Geology of the Northern Pennine Orefield

Volume 2 STAINMORE to CRAVEN

One of Britain's oldest industries, that of lead-mining, has left its characteristic imprint upon the beautiful hill country of the Yorkshire Dales, the geology and mineral deposits of which are described here. The deposits, exploited from Roman times on, occur in fissure veins and associated replacements of limestone in Carboniferous rocks belonging principally to the Asbian, Brigantian and Pendleian stages. The district has yielded about a million tonnes of galena, with much smaller quantities of copper and zinc ores. Providing the matrix of the deposits, fluorite occurs in restricted central areas; baryte, often accompanied by witherite, is more widespread, while calcite is generally present. Structurally, the stable Askrigg fault-block is flanked to the north by the Stainmore Trough and the Craven Basin respectively. The post-Carboniferous uplift of the block, probably under the influence of the density-deficient Caledonian granite underlying Wensleydale, produced the elaborate pattern of fractures that provided the vein channels.

Throughout the orefield there is a marked stratigraphical control of the mineralisation; in summarising the wealth of information, the horizon of each oreshoot is recorded. The effects of the mid-Pendleian unconformity become increasingly evident as it is traced towards the south-east, and it is shown that all the mineralisation lies beneath it except at Grassington and Ashfold Side. Variations in the deltaic and fore-delta cyclothems of the Brigantian and Lower Pendleian are traced and shown to have a bearing upon the incidence of oreshoots. Though mining in the orefield is now on a very small scale, the deposits have not been exhaustively explored in depth; in addition, substantial tonnages containing recoverable barytes, some with subordinate fluorite, remain in residues adjacent to the mines.

Plate 1
Section of a lead vein from Grassington High Moor, presented by His Grace the Duke of Devonshire to the 1851 Exhibition and now displayed in the Geological Museum, London SW7

Bands of galena alternate with fluorite and baryte. Width of vein 0.35 m

BRITISH GEOLOGICAL SURVEY

England and Wales

K. C. DUNHAM and
A. A. WILSON

CONTRIBUTORS

Mining geology
L. H. Tonks
J. R. Earp
D. J. C. Mundy
A. J. Wadge

Palaeontology
J. Pattison
W. H. C. Ramsbottom
A. R. E. Strank

Petrography
R. K. Harrison
B. R. Young

Geochemistry
E. Waine

Geology of the Northern Pennine Orefield

Volume 2 STAINMORE to CRAVEN

Economic memoir covering the areas of one-inch and 1:50 000 geological sheets 40, 41 and 50, and parts of 31, 32, 51, 60 and 61, New Series

1835 Geological Survey of Great Britain

150 Years of Service to the Nation

1985 British Geological Survey

Natural Environment Research Council

LONDON: HER MAJESTY'S STATIONERY OFFICE 1985

First published 1985

ISBN 0 11 884284 6

Bibliographical reference

DUNHAM, K. C., and WILSON, A. A. 1985. Geology of the Northern Pennine Orefield: Volume 2, Stainmore to Craven. *Econ. Mem. Br. Geol. Surv.*, Sheets 40, 41, 50, etc.

Authors

SIR KINGSLEY DUNHAM, DSc, FRS, FEng
Charleycroft, Quarryheads Lane, Durham DH1 3DY

A. A. WILSON, PhD
British Geological Survey, Keyworth, Nottingham NG12 5GG

Contributors

The late L. H. Tonks, MSc

J. R. Earp, PhD
South Chiltern, Orchard Coombe, Whitchurch Hill, Berkshire RG8 7QL

D. J. C. Mundy, PhD, J. Pattison, MSc, W. H. C. Ramsbottom, PhD, A. R. E. Strank and A. J. Wadge, PhD
British Geological Survey, Keyworth

R. K. Harrison, MSc, B. R. Young, MSc, and E. Waine, BSc
British Geological Survey, London

Other publications of the Survey dealing with the geology of this and adjoining districts.

Memoirs

Geology of the Northern Pennine Orefield: Volume 1, Tyne to Stainmore (2nd edition)
Geology of the country around Barnard Castle, Sheet 32
Geology of the country around Brough-under-Stainmore, Sheet 31

British Regional Geology
Northern England (4th edition)

Maps

1:625 000
Solid geology, south sheet
Quaternary geology, south sheet

1:50 000
Sheet 31 Brough-under-Stainmore (Solid, Drift)
Sheet 51 Masham (Solid, Drift *in press*)
Sheet 60 Settle (Solid *in press*)

1:63 360
Sheet 32 Barnard Castle (Solid, Drift)
Sheet 40 Kirkby Stephen (Solid, Drift)
Sheet 41 Richmond (Solid, Drift)
Sheet 50 Hawes (Solid, Drift)

Printed in the UK for HMSO
Dd 737385 C20 3/85

CONTENTS

PLATES

FIGURES

TABLES

PREFACE

The southern half of the Northern Pennine Orefield covers parts of eight New Series sheets of the British Geological Survey: 31 (Brough under Stainmore), 32 (Barnard Castle), 40 (Kirkby Stephen), 41 (Richmond), 50 (Hawes), 51 (Masham), 60 (Settle) and 61 (Pateley Bridge). The primary six-inch survey was carried out between 1868 and 1883, the surveyors principally engaged in the mineralised areas being G. Barrow, C. T. Clough, J. R. Dakyns, C. E. de Rance, J. G. Goodchild, W. Gunn and J. Lucas. The results were published as hand-coloured one-inch sheets between 1889 and 1893. Care was taken, in collaboration with the mineral owners and mine operators, to represent accurately the positions of the mineral veins, but only limited information was collected about the nature of the deposits, which received only brief description in the two memoirs published at this time, Mallerstang (Sheet 40) and Ingleborough (Sheet 50). During World War I R. G. Carruthers briefly examined the lead deposits, his report being published in 1923. As World War II approached, a thorough re-examination of this part of the orefield was put in hand, under the supervision of Mr T. Eastwood. In 1939 Mr L. H. Tonks, starting from the Pateley Bridge–Grassington area, began to work northward into Upper Wharfedale; in 1941 Dr J. R. Earp undertook the survey of mineral deposits in those parts of Swaledale, Arkengarthdale and Wensleydale falling within Sheet 41. Later in the same year Dr K. C. (now Sir Kingsley) Dunham took over from Mr Tonks, remapping part of Sheet 61 on the 1:5000 scale, continuing the mineral survey northwards through Sheet 50 to Sheet 40. Results for Greenhow Hill, the only area which appeared at the time to offer promise of further subsurface development, were published in collaboration with Dr C. J. (now Sir James) Stubblefield as palaeontologist in 1945. Attention was drawn to wastes worthy of retreatment for barium minerals in a Wartime Pamphlet in 1945, and for fluorspar in a revision of the special report on this mineral in 1952, but the preparation of the full economic memoir could not begin for pressure of other work.

A full six-inch revision of the geology of four of the sheet-areas, 31, 32, 51 and 60 has now been completed.

Maps on the one-inch scale (32) and the 1:50 000 scale (31) were published in 1969 and 1974, with descriptive memoirs respectively by D. A. C. Mills & J. H. Hull, and I. C. Burgess & D. W. Holliday. Meanwhile colour-printed editions of sheets 40, 41 and 50, compiled respectively by I. C. Burgess, D. A. C. Mills and A. A. Wilson, were issued in 1972, 1970 and 1971, incorporating new information from both official and private sources. The revision of Sheet 51, mainly by Dr Wilson, is approaching publication.

The preparation of this volume of the memoir has been undertaken by Sir Kingsley Dunham since his retirement, at the end of 1975, from the Institute of Geological Sciences, with the collaboration of Dr A. A. Wilson who has been responsible for the stratigraphical chapters, and who has measured many new sections for the purpose. The text has been edited by Messrs J. G. O. Smart and I. C. Burgess. This volume will constitute, in addition to its function as a summary of the available data about the mineral deposits, an explanation of the stratigraphy, particularly for the Kirkby Stephen, Richmond and Hawes sheets.

Gratitude is expressed to the North Yorkshire County Record Office, the Trustees of the Chatsworth Settlement, the Earby Mines Research Group and the Department of Mining, Leeds University, who between them hold most of the mine plans not in the care of the Health and Safety Executive. The Keeper of Minerals of the British Museum (Natural History) kindly permitted a thorough examination to be made of the main and Russell collections, to supplement data obtained from the Mineral Inventory collection of the Geological Museum. Acknowledgement is also gratefully given to the many individuals who have assisted in various ways, including Professor A. C. Dunham, Dr G. A. L. Johnson, Dr D. Moore, Dr P. F. Hicks, Dr J. G. Holland, Dr A. T. Small, Dr Arthur Raistrick, Dr F. W. Smith and the late Mr J. H. Clay among many others. For access to confidential information, including borehole sections and samples, the assistance of the following companies is gratefully acknowledged: Cominco, Lonrho, Monsanto and Tilcon Ltd.

During the preparation of this memoir, the Geological Survey undertook, as part of its Mineral Reconnaissance Programme, geophysical and geochemical surveys of parts of the orefield on behalf of the Departmant of Industry. Mr D. Ostle was latterly in charge of this work, and other officers concerned have included A. J. Wadge, J. H. Bateson, J. D. Cornwell, A. D. Evans, J. M. Hudson and D. J. Patrick. Their reports are available on an open-file basis.

G. M. Brown
Director

British Geological Survey
Keyworth
Nottingham NG12 5GG

3 January 1985

GEOLOGICAL SUCCESSION

The formations represented on the one-inch and 1:50000 geological maps and sections are summarised below:

SUPERFICIAL FORMATIONS

Quaternary

Peat

Alluvium

River Terraces

Glacial Sand and Gravel

Morainic Drift and Boulder Clay

Unconformity

SOLID FORMATIONS

			Approximate thickness m
Permian			
Upper Permian	Magnesian Limestone	dolomite and dolomitised limestone, marls	75
Lower Permian		basal breccias and sand	2

Unconformity

Carboniferous			
Silesian (Westphalian)	mudstone and sandstone outside the limits of the orefield		
Silesian (Namurian)			
Yeadonian			77
Marsdenian			40
Kinderscoutian	Libishaw Sandstone Cayton Gill Shell Bed		98
Chokierian	Upper Follifoot Grit	sandstone, shales	28
Arnsbergian	Lower Follifoot Grit	sandstone	
	Scar House Beds	shale with sandstone	
	Shunner Fell-Colsterdale Limestone		
	Pickersett Edge-Red Scar Grit	coarse sandstone and shales	
	Fossil Sandstone, Hearne Beck Limestone		
	Upper Howgate Edge Grit	sandstone	
	Nidderdale Shales		
	Mirk Fell Ironstone-Cockhill Limestone		184
Pendleian	Lower Howgate Edge-Grassington Grit	coarse sandstone	
	Upper Bowland Shales		

Unconformity

Pendleian	Stonesdale, Crow, Little and Main cyclothems	sandstones shales, cherts, limestones	140
DINANTIAN (VISÉAN)			
Brigantian	Underset, Three Yard, Five Yard, Middle, Simonstone, Hardrow Scar, Gayle and Hawes cyclothems Lower Bowland Shales	each cyclothem when fully developed shows a descending sequence of coal, seatearth, sandstone, sandy shale, shale, chert (in Underset only), limestone	430
Asbian	Danny Bridge, Kingsdale, Cove (in part), Garsdale (in part) limestones; Thorny Force Sandstone; Hargate End, Greenhow, Stump Cross and Timpony limestones at Greenhow		245
Holkerian	Fawes Wood, Kilnsey (in part), Horton, Cove (in part), Garsdale (in part) limestones		110
Arundian	Ashfell Sandstone; Tom Croft Limestone; Kilnsey Limestone (in part); Penny Farm Gill Dolomites (in part)		113
Chadian	Penny Farm Gill Dolomites (in part); Marsett Sandstone; Raydale Dolomite; Nor Gill Sandstone; Sedbergh Conglomerate		148
DINANTIAN (TOURNAISIAN)			
Ivorian	Shap Conglomerate, Pinskey Gill Beds		90

Major unconformity ~~~~~~

Silurian			
LUDLOW		sandstone turbidites	250
		siltstones	250
	Studfold Formation	greywacke turbidites	100
	Horton Formation	siltstones	400
	Arcow Formation (in part)	calcareous mudstones	6
WENLOCK	Arcow Formation (in part)	calcareous mudstones	12
	Austwick Formation	greywacke turbidites and mudstone	500
LLANDOVERY	Browgill Beds	fine turbidites	30
	Skelgill Beds	mudstone	?15
Ordovician			
ASHGILL		mudstones and siltstones	160
ARENIG	Ingleton Group	turbidites, siltstones and mudstones	762
INTRUSIVE ROCKS	Wensleydale Granite, ?intrusive into Lower Palaeozoics Quartz-microdiorite, intrusive into Ingleton Group		
MINERAL VEINS	Principally cutting Carboniferous rocks beneath the intra-Pendleian unconformity; but in Wharfedale and Nidderdale cutting rocks up to Kinderscoutian in age		

CHAPTER 1

Introduction

GEOGRAPHY

The Northern Pennine Orefield, as defined in Dunham (1948), extends southwards from the Roman Wall on the Whin Sill escarpment north of the main Tyne valley, to the Craven limestone uplands between Pateley Bridge and Settle. It is divided into two almost symmetrical halves by the pass of Stainmore, between Bowes in Durham and Brough in Cumbria. This volume is devoted to the southern half of the field, lying entirely in Yorkshire apart from a small area of Cumbria in the north-west (formerly in Westmorland) and a smaller northern area in Durham. The region where the ores are found forms part of an uplifted and deeply dissected plateau, with its western margin running from Stainmore near Kirkby Stephen through the hills immediately west of Upper Edendale, such as Clouds (468 m), Wild Boar Fell (708 m) and on through Baugh Fell (676 m), Aye Gill Pike (687 m) and Great Coum (686 m). There is, however, no marked topographic feature comparable with the escarpment bounding the northern region, for the great rounded mass of the Howgill Fells lies to the west of the present area. In the south-west part of the plateau, the striking mountains of Whernside (737 m), Ingleborough (723 m) and Pen-y-ghent (762 m) remain, but the largest and most inaccessible part of the upland plateau lies around Great Shunner Fell (713 m) where rise the rivers Swale and Ure. The most extensive mineral deposits have been worked in Swaledale and its northern tributary, Arkengarthdale, along the Swale–Ure watershed and to a lesser extent in Wensleydale (the valley of the Ure) itself. The orefield then continues in upper Wharfedale and over the Wharfe– Nidd interfluve between Hebden and Pateley Bridge. The plateau slopes gently eastward until, beyond Richmond, Leyburn and Masham it merges with the Vale of York which lies far beyond the eastern limit of known mineralisation. Apart from the small deposits that continue along the southern margin of the plateau almost as far west as Settle, the south-western quarter in general is almost devoid of such deposits. This particularly applies to the area covered by the Hawes (40) Sheet although the same rock formations continue there; its stratigraphy is included in the description to be given here.

Arthur Raistrick in his numerous historical and descriptive works (to be referred to in the next section) has preferred to distinguish the North Swaledale–Wensleydale area as one separate orefield, and Wharfedale–Nidderdale as another. For purposes of geological investigation, we wish to emphasise the essential unity of the southern half of the Northern Pennines, which is as striking as that of the northern half, described in Dunham (1948).

Each of the Yorkshire dales has its own characteristic scenery. Swaledale is the narrowest, with steeper sides than the others which, below Reeth, become extensively wooded. Arkengarthdale at first follows the escarpment of Fremington Edge (Plate 2) north-west from Reeth, then opens out above Langthwaite into a broader river basin, through which a road leads to Tan Hill, a former coalmining centre. From Keld in upper Swaledale a spectacular pass, Buttertubs, so named from the several large sink-holes in the Main Limestone, leads over to Wensleydale. This valley, parallel in its east–west course to Swaledale, is nevertheless much wider and accommodates a discontinuous train of drumlins. All three dales so far mentioned show fine step-featuring marking the outcrops of hard beds in the Carboniferous sequence, particularly limestones. The long runs of these features give an erroneous impression that the beds are flat, and that the region is structureless.

Wharfedale with its general north–south direction and its large west-side valleys of Langstrothdale and Littondale differs from the more northerly dales in that the middle and lower slopes of the valley sides expose steep crags and cliffs of white limestone; the same rock predominates in the walls that criss-cross the countryside, and is exposed in wide griked pavements in the Craven uplands, particularly westward from Grassington.

Lead mining has been practised for two millennia in the dale country, leaving its mark in the gashes produced by a form of hydraulic mining known as 'hushing' (Plate 2), in the remains of innumerable bell-pits and shallow shafts, and the debris from levels, smelt-mills and dressing floors. Very few of the former buildings remain intact. Generally, the remains have already merged into the scenery, to which they impart a certain character; and only in a few places does industrial wilderness, such as can be seen around the Hungry Hushes and the Octagon smelt-mills in Arkengarthdale, remain as an eyesore to some. It may be assumed that men were attracted into the area, especially from late mediaeval times on, because it gave them scope to be their own masters, to work as individuals or in small partnerships. Many were also smallholders. Deposits were found and worked in very remote places; smelt-mills grew up near villages in the valleys, and ore was carried to them by pack-horse. The enlargement of the mines later called for capital which was provided by investors, or in some cases, under company organisation, by the lords of the royalties. The human side of the lead mining industry has been interestingly portrayed, with accurate technological detail, by G. T. Armstrong in his novel *Adam Brunskill* (1952); there is factual foundation for some of the inter-company rivalry that forms part of the story.

The motive power for the mines, once they began to be equipped, was water, ingeniously used to operate large wheels on surface and underground for winding, pumping, crushing and driving the bellows of the smelting hearths. A few early steam engines were erected, but several were abandoned as too expensive, in favour of water. The horse was also used for winding by means of whims or gins, and was the means of transport in those long underground tunnels where rails were installed. Virtually none of the ore was sent outside the field for smelting; bole or baal hills, sited in windy places and burning charcoal, were used until the end of the sixteenth century, after which smelting hearths, coupled to long flues up the fell sides and equipped with bellows for forced draught, were employed. Underground, hundreds or perhaps thousands of kilometres of tunnels were driven by hand labour; compressed air had only just been introduced when the late nineteenth century decline in mining, from which the industry has never recovered, set in. The existing workings are nevertheless so extensive as to raise the question whether modern methods of prospecting and mining could revive the industry, and as base-metal and non-metallic mineral resources become scarcer in the future, the question is likely to be repeated. In the present century there have been limited attempts at a revival at Greenhow, and attention has been given to the recovery of fluorspar, baryte, witherite and calcite from the wastes and unworked ground left by the lead miners; on a small scale this is still continuing. Of metals other than lead, there has been a small production of zinc as carbonate and hydroxysilicate (calamine) from Malham and near Redmire. For three short periods, copper was mined near Richmond and near Kirkby Stephen, and small amounts, far below workable grade, are widespread in the western part of the region. The lead ores contain silver, but not in sufficient quantity to render the metal worth recovering; a few hundred ounces were produced during tests at several places during the last century, but the uneconomic grade of the silver ores here contrasts strikingly with that in the northern half of the Northern Pennine Orefield.

Since 1954, the greater part of the present region has been constituted the Yorkshire Dales National Park. The Park's northern boundary follows the new Durham–Yorkshire boundary along the watershed between the Arkengarthdale and Stainmore drainage, south of Stang Forest, and extends

Plate 2 Turf Moor Hush and Arkengarthdale viewed from the head of the hush

The prominent cliff is in Main Limestone faulted against mudstones on the left of the hush, overlain by scree from the Ten Fathom Grit. In the distance is Fremington Edge capped by Richmond Cherts and the outlier of Calver with cliffs in Main Limestone (right).

as far south as Grassington; much of the orefield therefore comes within the jurisdiction of the Park authorities.

Except at the old mining village of Greenhow Hill, the population is restricted to the valleys which contain a series of beautiful sandstone-built villages in the northern dales, with limestone becoming more important in Wharfedale. Settle, Grassington and Pateley Bridge are regional centres and, for many visitors, Reeth in Swaledale also fulfills this role. Kirkby Stephen, Richmond and Leyburn are peripheral to the orefield but have good road access to it. From a beginning with tracks made by the early lead miners, the field now has an excellent network of roads, including

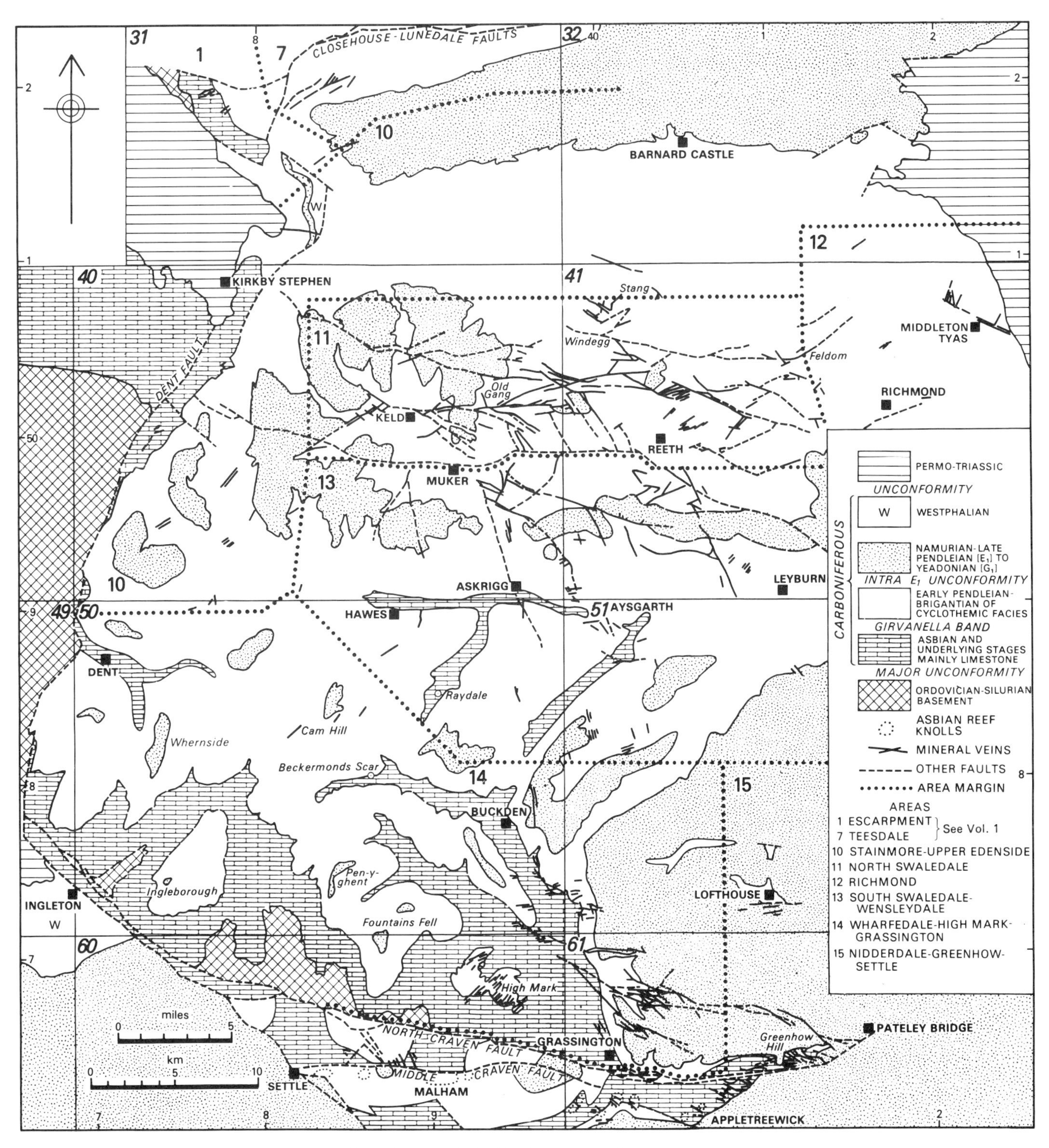

Figure 1 Sketch map of the Northern Pennine Orefield (Stainmore to Craven) showing the subdivisions of the field and the broad divisions of the Carboniferous sequence

many that cross the watersheds at altitudes over 650 m. Only one railway remains, the Leeds to Carlisle line, which crosses the west part of the region from Settle to Kirkby Stephen; Wensleydale was formerly served by a line from Leyburn through Hawes, but this is no longer in use. In the south-eastern part of the region, Nidderdale contains three substantial reservoirs belonging to the Yorkshire Water Authority, Angram, Scar House and Gouthwaite; there is a fourth in the Grimwith valley, at present being enlarged.

Around Ingleborough, Malham and in upper Wharfedale, the limestone contains extensive cave-systems and other karst phenomena, which continue to attract cave explorers. The dangers of this form of recreation need no emphasis here, but it is worth mentioning that although some of the old mining workings are still accessible, these are, in general, more dangerous than the caves because of the existence of unstable ground and of winzes in the bottoms of tunnels. The chief recreational attraction of this beautiful area is the opportunity it offers to the fell-walker.

SUBDIVISIONS OF THE FIELD

The results of this geological investigation of the southern half of the Northern Pennine Orefield are presented in two parts. The first gives a description of the stratigraphy and structure of the region, followed by a general description of the mineral deposits and a discussion of their possible origins. In the second part, the region is divided into six main areas numbered 10 to 15, continuing the scheme in the Northern Region (1 to 9). Within each area a detailed account is given of the oreshoots and the extent to which they have been mined. Figure 1 gives an overall view of the region and shows the boundaries of the six areas. Of the eight New Series geological maps involved in this study, full revision of Brough (31) and Barnard Castle (32) has been completed and both have been published, the former at 1:50 000, the latter at 1:63 360. Kirkby Stephen (40), Richmond (41) and Hawes (50) have appeared in new provisional one-inch editions. Masham (51) and Settle (60) revision surveys are complete and map production is in progress. The mineralised part of Pateley Bridge (61) was remapped in 1940–42, at 1:5000 scale (Dunham and Stubblefield, 1945).

HISTORY OF MINING

Raistrick has emerged as the leading authority on the history of this rural industry in Yorkshire. His numerous works (1927, 1930, 1955, 1967, 1973, 1975 and papers listed in chapters 9 to 13) and his joint book with B. Jennings (1965) give full references to the scattered earlier literature. Jennings (1959, 1967), J. M. Dickinson (1970), T. R. Hornshaw (1975); and Fieldhouse and Jennings (1978) are important recent contributors, and R. T. Clough (1962, 1980) has described and figured the former smelt-mills. The following brief summary, based largely on these works, is intended to provide background necessary for the detailed geological descriptions of the mines.

The Brigantes, the Belgic tribe that inhabited the Pennines at the time of the Roman invasions were skilled in metal work and some pre-Roman settlements have yielded fragments of lead ore, probably from outcrops. The first positive evidence of mining follows the Roman occupation and is based upon the discovery, at Heyshaw Bank, 4 km SE of Greenhow, of two pigs of lead marked with the stamp of the Emperor Domitian, to which the date AD 81 can be assigned (Kirkshaw, 1735; Raistrick, 1930); also, according to local tradition at Nussey Knot, Greenhow, one pig with the mark of Trajan (AD 98) (Lucas, 1885); and at Hurst, north of Swaledale, one pig of Hadrian (AD 117–138). One of the Heyshaw Bank pigs still exists, at Ripley Castle. No proved mine-workings of this period are known, though the small Jackass and Sam Oon levels at Greenhow have been thought to be Roman. Nothing is known of mining in the Dark Ages, but it may have continued in a small way, for it is recorded that St Wilfred protected the roof of York Minster with leaden sheets (William of Malmesbury, 690) and the widespread use of the term Groove for a mine is almost certainly Danish in origin. During the monastic period that followed the Norman Conquest, mining was being carried on, as shown by state and abbey records. In 1145 Jervaulx Abbey had a grant to dig iron ore and lead ore in Count Alan of Brittany's Forest of Wensleydale, but the Count reserved for his own use mines at Grinton and in Arkengarthdale. There is also a reference in the twelfth century, in 26 Henry II which shows that lead was sent from Swaledale to Waltham Abbey. Thereafter monastic records indicate that Bylands and Fountains abbeys had holdings at Greenhow, Bolton Priory in Wharfedale and Rievaulx Abbey in upper Swaledale. The practice of leasing ground to partnerships of miners had already begun. Reduction of the galena to lead metal was carried out in charcoal-burning boles, placing a demand for timber on the estates, but by 1445 what must have been one of the earliest smelt-mills with bellows was working on Fountains Abbey property at Brimham, outside the orefield. After the dissolution of the monasteries under Henry VIII, Fountains property passed to the successive owners of Bewerley, while the Bylands estates were purchased in 1549 by Sir John Yorke, in whose family they still remain. The Appletreewick Manor, including Burhill and Craven Moor, which had belonged to Bolton Priory since its purchase from James de Eston in 1300, contained mines which though they had been worked before 1548, were returned as of *nil* value at the Dissolution. This manor also passed after several resales to Yorke. In 1564 Elizabeth I brought skilled miners from the Clausthal area of Germany to her mines near Keswick in Cumberland and they introduced the systematic use of water power and other techniques, well illustrated in the contemporary *De Re Metallica* of Agricola (Georg Bauer, 1556). The impetus to mining spread far beyond the limits of Cumberland. In the reign of James I the important Grassington mines, on former Bolton Priory property, appear first in history under the ownership of the Earl of Burlington; they have remained with his descendants the Dukes of Devonshire, by whom they were vigorously developed, together with other parts of Wharfedale. In the early years of the seventeenth century, miners were brought from the Duke's Derbyshire estates to work these mines and with them came a system of Laws and Customs for regulating the leasing and working based on the ancient customs of Derbyshire, including the setting up of a

Barmote Court and the appointment of barmasters (Raistrick, 1936). Farther north, Leland was able to record in 1546 that Lord Scrope was searching for lead two miles from Bolton Castle in Wensleydale, and indeed the estates of the Lords of Bolton were to become the scene of active mining. At about the time of Leland's visit, another important figure appeared on the northern scene: Thomas Lord Wharton bought the manors of Muker and part of Healaugh from the Crown in 1544–46, covering much of upper Swaledale. By 1601 this property had passed to his grandson Philip who with his agent Philip Swale played a significant part in the promotion of Swaledale mining. On the north side of the dale the larger mines of the North Swaledale Mineral Belt were being exploited: Beldi Hill, Lownathwaite, Old Gang and Friarfold; small partnerships carried on the operations by means of opencuts and shafts, many of which were only 1 m in diameter, but reached depths over 100 m. In the 1670s, however, Lord Wharton decided to dispossess the small men, apparently with some compensation, and to organise the mines on a larger scale, no doubt influenced by Hochstetter's methods at Keswick. Already by 1685, water was being pumped from the 35-Fathom Level at Lownathwaite, and by 1705 this mine, with Old Gang, was producing 440 tons of lead per year (Raistrick and Jennings, *op. cit.*, p. 155). In 1715 Lord Wharton, like the Earl of Derwentwater (Dunham, 1948, p. 142) was involved in the rebellion and lost his lands by attainder. They were purchased from the trustees for the Crown by Lady Jane Coke on whose death in 1760 they passed to Anne Draycott, her niece. The initials by which the A.D. Mines were known until they finally closed during the present century probably came from this lady[1]; they were in use long before the A.D. Mining Company as such was formed in 1873. In 1656 Dr J. Bathurst, physician to the Lord Protector, had bought the Arkengarthdale royalty, and here the C.B. Mines (named for his grandson Charles) flourished and became the other main producers in the North Swaledale Belt. Eventually owned by Sir Thomas Sopwith, the C.B. Mines also continued in operation until the present century when Faggergill closed in 1912. Between the properties held by the A.D. and C.B. interests, high up on the Hard Level Gill–Arkengarthdale watershed lay the small but interesting Surrender property. Not enough is known of its history, but it had been penetrated before 1680, and it seems to have been one of the richest parts of the belt, with workings reaching over 656 ft (200 m) below surface. The period covered by the operations of all these mines witnessed great changes as they were worked to greater depths. The opencuts, hushes and shallow shafts of the earlier days gave place to development from long adits driven from the valley sides, for example Mould's Level (probably started in 1682), Parkes (1746), Hard Level (1785), Gillfield and Cockhill levels at Greenhow (1782) and Duke's Level at Grassington (1790). The extension of these and other long adits took place during the period from 1795 to 1812, when lead prices reached the highest level they attained before World War II (Dunham, 1944). In these years, capital was attracted into the industry from business and professional men in the towns and industrial areas surrounding the Northern Pennines; names such as Hutchinson, Cleaver, Wood and Horner began to appear in the records in the south, Overend, Tennant (somewhat earlier) and Place in Wharfedale, Chaytor in Wensleydale, Jaques, Knowles, Tomlin and the Bradleys in Swaledale, among many others, some of whom had a long connection with the mines. Many of the deep adits proved to be below the limit of well mineralised ground, though their value in draining and ventilating workings at higher levels is unquestionable. The last one to be driven, Sir Francis Level in Gunnerside Gill, started in 1864, eventually gave access to an oreshoot which was worked beneath the adit, but by this time the falling prices had made the working uneconomic. This was an interesting case where the royalty owner, Sir George Denys (a descendant of Anne Draycott) took a large share of the risk, while at Grassington after the turn of the century, the royalty owner—the Duke of Devonshire—provided support for the whole operation. Of the two great concerns that carried on lead mining operations in the northern region of the Northern Pennines during the eighteenth and nineteenth centuries, only the London Lead Company penetrated into the southern region. In 1733 the company acquired the already considerably-developed mines on the south side of Swaledale at Whitaside, Harkerside and Grinton; soon afterwards they were reconstructing the smelt mill at Spout Gill, where the mine was successfully worked at least until 1768, in one year producing, unless the report is exaggerated, 2000 tons of lead. The company also leased the Marrick mines and improved the smelt mills by installing reverberatory furnaces. Their operations also extended in 1734 into Wensleydale. Though the documentation in the southern region is much less satisfactory, it remains clear that none of their mines here compared with their properties in Alston Moor and Teesdale.

In the region generally, many smaller mines were worked other than in the main producing areas so far mentioned. Some had periods of success; others supported three or four men at a time when it was possible to live by mining a few tons of lead ore per year; others were prospects supported by the larger mines. It is a normal picture for a long-established metalliferous field. At all the mines, the crude ore was dressed on the spot, or where the nearest water was available, women and children often assisting their menfolk. Many of the small tailings heaps are so remote that even the simplest mechanical aids cannot have been available. In the eighteenth century, powered crushers came into use, and in the nineteenth, a few mines had buddles and powered jigs; but it is difficult to avoid the conclusion that tens of thousands of tons of tailings are the result of hand labour, including hand jigging which was still being practised in 1912. The only ore known to have been carried far from the region for smelting was that taken to Bollihope in Weardale by the London Lead Company. Within the present region at least 56 smelting houses were in operation for some part of the period 1600 to 1900, their outputs ranging from as little as 30 to 40 tonnes of metal per year up to 2500 tonnes or more.

The last two decades of the nineteenth century saw a marked fall in the price of lead, probably associated with the development of large mines in Spain and in the Middle West of the United States of America. The lead mining industry in

[1]According to a document by B. Jennings in the North Yorkshire County Record Office ZLB/41.

Yorkshire and Westmorland, already in decline after 1860, did not survive the depression. A revival of interest, resulting from the discovery of ore-bearing veins in an aqueduct tunnel driven beneath the once-prosperous mining district of Greenhow Hill, failed to re-establish the industry and the only mining since then has been on a small scale, with fluorspar, baryte and, in one place, calcite as its objectives.

Reasonably complete statistics of mineral production are available only from the middle of last century up to 1939 (Hunt, 1848a and b; Geological Survey, 1848–1881; Home Office, 1882–1919; Mines Department, 1920–1939). Although certain estates and companies were able to supply information for years prior to 1845, this proved to be very much less complete than that for the northern region (Dunham, 1944). In the present region, the statistical information mainly covers the period of decline, and for this reason, tentative estimates designed to fill in the missing data in broad terms have been included here. The basis for these varies from area to area, and is explained in chapters 8 to 13; in all cases both geological and mining evidence has been taken into consideration. While it is believed that these approximate the right orders of magnitude, a high degree of confidence should not be attached to them. The average recovery of lead from the ore concentrates during the period covered by the statistics improved from as low as 60 per cent to over 70 per cent as a result of the introduction of mechanised ore dressing, from the beginning of the nineteenth century on.

Discussion of the mineral deposits of the Stainmore to Craven region first appears in the geological literature in the classic work of J. Phillips (1836), whose stratigraphy of the Mountain Limestone and Yoredales enshrines knowledge gained through centuries of lead mining. He recognised with clarity the phenomenon of 'bearing beds', the stratigraphical control of vein-oreshoots. This subject was developed further by L. Bradley (1862) who as a result of his inquiries among all the mine operators in Swaledale preserved much valuable information. The primary geological survey carried out in 1870–1880 led to the publication of hand-coloured versions of all the one-inch maps shown on Figure 1, but only two memoirs, Ingleborough (Sheet 50) in 1890, and Mallerstang (Sheet 40) in 1891. A partly-completed manuscript for Richmond (Sheet 41) also exists. A brief re-examination of the deposits during World War I led to the Special Report by Carruthers and Strahan (1923). The more extended resurvey of the deposits on which the present account is based led to publication of results on barium minerals (Dunham and Dines, 1945) and fluorspar (Dunham, 1952), but full geological detail has been published only for Greenhow (Dunham and Stubblefield, 1945). The extensive geological literature up to 1958 has been summarised (Dunham, 1959) but references to these papers, and to more recent contributions, are included here in the appropriate chapters. KCD

Table 1 Production of lead concentrates in tonnes, Northern Pennines, Southern Region

Area		Period	Recorded	Estimated additional	Total
10	Stainmore-Mallerstang	1854-1884	1 032	3 000	4 032
11	North Swaledale	1696-1700, 1783-1913	339 612	215 000	554 612
12	Richmond	1864-1874	1 305	1 016	2 321
13	South Swaledale-Wensleydale	1851-1893	46 192	84 000	130 192
14	Wharfedale, Grassington	1735-1886	105 990	73 000	178 990
15	Nidderdale, Greenhow, Settle	1706-1936	70 855	75 250	146 105
		Total	564 986	451 266	1 016 252

Table 2 Production of copper concentrates, fluorspar, barytes, 'calamine' and calcite in tonnes

Area		Period	Copper ore	Fluorspar*	Barytes*	'Calamine' *Calcite*
10	Stainmore	to 1900	500†	—	—	—
11	North Swaledale	to 1975	—	2 500†	35 000†	—
12	Richmond	1742-1912	6593‡	—	—	—
13	Wensleydale	to 1975	—	5 000†	—	500†
14	Grassington	to 1964	—	10 545	§	—
15	Greenhow-Settle	to 1975	500†	121 920	3 000†	5 000† *720*
		Total	7 593	139 965	38 000	5 500 *720*

* The terms fluorspar and barytes as used in this memoir refer respectively to commercial concentrates of the minerals fluorite and baryte.
† Estimated from history and scale of operations.
‡ Based on partial statistics.
§ Some recent production.

REFERENCES

AGRICOLA, G. (BAUER, G.) 1556. *De Re Metallica* Libri XII. (Basel: Froben.) English translation by HOOVER, H. C. and HOOVER, L. H., 1950. (London: Constable.) 638 pp.

ARMSTRONG, G. T. 1952. *Adam Brunskill.* (London: Collins.) 574 pp.

BRADLEY, L. 1862. *An inquiry into the deposition of lead ore in the mineral veins of Swaledale, Yorkshire.* (London: Stanford.) 40 pp.

CARRUTHERS, R. G. and STRAHAN, A. 1923. Lead and zinc ores of Durham, Yorkshire and Derbyshire, with notes on the Isle of Man. *Spec. Rep. Miner. Resour. Mem. Geol. Surv. G. B.*, Vol. 26.

CLOUGH, R. T. 1962. *The lead smelting mills of the Yorkshire dales.* (Leeds: Clough.) 188 pp.

— 1980. *The lead smelting mills of the Yorkshire dales and Northern Pennines.* 2nd edit. (Keighley: Clough.) 331 pp.

DAKYNS, J. R., TIDDEMAN, R. H., GUNN, W. and STRAHAN, A. 1890. The geology of the country around Ingleborough, with parts of Wensleydale and Wharfedale. *Mem. Geol. Surv. G. B.*

— RUSSELL, R., CLOUGH, C. T. and STRAHAN, A. 1891. The geology of the country around Mallerstang, with parts of Wensleydale, Swaledale and Arkendale. *Mem. Geol. Surv. G. B.*

DICKINSON, J. M. 1970. The Greenhow lead mining field (A historical survey). *North. Cavern Mine Res. Soc.*, Individ. Surv. Ser. No. 4.

DUNHAM, K. C. 1944. The production of galena and associated minerals in the Northern Pennines; with comparative statistics for Great Britain. *Trans. Inst. Min. Metall.*, Vol. 53, pp. 181–252.

— 1948. The geology of the Northern Pennine Orefield: Vol. I, Tyne to Stainmore. *Mem. Geol. Surv. G. B.*, 1st edit.

— 1952. Fluorspar. *Spec. Rep. Miner. Resour. Mem. Geol. Surv. G.B.*, Vol. 4, 4th edit., pp. 65–79.

— 1959. Epigenetic mineralization in Yorkshire. *Proc. Yorkshire Geol. Soc.*, Vol. 32, pp. 1–29.

— and DINES, H. G. 1945. Barium minerals in England and Wales. *Wartime Pamphlet Geol. Surv. G. B.*, No. 46.

— and STUBBLEFIELD, C. J. 1945. The stratigraphy, structure and mineralisation of the Greenhow mining area, Yorkshire. *Q. J. Geol. Soc. London*, Vol. 100, pp. 209–268.

FIELDHOUSE, R. and JENNINGS, B. 1978. *A history of Richmond and Swaledale.* (London: Phillimore.) 520 pp.

HORNSHAW, T. R. 1975. Copper mining in Middleton Tyas. *North Yorkshire County Record Office*, Pub. No. 6.

JENNINGS, B. 1959. The lead mining industry of Swaledale. (Unpublished M.A. thesis, University of Leeds.)

— (Editor). 1967. *A history of Nidderdale.* (Huddersfield: Advertiser Press.) 504 pp; 1983. 2nd edit. (York: Sessions.)

KIRKSHAW, S. 1735. A letter concerning two pigs of lead, found near Ripley. *Philos. Trans.*, Vol. 41, p. 560.

LELAND, J. 1546. *Itinerary* (The itinerary of John Leland in or about the years 1536–43). TOMLIN SMITH, L. (Editor). Vol. 1, p. 79. (London: George Bell.) 352 pp.; Vol. 5, p. 140, 1910, (London: George Bell.) 352 p.

LUCAS, J. 1885. In *Old Yorkshire.* WHEATER, W. (Editor). (London: Hamilton Adams.) 324 pp.

MILLS, D. A. C. and HULL, J. H. 1976. Geology of the country around Barnard Castle. *Mem. Geol. Surv. G. B.*

PHILLIPS, J. 1836. *Illustrations of the Geology of Yorkshire*; Part II. The Mountain Limestone District. (London: John Murray.) 253 pp.

RAISTRICK, A. 1927. Lead mining and smelting in West Yorkshire. *Trans. Newcomen Soc.*, Vol. 7, pp. 81–97.

— 1930. A pig of lead with Roman inscriptions. *Yorkshire Archaeol. J.*, Vol. 30, pp. 181–183.

— 1936. 'Rare avis in terris'; The Laws and customs of lead mines in West Yorkshire. *Proc. Univ. Durham Philos. Soc.*, Vol. 9, pp. 180–190.

— 1955. *Mines and miners of Swaledale.* (Clapham: Dalesman.) 92 pp.

— 1967. *Old Yorkshire Dales.* (Newton Abbot: David and Charles.) 199 pp.

— 1973. *Lead mining in the Mid-Pennines.* (Truro: Bradford Barton.) 172 pp.

— 1975. *The lead industry of Wensleydale and Swaledale.* Vol. 1. The Mines; Vol. 2. The Smelting mills. (Buxton: Moorland.) 120 pp. each volume.

— and JENNINGS, B. 1965. *A history of lead mining in the Pennines.* (London: Longmans.) 347 pp.

— and JENNINGS, B. 1965. A history of lead mining in the Pennines. (London: Longmans.) 347 pp.

VARVILL, W. W. 1920. Greenhow Hill lead mines. *Min. Mag.*, Vol. 22, pp. 275–282.

WARD, J. 1756. Considerations of a draught of two large pieces of lead with Roman inscriptions upon them, found in Yorkshire. *Philos. Trans. R. Soc.*, Vol. 49, pp. 686–700.

WILLIAM OF MALMESBURY. *c.*690. *De Gestis Pontificum Anglorum.* Rolls Series, London, 1870.

Statistical data

DUNHAM, K. C. 1944. (*q.v.*)

ESTATE RECORDS Chatsworth Estate (per Dr Raistrick); Bewerley Estate (per Mr D. Gill).

FIELDHOUSE, R. and JENNINGS, B. 1978. See Appendix 5.

GEOLOGICAL SURVEY 1848–1881. Mineral statistics of the United Kingdom. *Mem. Geol. Surv. G. B.*

HOME OFFICE 1882–1919. Mineral statistics of the United Kingdom. (London: HMSO.)

HUNT, R. 1848. Produce of lead ore and lead in the U.K. for the years 1845 and 1846. *Mem. Geol. Surv. G. B.*, Vol. ii, pp. 703–706.

— 1848b. Produce of lead ore and lead for the year 1847. *ibid.*, pp. 707–710.

INSTITUTE OF GEOLOGICAL SCIENCES 1939–1978. Estimates by Mineral Statistics and Economics Unit.

MINES DEPARTMENT. 1920–1938. Mines and Quarries. General report with statistics. (London: HMSO.)

CHAPTER 2

The Lower Palaeozoic foundation

The Lower Palaeozoic rocks are seen only on the southern fringe of the area in a chain of inliers extending from Ingleton eastwards to Gordale Beck. Dinantian rocks rest on the folded earlier strata with conspicuous unconformity especially well seen at Thornton Force [SD 6948 7533] and Combs Quarry [SD 8002 7016]. The oldest strata, the Ingleton Group (probable Arenig age), are tightly folded on a WNW–ESE trend and unconformably overlain by Ashgill–Ludlow age sediments, generally in more open folds. These latter folds plunge at 10°–15° to the east tending to bring in successively higher beds in that direction. Slaty cleavage parallel to the fold axes is developed at many levels in the Lower Palaeozoic rocks.

Farther north, Bott and Masson Smith (1957) and Bott (1961) have delineated a belt of magnetic strata, hidden beneath Dinantian rocks, extending east-south-eastwards from the Ribblehead into Langstrothdale (upper Wharfedale). Recently this belt has been tested by drilling at Beckermonds Scar in Langstrothdale. North of the Wensleydale Granite there is no firm knowledge of the basement until one reaches the Cross Fell Inlier. Immediately west of the Dent Fault lie Ordovician and Silurian rocks in Barbon Fell and the Howgill Fells. It seems likely that strata of the same age might also occur in basement rocks on the eastern margin of the Askrigg Block.

ORDOVICIAN

Arenig Series

INGLETON GROUP This group of greywackes and slates is about 2500 ft (762 m) thick (Dunham and others, 1953). It is seen in inliers at Horton and notably at Ingleton where there are exposures in the ravines of the River Greta and Kingsdale Beck. The harder greywacke bands make spectacular waterfalls like Pecca Falls and also stood up as low ridges on the pre-Carboniferous landscape.

The lithologies in the Ingleton Group range from cleaved mudstones through to very coarse greywacke, the 'Ingleton Granite' of the quarry men. Rastall (1906), in studying the petrography of the greywackes, noted a number of igneous and metamorphic clasts, besides much quartz, feldspar, chlorite and locally abundant heavy minerals. Veining by quartz, chlorite and carbonate was recorded by Rastall (1906) and by Dunham and others (1953).

Leedal and Walker (1950) have proved several major folds in the Ingleton sections by plotting inverted repetitions of strata using sedimentary structures, particularly eroded tops of convoluted and cross-laminated beds. In view of the very steeply inclined tight folding in the Ingleton Group, coupled with the presence of quartz-chlorite veins which are rare in higher beds, a tentative Precambrian age has previously been assigned to the group. The absence of macrofossils, apart from a problematical trace fossil (Rayner, 1957) has supported this view. More recently rubidium-strontium dating has been applied to micas in greywacke samples by O'Nions and others (1973) who suggest a probable late Cambrian or early Ordovician age for these beds in the Horton area. Attempts to find microfossils in the outcrops at Ingleton and Horton have been unsuccessful.

The disposition of the Ingleton Group in Ribblesdale, on the north side of the Silurian and Ordovician rocks of the inliers, suggests that there is an extensive outcrop of these rocks under Ingleborough (Leedal and Walker, 1950, fig. 5). A borehole inclined at 15°–28° to the vertical was sunk by IGS in 1976 at Beckermonds Scar [SD 8635 8016] in Langstrothdale (Wilson and Cornwell, 1982; Berridge, 1982). It penetrated a rod length of 853 ft (260 m) of steeply dipping siltstones, mudstones and greywackes directly underlying Dinantian rocks. The lithologies resemble those in the Ingleton Group, but do not include the very coarse 'Ingleton Granite' lithology. Veins of quartz and chlorite occur as in the Ingleton Group of Ingleton, with the addition of pyrite, chalcopyrite and magnetite.

Samples were tested for acritarchs at 24 levels. One of them at a depth of 1299 ft (396.20 m) has yielded the following forms to Mr R. E. Turner: '*Archaeohystrichosphaeridium*' cf. *pungens*, *Veryhachium trispinosum*, *V. lairdi*, *Polygonium* cf. *gracilis*, *Pirea* cf. *dubia* and *Michrystridium sp*. The suggested age of this first assemblage from probable Ingleton Group rocks is Arenig and is in reasonable agreement with the radiometric work of O'Nions and others (1973). Equivalent beds lie within the Skiddaw Slates of the Lake District and Manx Slates of the Isle of Man.

Beckermonds Scar Borehole was drilled on a magnetic anomaly. Grains of magnetite are abundant at many levels in the presumed Ingleton Group and evidently contribute greatly to the magnetic anomaly. At least some of the magnetite seen in thin section is detrital in origin (Berridge, 1982) but it is also present in veins. The magnetite is very pure and lacks lamellae of ilmenite and ulvospinel. This is borne out by the high Curie temperature, which is related to the high purity of the magnetite (Wilson and Cornwell, 1982).

Recent work in the Lake District, summarised by Wadge (1978), indicates that there is a major unconformity beneath the Borrowdale Volcanic Group (Llandeilo to middle Caradoc age), which rests on the Eycott Group (higher Arenig – lower Llanvirn) and Skiddaw Group (Arenig). There is also a second unconformity between these beds and the overlying Caradoc sediments. Some of the folding in the pre-Borrowdale Volcanic Group strata clearly predates the deposition of the Borrowdale Volcanic Group (Soper and Moseley, 1978). Similarly the Ingleton Group in the present area is much more tightly folded than the Ashgill and younger rocks and some of the deformation is of pre-Caradocian age, as witnessed by the marked unconformity at Douk Ghyll.

Ashgill Series

These beds are usually faulted against rocks of the Ingleton Group except in Douk Ghyll [SD 8138 7237] near Horton where they rest unconformably upon them (O'Nions and others, 1973). Exposure is chiefly in anticlinal cores in the Horton – Austwick inliers and in a narrow belt alongside the North Craven Fault, near Ingleton. The fuller succession at Cautley, west of the present area, is only represented at Horton – Austwick by three short sections (Ingham and Rickards, 1974, fig. 7), parts of the Hirnantian, Rawtheyan and Cautleyan stages. Mudstones with calcareous nodules containing a shelly fauna with trilobites, brachiopods and corals covering parts of the Cautleyan and Rawtheyan stages are overlain by siltstones with thin volcanic tuffs (E. W. Johnson, personal communication); these are correlated with the Cautley Volcanic Formation of Rawtheyan age in the Cautley and Dent inliers (Ingham and Rickards, 1974, p. 35). There is an erosive break in the sequence overlain by the Wharfe Conglomerate, followed by banded siltstones, all of Hirnantian (highest Ashgillian) age.

SILURIAN

Llandovery Series

SKELGILL BEDS AND BROWGILL BEDS These are best exposed in Crummackdale where Skelgill Beds, ill-exposed black graptolitic shaly mudstones, are overlain by up to 100 ft (30 m) of green, platy mudstones, the Browgill Beds. Relations in the higher part of the Skelgill Beds were clarified by trenching at Hunterstye [SD 7814 7131] (King and Wilcockson, 1934) which revealed strata now referred to the Idwian and Fronian stages. Browgill Beds in the trench and at a nearby outcrop yielded *'Phacops' elegans*. A second trench south of Sowerthwaite Farm [SD 7759 6984] showed the Browgill Beds had thinned to a few feet in thickness.

Wenlock Series

AUSTWICK FORMATION This comprises up to 1640 ft (500 m) of greywacke sandstones alternating with siltstones (E. W. Johnson, personal communication). Turbidite units with numerous sole markings and averaging 2 ft 4 in (0.70 m) in thickness make up much of the sequence (McCabe and Waugh, 1973). The turbidite sandstones decrease in thickness and in numbers northwards in Crummackdale and Ribblesdale. Four out of a possible seven graptolite zones have been located in the siltstone (King and Wilcockson, 1934; Ingham and Rickards, 1974, fig. 10). Strata in upper Crummackdale showing liesegang rings (the 'Moughton Whetstones') are now thought to be in the Austwick Formation (McCabe, 1972).

ARCOW FORMATION This formation was named by McCabe (1972) and consists of up to 59 ft (18 m) of bioturbated calcareous mudstones with thin silty laminations and a fauna of trilobites, orthocone nautiloids and bivalves. A similar lithology has been found in the Monograptus ludensis Beds (highest Wenlock age) in North Wales, the Lake District, central Europe, Scandinavia and Nevada (P. T. Warren *in* discussion of McCabe and Waugh, 1973). The highest beds of the Arcow Formation are placed in the Ludlow Series (Ingham and Rickards, 1974, fig. 10).

Ludlow Series

HORTON FORMATION Some 1300 ft (400 m) of monotonous siltstones, quarried for roadstone, make up this formation. The siltstones tend to be laminated and better bedded in the lower part; fossils are not common, but Marr (1887) and Hughes (1907) recorded graptolites (forms usually associated with the *M. nilssoni-scanicus* - fauna) corals and orthocone nautiloids.

STUDFOLD FORMATION This comprises some 330 ft (100 m) of greywacke turbidites restricted to the eastern end of the Horton inlier and to Silverdale.

Dr E. W. Johnson (personal communication) notes that in the Silverdale area, north-east of Stainforth, 820 ft (250 m) of laminated siltstones similar in lithology to the Horton Formation occur above the Studfold Formation, here only 131 ft (40 m) thick. As yet they have yielded no fauna. Overlying these siltstones Dr Johnson reports at least 820 ft (250 m) of coarse-grained well-bedded turbiditic sandstones. AAW

WENSLEYDALE GRANITE

Research on the concealed foundation rocks of the Askrigg Block was commenced by Whetton, Myers and Watson (1965), whose gravity traverse from Kettlewell in Upper Wharfedale to Aysgarth in Wensleydale revealed a negative Bouguer anomaly, which they considered to form a closed low, lying 2 to 4 miles (3.2 to 6.4 km) south of Wensleydale and elongated east – west. This they interpreted as a Pre-Cambrian core of low-density crystalline rocks. The magnetic survey of Bott (1961) led him to postulate a granitic core, as well as a belt of magnetic rocks between this core and the Craven faults. The gravity survey completed as far north as Wensleydale by Myers and Wardell (1967) was interpreted by them as consistent with Bott's interpretation rather than the earlier one and Bott (1967) was able to delineate a small batholith with long axis east – west, 5¼ × 4 miles (8.4 × 6.4 km) in extent, with a single cupola in the vicinity of Semer Water. This was tested in 1973 by the Raydale Borehole at a site [SD 9026 8474] chosen after more detailed geophysical measurements by J. M. Gillingham. A preliminary account has been published by Dunham (1974) based on the log of the borehole by I. C. Burgess and contributions from BGS specialists. At the base of the Dinantian, 12.4 ft (3.78 m) of sandstone and conglomerate occur, with dolomitic siltstone bands and granite debris becoming conspicuous towards the bottom. Calichified granite was entered at 1624.2 ft (495.05 m) and drilling was continued for a further 333.75 ft (101.73 m).

The granite is a pink, mainly non-porphyritic and unfoliated medium-grained rock, superficially resembling the Ennerdale intrusion in the Lake District rather than the granites of Skiddaw, Eskdale or Shap. It differs markedly from the coarser, strongly-foliated Weardale Granite (Dunham and others, 1965). Dr N. G. Berridge finds that the principal constituents are pink microperthite, quartz, greenish cream albite and chlorite, the latter mineral partly after biotite, partly filling veinlets. The virtual absence of muscovite also contrasts with the Weardale rock, though a little sericite is present as an alteration-product. The amount of ferromagnesian material is low. Accessories include zircon, pyrite, calcite, fluorite, sphene, leucoxene and epidote. Pseudomorphs in a clay mineral and leucoxene may represent former amphibole.

The chemical composition of the granite has been investigated by J. G. Holland (in preparation), who analysed 35 samples from the solid rock and 24 from the weathered capping. In Table 3 his preferred averages are compared with the average result (Holland and Lambert, 1970) for the Weardale Granite. The Wensleydale Granite, in the 101 m penetrated, appears chemically homogeneous, but has probably been subjected to some deuteric alteration throughout, producing mainly chlorite. The kind of alteration that would be expected if brines of the type responsible for the Pennine mineralisation had penetrated it (see p. 103) is not seen in the core as there is little evidence of added potash, although logs by N. G. Berridge indicate that the granite core is sporadically affected by ferric oxide and quartz vein mineralisation throughout.

Rb:Sr and K:Ar age determinations were carried out by the BGS Isotope Unit under Dr Snelling. Whole-rock Rb:Sr yielded an isochron leading to an age of 400 ± 10 Ma. This is slightly later than the maximum figure obtained for the Weardale Granite by Holland and Lambert (*op. cit.*) but, within the limits of experimental error, it appears that both intrusions belong to the post-tectonic Caledonian suite, along with many Scottish granites and probably those of the Lake District. An apparent age of 300 Ma was obtained from the K:Ar measurements and the discrepancy is considered to indicate that the Wensleydale Granite was affected by a thermal event at about that time. It is interesting to note that independent evidence (p. 102) has now made it possible to substantiate this conclusion.

It is possible that the Arenig rocks encountered in the Beckermonds Scar Borehole lie within the thermal aureole of the Wensleydale Granite, since they are veined with magnetite, prob-

Table 3 Analyses of granite-bearing sediment and solid Wensleydale Granite compared with Weardale Granite

	1	2	3
Main constituents per cent			
SiO_2	46.60	74.89	72.18
TiO_2	0.12	0.18	0.16
Al_2O_3	6.45	13.22	15.57
Fe_2O_3	1.90*	1.11*	0.27
FeO			0.78
MgO	8.14	0.18	0.50
CaO	13.65	0.45	0.86
Na_2O	0.84	3.30	2.93
K_2O	3.95	5.59	5.96
$H_2O > 105°C$	0.44	0.23	0.69
$H_2O < 105°C$			0.22
P_2O_5	0.049	0.034	0.17
MnO	0.51	0.034	0.07
CO_2	17.34	0.41	—
S	0.04	0.01	
Total	100.03	99.64	100.36
CIPW norm (modified for column 1)			
Q	16.7	33.5	29.8
Cor	0.8	1.5	2.8
Or	23.6	33.0	35.6
Ab	7.1	27.7	24.6
An	0.4	0.8	4.2
Hyp		1.1	2.7
Dolomite	26.6	—	—
Cc	10.6	0.5	—
Trace elements parts per million			
Zr	76	120	52
Sr	39	18	106
Rb	111	207	423
Ba	319	179	170
Y	121	64	
Ce	57	94	
La	20	36	
Cu	87	6	10
Ni	4	1	16
Zn	50	17	45
Pb	41	18	37
Cr	9	4	
U	1.5	5.5	10.2†
Th	11.6	19	22.5†

1 Preferred mean of 28 samples of dolomitic sediment enclosing weathered granite, Raydale Borehole, range 1620½ to 1640 ft (493.9 to 499.9 m)

2 Preferred mean of 32 samples of solid Wensleydale Granite, Raydale Borehole, range 1641½ to 1967 ft (500.3 to 599.5 m)

3 Mean analysis of Weardale Granite, Rookhope Borehole, range 1672 to 2642 ft (509.6 to 805.3 m)

Analyses by J. G. Holland, University of Durham (Holland, personal communication, 1980 [1, 2]; Holland, 1967, and Holland and Lambert, 1970 [3])

* Total iron as Fe_2O_3

† Data supplied by the Geochemistry and Petrology Division of IGS, 1969-70

ably from remobilisation of detrital grains (Berridge, 1982), and the siltstones containing the magnetite are spotted. Professor A. C. Dunham made six electron microprobe analyses of magnetites; total Fe recalculated on Carmichael rules shows an average of 68.6 per cent Fe_2O_3, 29.7 per cent FeO. The magnetite was found to be free from ilmenite and other lamellae in reflected light, but one analysis showed 0.4 per cent TiO_2.

The Ordovician sediments in the Beckermonds core have been invaded by quartz-microdiorite between 1078½ and 1130 ft (328.72 and 344.37 m) depth. Dr Berridge reports that this rock consists of heavily sericitised plagioclase in stumpy prisms of 0.3 to 0.6 mm, euhedral chloritised hornblendes and a groundmass of felsitic character, with some late potassic feldspar. The grain-size decreases towards the contacts with the Arenig greywacke. It is assumed that this is a minor intrusion associated with the Wensleydale Granite.

The enhancement of trace-element values indicated by Dr Holland's analyses of the sediment overlying the Wensleydale Granite as compared with the values in the granite is interesting, but it is not clear what bearing this may have upon the concentration of these elements into the mineral deposits of the orefield.

KCD

REFERENCES

Bott, M. H. P. 1961. A gravity survey off the coast of north-east England. *Proc. Yorkshire Geol. Soc.*, Vol. 33, pp. 1–20.

— 1967. Geophysical investigations of the Northern Pennine Basement rocks. *Proc. Yorkshire Geol. Soc.*, Vol. 36, pp. 139–168.

— and Masson Smith, D. 1957. The geological interpretation of a gravity survey of the Alston Block and the Durham coalfield. *Q. J. Geol. Soc. London*, Vol. 113, pp. 93–117.

Berridge, N. G. 1982. Petrography of the pre-Carboniferous rocks of the Beckermonds Scar Borehole in the context of the magnetic anomaly at the site. *Proc. Yorkshire Geol. Soc.*, Vol. 44, pp. 89–98.

Dunham, K. C. 1974. Granite beneath the Pennines in north Yorkshire. *Proc. Yorkshire Geol. Soc.*, Vol. 40, pp. 191–194.

— Dunham, A. C., Hodge, B. L. and Johnson, G. A. L. 1965. Granite beneath Viséan sediments with mineralisation at Rookhope, northern Pennines. *Q. J. Geol. Soc. London*, Vol. 121, pp. 383–417.

— Hemingway, J. E., Versey, H. C. and Wilcockson, W. H. 1953. A guide to geology of the district round Ingleborough. *Proc. Yorkshire Geol. Soc.*, Vol. 29, pp. 77–115.

Holland, J. G. 1967. Rapid analysis of the Weardale Granite. *Proc. Yorkshire Geol. Soc.*, Vol. 36, pp. 91–114.

— (in preparation). The geochemistry of the Wensleydale Granite.

— and Lambert, R. St. J. 1970. Weardale Granite. Pp. 103–118 in *Geology of Durham County*. G. Hickling (Editor). *Trans. Nat. Hist. Soc. Northumberland, Durham and Newcastle upon Tyne,* Vol. 41.

Hughes, T. McK. 1907. Ingleborough. Part 4. Stratigraphy and palaeontology of the Silurian. *Proc. Yorkshire Geol. Soc.*, Vol. 16, pp. 45–74.

Ingham, J. K. and Rickards, R. B. 1974. Lower Palaeozoic Rocks. Pp. 29–44 in *The geology and mineral resources of Yorkshire.* Rayner, D. H. and Hemingway, J. E. (Editors). (Leeds: Yorkshire Geological Society.) 405 pp.

King, W. B. R. and Wilcockson, W. H. 1934. The Lower Palaeozoic rocks of Austwick and Horton-in-Ribblesdale, Yorkshire. *Q. J. Geol. Soc. London,* Vol. 90, pp. 7–31.

LEEDAL, G. P. and WALKER, G. P. L. 1950. A restudy of the Ingletonian Series of Yorkshire. *Geol. Mag.*, Vol. 87, pp. 57–66.

MCCABE, P. J. 1972. The Wenlock and Lower Ludlow strata of the Austwick and Horton-in-Ribblesdale inlier of north-west Yorkshire. *Proc. Yorkshire Geol. Soc.*, Vol. 39, pp. 167–174.

— and WAUGH, B. 1973. Wenlock and Ludlow sedimentation in the Austwick and Horton-in-Ribblesdale inlier, north-west Yorkshire. *Proc. Yorkshire Geol. Soc.*, Vol. 39, pp. 445–470.

MARR, J. E. 1887. The Lower Palaeozoic rocks near Settle. *Geol. Mag.*, Dec. 3, Vol. 4 (24), pp. 35–38.

MYERS, J. O. and WARDELL, J. 1967. The gravity anomalies of the Askrigg Block south of Wensleydale. *Proc. Yorkshire Geol. Soc.*, Vol. 36, pp. 169–173.

O'NIONS, R. K., OXBURGH, E. R., HAWKESWORTH, C. J. and MACINTYRE, R. M. 1973. New isotopic and stratigraphical evidence of the age of the Ingletonian: probable Cambrian of northern England. *Q. J. Geol. Soc. London*, Vol. 129, pp. 445–452.

RASTALL, R. H. 1906. The Ingletonian Series of West Yorkshire. *Proc. Yorkshire Geol. Soc.*, Vol. 16, pp. 87–100.

RAYNER, D. H. 1957. A problematical structure from the Ingletonian rocks, Yorkshire. *Trans. Leeds Geol. Assoc.*, Vol. 7, pp. 34–42.

SOPER, N. J. and MOSELEY, F. 1978. Structure. Pp. 45–67 in *The geology of the Lake District.* MOSELEY, F. (Editor). (Leeds: Yorkshire Geological Society.) 284 pp.

WADGE, A. J. 1978. Classification and stratigraphical relationships of the Lower Ordovician rocks. Pp. 68–78 in *The geology of the Lake District.* MOSELEY, F. (Editor). (Leeds: Yorkshire Geological Society.) 284 pp.

WHETTON, J. T., MYERS, J. O. and WATSON, I. J. 1965. A gravimeter survey in the Craven district of north-west Yorkshire. *Proc. Yorkshire Geol. Soc.*, Vol. 30, pp. 259–287.

WILSON, A. A. and CORNWELL, J. D. 1982. The IGS borehole at Beckermonds Scar, North Yorkshire. *Proc. Yorkshire Geol. Soc.*, Vol. 44, pp. 59–88.

CHAPTER 3

Carboniferous rocks: General account

INTRODUCTION

The mineral deposits of the Northern Pennines are effectively confined to the Carboniferous rocks and can only be understood in relation to the stratigraphy of this system. This chapter gives a general survey of the extensive literature in terms of classification, stratigraphical palaeontology, patterns and problems of sedimentation; detailed discussion by stages and formations follows in chapters 4 and 5.

Phillips (1836) in a scholarly work was responsible for the first description of the Carboniferous of the Yorkshire Dales; he recognised that basement beds resting on the Lower Palaeozoic rocks were followed by massive limestone formations (the Great Scar Limestone) succeeded upwards by repeated alternations of limestone, shale and sandstone (his Yoredale Beds), in turn overlain by the coarse sandstones, shales and thin limestones of the 'Millstone Grit'. This classification was accepted by the official surveyors of a century ago, who found that the relatively thin Yoredale limestones could be mapped accurately over very wide areas. Their six-inch mapping has for the most part stood the test of time, as the re-issued and revised one-inch and 1:50 000 maps show. They did not, however, subdivide the thick Great Scar Limestone; this work remained for the monumental labour of Garwood and Goodyear (1924), using expertise built up during years of research in north-west England (Garwood, 1913). By means of a combination of fosil bands and faunal zones, they showed how the Carboniferous deposits lapped up against and eventually submerged the Lower Palaeozoic landscape. Three stratigraphers trained originally by Garwood developed his approach: Chubb and Hudson (1925) delineated the unconformity (to which we attach some importance) at the base of the 'Millstone Grit' in Wharfedale, and subsequent years saw important papers from Hudson (notably 1924, 1941, 1944) and Turner (1927, 1959, 1962). Wartime mapping directed towards mineral resources led to the reclassification of the limestones below the sub-Grassington Grit (intra-Pendleian) unconformity at Greenhow by Dunham and Stubblefield (1945) and the establishment of the age of the top of the grit on the goniatite time-scale.

Post-war university research led to the establishment of an acceptable position at the base of the Great-Main Limestone for the major division between the Dinantian and Namurian as defined on the European mainland (Johnson *in* Mills and Hull, 1976; Johnson, Hodge and Fairbairn, 1962). Other university research projects, particularly at Leeds and Durham, contributed significantly to the stratigraphical picture, using sedimentological, petrological and palaeontological techniques (Wells, 1955, 1957; Rowell and Scanlon, 1957a, b; Moore, 1958, 1959, 1960; Wilson 1960 a, b; Wilson and Thompson, 1959, 1965). Wells's (1955) discussion of the genesis of the Richmond Cherts and Moore's exhaustive study of the Yoredale facies beds in their type area with a suggested reconstruction of their environment of deposition are particularly noteworthy. Recent attempts have been made to divide up or correlate the 'Great Scar Limestone' by Schwarzacher (1958), Doughty (1968, 1974) and Waltham (1971). Burgess and Mitchell (1976) have demonstrated in detail how the beds near the Asbian – Brigantian boundary split northwards into the Stainmore trough. The results of the resurvey of the Brough and Barnard Castle sheets (31 and 32) on the northern fringe of the area have been published (Burgess and Holliday, 1979; Mills and Hull, 1976). The latest trend in university research is to study the sedimentology of a limited set of beds over a wide area, as for example, the Middle, Underset and Crow limestones (Cousins, 1977, unpublished thesis).

The first major review of the geology of the area was in Kendall and Wroot's popular account (1924). In recent years the Dinantian geology has been covered by review papers by Rayner (1953) and Ramsbottom (1974). The recent syntheses of Dinantian and Silesian stratigraphy by Geological Society of London specialist committees are a useful guide to correlation within the area and further afield, (George and others, 1976; Ramsbottom and others, 1978).

Classification of the Dinantian

The earliest detailed attempts to divide up the sequence by Garwood (1913) and Garwood and Goodyear (1924) were based on macrofossils (Figure 2). Corals and brachiopods were used to erect a scheme of zones and subzones which were mapped on the ground. Whilst sandstones were also surveyed, the limestone lithologies received little attention. The first attempt in areas in and close to the present ground to divide up the massive limestone sequence primarily on lithostratigraphical grounds was by Dunham and Rose (1941) in Furness, and by Dunham and Stubblefield (1945) at Greenhow. More recently Burgess and Holliday (1979) have divided the Dinantian succession at Brough into the Orton Group and Alston Group, the latter being equivalent to the Lower plus Middle Limestone groups of Dunham (1948) in the Alston area.

A recent refinement for purposes of correlation is the subdivision of the Dinantian into six stages (George and others, 1976). Six stratotype localities, one for the base of each stage are defined. The Asbian Stage with its stratotype at Little Asby Scar, just outside the present area, closely corresponds with the Lower Limestone Group of Dunham (1948). The Brigantian Stage, with its stratotype at Janny Wood on the River Eden in the present area, corresponds with the Middle Limestone Group, but the base is drawn at the base of the Peghorn, rather than the Smiddy Limestone. Schemes of Dinantian correlation using foraminifera and conodonts are in the process of being refined and some of the results for the former are given in the present work.

The Yoredale Beds of Phillips (1836), redefined as a Series by Moore (1958) is only used in a facies connotation in this account; the Brigantian Stage of the Dinantian is wholly of Yoredale facies except for the lower beds in the south of the area, but in the north the facies appears below the base of this stage and extends up into the Namurian.

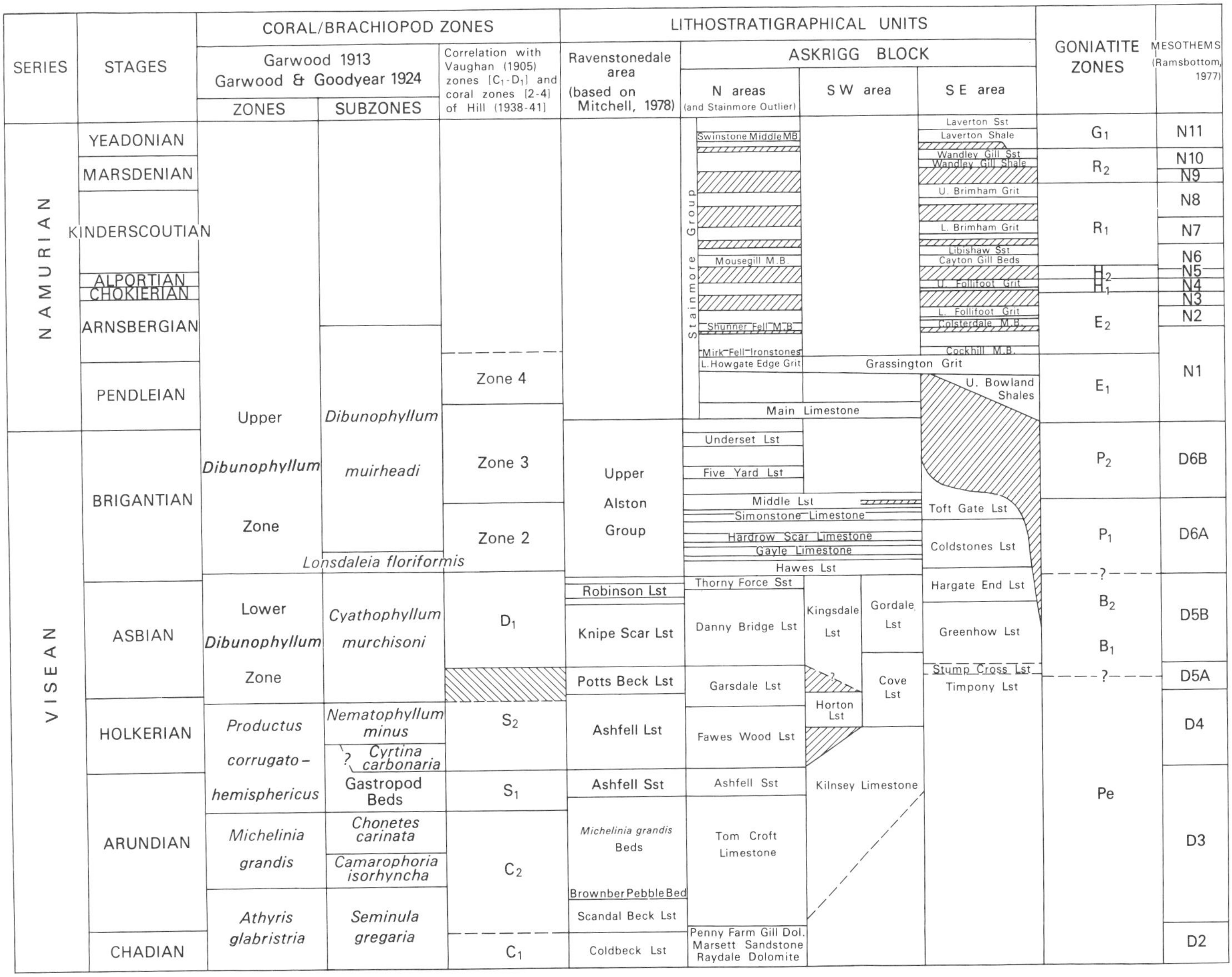

SERIES	STAGES	CORAL/BRACHIOPOD ZONES: Garwood 1913, Garwood & Goodyear 1924 — ZONES	SUBZONES	Correlation with Vaughan (1905) zones (C_1-D_1) and coral zones (2-4) of Hill (1938-41)	LITHOSTRATIGRAPHICAL UNITS: Ravenstonedale area (based on Mitchell, 1978)	ASKRIGG BLOCK: N areas (and Stainmore Outlier)	SW area	SE area	GONIATITE ZONES	MESOTHEMS (Ramsbottom, 1977)
NAMURIAN	YEADONIAN					Stainmore Group; Swinstone Middle MB		Laverton Sst; Laverton Shale	G_1	N11
	MARSDENIAN							Wandley Gill Sst; Wandley Gill Shale	R_2	N10; N9
	KINDERSCOUTIAN					Mousegill M.B.		U. Brimham Grit; L. Brimham Grit; Libishaw Sst; Cayton Gill Beds	R_1	N8; N7; N6
	ALPORTIAN							U. Follifoot Grit	H_2	N5; N4
	CHOKIERIAN								H_1	N3
	ARNSBERGIAN					Shunner Fell M.B.		L. Follifoot Grit; Colsterdale M.B.	E_2	N2
	PENDLEIAN			Zone 4		Mirk Fell Ironstones; L. Howgate Edge Grit; Main Limestone	Grassington Grit	Cockhill M.B.; Grassington Grit; U. Bowland Shales	E_1	N1
VISEAN	BRIGANTIAN	Upper *Dibunophyllum* Zone	*Dibunophyllum muirheadi*; *Lonsdaleia floriformis*	Zone 3; Zone 2	Upper Alston Group	Underset Lst; Five Yard Lst; Middle Lst; Simonstone Limestone; Hardrow Scar Limestone; Gayle Limestone; Hawes Lst		Toft Gate Lst; Coldstones Lst	P_2; P_1	D6B; D6A
	ASBIAN	Lower *Dibunophyllum* Zone	*Cyathophyllum murchisoni*	D_1	Robinson Lst; Knipe Scar Lst; Potts Beck Lst	Thorny Force Sst; Danny Bridge Lst; Garsdale Lst	Kingsdale Lst; Gordale Lst; Cove Lst	Hargate End Lst; Greenhow Lst; Stump Cross Lst; Timpony Lst	?; B_2; B_1; ?	D5B; D5A
	HOLKERIAN	*Productus corrugato-hemisphericus*	*Nematophyllum minus*; ? *Cyrtina carbonaria*; Gastropod Beds	S_2; S_1	Ashfell Lst; Ashfell Sst	Fawes Wood Lst; Ashfell Sst	Horton Lst; Kilnsey Limestone		Pe	D4
	ARUNDIAN	*Michelinia grandis*	*Chonetes carinata*; *Camarophoria isorhyncha*	C_2	*Michelinia grandis* Beds; Brownber Pebble Bed; Scandal Beck Lst	Tom Croft Limestone				D3
	CHADIAN	*Athyris glabristria*	*Seminula gregaria*	C_1	Coldbeck Lst	Penny Farm Gill Dol.; Marsett Sandstone; Raydale Dolomite				D2

Figure 2 Classification of the Viséan and Namurian

Classification of the Namurian

The time-honoured division into Millstone Grit and Carboniferous Limestone has carried a varying connotation in different areas (Ramsbottom and others, 1978, fig. 1), and it is not proposed to use these terms. In terms of lithostratigraphy the Namurian rocks of the present area all belong to the Stainmore Group (Burgess and Holliday, 1979). In chronostratigraphical terms Namurian beds are divided up into seven stages which have grown up gradually following an early proposal by Bisat (1928). Within these stages a series of zones and subzones have been defined, their limits being taken at the base of the marine band containing each characteristic fossil. The goniatite record on the Askrigg Block is, however, less complete than in the Craven Basin and there is consequently a greater margin of error in ascribing a given set of beds to a particular zone (Figure 3).

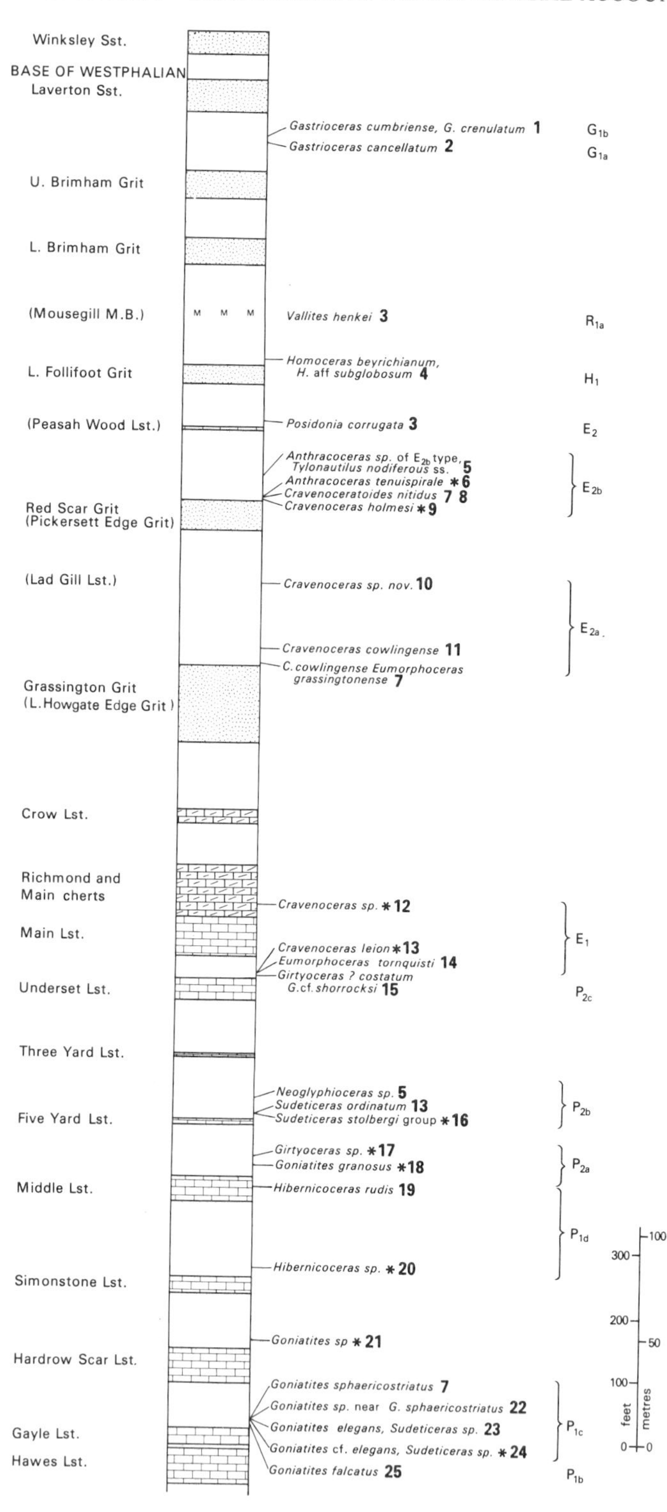

Figure 3 Goniatites and other macrofossils of value in goniatite zonation collected from the district. The stratigraphical column up to and including the Crow Limestone is based on the succession in upper Swaledale and upper Wensleydale. The higher part is based on Nidderdale and lower Wensleydale, but some units from the north-west of the district which have yielded stratigraphically significant fossils are inserted, with their names in brackets

CONTROL OF DEPOSITION BY BLOCKS AND BASINS

Following the work of Marr (1921) the Pennine country between the Craven and Stublick faults was regarded as a rigid block, little affected by large scale folding and faulting and consisting of gently dipping sediments (Figure 15). However, Bott (1961), on the basis of a gravity survey, suggested that Lower Carboniferous and possible Devonian rocks beneath Stainmore were about 10 000 ft (3048 m) thick and recent syntheses (Johnson, 1967; Taylor and others, 1971, fig. 17) indicate a belt of trough sedimentation extending westwards to Ravenstonedale separating the Alston and Askrigg blocks. Dinantian sedimentation on these blocks started in late Arundian and Holkerian times and the beds are much thinner than in adjacent areas of trough sedimentation.

The concept of the Stainmore Trough is important to the present study since it affects the northern half of the present area. A near doubling of Dinantian rock thicknesses between the Beckermonds and Raydale boreholes, with beds progressively onlapping to the south, is strong indication of northward thickening into this trough (Figure 4).

Bott (1967) had suggested, however, that the Stockdale Disturbance lying farther north in Swaledale was the southern limit of the Stainmore Trough. The Brigantian isopachs (Figure 10) and the evidence of the above two boreholes suggest, in fact, that great thickening begins just north of Beckermonds Scar Borehole and persists and accelerates somewhat across the Stockdale Fault towards the axis of the Stainmore Trough (Figure 4) where recent geophysical work by Swinburn (p. 72) suggests stratal thicknesses up to 11 350 ft (3460 m).

Bott (1961) suggested that the Butterknowle Fault belt formed the northern limit of the Stainmore Trough in eastern County Durham. There is now strong confirmation of the eastern extension of the Stainmore Trough from a deep, largely rock-bitted borehole at Seal Sands, Middlesbrough [NZ 538 238]. Strata ranging in age from early Pendleian or late Brigantian down to probable Arundian total a remarkable 11 295 ft (3443 m) in thickness, equalling the maximum thickness for the same beds in the west of the Northumberland Trough, north of the Alston Block (Taylor and others, 1971, pl. 5, J). The succession at Seal Sands

Key to Figure 3
1 Wilson and Thompson (1965, p.213); 2 Ramsbottom and others (1978, p.29); 3 Owens and Burgess (1965, pp.23–24); 4 Wilson (1960b, p.437); 5 Ramsbottom *in* Rayner and Hemingway (1974, pp.70, 85–86); 6 p.23*; 7 Dunham and Stubblefield (1945, pp.238, 240, 257); 8 Wilson and Thompson (1959); 9 p.66*; 10 Burgess and Ramsbottom (1970); 11 Hudson (1941, p.26); 12 New record in mudstone above Main Chert, Swinner Gill [NY 9118 0119]; 13 Figure 8* and Wilson (1960a, pp.293, 297); 14 O'Connor (1964, p.68). The specimens of *Eumorphoceras pseudobilingue* from the area and horizon cited by Hudson (1941) and Black (1950) are apparently lost; 15 Johnson and others (1962, p.439); 16 p.44*; 17 p.42*; 18 p.22*; 19 Hicks (1959, p.40); 20 p.39*; 21 p.38*; 22 Bisat (1957, p.124); 23 Redetermined by W. H. C. Ramsbottom from Bisat (1924, p.45); 24 p.31*; 25 Moore (1958, p.99)

* Denotes new record with reference to the cited page in this volume

comprises some 4000 ft (1219 m) of beds of Yoredale facies, also known in cored boreholes through nearly 1000 ft (305 m) of strata in the northern suburbs of Middlesbrough. These rest on 7380 ft (2249 m) of mudstones, siltstones and sandstones with a few thin limestones. Dr W. H. C. Ramsbottom notes that specimens taken from a core sample at a depth of 13 495 to 13 507 ft (4113–4117 m) near the bottom of the borehole contain archaediscids and are therefore of post-Chadian age (probably late Arundian). Dr B. Owens reports that the miospores confirm a Pu Zone (Chadian to Holkerian) age.

Information about the eastern margin of the Askrigg Block under Permian and later strata is derived solely from boreholes since the most easterly exposures do not show any marked thinning suggestive of the proximity of a block margin. Kent (1966, figs 1, 2) tentatively places the boundary north-east of Northallerton between the Harlsey and Cleveland Hills boreholes. The most easterly, and lowest stratigraphical penetration, the Cleveland Hills Borehole, is placed by Kent beyond the eastern limit of the Askrigg Block. From Fowler's detailed account (1944) of this borehole it is apparent that the upper half of the Carboniferous sequence, ranging from the Richmond Cherts down to about the Middle Limestone, is broadly similar to that at Richmond. Several short cores yielded typical Yoredale facies fossils. The Yoredale facies beds rest on over 1000 ft (305 m) of mudstone reminiscent of the Bowland Shales from which two short cores were taken, each with goniatites (P_{1a}–P_{1b}). Near the base of the borehole a 95 ft (29 m) thick limestone yields brachiopods of late Asbian aspect, broadly similar to those in the Draughton Limestone of the Craven Basin. A slightly lower limestone yields *Productus redesdalensis* and *'P.' hemisphaericus*, forms found in the early part of the late Asbian of the Northumberland Trough. The Harlsey Borehole, thought by Kent to be on the block, penetrated only Yoredale facies beds which also probably included Richmond Cherts. Limestones below the supposed Richmond Cherts are thicker than at Cleveland Hills, but the penetration of the succession is much less.

The southern margin of the Askrigg Block, located on the Middle Craven Fault, is the site of rapid facies change across a marginal reef belt (Figure 6). In the east, near Appletreewick, the Middle Craven Fault seems discontinuous, but the zone of facies change must pass south of the block facies area of Trollers Gill [SE 066 617]. Farther east the Aldfield Borehole (Falcon and Kent, 1960) lies on the Askrigg Block; but the unusually thin Namurian sequence may result from contemporaneous effects close to the concealed eastern continuation of the Craven line. The Ellenthorpe Borehole [SE 4233 6703] near Boroughbridge shows a thick Dinantian sequence which seems to be in a basin (Kent, 1966, fig. 1).

Close to the tectonically active southern margin of the block, rocks of Holkerian – Asbian age are particularly thick, as at Greenhow, and conditions must have been unusually favourable for carbonate accumulation. South of the boundary a thickness of up to 6525 ft (1989 m) of Dinantian sediments accumulated in the Craven Basin, including many wackestones and calcareous mudstones ranging from Tournaisian to Asbian in age. These contrast with the thin and dominantly pure carbonate sediments on the Askrigg Block, with a late Arundian – Holkerian onset of sedimentation. In the Brigantian, too, there is a profound facies change from the Yoredale sediments of the block, with their characteristic brachiopod faunas, to the goniatite bearing Bowland Shales of the Basin. The southward transition from Yoredale to Bowland Shale sedimentation is more gradual in beds of late Brigantian (P_2) age than has perhaps been realised. Owing to the virtual failure of the Three Yard and Underset limestones in the south of the block around Fountains Fell and Ingleborough, much of the Yoredale facies beds consist of mudstone and sandstone, albeit largely without goniatites, and a relatively minor facies change is needed to effect the transition into Bowland Shales with their numerous intercalated sandstones (Earp and others, 1961, pl. 7). AAW

The complex movements along the Craven line are best known in relation to Asbian and later strata. In early Asbian times a continuous reef belt became established on the southern edge of the Askrigg Block, with upstanding knolls south of this near Cracoe (see Rayner, 1953 and Ramsbottom, 1974). The continuous reef belt became broken into a number of blocks by pre-Namurian fault movements associated with the Middle Craven Fault (Hudson, 1930, 1932). Prior to this faulting, however, the reef belt experienced several episodes of uplift, resulting in local unconformities in late Asbian (Upper B_2) and a distinct faunal change in P_{1a} times (Mundy, 1978). The latter change reflects re-establishment of a reef fauna after a short hiatus. An erosive episode in P_{1a-b} is marked by the accumulation of reef-derived boulder-beds and conglomerates south of the reef belt, as for example in the Cracoe area. These debris beds are conformable with overlying Bowland Shales of P_{1a} age in Skelterton Beck (Black, 1940). This erosive episode, although due to a major uplift of the reef belt, cannot as yet be linked with proven fault movements.

The main movements of the Middle Craven Fault may, however, be associated with later erosive episodes which are indicated by boulder beds in Bowland Shales of P_2 (Black, 1957) and especially P_2 – E_1 age (Dixon and Hudson, 1931; Hudson, 1944). It was probably during high P_2 times (latest Brigantian) that the pre-Grassington Grit fault of Waterhole North Vein at Greenhow (Figure 17) had its main movement, although its initiation was during deposition of the sandstones of the Hardrow Scar Cyclothem (P_1). The south facing escarpment in the Malham – Bordley area, created by down-south movements of the Middle Craven Fault, was probably fully developed during this P_2–E_1 event, and subsequent Bowland Shales of Pendleian (E_1) age overlapped this escarpment together with the down-faulted reef topography to the south (Hudson, 1944). A similar overlap of a pre-E_1 escarpment took place north of the North Craven Fault in a belt from Grassington (Black, 1950) to Grimwith (Wilson, Dunham and Pattison, in preparation). Walker (1967) suggested that the pre-E_1 escarpment was not the result of faulting associated with pre-Bowland Shale uplift but of shoaling of the lower Yoredales near the edge of the Askrigg Block, together with the effect of differential subsidence and sediment starvation of the Craven Basin. Whilst the crinoidal mounds in the Middle Limestone of Grassington

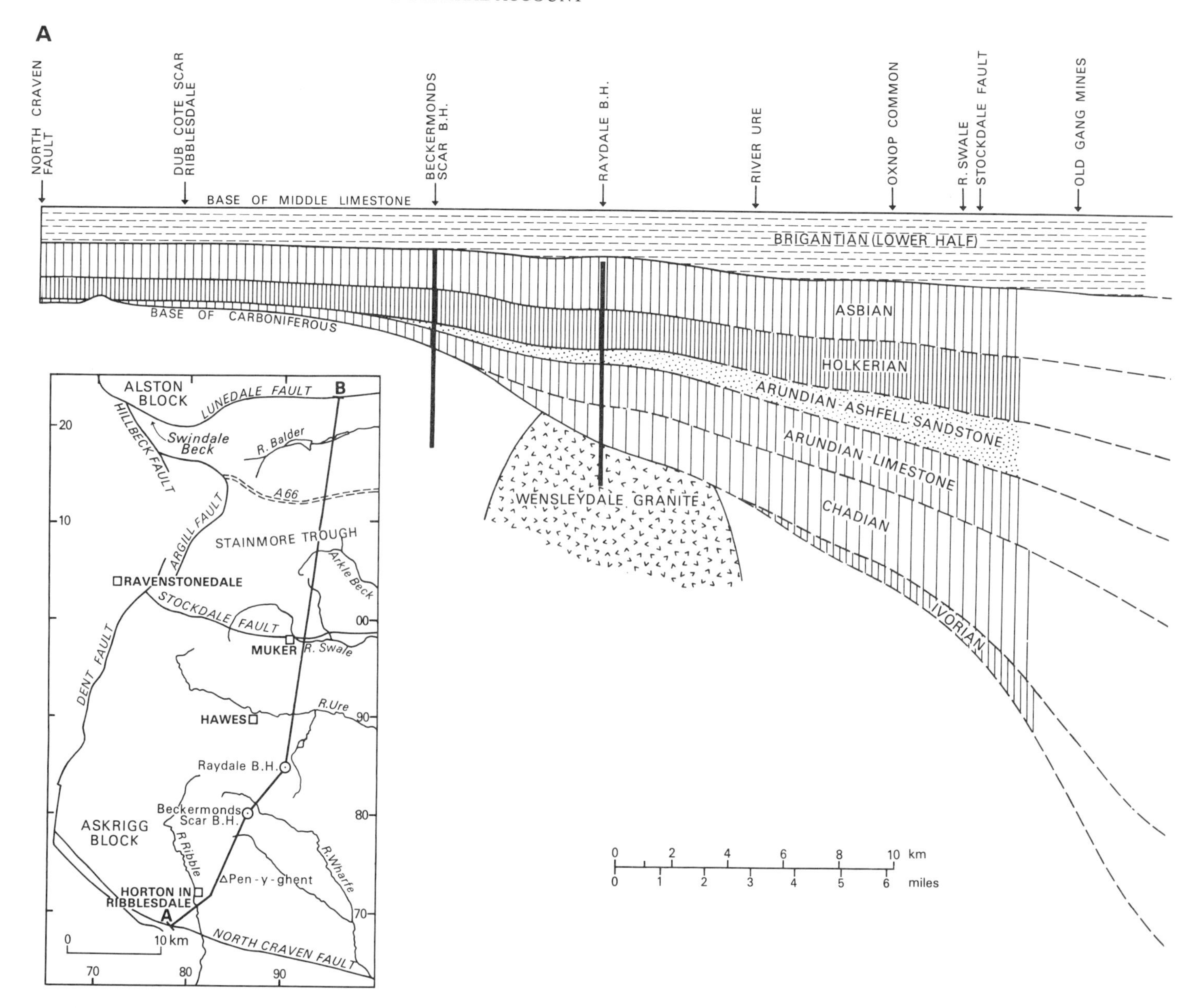

Figure 4 Section to illustrate the stratigraphy of the Askrigg Block and Stainmore Trough.

Dashed lines denote uncertainty, and the area with ornament omitted indicates the zone of greatest uncertainty

could represent a marginal facies, the attendant thickening of the limestone is not solely a block-edge phenomenon as it also runs north into lower Wensleydale. A theory involving faulting and erosion seems to fit more of the evidence.

Despite detailed surveys in the Settle – Malham area a fundamental problem remains. Exposures of the portion of the Lower Bowland Shales of P_1 age and beds of Yoredale facies known to be equivalent on goniatite evidence are always separate and there is no visible interdigitation of the two facies. It is possible that abrupt change in water depth over the reef belt could have effectively confined the two types of sedimentation. Alternatively in the light of evidence of uplift of the reef front prior to Bowland Shales sedimentation, it is possible that a physical barrier may have separated the two areas.

DJCM

The latest known effects on sedimentation along the Craven line relate to a great southward thickening of Millstone Grit facies strata which took place over a belt several kilometres wide. At Pateley Bridge A. T. Thompson (personal communication) has noted dramatic southward thickening of the Lower Brimham Grit from 30 to 400 ft (9–122 m) across the North Craven Fault. This is indicative of continued movement during Kinderscoutian sedimentation.

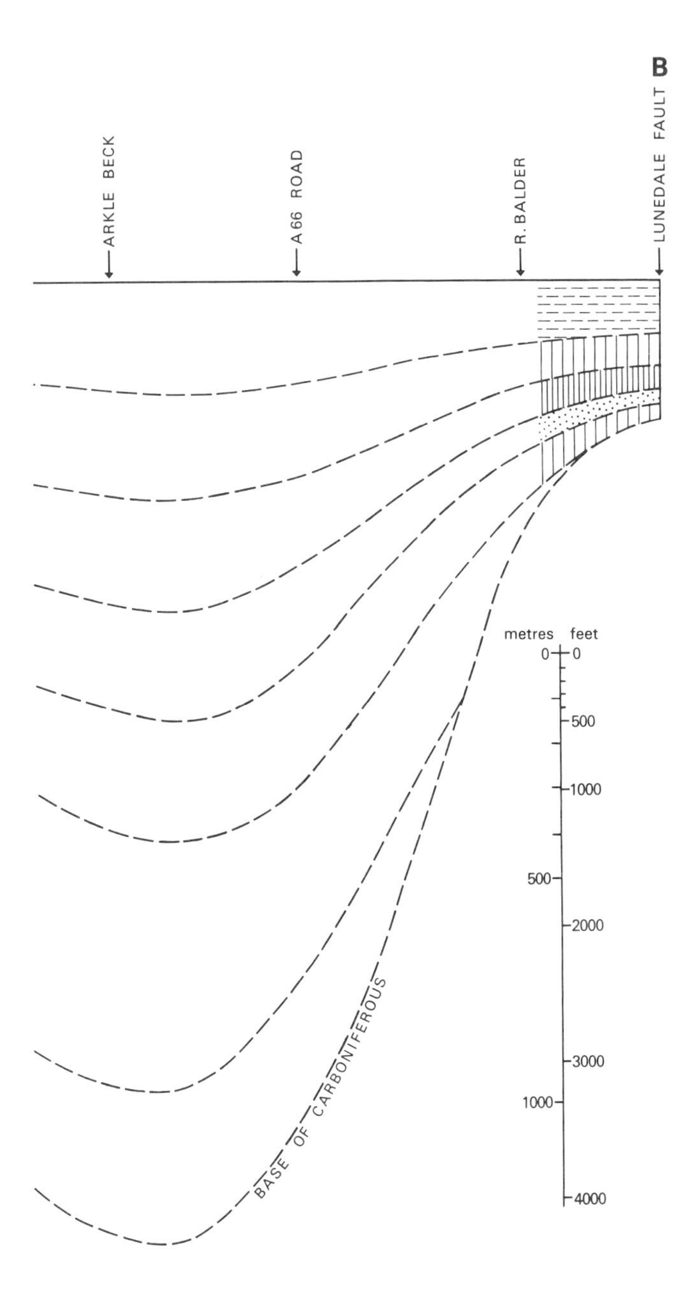

The western edge of the Askrigg Block is commonly taken at the Dent line, but there is no common agreement as to whether it is flanked by another block or a trough. Johnson (1967, fig. 1) represents the Askrigg Block as continuous with the Lake District Massif, whilst George (1958, fig. 21) and Kent (1966, fig.1) show a trough extending up the west side to Ravenstonedale. The present study has shown that, in relation to the Stainmore Trough, the northward thickening in most of the Brigantian cycles coincides with that in the rest of the underlying Dinantian. Thus the considerable westerly thickening in the Brigantian cycles noted by Dr Hicks (personal communication) could well indicate corresponding thickening in the lower beds. Garwood (1913) has mapped beds of the *Seminula gregaria* Subzone (Chadian) in the Kendal outliers, and these are interpreted by George as situated on the western edge of a trough extending from Morecambe Bay to Ravenstonedale. Garwood's mapping (1913) shows only Arundian limestone resting on basement conglomerates close to Kirkby Lonsdale, however. The width of outcrop in which one can fit Tournaisian rocks between Kendal and Kirkby Lonsdale is small, but it is just possible that such rocks, now removed by erosion, once occurred in a narrow trough a few miles west of the Dent Line.

AAW

PATTERNS OF CARBONIFEROUS SEDIMENTATION

The boreholes at Raydale and Beckermonds Scar, and surface exposures, show that there is an onlapping sequence on the south side of the Stainmore Trough (Figure 4). The earliest, Ivorian, strata are restricted to the far north in Ravenstonedale and beds of probable Chadian age in the Raydale Borehole succession also wedge out southwards, being absent in Beckermonds Scar Borehole and at crop in the south. The Askrigg Block thus, initially, stood up as an island with progressively higher Dinantian horizons lapping on to it. The history of sedimentation can be divided up into episodes as follows.

Prelude to the main carbonate deposition

In Ivorian times, at Ravenstonedale, dolomites of probable intertidal origin followed by fluviatile sandstones were laid down at the western head of the Stainmore Trough. It may be that these beds change into deeper water facies to the east, along the trough. The Askrigg Block to the south stood up as an island whilst Tournaisian or earlier fan conglomerates were laid down in the Sedbergh area.

Main period of carbonate deposition

In Chadian times, at Ravenstonedale, limestones and dolomites carry a restricted fauna perhaps indicative of hypersalinity. Southwards, in Garsdale, thick palaeosols occur in the Nor Gill Sandstones and overlying dolomites of probable intertidal origin were deposited along with fluviatile sandstones. The whole picture is of very shallow water or emergence and there seems to be little indication here of the two Chadian marine transgressions postulated by Ramsbottom (1977). Algal beds at Ravenstonedale could, however, mark an end-Chadian regression of sea level (Mitchell, 1978, amended by subsequent work).

Arundian times saw the start of a period of uninterrupted limestone deposition over much of the area, containing the first prolific coral fauna, notably in the lower half of the sequence. In the south, Arundian seas lapped against irregularities in the old landscape fringed by beach deposits, namely conglomerates and calcareous sandstones.

Towards the close of Arundian times, the Ashfell Sandstone was laid down on the south side of a land ridge stretching across from the Lake District to the Alston Block. Deposition was in a moderate energy environment with current ripple lamination and some cross-bedding. It was interrupted by several periods of limestone deposition. Further south, grainstones, well-washed limestones from a higher than usual energy environment, occur high in the Kilnsey

Limestone at the approximate level of the Ashfell Sandstone. It is possible that all these clastic beds and grainstones may result from the late Arundian regression of sea level as postulated by Ramsbottom (1973).

Continuous limestone deposition resumed in Holkerian times punctuated by some brief periods of shallow water, locally marked by scattered porcellanous calcilutite bands. Corals, notably *Lithostrotion martini*, flourished.

In late Holkerian – early Asbian times extensive signs of shallowing of the water were particularly pronounced in the centre of the area, possibly in part as the result of an end-Holkerian regression (Ramsbottom, 1973) but also spanning early Asbian beds in Beckermonds Scar Borehole. Siltstones and mudstones occur interbedded with up to ten calcilutite bands, locally with fenestral fabric, pointing to an intertidal origin. Thin coals occur, one of them on a seatearth mudstone. Limestone faunas in the central area include rhynchonelloids, whilst corals are usually absent. Southwards there are only one or two calcilutite bands and there is less sign of very shallow water.

Early Asbian times were marked by nearly continuous carbonate deposition with good growth conditions for *Lithostrotion martini*. In the north and central areas a characteristic alternation of dark and paler limestones occurs and these pass southwards into the light grey lower beds of the Kingsdale Limestone. The southern edge of the Askrigg Block at Greenhow had optimum conditions for very thick carbonate deposition in early Asbian times. A sandstone within the limestone sequence at Ravenstonedale perhaps marks the regression of the sea that took place at the end of the early Asbian, as suggested by Ramsbottom (1977), but the core of Raydale Borehole does not seem to show evidence of regression at this level.

The late Asbian is marked by a long period of limestone deposition, favouring some coral growth, punctuated by several periods of very shallow water marked by karstic or bioturbated surfaces and rare coal seams. Distinctive unfossiliferous pyritic pale grey mudstones are probably the product of ash-falls from distant volcanic eruptions. The basal beds in cycles 3 and 4 of Schwarzacher (1958) are locally cross-bedded grainstones, the product of a high energy depositional environment. Near the southern edge of the Askrigg Block the clay bands seem to have been largely washed away by sea bottom currents.

The special reef facies of algal mudstones with breccias accumulated in knolls or mounds along the Craven line in Asbian times.

Ramsbottom (1973, 1977) has proposed a theory of eustatic control to explain certain lithological and faunal variations in Dinantian strata. Worldwide regressions of sea level are thought to have caused widespread faunal exterminations followed by evolutionary bursts during the next transgression. The phases are also to be identified on specific regressive phase lithological criteria, especially in stable block areas. Originally six mesothemic cycles were proposed (1973) but a more recent diagram shows eleven (1977). The remarks in the foregoing paragraphs indicate that there is lithological evidence in beds of Viséan age for most of the regressions, but in some cases over only part of the area. George (1978), in particular, is sceptical about the effectiveness of the eustatic control theory as a unifying principle since he believes local tectonic movements tend to mask the effects of regressions of sea level when and if they occur. One of the difficulties in applying the theory to lithological characters lies in areas where regressive criteria are abundant, as in the Northumberland Trough, thus inhibiting a widespread and at the same time unequivocal application of the sedimentary criteria (George, *op. cit.*, pp. 244–245 and the ensuing discussion).

Period of Yoredale Facies deposition

The long period of carbonate deposition was interrupted just north of the present area by deep channelling of underlying limestones beneath a cross-bedded sandstone which persists southwards to Wensleydale. This sandstone is thought to mark the end-Asbian regression by Ramsbottom (1973) and heralds the incoming of another facies—the Yoredales. Broadly speaking the Yoredales consist of eleven cyclic repetitions of limestone, mudstone and sandstone, on the lines detailed by Hudson (1925), from the Hawes up to the Crow Limestone.

The most persistent members of the Yoredale cyclothems are the limestones indicative of widespread periodic submergence under extensive shallow shelf seas. The main exceptions are the Three Yard and Underset limestones which become patchy and impoverished in the south of the Askrigg Block, despite the fact that they can be traced northwards across both the Alston Block and Northumberland Trough (Johnson, 1960, fig. 3). The Middle Limestone also is split by clastic interbeds in the north-west.

Sudden local variations in thickness due to biological 'build-ups' are seen in the Gayle, Hardrow Scar, Middle, Underset and Main limestones and Richmond Cherts. The most common feature is a mound-like accumulation of crinoid debris, frequently whole stems, with or without a central core of calcilutite. Original depositional dips are seen, though the crinoid debris rocks are in many cases poorly bedded. Associated fossils include blastoids in the Middle Limestone (Joysey, 1955) and *Gigantoproductus* in the Main Limestone.

The clastic interbeds from the Hawes up to the Three Yard Limestone are thinnest in the south-east where only those of the Hardrow Scar cyclothem are thick and contain sandstone. The sequence consists mostly of limestone with thin fossiliferous marine mudstones. Farther north, beds thicken greatly and sandstones come in progressively, with evidence for a number of minor cycles in the strata of the Hardrow Scar and Simonstone cyclothems. North of the River Ure, seatearths are present at many horizons showing that though the pile of sediments was very thick they frequently were built up above sea level.

Moore (1958, 1959, 1960) interpreted the Yoredale facies as the products of a deltaic environment, with recognisable distributaries and interdistributary deposits interrupted by migration of the delta allowing the sea in and limestone to be deposited. The present study bears out much of this synthesis, though it is arguable whether mere shifts in the delta pattern are enough to accommodate some of the more extensive limestones. They would however readily explain the splitting of the Three Yard and Middle limestones by clastic partings in the present area. A more widespread mechanism is perhaps needed to explain some of the major extensive marine incursions. Broadly the mechanisms fall into two main types. The first is based on global variations in sea level which may have caused cyclicity in widespread areas of the northern hemisphere from the USA to the Urals (Wells, 1960; Ramsbottom, 1973). The second is a tectonic explanation, of which several variants have been advanced (e.g. Weller, 1930, 1956; Dunham, 1950). The most recent, the mantle flow theory of Bott and Johnson (1967), links isostatic uplift of a denuding mountain range with contemporary sinking of adjoining basins by a single mechanism.

Cherts occur at four main levels in strata of late Brigantian – early Pendleian age in Swaledale, with a tendency to occur just above limestone formations. Each successive cherty formation is in some degree distinguished from the others by faunal or lithological characteristics. The lowest, the Underset Chert, is commonly a blocky greyish black rock with few fossils. The Main Chert is chiefly distinguished by abundant *Latiproductus latissimus* at many outcrops. The Richmond Cherts, the thickest formation at 130 ft (39.60 m), is locally very variable in lithology with sudden lateral passage into crinoidal grainstones. The lowest beds in Swaledale are commonly platy silty cherts with abundant *Zoophycos*, an arcuate trace fossil. The highest cherty formation, the Crow Limestone, can contain cherty limestones and cherts at any level, which weather in parts of Swaledale to a white block-field. A fifth chert rich in bryozoa and small productoids underlies the Underset Limestone in the vicinity of Buckden Pike, south of the main area of chert formation.

A common lithology in these five formations is a blocky streaky textured chert which consists of an intimate mixture of cryptocrystalline silica and calcite, thought by Wells (1955) to be a primary precipitate. The sharp top and bottom contacts of these cherts against unaltered limestones also suggest a primary origin. A second type of chert includes a variable amount of crinoid debris and shows gradations to pure limestone. Much of the Crow Limestone of the Richmond area is in this category and was thought by Hey (1956) to be due to early diagenetic alteration of limestones. Silicified colonies of *Diphyphyllum* in the Red Beds of the Richmond Cherts are also, probably, in diagenetically altered beds. The glassy Marske Chert Bed, also in the Red Beds, is rich in chalcedony and has a secondary origin according to Wells (1955). Its contacts are gradational, in line with a secondary origin. Locally, near Bowes, the Main Chert is also glassy and of secondary origin, according to Wells.

The presence of four successive cherts in the same geographical area suggests a continuing silica source, possibly silica rich springs on the sea bed or a river with a high silica content. Usually silica is only present in minute quantities in sea water but is known to be fixed nowadays by diatoms, radiolaria and sponges. Mr Pattison notes that the abundant spicules in the Richmond Cherts display the characteristic symmetry of siliceous sponges and that the mooring spicules of *Hyalostelia* are also siliceous. Wells (1955) inclined to the view, on petrographic grounds, that most of the spicules were originally calcareous, but it seems that he may have been mistaken. He favoured an incompletely understood process of inorganic co-precipitation of silica and calcium carbonate. It remains possible that both organic and inorganic processes were involved.

Around Lofthouse, in Nidderdale, where the Grassington Grit Formation comes to rest on progressively older horizons of Brigantian age in a southerly direction, cherty limestones are present in the Five Yard Limestone, a formation free of cherts in the country to the north. It seems likely that some of the silica in the Richmond Cherts, which underlie large areas of the pre-Grassington Grit erosion surface to the north, has been redistributed.

The Millstone Grit facies

The base of the Grassington Grit Formation is marked by a widespread but generally very gentle unconformity. The beds above the break are very different from the Yoredale facies strata beneath them, since limestones are virtually absent in a succession of sandstones and mudstones. The most characteristic strata are cross-bedded, fluviatile, feldspathic sandstones wth a number of seatearth horizons, the latter pointing to periodic times of emergence with local development of coal swamps. Distributary channels occur at several levels, particularly in the Pendleian of the Stainmore Trough. Studies of the petrography and cross-bedding of the sandstones suggest derivation from the north and north-east (Gilligan, 1920; Walker, 1955; Rowell and Scanlon, 1957a, fig. 11). Turbidites laid down in front of the delta occur in the Scar House Beds. In Kinderscoutian times several shelly sandstones without cross-bedding were laid down in tranquil waters in shallow shelf seas.

At rare intervals goniatite mudstones extended over the area (Figure 3). These are much less numerous than in the Craven Basin and their rarity has led to the hypothesis (Ramsbottom, 1977, fig. 8) that the Namurian consists of eleven mesothems and that the lower, regressive elements are chiefly represented on the Askrigg Block by gaps in the succession, notably at seatearth horizons. Since seatearths are also present in much of the basin succession around Huddersfield the hypothesis is hard to prove on this criterion, but it postulates a mechanism which could, for instance, explain the absence of Alportian goniatites on the Askrigg Block. In boreholes at Greenhow the Upper Follifoot Grit contains much seatearth so there were undoubted long periods of emergence at this level. Indeed so numerous and thick are these seatearths here and in Colsterdale that one wonders if they might equate with the fundamental break between the Bashkirian and underlying Serpukhovian stage of the USSR (this is the proposed contact of the upper and lower division of the Carboniferous, discussed by Ramsbottom, 1977, p. 287, Aisenverg and others, 1979 and Rotai, 1979). AAW

STRATIGRAPHICAL PALAEONTOLOGY

This section is largely confined to describing the macrofossil succession in the shelf-sea sediments of the Askrigg Block but incorporates the results of recent work on the Holkerian and Asbian foraminifera and the goniatites of the Craven Reef Belt. It is based on previously published work as well as systematic collections from Raydale and Beckermonds Scar boreholes in the centre of the district, less comprehensive sampling from boreholes in the Greenhow/Pateley Bridge area, large collections from surface localities in the Pendleian chert beds around Swaledale and fossils yielded by a few selected sections elsewhere.

Pre-late Arundian

The oldest firmly diagnostic Carboniferous macrofaunas recorded from the Askrigg Block are of late Arundian age but macrofossils which are probably older have been found at several places in the district.

From the Penny Farm Gill Dolomite in Hebblethwaite Gill, Garwood (1913, pp. 524–525) recorded cf. *'Rhynchonella' fawcettensis.* Shells, including forms doubtfully referred to both that species and *Linoprotonia globosa,* were also found in a sandy limestone within the same formation between 1207 and 1224 ft (368 and 373 m) in the Raydale Borehole. Both finds suggest a correlation with the lower part of the Scandal Beck Limestone in the Ravenstonedale area, although a fauna of this nature is probably more indicative of facies than horizon. The beds below the Tom Croft Limestone in the Raydale Borehole also include several bands with '*Modiola*-phase' faunas. In one of these, about 30 ft (9 m) higher than the above-mentioned brachiopod limestone, mudstone containing *?Viviparus carbonarius* is overlain by an argillaceous limestone with *Modiolus latus* and ostracods. This sequence recalls that described by Garwood (1922) at the local base of the Carboniferous near Horton-in-Ribblesdale.

Further probable early Arundian macrofossils collected from the district include the possibly derived specimens of *Stenoscisma isorhyncha* recorded by Garwood and Goodyear (1924, p. 191) from grey limestones overlying the basal Carboniferous unconformity at Thornton Force, Ingleton and Nappa Scar, Austwick. The species is characteristic of the early Arundian Red Hill Oolite of south Cumbria. The lower part of the Tom Croft Limestone, between 1017 and 1148 ft (310 and 350 m) in the Raydale Borehole contained, among a sparse marine fauna, a coral form doubtfully assigned to *Lithostrotion caswellense*, the type locality of which is in rocks of early Arundian age in South Wales.

Late Arundian

A prolific and distinctive coral/brachiopod fauna similar to that of the late Arundian Dalton Beds in south Cumbria has been recorded in several parts of the district. The strata yielding this fauna include the upper beds of the Tom Croft Limestone in the River Clough and the Raydale and Beckermonds Scar boreholes as well as the lowest beds of the Kilnsey Limestone seen at Twistleton (Chapel-le-dale) and Mill Scar Lash in Wharfedale. The most common species are *Clisiophyllum ingletonense*, *Haplolasma subibicina*, *Lithostrotion martini*, *Michelinia megastoma*, *Palaeosmilia murchisoni*, *Siphonophyllia garwoodi*, *Syringopora spp.*, *Athyris expansa*, *Composita ambigua*, *Linoprotonia spp.* and *Straparollus spp.* Only a few doubtful examples of Garwood's subzonal fossil, *Delepinea carinata*, have been found and it appears probable that the species is less common here than in areas to the north-west.

In the south-western parts of the district these richly fossiliferous beds are overlain by further dark limestones within the Kilnsey Limestone which contain no stratigraphically diagnostic macrofossils. However, Dr A. R. E. Strank, reporting on the foraminifera from Kilnsey Crag, states that the lowest 23 ft (7 m) of the section there has yielded several *Ammarchaediscus*, *Glomodiscus* and *Rectodiscus* but no diagnostic Holkerian genera, whereas immediately above that level the incoming of *Koskinotextularia sp.* followed higher up by *Endothyranopsis crassa*, *Nevillella sp.*, *Nibelia nibelis* and nodose archaediscids, all post-Arundian forms, indicates that the Arundian – Holkerian boundary coincides approximately with the change from very dark to less dark limestones at this horizon. The accompanying *Brunsia spirillinoides*, *Eostaffella parastruvei*, *Glomospiranella dainae* and *Septabrunsiina sp.* form a characteristic Holkerian foraminiferal assemblage.

The stage boundary has not been precisely located elsewhere in these south-western areas but, on the basis of Dr Strank's evidence, it seems probable that it lies within the Gastropod Beds of Garwood and Goodyear (1924). Consequently, beds taken to be the latest Arundian strata there and their probable part-correlative, the uppermost Tom Croft Limestone between 705 and 755 ft (215 and 230 m) depth in the Beckermonds Scar Borehole, are dark limestones containing a fairly abundant but monotonous fauna including *L. martini*, *Composita ficoidea* and *Linoprotonia hemisphaerica*.

In the northern part of the district the youngest Arundian strata are the interbedded clastic rocks and dark limestones thought to be the attenuated local equivalent of the Ashfell Sandstone in the Ravenstonedale area. In the Raydale and Beckermonds Scar boreholes these limestones contain abundant *L. martini*, among which diphymorphic corallites are common, while the shales have yielded a bivalve/ostracod fauna. The bivalves include *Aviculopecten eskdalensis*, *A. subconoideus*, *Leiopteria hendersoni* and *Pteronites angustatus*, all long ranging forms in the Carboniferous rocks of the Northumberland Trough, from probable Arundian up to early Asbian age.

Holkerian – early Asbian

The Holkerian and early Asbian are discussed together, in view of difficulties which have arisen in locating the boundary between the two.

As noted above, in the southern part of the district the Holkerian base has not been located accurately using macrofauna, although the presence of cerioid *Lithostrotion* indicates that at least the highest beds of the Kilnsey Limestone are of Holkerian age. In the northern part, the macrofaunas support a correlation of the Arundian – Holkerian boundary with the top of the Ashfell Sandstone. The lowest beds of the Fawes Wood Limestone in the Raydale and Beckermonds Scar boreholes contain *Lithostrotion sociale* (together with *Diphyphyllum smithii* in the former), a characteristic they share with the earliest Holkerian rocks in the Cartmel peninsula, at and near the stage stratotype.

Davidsonina [*Cyrtina*] *carbonaria* and *Lithostrotion* [*Nematophyllum*] *minus* are the eponymous species of Garwood's two subzones (1913) in northern England within the strata now referred to the Holkerian. Although beds with *D. carbonaria* underlie others with abundant *L. minus* over much of the district, use of the subzones is no longer tenable for two reasons. Firstly, *D. carbonaria* ranges from the earliest Holkerian to early Asbian and, at Twistleton, the only examples of the species found during the course of this work occurred with *L. minus*. It has been suggested that strata of *D. carbonaria* Subzone age are missing from some southern parts of the block (Garwood and Goodyear, 1924, p. 194; Turner, 1963, p. 162). Secondly there are about 164 ft (50 m) of Holkerian strata above the highest beds with abundant *L. minus* as shown by the Raydale and Beckermonds Scar boreholes.

Holkerian macrofaunas in general are characterised more by the absence of Arundian or Asbian fossils rather than the presence of specifically Holkerian forms. Fossils common in strata of this age in the district are *L. martini*, *Syringopora geniculata*, *Composita spp.*, *Linoprotonia spp.* (notably *L. hemisphaerica* and *L. ashfellensis*), *Megachonetes papilionaceus* and gastropods, commonly including large specimens of *Macrochilina*. The probable regressions of the sea towards the end of the stage resulted, in the southern part of the district, in shallow clear waters where calcite-mudstones with *Composita*/gastropod faunas were deposited, and farther north, as shown by the lower beds of the Garsdale Limestone, in the introduction of fine terrigenous sediments accompanied by a fauna of small brachiopods including rhynchonelloids and *Productus garwoodi*.

Strata of the early Asbian mesothem postulated by Ramsbottom (1977, fig. 10) include the Potts Beck Limestone of the Ravenstonedale area, up to 328 ft (100 m) thick, and the base of which defines the Holkerian/Asbian boundary. Early Asbian strata are clearly thinner to the south of the Ravenstonedale area but are difficult to distinguish from those of Holkerian age as their macrofaunas and microfaunas have several features in common. At the stratotype section, the lowest 66 ft (20 m) of Asbian strata contain the thickly-ribbed *Gigantoproductus sp.* of the *G. maximus* group characteristic of the stage, but the most abundant species are *Productus garwoodi* and *Spirifer bisulcatus*, constituents of the faunas in the brachiopod beds common to both late Holkerian and early Asbian strata in the Ravenstonedale/Stainmore Trough and the Alston Block. Other faunal characteristics of the Holkerian persisting into the Potts Beck Limestone are the abundance of tabulate corals, *L. martini* and bellerophontids as well as the presence of *D. carbonaria*. The formation, however, also contains *Daviesiella llangollensis*, together with the diagnostic post-Holkerian species *Dibunophyllum bourtonense* and *Lithostrotion junceum*. The first-named species has been considered diagnostic of an early Asbian age (George and others, 1976, pp. 32–34, 40). However, Cope (1940, p. 207) recorded the range of *D. llangollensis* in the Wye Valley section, North Derbyshire, as being 'from the top of the Daviesiella Beds and the base of the Chee Tor Beds (S_2–D_1)'. Dr Strank's studies of the foraminifera from the Wye Valley suggest a Holkerian age for the Daviesiella Beds (= Woo Dale Beds) and an Asbian age for the Chee Tor Beds. From this it is clear that *D. llangollensis* is characteristic of beds spanning the Holkerian – Asbian boundary but is not diagnostic of the early Asbian alone.

On the northern edge of the Askrigg Block, the Holkerian – Asbian boundary should lie within the Garsdale Limestone, as the succeeding Danny Bridge Limestone is of late Asbian age. However, the relevant strata are lacking in diagnostic species. In the River Clough, *D. llangollensis* occurs with *D. carbonaria* in the topmost beds, indicating a possible correlation with the lowest beds of the Potts Beck Limestone.

In the south-western part of the district, the fauna of the lowest 66 ft (20 m) or so of the Kingsdale Limestone differs from those higher in the same formation and has much in common with that in the underlying Holkerian, thus paralleling the faunal succession in the Ravenstonedale area below, in, and above the Potts Beck Limestone. The most common macrofossils in these beds are *Axophyllum vaughani*, chaetetids, *L. martini*, *S. geniculata*, athyrids, *Gigantoproductus maximus* group, *Linoprotonia spp.*, *M. papilionaceus* and gastropods. A doubtful specimen of *D. llangollensis* was also found. The same species has been recorded at Horton-in-Ribblesdale by both Mr P. F. Dagger and Dr W. H. C. Ramsbottom.

In the Greenhow area, the equivalent, transitional beds may be the Stump Cross Limestone and part of the Timpony Limestone. Macrofaunas from below 1004 ft (306 m) in the Coldstones No. 4 Borehole included the species *L. minus*, *L. sociale* and *Linoprotonia ashfellensis*, a fauna of Holkerian aspect, about 300 ft (91 m) below the base of the Asbian Greenhow Limestone. Diagnostic post-Holkerian corals were also recorded by Dunham and Stubblefield (1945) from both formations (*L. junceum* in the Stump Cross Limestone and *L. pauciradiale* in the Timpony).

Late Asbian

Common fossils throughout the late Asbian rocks of the Askrigg Block (Figure 5) include *Koninckopora inflata*, *Dibunophyllum bourtonense*, *L. martini*, *Palaeosmilia murchisoni*, *Syringopora spp.*, athyrids, *Gigantoproductus spp. maximus group*, *Linoprotonia hemisphaerica* and *Megachonetes spp*. Some geographical variation in the faunas can be distinguished, notably the increasing abundance of *L. junceum*, gastropods and bivalves northwards and of *Davidsonina septosa* southwards. The only vertical changes observed are the relative abundance of the faunas in the middle part of the late Asbian (coinciding approximately with the range of *D. septosa*) and the incoming of some Brigantian forms before the end of the stage. JP

The age of the reef limestones along the south side of the Middle Craven Fault has been discussed by several authors including Garwood and Goodyear (1924) and Hudson (1930), who referred them to D_3 and S_2/D_1 respectively, and Bisat (1924, 1934), Hudson (1938) and Hudson and Cotton (1945) who assigned them to the goniatite zones B_2 to P_{1a}. Bond (1950) proposed some lithostratigraphical units related to the goniatite zones but they are here considered to be of only limited use in such varied and complicated facies.

Goniatites recently collected from the Craven reef limestones allow the recognition of the following subdivisions:

P_{1a}	*Goniatites crenistria* s.s
Upper B_2	*Goniatites schmidtianus* *Goniatites moorei* *Bollandites castletonensis*
Lower B_2	*Goniatites wedberensis* *Goniatites hudsoni*

The accepted relationship of the goniatite zones to the coral/brachiopod zones and of the latter to the stages of George and others (1976) is $B_2 = D_1$ (Asbian) and P_{1a-b} = early D_2 (early Brigantian) (see Mitchell *in* Stevenson and Gaunt, 1971, p. 152) but the base of P_{1a} has not been precisely located in chronostratigraphical terms.

WHCR, DJCM

Foraminifera of the Holkerian and Asbian stages

On the northern edge of the Askrigg Block studies of the foraminifera from the River Clough and from the boreholes at Raydale and Beckermonds Scar supplement the macrofaunal evidence outlined above. The Fawes Wood Limestone and all but the highest beds of the Garsdale Limestone yield a typically prolific Holkerian assemblage including *Archaediscus pulvinus*, *A. stilus*, *Brunsia spirillinoides*, *Endothyra blatoni*, *Eostaffella parastruvei*, *E. mosquensis*, *Globoendothyra insigna*, *Glomospiranella latispiralis*, *Holkeria avonensis*, *H. topleyensis*, *Koskinotextularia cribriformis*, *Nibelia nibelis*, *Septabrunsiina sp.*, *Spinobrunsiina lexhyi* and *S. tynanti*, together with the algae *Girvanella densa* and *Koninckopora inflata*.

The topmost 30 to 60 ft (9–18 m) of the Garsdale Limestone have a similar, but more restricted assemblage including *Archaediscus tataricus*, *Bogushella ziganensis*, *Brunsia spirillinoides*, *Earlandia moderata*, *Eostaffella parastruvei*, *Globoendothyra sp.*, *Glomospiranella dainae*, *Koskinotextularia cribriformis*, *Nibelia nibelis*, *Nudarchaediscus spirillinoides*, *Omphalotis sp.*, *Plectogyranopsis ampla* and *Pseudoammodiscus volgensis*, with the algae *Koninckopora inflata* and *K. mortelmansi* and various calcispheres. This fauna also is of Holkerian aspect, but is identical to that found in the lowest few metres of the early Asbian Potts Beck Limestone above the Asbian boundary stratotype at Little Asby Scar. The similarity of the Holkerian and earliest Asbian faunas is typical in sequences of this area where the characteristic early Asbian *Vissariotaxis compressa* and the Holkerian *Dainella holkeriana* have not been found, and for this reason it is not possible, at present, to locate the Holkerian – early Asbian boundary accurately in most sections. It is also of interest to note that fine-grained or porcellanous limestones (Figure 5) with an unmistakable fauna of abundant and varied calcispheres are found near, or at the Holkerian – Asbian boundary in many British sequences; for instance, similar beds have been noted by Somerville (1979) in the

Ty-Nant Limestones of Llangollen, North Wales. The Danny Bridge Limestone yields a diagnostic late Asbian assemblage including *Archaediscus cyrtus*, *A. inflexus*, *A. itinerarius*, *Asperodiscus sp.*, *Bogushella ziganensis*, *Climacammina* aff. *simplex*, *Cribrospira panderi*, *Cribrostomum lecomptei*, *Endostaffella fluctata*, *E. fucoides*, *Endothyra phrissa*, *E. spira*, *E. uva*, *Endothyranopsis crassa*, *Forschia subangulata*, *Forschiella prisca*, *Koskinobigenerina sp.*, *Lituotubella magna*, *L. glomospiroides*, *Millerella infulaeformis*, *Neoarchaediscus occlusus*, *Nudarchaediscus concinnus*, *Omphalotis minima*, *Plectogyranopsis ampla*, *P. settlensis*, *Pseudoendothyra sublimis* and *Pseudolituotuba wilsoni*.

In the southern part of the district, a number of boreholes examined, from Settle to Geenhow, yield a diagnostic Arundian foraminiferal assemblage including *Eblanaia michoti*, *Glomodiscus miloni* and *Rectodiscus sp.* A late Asbian fauna including *Cribrostomum lecomptei*, *Cribrospira panderi* and *Endothyra spira* is also readily recognised. However, the succession between the top of the Arundian and the base of the late Asbian usually contains a poorly developed, restricted assemblage in which the foraminifera are commonly stunted and also associated with abundant *Koninckopora inflata*. Only one borehole, at Silverdale [SD 8435 7144], yielded characteristic Holkerian faunas, in two-metre thick bands at both bottom and top of this facies. This restricted fauna is believed to represent a somewhat deeper-water environment than the shelf to the north, which yields prolific Holkerian faunas throughout. As noted above, Holkerian microfaunas persist into basal Asbian strata at the stratotype, so that the base of the early Asbian cannot be precisely located from this information. The facies is represented by the Cove and Horton limestones in the west, and by the Timpony and Stump Cross limestones in the Greenhow area. Late Asbian assemblages are found in the Gordale, Greenhow and Hargate End limestones. ARES

Brigantian

The Asbian – Brigantian boundary is the easiest of the Dinantian stage boundaries to correlate on the Askrigg Block because the stratotype is within the district (Janny Wood near Kirkby Stephen) and is drawn at a readily recognisable horizon, i.e. at the base of a limestone (the Peghorn/Hawes) which contains a distinctive biostrome (the Girvanella Nodular Band of Garwood, 1913).

Reference to the goniatite zonal scheme established by Bisat (1924 and other papers) and Moore (1936, 1941) in the Craven lowlands is one of several methods to have been applied to the biostratigraphical subdivision of the Brigantian rocks in the district (see Ramsbottom, 1974, fig. 25). Others were proposed by Hudson (1926, 1929a) based on the vertical distribution of rugose corals, especially species of *Orionastraea*, by Hill (1938) who divided what is now the Brigantian into two coral zones and by Hallett (1970) who described the ranges of foraminifera and algae. Ramsbottom (1977, fig. 10 and personal communication) assigns Brigantian strata to two mesothems (D6A and D6B) with their common boundary in northern England at the top of the Single Post Limestone.

The obvious way to divide Askrigg Block Brigantian strata is into 'Yoredale' cyclothems and their constituent phases. Both a description of the way faunas change with lithology in 'Yoredale' rocks and a detailed account of the faunas collected from each cyclothem in the district are beyond the scope of this work. The former was discussed by Hudson (1924), Johnson and Dunham (1963) and Pattison (*in* Burgess and Holliday, 1979) and the vertical distribution of the many macrofossils recorded from the Brigantian rocks of the Askrigg Block is well-documented both by previous authors, e.g. Garwood and Goodyear (1924), Hudson (1925), Miller and Turner (1931), Dunham and Stubblefield (1945), Moore (1958) and Wilson (1960a), and in the faunal lists accompanying the stratigraphical details in Chapter 4. Few of these fossils have proved to be of stratigraphical value but some groups are more useful than others and they are discussed below. JP

Several goniatites have been found in the Brigantian rocks of the district during the present study and their identifications and the goniatite zonation are included in Figure 3. Goniatites of the P_{1b} and P_{1c} zones are well represented by fragmentary specimens from the mudstones between the Gayle and Hardrow Scar limestones. It should be noted that these were erroneously placed below the Gayle Limestone in Ramsbottom, 1974, fig. 25, p. 70. The record of *Goniatites granosus* from the shales over the Middle Limestone parallels a similar record from the same horizon (i.e. above the Scar Limestone) at Bow Lees near Middleton in Teesdale by Rayner (1953). The Bow Lees specimens are comparable with examples of *G. granosus* from the second goniatite horizon above the base of P_2 in the typical section at Little Mearley Clough, Pendle Hill. This implies that the base of the P_2 Zone is below the base of the Scar Limestone member of the Middle Limestone, and it could well lie between the Single Post and the Cockleshell Limestone members, a horizon of widespread withdrawal of the seas from the area. Thus the goniatite zonal boundary at the base of P_2 probably coincides with the coral zones 2/3 boundary of Hill (1938), the new faunas of the P_2 goniatite zone and the Zone 3 corals entering with the transgression at the base of the D6B Mesothem. WHCR

Among the colonial corals, *Lithostrotion spp.* and related diphymorphic forms flourish in the lowest part of the Brigantian but decline from the Simonstone Limestone cyclothem upwards until only small, fasciculate species with diphymorphic tendencies remain at the top of the stage. *Orionastraea* occurs in the Hardrow Scar and Simonstone and the lower part of the Middle Limestone but is otherwise a rare fossil. The Lonsdaleiids range throughout but are most common in the Gayle and Hardrow Scar limestones while *Palaeosmilia regia* is found in the same two limestones as well as the Hawes, Simonstone and Scar. Caninioid corals are seldom common, especially above the Hardrow Scar Limestone, but the clisiophylloids, of which *Aulophyllum fungites*, *Clisiophyllum keyserlingi* and *Dibunophyllum bipartitum* are the most abundant, are present in all the major limestones. Zaphrentoid coral faunas, for which some correlative value was claimed, were described from the shales overlying the Middle and Simonstone limestones by Hudson and Fox (1943) and Black (1952) respectively but the latter author did suggest that the stratigraphical usefulness of zaphrentoids is limited by the large collections required to determine the morphological range within each species.

A similar proviso applies to the use of gigantoproductoid brachiopods as stratigraphical tools but this limitation is offset by their widespread abundance in most of the Brigantian limestones. Their vertical distribution in the Brigantian rocks of the district was described by Pattison (1981) and can be summarised as follows: In contrast to the predominance of the non-fluted, coarsely-ribbed forms *Gigantoproductus maximus* and *G. semiglobosus* in the Asbian, around the beginning of the Brigantian there was a proliferation of gigantoproductoid forms including thick-shelled, coarsely-fluted brachiopods related to *G. giganteus*, large, thin-shelled species comparable with *G. gigantoides* and latissimoid shells mostly referrable to *Latiproductus latissimus* s.l. Particularly common in the lowest three or four cyclothems of the stage are *G. edelburgensis*, *G. gigantoides*, *G. inflatus*, *G. okensis* and *Semiplanus semiplanus*. An abundance of *Gigantoproductus gaylensis* (a very large, thin-shelled form) and *Linoprotonia dentifer* is characteristic of the Gayle and Hardrow Scar limestones respectively while *G. edelburgensis* is most common in the Hardrow Scar. The number of different forms decreases from the Simonstone Limestone upwards. Gigantoproductoids are abundant in the upper part of the Middle Limestone but those seen could all be referred to *G. giganteus* or *G. varians*. From the Five Yard Limestone to the Main the most common species are *G. elongatus* and *G. expansus*.

Other Brigantian macrofossils of some stratigraphical value include the thin-shelled bivalves *Posidonia becheri* and *P. corrugata* found respectively in the P_1 and P_2 goniatite zones, nautiloids of the genus

Catastroboceras which, at least within the district, appear to become common only above the Five Yard Limestone, and the distinctive gastropod *Tropidodiscus* which is particularly characteristic of the shales in the Gayle and Hardrow Scar cyclothems. Dunham and Stubblefield (1945, p. 257) proposed that the point in its growth at which the productoid *Eomarginifera tissingtonensis* s.l. developed a median fold could be a factor of stratigraphical significance. The evidence from the present work is inconclusive but unpromising. What is clear is that the species is common in the Gayle to Simonstone cyclothems but is superseded by the broader, finer-ribbed *E. longispina* group in the later Brigantian rocks.

Pendleian

The base of the Pendleian Stage, coinciding with that of the Namurian Series, is defined at the base of the strata containing the earliest occurrence of the goniatite *Cravenoceras leion*. In most of the district this horizon is within strata of Yoredale facies containing few goniatites. The case for putting the boundary between the Underset and Main Limestones or their equivalents was presented by Johnson and others (1962). Subsequently the lower boundary of the Namurian in northern England has been taken at the bottom of the Great (= Main) Limestone because that is the nearest mappable horizon to the base of the series, although diagnostic E_{1a} goniatites collected from shales above the Underset Limestone indicate that the true base is lower (Figure 3).

The Main Limestone in the northern areas of the block contains a coral/brachiopod fauna comparable in richness with those in the early Brigantian limestones. Collections from Swinner Gill in Swaledale demonstrate the presence of two fossil bands described further north by Johnson (1959), namely the *Chaetetes* Band at the base and the Frosterley Band 23 to 16 ft (7–5 m) below the top of the limestone. The former yielded abundant *Chaetetella depressa*, *Derbyia sp.* and *Martinia glabra* while in the latter *Dibunophyllum bipartitum* and latissimoid gigantoproductoids were found to be common.

Fossils from the Pendleian rocks above the Main Limestone in the north-western corner of the Askrigg Block were listed by Rowell and Scanlon (1957a). For the present work, collections in the Swaledale area from the Main, Richmond and Crow cherts have yielded large faunas which are possibly the richest to have been found in British Pendleian rocks. As a whole they are characterised by a rarity of corals and an abundance of brachiopods, sponge remains and *Zoophycos*. As in other rocks of this age in northern England, the most common brachiopods are the coarsely-ribbed forms *Buxtonia spp.*, *Eomarginifera spp. longispina* group, *Pleuropugnoides greenleightonensis* and *Spirifer trigonalis*. The varied fauna of the Main Chert is notable for the profusion of latissimoid gigantoproductoids and it also includes many brachiopod and bivalve species of which *Brachythyris integricosta*, *Productus carbonarius*, *Aviculopecten interstitialis*, *Edmondia expansa* and *Limipecten dissimilis* are among the most common. The Richmond Cherts faunas are discussed on p.54. Those of the Crow Chert are comparable in that they largely consist of brachiopods; they include the species *Antiquatonia sulcata*, *Echinoconchus punctatus* and *Overtonia fimbriata*.

In the south-eastern parts of the district the mudstones at the base of the Pendleian succession are the local equivalent of the Upper Bowland Shales. Zonally useful fossils collected from them include those recorded by Garwood and Goodyear (1924, p. 246) and Black (1950, pp. 35–36) in addition to the *Eumorphoceras pseudobilingue* and *Posidonia membranacea* found at Grimwith during the present work. Shale bands within the overlying Grassington Grit Formation have yielded *Lingula*, ostracods and a few molluscs from boreholes in the Greenhow area; other fossils recorded from the formation were noted by Wilson (1960a, p. 306).

Arnsbergian

The diagnostic goniatite of the basal marine band in the Arnsbergian is *Cravenoceras cowlingense* which is present in the Cockhill Marine Band in the south-eastern part of the Askrigg Block as well as in the Mirk Fell Ironstones of the north-western corner. The Cockhill Marine Band (E_{2a}), together with the Colsterdale Marine Beds (E_{2b}), was described in detail by Wilson and Thompson (1959). In the Greenhow area the former consists of a limestone, with abundant *C. cowlingense* and anthracoceratids, and an overlying fossiliferous shale, in which bivalves, including *Posidonia corrugata* and a large pectinoid referred to *Obliquipecten* aff. *costatus*, are common. No *Eumorphoceras* was found but the type locality of *E. grassingtonensis* (Dunham and Stubblefield, 1945, p. 258) is in the Cockhill Limestone of Bolton Gill, about 8 km W of Greenhow. The *C. cowlingense* in the Mirk Fell Ironstones occurs in a band together with nuculoids, about 10 ft (3 m) below which is another fossiliferous bed, containing zaphrentoid corals (Hudson, 1941).

The only other strata of Arnsbergian age in the district to have yielded diagnostic goniatites are the Lad Gill Limestone and the Colsterdale Marine Beds. The former contains *Cravenoceras sp. nov.* aff. *cowlingense* and has been tentatively correlated with the *Eumorphoceras bisulcatum* Marine Band (E_{2a}) (Burgess and Ramsbottom, 1970). The latter are up to 40 ft (12 m) thick in the Greenhow area where they consist of a lower shale in which *Posidonia corrugata* and *Posidoniella variabilis* are common, the Colsterdale Limestone containing *Cravenoceratoides nitidus* and an overlying shale which has yielded abundant *Anthracoceras* including *A. tenuispirale*. A comparison with the fossils recorded by Yates (1962, pp. 374–375) at Slieve Anierin in Ireland suggests that both her *Eumorphoceras leitrimense* and *Anthracoceras tenuispirale* faunas in the *Cravenoceratoides nitidus* Zone (E_{2b}) are represented. In the Aiskew Bank Farm and Croft House boreholes the Colsterdale Marine Beds are thicker (c. 72ft, 22 m) and contain *Cravenoceras holmesi*. The fauna of the Shunner Fell Marine Beds in the north-western corner of the district, which are the probable equivalent of the Colsterdale Marine Beds, consists mostly of brachiopods and zaphrentoid corals together with *Anthracoceras*.

In the Fewston area, about 8 miles (13 km) south-east of Greenhow, Wilson (1977, pp. 10–11) recorded E_{2c} fossils from an horizon just above the Lower Follifoot Grit thus indicating that the same beds in the Greenhow area are probably of Arnsbergian age although no diagnostic fossils have been found locally.

Post-Arnsbergian

Within the district, post-Arnsbergian strata are confined to the south-eastern and north-western corners. Fossils collected from boreholes in the Greenhow area comprise Chokierian faunas, including *Homoceras sp.* and *Phestia* aff. *sharmani*, from the beds between the Follifoot grits, and large collections through the Kinderscoutian Cayton Gill Beds. The latter, with up to 56 ft (17 m) of marine fossil-bearing strata, yielded abundant brachiopods in the lower part, including *Derbyia gigantea*, *Productus carbonarius*, *Rugosochonetes laguessianus* and especially numerous *Schizophoria hudsoni* while the collections from the higher beds showed a more varied fauna with some of the same brachiopods (*P. carbonarius*, *R. laguessianus* and *S. hudsoni*) together with *Rhombopora sp.*, *Crurithyris urii* and several bivalve forms including *Parallelodon* cf. *tenuistria* and *Sanguinolites tricostatus*.

In the Stainmore Outlier Owens and Burgess (1965) recorded the goniatites *Vallites henkei* (R_{1a}) and *Gastrioceras* cf. *cumbriense* (?G_{1b}) from the Mousegill Marine Beds and Swinstone Middle Marine Band respectively. JP

PETROGRAPHY OF CARBONIFEROUS SEDIMENTARY ROCKS

Open-sea, fore-delta, deltaic and fluviatile sediments make up the Carboniferous succession in the orefield, but only a limited number of petrographical types are represented. Limestones, cherts, mudstones, siltstones, sandstones, seatearths and minor coals are many times repeated in the sequence, deviating little from a few recognisable norms. The following account is based upon 65 samples of limestone, 12 of chert and 40 of arenaceous rocks chosen to illustrate the principal lithologies. A sequential suite from the fine section in Brigantian and early Pendleian strata exposed in Greenseat Beck, near Muker, Swaledale (MR 34661–34881)[1] is included and the remaining collecting points range across the Askrigg Block from Hartley Fell in the north-west to Ingleborough, Fountains Fell and Whernside in the south and Greenhow in the south-east.

Limestones and cherts Within certain limits of variation in structure, texture and fossil assemblages, there is little variation between the specimens from the normal limestone facies (Hudson, 1924); most are bioclastic, medium- to fine-grained and relatively pure carbonate rocks with little or no admixture of terrigenous sediment. Apart from sporadic sedentary corals and crinoids, most of the limestones are formed of derived components (allochems), dominantly of biota with minor intraclasts and pellets (Folk, 1959). No true ooliths were detected. In general the allochems other than sporadic calcirudite crinoid stems, plates and brachiopods are moderately to well sorted and lie in the calcarenite size-range (2.0 to 0.06 mm). This is often the case in the Asbian limestones, but two samples (MR 34721–2) from Hartley Quarry, Kirkby Stephen are biomicrites with well-sorted, closely packed biota which include foraminifera and other fossils with very finely comminuted material in the silt to mud range. The general texture in the calcarenites is grain-supported. The grains are cemented with micrite (less than 0.004 mm), microsparite (0.004 to 0.1 mm) or less commonly, by coarsely crystalline sparry calcite. Finer allochems are commonly aligned to form laminations, which are emphasised in the less pure limestones by carbonaceous clay and other materials. The characteristically pale limestones of the Asbian contain less carbonaceous matter than the normally grey to dark grey limestones that occur throughout the Brigantian (with few exceptions) and persist into the early Pendleian. Laminations, however, occur in the Asbian rocks (for example, MR 34678) as well as in the Three Yard (MR 34730), a limestone between the Three Yard and Underset (MR 34651), Underset (MR 34729, 34690), and Lower Stonesdale Limestone chert (MR 34736).

In general terms the specimens examined represent the accumulation of organic remains with co-precipitation of calcium carbonate as aragonitic ooze evidently under low-energy conditions, though perhaps with sporadic stronger currents winnowing away the interstitial ooze to allow sparry calcite to form. However, because of post-depositional recrystallisation it is generally difficult to distinguish in some specimens between such primary pore-filling by calcspar and recrystallisation of micrite. Whether the micrite represents a paramorphic replacement of original aragonite mud as in Recent carbonate sediments, (e.g. Friedman, 1964) is unknown for ancient sediments, but the analogy may be valid.

The relation between mineralogy and depositional environment of such Recent carbonates (Friedman, *op. cit.*) has indicated that debris from reefs contains high-Mg calcite, whereas deep-water carbonate sediment contains low-Mg calcite. Subsequent processes of lithification erase these primary differences; low-Mg calcite replaces unstable aragonite and high-Mg calcite. This releases Mg-ions which may be one source of dolomite. However dolomite also forms penecontemporaneously with sedimentation in the intertidal zone and in deep-water marine conditions. In the present specimens dolomite occurs sporadically replacing micritic calcite as in an S_2 limestone (MR 34726) and the Upper Stonesdale Limestone (MR 34737). More extensive dolomitisation occurs as dolomite-calcisiltite in a thin limestone in shale beneath the Main Limestone (MR 34656), and in micritic dolomite (MR 34724). These are evidently replacements presumably of calcite; if the freeing of Mg^{++} from ubiquitous high-Mg calcite during lithification contributed to dolomite formation then a more evenly widespread replacement by dolomite might perhaps be expected rather than the patchy occurrence. Parankerite occurs in the Lower Stonesdale Limestone (MR 34739) replacing micritic matrix, where it is associated with carbonaceous, pyritic and phosphatic components.

Silicification varies from selective to complete, resulting in cherts. Selective replacement of carbonate biota occurs in several specimens, e.g. of corals (Underset Limestone, MR 34690) (beneath the Underset, MR 34651), and of micritic envelopes in the Lime Plate (MR 34711). Primary laminae too may be selectively silicified (Underset, MR 34651, the Faraday House Marine Band, MR 34740, Main Chert, MR 34783). Silicification is associated with glauconite and carbonaceous material (e.g. Crow Chert, MR 34718), and occurred under locally reducing conditions. Chalcedony is commonly associated with sponge spicules (*Hyalostelia sp.*) in the present specimens so a biogenic origin seems likely for part at least of the silica. In modern deep ocean deposits, the source (at least in nodular cherts) appears to be opaline skeletal silica of radiolarian tests and sponge spicules (Davies and Supko, 1973). Reference has already been made (p. 19) to the problem of the origin of the chert formations such as the Richmond Cherts. The present limited petrographic study offers no further new evidence.

Apart from sporadic dolomitisation and silicification most of the specimens are relatively pure limestones. Some ingress of resistate and clay detritus is shown, by for example, the *Spirifer* bed (Four Fathom Limestone: MR 34729) where terrigenous components are intercalated with shelly limestone laminae. Reorganisation of clay, carbonaceous material, sulphide and other insoluble residue is shown by microstylolites as in the Coldstones Limestone (MR 34637), Smiddy (MR 34635) and Simonstone Limestone (MR 34627). These microstylolites probably formed during diagenesis.

Arenaceous rocks Nearly all the samples examined from the Brigantian and early Pendleian represent interdistributary sheet deposits, forming, as Elliott (1975) has emphasised in his account of the Great cyclothem, the upper parts of coarsening-upward sequences. Quartzitic sandstones of fine sand grade (0.06 to 0.2 mm) predominate, with relatively hard, even-grained tough textures due to closely-packed resistate detrital grains, amongst which igneous quartz is predominant. With few exceptions these rocks are cemented by secondary silica moulded around grain cores; microstylolitic contacts are not conspicuous, suggesting that chemical deposition rather than pressure-solution was the principal factor during lithification. This has been noted in Upper Carboniferous sandstones also by Greensmith (1957). The sand grains exhibit a low degree of roundness, indicating first cycle deposition. A majority of the samples could be termed orthoquartzites in the sense of Krynine (1948), but they have not been recrystallised or metamorphosed, and are therefore classified here as quartzitic sandstones. Muscovite, where present, may sharply transgress both quartz individuals and their secondary overgrowths, indicating solid-state growth following silicification; secondary muscovite is also strewn along joint-planes. The sandstones are commonly light

[1]Numbers refer to the Museum Reserve Collection of the British Geological Survey.

grey (near N7) except where stained by secondary iron hydroxide. Some samples show thin bedding to primary lamination, and rarely, cross-lamination with grading. In these cases bedding planes are usually picked out with fine muscovite, carbonaceous flakes or by staining with limonite. Rarely clay pellets or galls occur, but carbonate either derived or precipitated is hardly ever found in spite of the important part played by limestone in the cyclothems. There are very sparse rock particles, including cherty quartz, fine siltstone with variable clay matrix and quartzitic siltstone; metamorphic quartz is generally very rare. Feldspar grains are no more than minor or even trace constituents; where present they include fresh to partly kaolinised orthoclase and fresh albite-oligoclase; the quantity is, however, far below that required (25 per cent) for arkose. Biotite or hydrobiotite and muscovite may be scattered along bedding planes. Primary matrix is sparse, occurring only as thin intergranular pellicle coatings of clay material perhaps augmented by kaolinite dust transported in during deposition, rather than resulting from weathering. The main constituents of the quartzitic interdistributary sandstones suggest a first-cycle granitic source, from which deep weathering has eliminated most of the feldspar.

An essentially granitic source is also suggested by the heavy detrital minerals. These were first studied by J. A. Butterfield (1939a,b) and the concentrates from the present suite largely confirm his findings. There is a simple association of first-cycle colourless zircons, pleochroic tourmalines and sparse apatite. Rutile is probably primary, but the other titanium polymorphs present (brookite, anatase, 'leucoxene') are likely to be authigenic. The zircons, carrying fluid and solid inclusions, are mostly angular or broken, the grains seldom exceeding 0.2 mm but occasionally reaching 0.5 mm. Rounding or abrasion is rare. Tourmaline, in euhedra or angular broken grains shows pleochroism characteristic of a granite (olive-green, yellow-brown or black to colourless or pale brown) or granite-pegmatite (blue to colourless) provenance. Garnet is rare and was found in only six samples. Rutile occurs as euhedra or broken crystals averaging 0.1 mm, but geniculate twins are also present. Hornblende is rare as green cleavage particles and opaques include ubiquitous 'leucoxene' and rare iron-titanium oxides.

Sandstones forming the upper parts of the following cyclothems are represented by the foregoing description: Gayle (MR 34681, 34697, 34699); Hardrow Scar (the Dirt Pot Grit) (MR 34628, 34645, 34679); Simonstone (MR 34677, 34700); Single Post (MR 34648, 34676); Middle (MR 34649, 34674, 34703); Five Yard (MR 34655, 34671–2, 34694, 34705–6); Three Yard (the Nattrass Gill Hazle) (MR 34708); Underset (MR 34650, 34666, 34731); Main (the Coal Sills Sandstone) (MR 34713); Little (the Faraday House and Uldale sills, and the Ten Fathom Grit) (MR 34715, E 53855[1], MR 34717, 34661, 34686, 34688).

A distributary sandstone cutting the Three Yard Cyclothem (MR 34624, 34669, 34727–8) contrasts with the sheet sandstones so far described in including subarkose of closely-packed poorly-sorted igneous quartz, subordinate cloudy orthoclase, albite, cryptoperthite and a little chert, muscovite, polygranular metamorphic quartz. Muscovite is concentrated along laminae and there is conspicuous intergranular kaolinite and iron hydroxide. In one sample the detrital grains are separately enclosed in pellicles of clay micas and kaolinite.

Above the intra-E_1 unconformity the Millstone Grit facies makes its appearance and the style of arenaceous sedimentation represented by most of the major sandy units is markedly different from that of the Yoredale facies sheet-sandstones. The Lower Howgate Edge Grit (MR 34693) and its postulated equivalent farther south, the Grassington Grit (E 19914–7) are pebbly rocks in which igneous or metamorphic quartz provides most of the coarse clasts. However, subordinate microcline, orthoclase and sodic plagioclase are present, and there is a notable amount of kaolinitic material presumably derived from weathering, both at source and locally, of the feldspars. Shreds of mica are present but rarely abundant. The heavy mineral suite is similar to that already described. The general impression gained is of their closer proximity to the source area, or of greater elevation of that area, whether it be to the north as Gilligan (1920) or to the south as Evans and others (1968) postulated for the Millstone Grit of Yorkshire and Macclesfield respectively. Nevertheless, the proportion of unaltered feldspar remains low, and a nearby source is unlikely. Quartzitic sandstones also recur in this facies; for example, the Mirk Fell Ganister (MR 34738), according to the mapping of Rowell and Scanlon (1957a) the lateral equivalent of the Lower Howgate Edge Grit, is such a rock and the type strikingly makes up the Follifoot grits (E 19922–3) which are medium-grained quartzites. On the other hand the Red Scar Grit (E 19920–1) is coarse with pebbles of quartz and quartzite and though the proportion of fresh feldspar is low, the rock is often noticeably kaolinitic. Sufficiently fresh material has not been obtained to show if the reddish weathering of this rock is due to oxidation of primary pyrite or siderite in it. The sliced specimen from the equivalent Pickersett Edge Grit (MR 34743) is quartzitic sandstone, but in places much coarser material is also present at this horizon.

The petrographic investigation of the arenaceous rocks has failed to reveal any systematic variation with stratigraphical position other than the greater abundance of coarse pebbly grits of fluviatile origin above the intra-E_1 unconformity. However, the channels cutting through the Brigantian and early Pendleian strata (as indicated by Moore (1958, 1959) also contain rocks of similar lithology, coarser and much less uniform than the sheet sandstones which typify the cyclothems of the Yoredale facies. RKH

Argillaceous rocks The petrographical investigation, even of small numbers of samples of the shales, mudstones, siltstones and striped 'grey beds' that make up the remainder of the rocks in the Carboniferous sequence did not appear likely to be sufficiently profitable to justify it in the present connotation; as will emerge later, these rocks are unfavourable to mineralisation on account of the physical properties (in failing to promote clean fracturing) and their impermeable character. Study of typical shales from the Yoredale facies around Moor House (Johnson and Dunham, 1963) showed that they are composed of illite, chlorite, interlayer illite-chlorite and subordinate kaolinite, sometimes with kandite, together with fine detritus, chiefly of quartz. In the non-marine and sometimes in the marine shales, siderite becomes an important constituent, chiefly in the form of diagenetic nodules.

Conclusion

It is a reasonable conclusion that prior to tectonic fracturing, the entire pile of Carboniferous sediments hereabout possessed a low permeability; the limestones, because during diagenesis all available open space was filled with late carbonate; the cherts and the quartzitic sandstones because of extensive diagenetic silification; and the shales and mudstones because, once consolidated, they became low-permeability rocks. Only the coarser grits in the channels and fluviatile spreads perhaps retained some porosity. There is, as will be shown, no evidence to suggest that either these or the sheet sandstones acted as widespread aquifers for the mineralising fluids. On the contrary, tectonic fracturing of the sedimentary pile is the key to the start of the mineralising process. KCD

[1]Numbers prefixed by E refer to the English Sliced Rock collection of the British Geological Survey.

CHAPTER 4

The country rock: Dinantian stratigraphy

The Lower Carboniferous or Dinantian includes the most widely-exposed group of rocks in the Northern Pennine Orefield south of Swaledale, where the Main Limestone at the base of the Upper Carboniferous is high on the hillsides. The main valleys cut deeply into the Dinantian, but fail to reach the base except in the south, between Austwick and Settle. In the compass of a chapter it is impossible to do full justice to the detail of the many excellent exposures, but by describing representative sections it is feasible to build up a picture of the stratigraphy of this large area. Lateral variations in thicknesses of individual units are represented both in the comparative vertical columns (Figures 5, 7 and 8) and by means of isopach maps (Figure 10).

The nomenclature of the subsystem adopted here follows that proposed by T. N. George and others (1976) on behalf of the Geological Society of London, and is set out in Figure 2. For comparison with former usage in the orefield, as established in Dunham (1948) it should be said that the Middle Limestone Group is assigned here to the Brigantian, the Lower Limestone Group to the Asbian, and the Basement Group of the Tyne to Stainmore region includes strata now assigned to the Holkerian and to earlier stages.

TOURNAISIAN

Rocks of proven Tournaisian age are exposed just outside the north-western edge of the area at Ravenstonedale and probably persist eastwards through the Stainmore Trough, but thin out to the south (Figure 4). The lowest strata, the Pinskey Gill Beds are 148 to 164 ft (45–50 m) of grey vuggy limestone, dolomitic limestones and dolomites interbedded with grey to dark grey calcareous mudstones which rest with marked unconformity on Silurian slates. Calcareous silty sandstones occur in the basal 33 ft (10 m). Miospores and conodonts suggest an Ivorian age for these beds (Johnson and Marshall, 1971; Holliday, Neves and Owens, 1979; Varker and Higgins, 1979). The overlying Shap Conglomerate, of probable Ivorian age, consists of fine or medium-grained sandstones with many small rock fragments and bands of conglomerate, interbedded with siltstones and mudstones. These beds are thought by Holliday, Neves and Owens (1979) to be of fluvial origin.

In the Sedbergh area some 700 ft (233 m) of crudely bedded reddish brown conglomerates fill up irregularities in the old landscape (Butterfield, 1920). The age of these Sedbergh Conglomerates is uncertain and they could include Devonian and Tournaisian strata.

CHADIAN

Beds of this probable age are restricted to the northern half of the area and are only seen at surface in an inlier at Penny Farm Gill [SD 698 932] and Nor Gill (Figure 5) [SD 699 933]. A somewhat similar succession 470 ft (143 m) thick is now recorded in the Raydale Borehole (Burgess, b, in preparation). There is further northwards thickening to about 1000 ft (304 m) at Ravenstonedale, and it is likely that these thick beds extend eastwards in the Stainmore Trough under the north of the present area.

At Penny Farm Gill the Sedbergh Conglomerates are overlain by Nor Gill Sandstone, 31 ft (9.45 m) of fine-grained sandstones and siltstones with many bands of 'cornstone', palaeosols with calcareous nodules, in the upper beds. Higher in the sequence the Penny Farm Gill Dolomite, probably of part-Chadian, part-Arundian age and totalling 166 ft (50.60 m) in thickness is well exposed (Figure 5, section 6). This formation consists of many alternations of sandy dolomites, dolomites and fine-grained sandstones thought to have been deposited in very shallow water (Burgess, a, in preparation).

Equivalent strata in Raydale Borehole are apparently much thicker since there are two separate dolomite formations (Figure 5, section 7). A lower formation, the Raydale Dolomite is closely similar to the Scottish cementstone facies. It is 102 ft (31 m) of thinly bedded and nodular dolomites with much interbedded siltstone and fine-grained sandstone. Miospores suggest a post-Tournaisian age (Owens *in* Burgess, b, in preparation). The overlying Marsett Sandstone consists of 200 ft (61 m) of reddish brown and greenish grey sandstones and conglomerates with rare bands of dolomite. The third and highest formation, 184 ft (56 m) thick, and probably of part-Chadian, part-Arundian age corresponds closely in lithology with the Penny Farm Gill Dolomite at its type locality.

Chadian strata at Ravenstonedale consist of two formations. The lowest are the Stone Gill Limestones, about 568 ft (173 m) of grey, well-bedded limestone with some dolomite bands. Near the base and top there are several porcellanous calcilutite bands (Holliday, Neves and Owens, 1979; Johnson and Marshall, 1971). Rugose corals are absent and shelly fossils tend to be in bands crowded with individuals of a few species like *Camaratoechia proava* (Garwood, 1913). The overlying well-bedded Coldbeck Limestones some 246 ft (75 m) in thickness have several mudstone partings and at least three algal bands, the basal one being the Spongiostroma Band.

ARUNDIAN

Limestones of Arundian age These are chiefly dark grey limestones with a distinctive coral-brachiopod fauna characterised by *Michelinia megastoma* and *Delepinea carinata*. Where the highest Arundian Ashfell Sandstone is present in the northern half of the area the approximate upper limit of the Arundian is easy to fix. Farther south where the sandstone is absent the upper limit is thought to be near the top of the Kilnsey Limestone, a distinctive unit which contrasts with the overlying light and very light grey limestone of the Cove Limestone member of the Malham Formation (Mrs L. C. Jones, personal communication). The colour of the Kilnsey Limestone is typically dark grey, with grey beds coming in near the gradational top. The limestones are fine- and medium-grained calcarenites with a tendency for grainstones to occur towards the top of the formation, at about the level where the sandstones come in to the north. In the lower beds of the Kilnsey Limestone the characteristic *Michelinia megastoma* fauna was recorded by Garwood and Goodyear (1924, pl. 20) in a belt between Kingsdale and Kilnsey on the southern fringe of the Askrigg Block. These beds rest directly on the eroded Lower Palaeozoic rocks and are locally banked against humps in the old landscape, as at Twistleton Scars, Chapel le Dale (Figure 5, section 3). Here Dr D. Moore (personal communication) has studied a hummocky terrain with a relief of 34 ft (10.40 m). At the base patches of conglomerate rich in

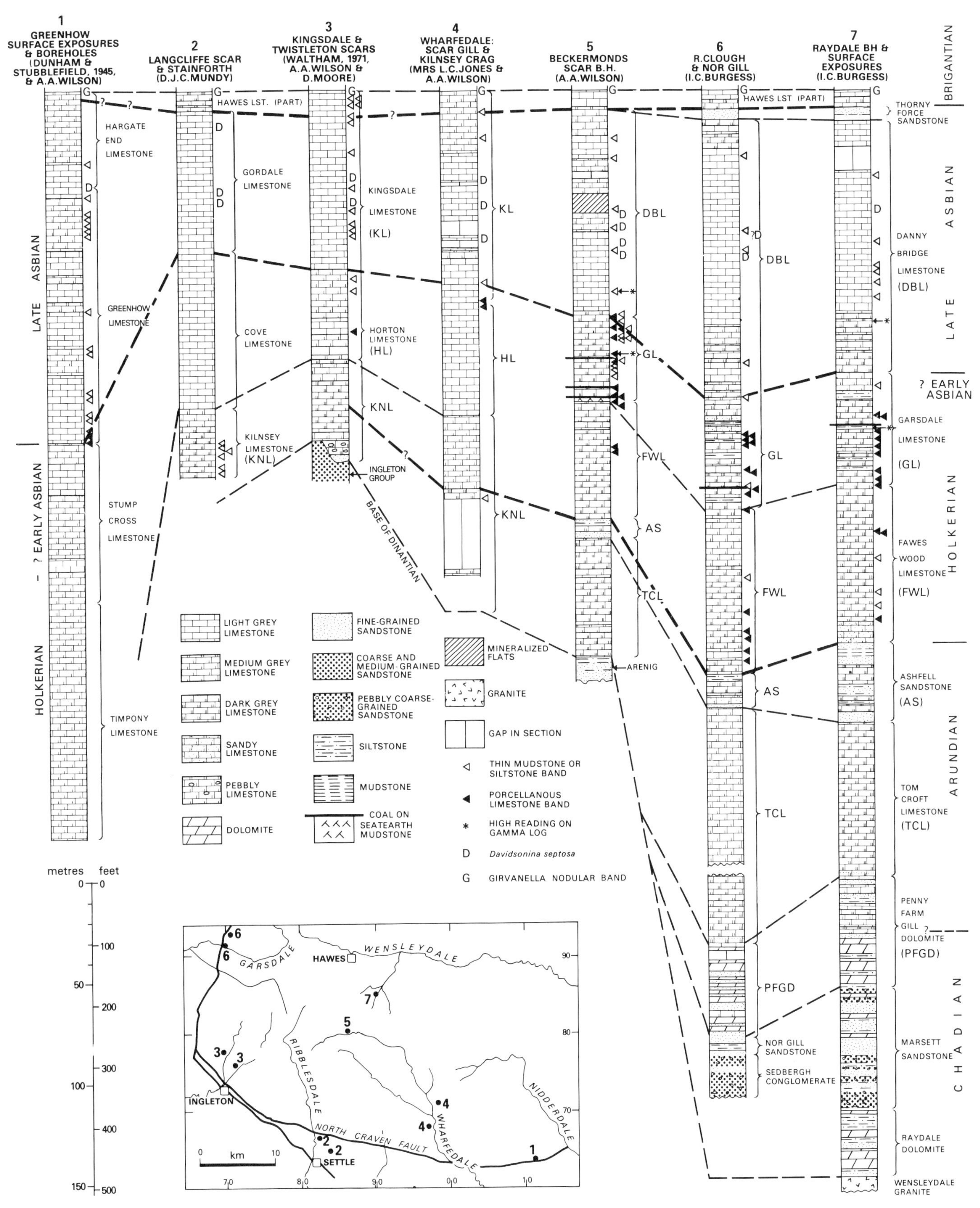

Figure 5 Comparative sections of Dinantian strata from the base of the Carboniferous up to the top of the Asbian

greywacke clasts are overlain by 7 ft (2.13 m) of calcareous sandstone. Medium- and fine-grained sparites with conglomerate bands follow; *Michelinia megastoma* and *Delepinea* are typical fossils collected here by Mr J. Pattison (see also p. 20). Higher grey and dark grey biosparites above the hump in the old landscape yield *M. megastoma* and *Siphonophyllia garwoodi*. At the top are grey, fine-grained, sparsely fossiliferous sparites with *Lithostrotion martini* and large gastropods. The horizon of these last beds on faunal evidence is that of Garwood's Gastropod Beds.

East of Kingsdale the upper sparsely fossiliferous part of the Kilnsey Limestone is thicker than at Kingsdale, attaining 120 ft (36.60 m) at Kilnsey Crag, Wharfedale. The *Michelinia* fauna occurs at the lowest exposure in the Wharfedale sequence (Figure 5, section 4) at Mill Scar Lash [SD 9796 6639] (Garwood and Goodyear, 1924).

Sections in the River Clough, Raydale Borehole and Beckermonds Scar Borehole show a basic similarity. Overlying the partly Arundian Penny Farm Gill Dolomite (p. 26) there is a thick run of Tom Croft Limestone (Burgess,a, in preparation), dominantly dark grey in colour and virtually free of mudstone partings. *Michelinia megastoma* and *Delepinea carinata* have been recorded in the Tom Croft Limestone of the Beckermonds Scar Borehole. *D. carinata* was collected by Garwood (1913) in Penny Farm Gill. The upper beds include *Lithostrotion martini*, particularly in the above borehole (p.20). Though these limestones are presumably equivalent in part to the Gastropod Beds these fossils are rare in the boreholes.

At Ravenstonedale most if not all of the partly dolomitic Scandal Beck Limestone (Ramsbottom, personal communication, based on foraminiferal evidence) and all the Michelinia grandis Beds are of Arundian age. These formations are fossiliferous limestones with mudstone partings, totalling some 850 ft (260 m) in thickness.

Ashfell Sandstone Towards the close of Arundian times a distinctive 500 ft (152 m) thick sandstone formation with four limestone interbeds was deposited in the Ravenstonedale area. The sandstones are fine, rarely medium-grained, with current-ripple laminations and rare cross-bedding. Convoluted bedding is well seen in the upper beds on Ashfell Edge. Thinning northwards towards the Shap area, they were thought to have been derived from a partly granitic landmass to the north stretching from the edge of the Lake District to the Alston Block (Turner, 1950, fig. 2). Their sudden thinning against this old landmass at the Swindale Beck Fault was demonstrated by Burgess and Harrison (1967, fig. 2).

The Ashfell Sandstone in the Raydale Borehole consists of sandstone and siltstone members with six limestone interbeds, the whole formation totalling 126 ft (38.40 m) in thickness (Burgess,b, in preparation). Further attenuation occurs in the River Clough (Turner, 1959) and Beckermonds Scar Borehole where there are three or four clastic intervals separated by limestones (Figure 5, sections 5, 6). In both boreholes silty mudstone bands contain species of *Pteronites* and *Leiopteria* (p.20).

In the south of the area the Ashfell Sandstone has not been noted, though on faunal grounds Mr Pattison has identified Gastropod Beds in the Twistleton Scars section which should be broadly their equivalent.

HOLKERIAN AND EARLY ASBIAN

Horton Limestone, Cove Limestone (in part) and equivalents Beneath the Kingsdale limestones in a belt of country between Ingleton and Wharfedale there is a distinctive light or very light grey limestone formation referred to S_2 by Garwood (1913). Though it forms a part of a conspicuous cliff at Malham Cove and Kilnsey Crag it generally tends to make extensive scree covered slopes. This is variably called the Horton Limestone (Ramsbottom, 1974) when its top is fixed at the highest porcellanous band or the Cove Member of the Malham Formation when its top is defined by the top of Schwarzacher cycle 2 (D. J. C. Mundy, personal communication).

In the River Clough, Beckermonds Scar and Raydale boreholes the succession is very different from that in the south. Instead of pale grey limestones dark grey beds dominate the sequence which falls into two formations newly named from the River Clough (Burgess,a, in preparation):

The Fawes Wood Limestone, which ranges in thickness from 265 ft (80.77 m) in the River Clough to 183 ft (55.80 m) in Beckermonds Scar Borehole, consists dominantly of dark grey and grey fine-grained grainstones and packstones with a few mudstone and siltstone bands. Though it is likely that most of these beds are higher than Garwood's Gastropod Beds they nevertheless contain many *Macrochilina*?, a large gastropod, besides *Lithostrotion martini*, present at several levels common to Raydale and Beckermonds Scar Boreholes (Pattison, personal communication).

The overlying Garsdale Limestone is 135 to 190 ft (41–58 m) of dominantly dark grey fine-grained wackestone with many interbeds of siltstone and mudstone; one to four thin coal seams and seven to ten bands of calcilutite, up to 4 ft (1.22 m) thick, also occur. Two of the latter show fenestral fabric in Beckermonds Scar Borehole, pointing to emergent conditions, as also evidenced by one of the coal seams which clearly rests on a seatearth. In Raydale and Beckermonds Scar boreholes a high gamma marker close to the middle of the formation serves to strengthen the correlation. In the River Clough the higher part of the Garsdale Limestone contains two bands with *Daviesiella llangollensis* (p. 21), a form not found in the boreholes though *Daviesiella*? was found near the base of the formation in Beckermonds Scar Borehole.

At Ravenstonedale the Ashfell Limestone, judged to be of Holkerian age, is a poorly bedded pale grey limestone with some sandy mudstones (Ramsbottom, 1974, p. 58). It equates with the Fawes Wood and perhaps part of the Garsdale Limestone. Overlying it is the Potts Beck Limestone (with *Daviesiella llangollensis*) the base of which at Little Asby Scar [NY 6988 0827] is taken as the Asbian stratotype (George and others, 1976).

Ramsbottom (1974, fig. 20) has indicated that beds with *D. llangollensis* thin out southwards on to the Askrigg Block. The precise degree and manner of this thinning is in some doubt as there is some evidence of a lateral change of facies to the south. It is possible that the eight cycles discovered by Jefferson (1980) close to the Holkerian – Asbian boundary will prove to be equivalent to the Garsdale Limestone further north which also falls into about eight cycles.

Another porcellanous band 42 ft (12.80 m) below the main porcellanous band at Twistleton Scars, probably lying within the group of beds studied further east by Jefferson, is likely to be equivalent to one of the lower calcilutites in the Garsdale Limestone. Recently Dr Ramsbottom has recorded *D. llangollensis* as far south as Horton in Ribblesdale, showing that beds with this fossil have not thinned completely.

At Greenhow no definite Arundian rocks are known and 640 ft (195 m) of strata are of Holkerian – early Asbian age (see p. 21 for the palaeontological evidence). These consist of two formations mapped by Dunham and Stubblefield (1945). The Timpony Limestone, the lowest formation exposed at Greenhow, consists of 380 ft (116 m) of grey and grey to dark grey, fine-grained limestones with some chert nodules in the upper beds. Coldstones No. 4 Borehole sunk to within 80 ft (24.38 m) of the base of the exposed succession confirms that no mudstone partings occur down to that level in the Timpony Limestone, but debris in dumps indicates that the lowest beds contain some mudstone bands (Dunham and Stubblefield, 1945). The absence of a long run of dark grey limestone in the borehole cores suggests that the dominantly dark grey Kilnsey Limestone of Arundian – Holkerian age is buried in the Greenhow Anticline and does not come to crop. The Stump Cross Limestone, overlying the Timpony Limestone and including the well known caverns, consists of 260 ft (79.25 m) of fine-grained buff and light grey limestones.

LATE ASBIAN

Kingsdale Limestone and equivalents The Kingsdale Limestone was first defined by Ramsbottom (1974) as the beds between the 'Porcellanous Band' and the base of the Girvanella Nodular Band. In the northern half of the area the approximate equivalent, the Danny Bridge Limestone (Burgess,a, in preparation), ranges from the top of the Garsdale Limestone to the base of the Thorny Force Sandstone. It corresponds with the beds of the D_1 Zone of Garwood (1913) and forms a characteristically stepped landscape with many karstic pavements. Overall thickness varies from 336 to 550 ft (102–168 m) and is greatest on the southern margin of the Askrigg Block in a belt from Giggleswick Scar to Greenhow.

Using clearly defined bedding planes, which are often related to the tops of major features, Schwarzacher (1958) divided up the Kingsdale Limestone into nine cycles. Doughty (1968) used joint density as an aid to the recognition of these cycles and maintained that each successive topographic feature is commonly built up of densely jointed grainstone which tends to form scree, overlain by less densely jointed finer-grained cliff-forming limestone. Using this method he remapped an area north of Settle and discounted the existence of several faults mapped by earlier workers (Doughty, 1974).

Recently McCabe (1975) and Mundy (discussion of Jefferson, 1979) have disagreed with several details of Doughty's mapping. Recent surveys between Giggleswick Scar and Kilnsey (D. J. C. Mundy and Mrs L. C. Jones, personal communication) have shown that features commonly split into two, especially in the south where there are no clay bands, and cannot be relied on for the purposes of defining each and every Schwarzacher cycle beyond reasonable doubt. Above Giggleswick Scar, for instance, there are at least sixteen separate features.

The most distinctive lithological characteristic of the Kingsdale Limestone, the clay bands, was first noted by Dunham and Stubblefield (1945) in the Greenhow Limestone, equivalent to the lower three-quarters of the Kingsdale Limestone. There are about 15 such bands at Greenhow, present in underground sections and boreholes (Wilson, 1974) and also in cave systems near Ingleton (Waltham, 1971) and in boreholes at Beckermonds Scar (Wilson and Cornwell, 1982) and Raydale (Burgess,b, in preparation). Typically pale grey, unfossiliferous and with scattered pyrite crystals weathering to limonite, these partings average 1 ft (0.30 m) in thickness and commonly rest on irregular palaeokarst surfaces. Some appear to separate the Schwarzacher cycles, whilst others occur within them. Recent boreholes near Greenhow and west of Malham show that the clay bands decrease in number towards the southern edge of the Askrigg Block and are locally absent. Out of 15 bands proved one mile to the north only five persist to Greenhow itself (Figure 5, section 1). The clay bands are rarely exposed, but tend to form lines of seepage in the mineralised area north of Kilnsey Moor (Mrs L. C. Jones, personal communication) and three bands are exposed in cliffs and stream sections in Wharfedale (Figure 5, section 4).

The Kingsdale limestones are light and medium grey, fine- and medium-grained, chiefly wackestones with some grainstones. At several levels probable bioturbated beds occur which weather as 'pseudobreccias'.

Schwarzacher's (1958) first cycle is easy to recognise where the Porcellanous Band is present but tends to be poorly exposed. In the Settle – Kilnsey area, where the Porcellanous Band is absent, cycle 1 is hard to define (D. J. C. Mundy, personal communication) and a new formation, the Gordale Limestone, is proposed to include beds from the top of cycle 2 (marked by the top of a major cliff) to the base of the Hawes Limestone. Cycle 2 forms a lengthy cliff in Wharfedale, notably around Kettlewell, and in Chapel le Dale, and its base appears to be marked by a clay band on a palaeokarstic surface.

Higher cycles are harder to tell apart with total certainty. The base of cycle 3 commonly includes cross-bedded grainstones, also present in the middle of cycle 4 (D. J. C. Mundy, personal communication). The richest band with *Davidsonina septosa* is thought to lie in cycle 5 but other such bands can occur in cycles 4 to 8. These biostromes tend to fail towards the southern margin of the Askrigg Block (D. J. C. Mundy, personal communication).

Schwarzacher's ninth cycle is apparently equivalent to the Robinson Limestone of the Alston Block and the north-western part

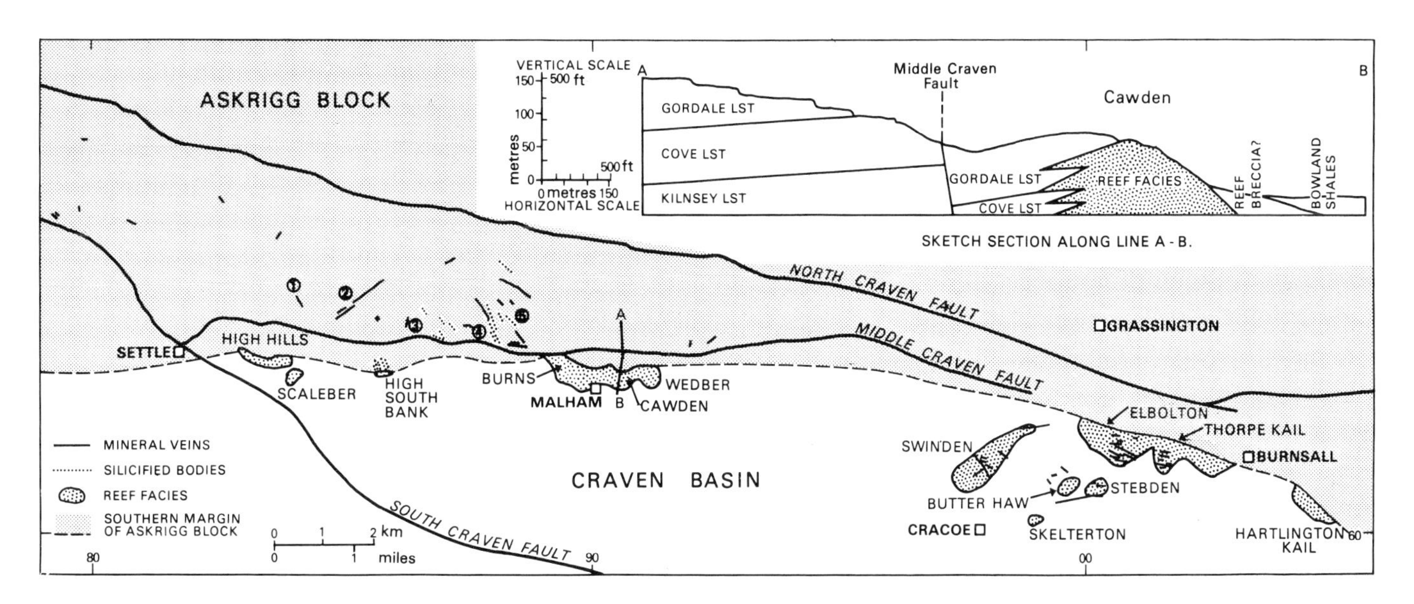

Figure 6 Stratigraphical and structural relationships of the Craven reefs

Key to numbered localities **1** Pikedaw veins **2** Malham 'calamine' **3** Great Scar veins **4** Settle Scar-Back Scar veins **5** Benscar veins

of the present region. Over much of the present area the mudstone parting below cycle 9 is absent or very thin and hence the cycle cannot everywhere be defined. In the north-west, however, the parting thickens, notably in the River Rawthey where 31 ft (9.45 m) of mudstone with thin limestone bands is exposed (Burgess and Mitchell, 1976). Farther north, in the mineralised ground at Fell End Clouds, the Robinson Limestone forms a strong feature into which an adit has been driven [NY 7418 0019]. The dumps are chiefly of mudstone with a little dark grey argillaceous limestone; levelling indicates a likely thickness for these beds of 10 ft (3 m). The section of Fell End Clouds, according to Schwarzacher (1958, fig. 2), ranges from cycle 1 to the base of cycle 8. This is puzzling since cycles 8 and 9 are particularly well exposed, and there is likely to have been a miscorrelation. There is considerable north-westward thickening of the mudstone beneath the Robinson Limestone into the Stainmore Trough where 80 ft (24.40 m) of beds including sandstone are exposed in Augill Beck (Burgess and Mitchell, 1976).

The equivalents of the Kingsdale Limestone at Greenhow have been divided into two formations by Dunham and Stubblefield (1945). The lower, the Greenhow Limestone, is dominantly light grey with some grey beds and totals 400 ft (122 m) in thickness. Up to 15 mudstone bands occur, in marked contrast to the underlying Stump Cross Limestone which is devoid of mudstones in Coldstones No. 4 Borehole. The higher formation, the Hargate End Limestone, is dominantly a dark grey or grey crinoidal limestone, 150 ft (45.70 m) thick, the base of which is taken at *Davidsonina septosa* which here only forms a single band, known at outcrop and in a borehole; only one impersistent mudstone band is known. It has not proved possible to recognise the identity of Schwarzacher cycles at Greenhow. AAW

The upper beds of the Cove Limestone, the Gordale Limestone and part of the Hawes Limestone pass southwards into the Craven Reef Belt which was established on the edge of the Askrigg Block (Figure 6). This belt, thought to have been a once continuous apron for at least a 14 mile (23 km) length has suffered faulting followed by inter-Carboniferous erosion (p. 15). The former reef apron undoubtedly contained algal boundstone frameworks as at High Hills, Cawden and Thorpe Kail but these were only subordinate components in these complex build-ups.

The hills of Stebden and Butter Haw show original quaquaversal dips of up to 35° and are evidently isolated mound shape build-ups which grew up south of the main reef apron. The palaeoecology of Stebden (over a bathymetric range of 110 m) has been studied by Mundy (1978). A wave-resistant organic framework constructed by stromatolitic algae, lithistid sponges, encrusting bryozoans, tabulate corals and attached pseudomonotid bivalves has been identified on the summit of the hill. It appears however that this framework does not extend into the core of the build-ups, but was a later event in its history. DJCM

BRIGANTIAN

Lower Bowland Shales These are dark grey shaly mudstones with thin limestone bands which locally thicken greatly to 950 ft (290 m) due to the incoming of the Pendleside or Nettleber Sandstone in parts of the Craven lowlands. The Lower Bowland Shales have not been recorded on the Askrigg Block, but their equivalence to Yoredale facies beds from the Gayle up to the Underset Limestone is demonstrated by the numerous goniatite bands ($P_{1a}-P_{2c}$). These equate with the rare goniatite records on the Askrigg Block (Figure 3, p. 22).

The account which follows describes the beds of Yoredale facies on the Askrigg Block and to the north, which include the equivalents of the Lower Bowland Shales.

Hawes Limestone This formation, first named by Moore (1958) ranges from 41 to about 80 ft (12.50–24.40 m) in thickness, averaging some 50 ft (15.25 m) over most of the area. When traced northwards on to the Alston Block it splits into two separate limestones, the Peghorn and Smiddy (Burgess and Mitchell, 1976). Though the clastic parting between these two limestones is absent in the present area, their equivalents can be recognised within the Hawes Limestone, which is generally divisible into three members. The lower two members equate with the Peghorn Limestone, the top one with the Smiddy. In the south-east, around Greenhow, the lowest member may be represented by the topmost beds of the Hargate End Limestone; the middle and top members form part of the Coldstones Limestone (Figure 7, section 22).

The lowest member of the Hawes Limestone is a medium grey medium-grained biosparite which commonly forms a steep feature in Wharfedale. Locally, as at Dow Cave [SD 9836 7432], Wharfedale, high current activity is shown by transgressive bedding planes. The upper beds locally contain large coral colonies, as for instance in the section at Aysgarth Force, Wensleydale [SE 0210 8885 to 0127 8870] which also shows clearly the tripartite division of the Hawes Limestone:

	Thickness		
	ft	in	m
GAYLE LIMESTONE	–	–	–
GAYLE SHALE (in Middle Force, Aysgarth)			
Mudstone with numerous bands of dark grey limestone up to 10 in (0.25 m) in thickness; *Antiquatonia sulcata*, *Gigantoproductus sp.*, *Latiproductus latissimus*, *Edmondia sp.*	8	3	2.50
HAWES LIMESTONE, above Girvanella Nodular Band			
Limestone, dark grey, fine-grained biosparite with numerous mudstone partings up to 4 in (0.10 m) thick, *Saccamminopsis fusulinaformis*, *Lithostrotion martini*	7	4	2.24
Limestone, grey, fine-grained sparite; well-bedded, becoming wavy downwards	19	8	6.00
HAWES LIMESTONE, Girvanella Nodular Band (in Aysgarth Force)			
Limestone, dark grey sparite; well-bedded; abundant *Linoprotonia dentifer*, also *Athyris expansa*; '*Girvanella*' near top and base	5	0	1.52
HAWES LIMESTONE, below Girvanella Nodular Band			
Mudstone, calcareous, with muddy limestone	1	0	0.30
Limestone, dark grey sparite, fine-grained; mudstone partings	4	3	1.30
Limestone, grey sparite; 'pseudobreccia'	1	4	0.40
Limestone, biosparite, fine-grained; *Diphyphyllum lateseptatum*, *Lithostrotion junceum*, *L. portlocki*, *Syringopora* cf. *geniculata, Palaeosmilia murchisoni*, *Gigantoproductus sp.*	2	2	0.66
Limestone, light grey biosparite; 'pseudobreccia' at top	4	3	1.30

An estimated 31 ft (9.45 m) of grey and light grey sparites lie between the lowest bed of this section and the underlying Thorny Force Sandstone.

The highest stratum of the lowest member of the Hawes Limestone was termed the White Post by Burgess and Mitchell (*op. cit.*). This commonly forms a pale-weathering karstic pavement, but in Wharfedale the internal rock colour is medium grey or dark grey as at Aysgarth, rather than the light grey seen farther north. The top of the White Post is heavily bioturbated in places and is likely to have been an emergence surface.

In the country from Dentdale, through Kingsdale to Greenhow, dark grey limestones up to 15 ft (4.57 m) in thickness underlie the White Post. In this area and in parts of Wharfedale the exact base of the Hawes Limestone is indefinite and is usually taken at a thin mudstone parting or major bedding plane.

The middle member of the Hawes Limestone is a well-bedded, dark grey, fine-grained, often richly shelly biosparite, with thin muddy partings. Within these beds, and commonly towards the top and base, the encrusting alga *'Girvanella'* (from genus *Osagia*, Johnson, 1959), is seen. These algal nodules are particularly abundant at Coldstones Quarry, Greenhow [SD 124 643] (Dunham and Stubblefield, 1945) and form a valuable datum in boreholes sunk in the area. Locally the algal nodules are rare and the abundant small brachiopods, notably *Linoprotonia dentifer*, are a more prominent part of the fauna.

The top member of the Hawes Limestone is a medium grey, medium-grained sparite, with dark grey beds close to the top. Fossils are rare except for a band with the large foraminifer *Saccamminopsis fusulinaformis*, which is commonly found in the top 10 ft (3 m) of limestone.

Beds above the Hawes Limestone These beds are generally thin except in the north and west (Figure 10, map B) and are referred to as the Gayle Shale (Hudson, 1925). Averaging 6 ft (1.80 m) in thickness in Wensleydale and Wharfedale, these are typically calcareous mudstones interbedded with several bands of muddy dark grey limestone. The undercut sections of picturesque waterfalls as at Aysgarth Lower Force and West Burton Force are developed at this level. To the west these beds thicken to 12 to 24 ft (3.66–7.32 m) in Kingsdale and Dentdale (Figure 7), but the precise base of the overlying Gayle Limestone is in places indefinite due to numerous mudstone partings. The greatest thickening is in the north-west, to a maximum of 125 ft (38 m), in the Stainmore Trough (Burgess and Mitchell, 1976). The beds around Brough include a sandstone, the Smiddy Ganister, but in the remainder of the present area of study sandstones are absent.

In the south-east of the area the Gayle Shale becomes very thin. In Darnbrook Beck [SD 885 714], for instance, it is marked only by a few thin argillaceous partings in limestone. At Greenhow the parting between the Gayle and Hawes members of the Coldstones Limestone is merely a muddy bedding plane, not identifiable with certainty (Figure 7, sections 22, 23).

Gayle Limestone This formation, first named by Hudson (1924), shows a marked tendency to split in the west of the area. Burgess and Mitchell (1976) have shown that two limestones on the Alston Block, locally separated by up to 56 ft (17.00 m) of mudstone and sandstone in the Brough area come closer together as they are traced southwards into Wensleydale. The section in the Birkett Cutting [NY 7734 0317] shows the Grain Beck Limestone overlain by 20 ft (6 m) of measures which include a 6 in (0.15 m) seatearth sandstone directly under the Lower Little Limestone (itself in two leaves) (Figure 7, section 1). Farther south, in Scotchergill, Dentdale [SD 7205 8717, Figure 7, section 5] and in parts of Kingsdale (Hicks, 1957), a calcareous sandstone occurs in the middle of the Gayle Limestone and is presumed to be at the level of the split between the Lower Little and Grain Beck limestones. In Cluntering Gill, Kingsdale [SD 7106 8023] the entire bottom half of the Gayle Limestone is a very sandy limestone with cross-bedding etched out by weathering (Figure 7, section 10). The upper half of the limestone carries a red colouration in this stream and many others around Whernside (Hicks, 1957).

In the central and eastern parts of the area the Gayle Limestone commonly consists of a single limestone with local thin mudstone partings which come in westwards, especially at the head of Dentdale (Figure 7, section 6). The Gayle Limestone is typically a grey or dark grey, medium or fine-grained biosparite, with up to three alternations in colour from dark to light, as at Aysgarth Falls. Due to the easily undercut Gayle Shale, the limestone forms picturesque waterfalls as at West Burton Falls and the Middle Force, Aysgarth. Typically the overhanging lower beds contain very large *Gigantoproductus*. Corals commonly occur and also, locally, algal overgrowths.

In the upper part of the Gayle Limestone in central Wensleydale and its tributary valley Raydale (Moore, 1958), a number of crinoid biostromes cause local thickening of the whole formation to a maximum of 78 ft 6 in (23.93 m). Burtersett is built on one such mound of grey friable grainstone, with an abundance of articulated crinoid stems and columnals, showing original depositional dips. Moore records large spiny productoids near its upper surface.

As the Gayle Limestone is followed southwards down Wharfedale, the mudstone partings which divide it from the Hawes and Hardrow Scar limestones become progressively thinner and the precise limits of the Gayle Limestone become harder to fix. This is especially so at Greenhow where some 23 ft 7 in (7.19 m) of the Coldstones Limestone are tentatively referred to the Gayle. These are both dark and light grey limestones and include a nodular lithology with muddy envelopes. Whereas the top of the Gayle is here defined by 4 ft (1.22 m) of mudstone, there is no discernible parting at the base (Figure 7). The same possibly applies at Grimwith where in No. 4 Borehole 36 ft (11.00 m) of medium dark grey micrite include several muddy limestone bands and nodular limestones with muddy envelopes but no definite basal mudstone parting (Figure 7, section 21). The Girvanella Band was not reached.

Beds above the Gayle Limestone These rocks are in two markedly contrasting facies. In the north and west, sandstones form a major part of a sequence up to 193 ft (59 m) in thickness (Figures 7; 10, map C). Together with the overlying Hardrow Scar Limestone these sandstones make spectacular waterfalls, notably Hardrow Force and Mill Gill Force in Wensleydale (Moore, 1958, p. 98). In the south and east, the beds are much thinner and consist, dominantly, of mudstones with one or more local limestone bands. In parts of Wharfedale and at Greenhow, the Gayle and Hardrow Scar limestones are as little as 4 ft (1.22 m) apart.

The beds directly overlying the Gayle Limestone are everywhere mudstones which have locally yielded *Posidonia becheri* and zonally useful goniatites at Mill Gill (Askrigg), in Dentdale and Wharfedale (Figure 3). They are overlain by unfossiliferous mudstones with some sandstones showing current-ripple lamination and trails, as in the Birkett Cutting section and Garsdale (Figure 7, sections 1, 4). The main group of sandstones in the upper half of the measures under discussion is particularly well seen in Garsdale along the banks of the River Clough [SD 7700 9032] where pronounced channelling is seen in the lower beds of the sandstone member. In upper Swaledale there are excellent sections in Ivelet Beck and Greenseat Beck (Figure 7, section 2) with current ripple lamination and trails seen at several levels, but no cross-bedding. Seatearths were not seen, but in Birkett Cutting two thin coals resting on seatearths were recorded (Figure 7, section 1).

Southwards the sandstone facies is split by mudstone partings, notably in Great Blake Beck, Dentdale [SD 7690 8547] and Backstone Gill, Kingsdale [SD 7067 7990] and there is an overall thinning of beds (Figure 7, sections 6, 10). In Crook Gill, Cray [SD 9331 7944] there is only 2 ft 2 in to 3 ft 5 in (0.65–1.05 m) of fine-grained sandstone sandwiched between thin bands of limestone and mudstone (Figure 7, section 12).

South-east of the line shown on Figure 10, map C, sandstones are absent and the sequence is dominated by mudstones, much of them fossiliferous. Around Aysgarth in Wensleydale, these mudstones contain corals, brachiopods and bivalves. A strong bed of limestone some 3 ft (0.90 m) in thickness is a feature of the Aysgarth area and occurs locally in Wharfedale, notably in Buckden Beck [SD 9495 7775] and Crystal Beck [SD 9136 7521], Littondale, in association with other thinner limestones.

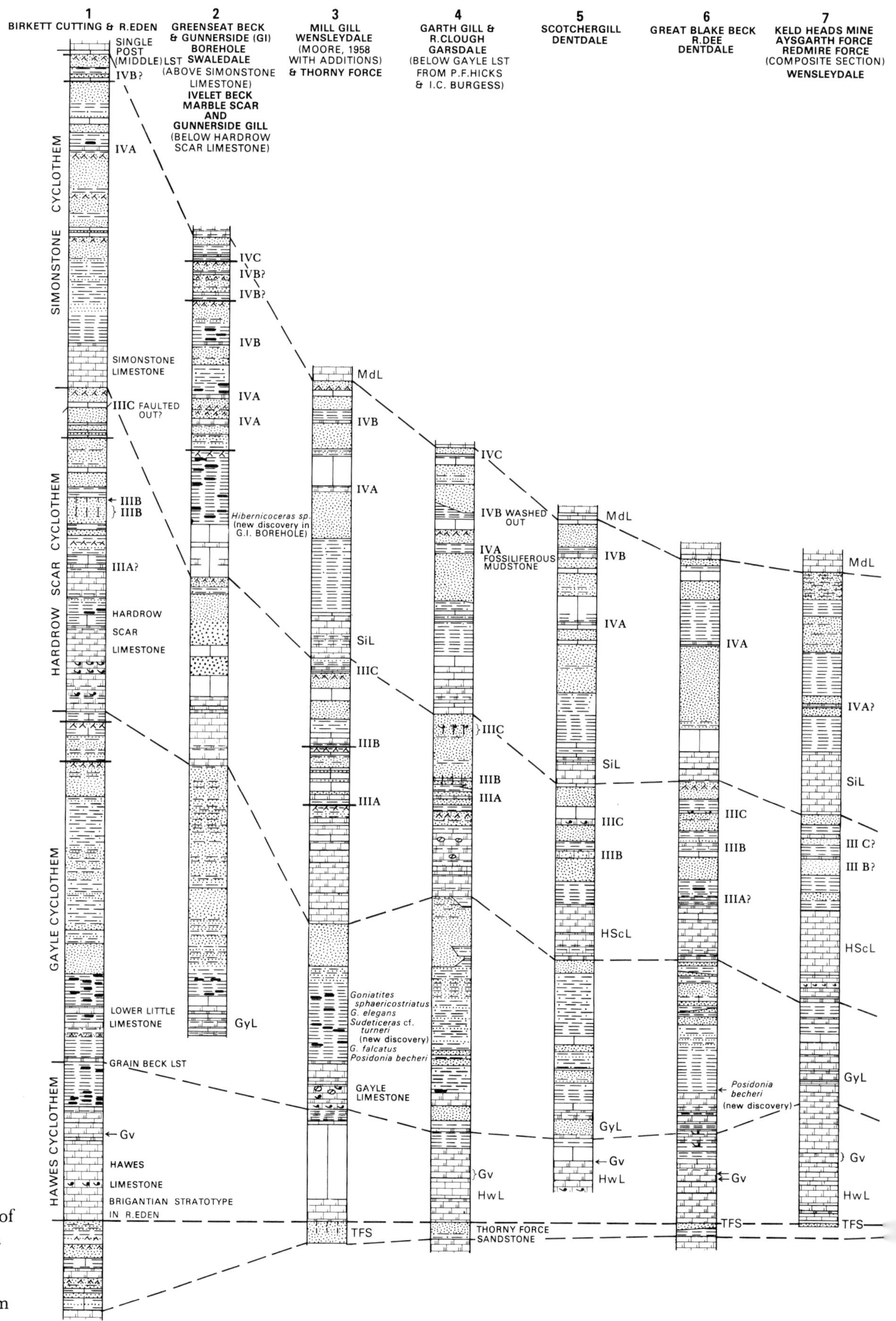

Figure 7
Comparative sections of Brigantian strata from the base of the Hawes up to the top of the Simonstone Cyclothem

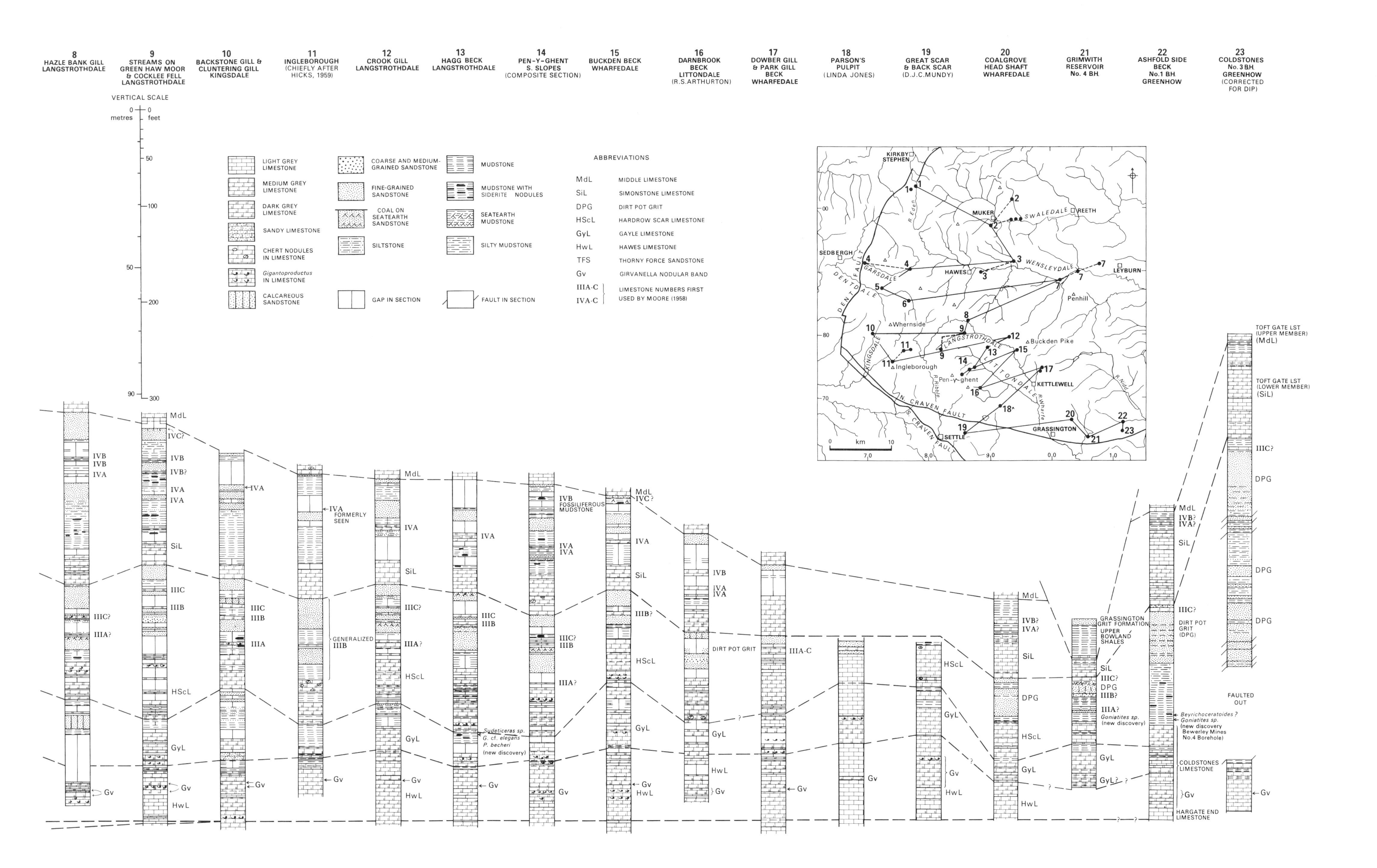

8 HAZLE BANK GILL LANGSTROTHDALE
9 STREAMS ON GREEN HAW MOOR & COCKLEE FELL LANGSTROTHDALE
10 BACKSTONE GILL & CLUNTERING GILL KINGSDALE
11 INGLEBOROUGH (CHIEFLY AFTER HICKS, 1959)
12 CROOK GILL LANGSTROTHDALE
13 HAGG BECK LANGSTROTHDALE
14 PEN-Y-GHENT S. SLOPES (COMPOSITE SECTION)
15 BUCKDEN BECK WHARFEDALE
16 DARNBROOK BECK LITTONDALE (R.S.ARTHURTON)
17 DOWBER GILL & PARK GILL BECK WHARFEDALE
18 PARSON'S PULPIT (LINDA JONES)
19 GREAT SCAR & BACK SCAR (D.J.C.MUNDY)
20 COALGROVE HEAD SHAFT WHARFEDALE
21 GRIMWITH RESERVOIR No. 4 BH.
22 ASHFOLD SIDE BECK No.1 BH GREENHOW
23 COLDSTONES No.3 BH. GREENHOW (CORRECTED FOR DIP)
VERTICAL SCALE
metres
feet
LIGHT GREY LIMESTONE
MEDIUM GREY LIMESTONE
DARK GREY LIMESTONE
SANDY LIMESTONE
CHERT NODULES IN LIMESTONE
Gigantoproductus IN LIMESTONE
CALCAREOUS SANDSTONE
COARSE AND MEDIUM-GRAINED SANDSTONE
FINE-GRAINED SANDSTONE
COAL ON SEATEARTH SANDSTONE
SILTSTONE
GAP IN SECTION
MUDSTONE
MUDSTONE WITH SIDERITE NODULES
SEATEARTH MUDSTONE
SILTY MUDSTONE
FAULT IN SECTION
ABBREVIATIONS
MdL MIDDLE LIMESTONE
SiL SIMONSTONE LIMESTONE
DPG DIRT POT GRIT
HScL HARDROW SCAR LIMESTONE
GyL GAYLE LIMESTONE
HwL HAWES LIMESTONE
TFS THORNY FORCE SANDSTONE
Gv GIRVANELLA NODULAR BAND
IIIA-C IVA-C LIMESTONE NUMBERS FIRST USED BY MOORE (1958)
KIRKBY STEPHEN
MUKER
SWALEDALE
REETH
SEDBERGH
GARSDALE
HAWES
WENSLEYDALE
LEYBURN
Penhill
DENT FAULT
DENTDALE
Whernside
LANGSTROTHDALE
Buckden Pike
KINGSDALE
Ingleborough
Pen-y-ghent
LITTONDALE
KETTLEWELL
R. Eden
R. Ribble
R. Wharfe
R. Nidd
N. CRAVEN FAULT
S. CRAVEN FAULT
SETTLE
GRASSINGTON
km
TOFT GATE LST (UPPER MEMBER) (MdL)
TOFT GATE LST (LOWER MEMBER) (SiL)
FAULTED OUT
GENERALIZED IIIB
IVA FORMERLY SEEN
IVB FOSSILIFEROUS MUDSTONE
DIRT POT GRIT
GRASSINGTON GRIT FORMATION
UPPER BOWLAND SHALES
Sudeticeras sp.
G. cf. elegans
P. becheri
(new discovery)
Goniatites sp. (new discovery)
Beyrichoceratoides ?
Goniatites sp. (new discovery Bewerley Mines No.4 Borehole)
DIRT POT GRIT (DPG)
COLDSTONES LIMESTONE
HARGATE END LIMESTONE

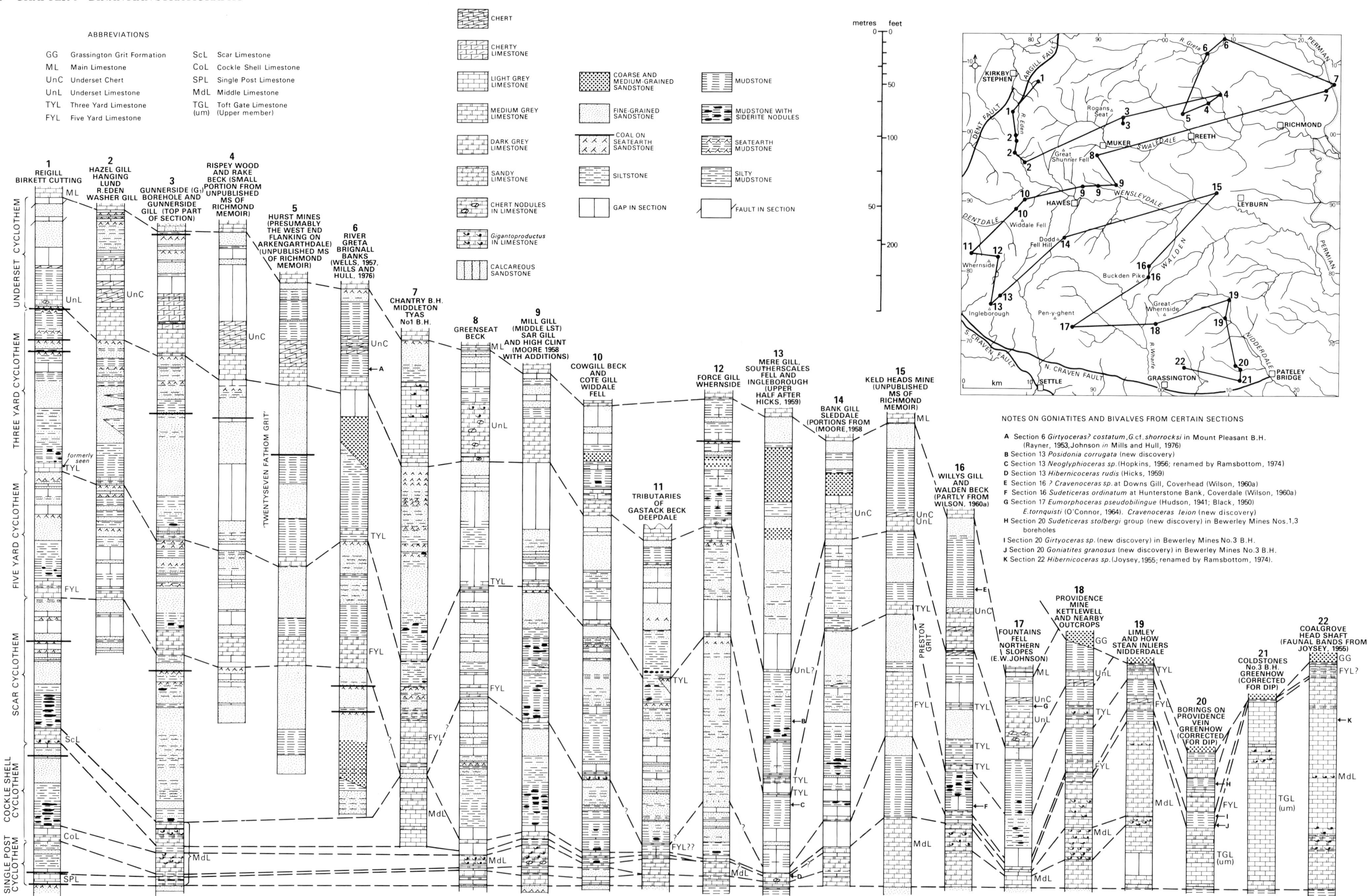

Figure 8 Comparative sections of Brigantian strata from the base of the Middle up to the top of the Underset Cyclothem

Further abrupt thinning of the shaly facies occurs southwards on Ingleborough (Hicks, 1959) and south of Starbotton in Wharfedale to as little as 6 in (0.15 m). There is some local thickening towards Greenhow where the mudstone parting is 4 ft (1.20 m) in thickness. At Grimwith, borehole records show great thickness variations from 9 ft 9 in to 30 ft (2.97–9.14 m) in a horizontal distance of 250 ft (76.20 m). The lesser thickness was logged in Grimwith No. 4 Borehole [SE 0601 6405] as follows:

	Thickness ft	in	m
Hardrow Scar Limestone	–	–	–
Mudstone, medium dark grey, shaly; fossils include *Fenestella* cf. *parallela*, *F.* cf. *plebeia*, *Penniretepora sp.*, *Sulcoretepora sp.*, *Avonia youngiana*, *Latiproductus latissimus* group, *Plicochonetes crassistrius*, *Hesperiella* aff. *thomsoni*, *Acanthopecten* cf. *nobilis*, *Edmondia* cf. *gigantea*, *Phestia brevirostris*	9	9	2.97
Gayle Limestone			

A section in Hagg Beck, Langstrothdale [SD 8975 7813] is also in the area of mudstone facies and, although incomplete, is notable for the discovery of the only goniatites in the Wharfedale area at this horizon (see below and Figure 7, section 13).

Hardrow Scar Limestone This formation was named the Hardraw Limestone by Phillips (1836) but modified by Moore (1958) to Hardrow Scar Limestone. In Wensleydale it forms a prominent and persistent feature marked by spectacular waterfalls at Hardrow Force and Mill Gill Force. Thickness ranges from a probable 21 ft (6.40 m) at Greenhow (where it forms the topmost part of the Coldstones Limestone) to 92 ft (28 m) in the Birkett Cutting section. Lithology is variable, with a large range of grain size and colour. Over much of the area the limestone is split into two members by a median mudstone parting; but where the parting is absent, notably in the more south-easterly outcrops, details of the variable lithologies are difficult to correlate from one section to another. The tripartite division of the limestone has been clearly illustrated in the south-west by Hicks (1959, fig. 3), and is here detailed for Hagg Beck in Langstrothdale to bring out the salient lithological and faunal characteristics. The section includes some beds of the Gayle Cyclothem.

		Thickness ft	in	m
Hardrow Scar Limestone				
Limestone, grey biosparite, fine- grained; well-bedded; with *Dibunophyllum sp.*, *Lithostrotion junceum*, *L. maccoyanum*, *L. pauciradiale*, *Lonsdaleia floriformis floriformis*, *Syringopora* cf. *catenata*, *S.* cf. *geniculata*, *Gigantoproductus* cf. *moderatus*, *G. okensis*, *Latiproductus sp.*, *Linoprotonia dentifer*, *Spirifer striatus*		18	0	5.49
Mudstone, dark grey, with three bands of dark grey limestone up to 1 ft 6 in (0.45 m) in thickness; collection from the top 2 ft 10 in (0.85 m) of mudstone yielded the following abundant forms: *Aulophyllum fungites*, *Gigantoproductus edelburgensis*, ostracods; other forms include *Fenestella* cf. *parallela*, *Polypora verrucosa*, *Eomarginifera tissingtonensis cambriensis*, *Latiproductus latissimus*, *Productina margaritacea*, *Spiriferellina perplicata*, *Aviculopecten interstitialis*, *Limipecten dissimilis*, *Palaeolima simplex*, *Pernopecten concentricum*, *P.* cf. *sowerbii*, *Posidonia sp.*, *Eocaudina sp.*, *Paleochiridota sp.*		17	3	5.26
Limestone, dark grey, biosparite, fine- grained; well-bedded, with a few thin mudstone partings, notably near the base; collection 2 to 5 ft (0.60–1.50 m) above base included *Diphyphyllum lateseptatum*, *Lithostrotion junceum*, *Pseudozaphrentites sp.*, *Eomarginifera setosa*, *Gigantoproductus* cf. *moderatus*, *G. inflatus*, *G.* cf. *okensis*, *Latiproductus latissimus*, *Overtonia fimbriata*		31	0	9.45
Beds of Gayle Cyclothem				
Mudstone, dark grey, siderite nodules; collections from the uppermost 5 ft (1.50 m) yielded the following common forms: *Latiproductus latissimus* and *Megachonetes dalmanianus*; other fossils include *Amplexizaphrentis enniskilleni*, *Aulophyllum fungites*, *Fenestella* cf. *arthritica*, *Fistulipora incrustans*, *Eomarginifera tissingtonensis cambriensis*, *Gigantoproductus gigantoides* group, *Rhipidomella michelini*, *Rugosochonetes hardrensis*, *Retispira sp.*, *Edmondia maccoyii*, *Pernopecten concentricum*, *Posidonia becheri*, *Sanguinolites plicatus*, *Eocaudina sp.*; the middle 3 ft 3 in (1 m) yielded *Globosochonetes* aff. *parseptus*, *Martinia sp.*, *Posidonia becheri*, *Goniatites* cf. *elegans*, *Sudeticeras sp.*, *Coleolus* cf. *carbonarius*; the basal 3 ft 3 in (1 m) yielded *Spirorbis sp.*, *Crurithyris urii*, *Megachonetes dalmanianus*, *Mourlonia sp.*, *?Pernopecten concentricum*, ostracods		11	6	3.50
Gap to highest Gayle Limestone exposure	about	10	0	3.05

The lower member of the Hardrow Scar Limestone, where it is separately recognisable, is a grey to dark grey biosparite with local mudstone partings near the base. It varies from thinly to thickly bedded and contains local rich faunal bands as follows: sections in Wharfedale, including Hagg Beck, contain *Gigantoproductus* some 10 ft (3 m) above the base of the limestone and in Wensleydale *Linoprotonia dentifer* is abundant at the top (the *hemisphericus* bed of Hudson, 1924). Moore (1958) also records an orthotetoid bed at the top of the lower member in Wensleydale. In the Ingleborough area Garwood and Goodyear (1924, p. 204) noted an extensive fauna rich in bivalves preserved in splintery limestone near the middle of the lower member. Hicks (1959, fig. 3) has subsequently found it on Whernside.

The mudstone member which divides up the limestone near the middle ranges in thickness from nil locally to a maximum of 19 ft (5.80 m) in the Birkett Cutting section. It consists of dark grey mudstone, with several thin limestone bands present in Hagg Beck and other sections. Especially well developed in Kingsdale and on Ingleborough, it thins northwards to under 3 ft (0.90 m) in Garsdale. Although thick again farther north in the Birkett Cutting it is absent in the Brough area (Burgess and Holliday, 1979). In Wensleydale (Moore, *op. cit.*) the parting averages 1 ft (0.30 m) in thickness, and is thin or absent in the south-east, between Fountains Fell, Greenhow and Buckden, and also in Swaledale. The mudstone member contains a bivalve fauna in the south-west (Hicks, 1959) which was also collected in Hagg Beck, Langstrothdale (see above).

The upper leaf of the Hardrow Scar Limestone in the south of the area commonly contains nodules of black chert in dark grey, well bedded biosparite, and is especially notable for the great abundance of its compound corals, notably *Lithostrotion junceum* and productoid brachiopods (e.g. Hagg Beck, and Pen-y-ghent Gill, SD 853 737). Farther north, in Wensleydale and Mallerstang, medium grey, coarse-grained biosparites also occur. Algae are locally found at top of the limestone in Wensleydale (Moore, *op. cit.*) and *Gigantoproductus* is abundant in a stream on Hazle Bank Edge, Langstrothdale. [SD 8658 8215] (Figure 7, section 8).

In areas where the median mudstone parting has not been seen, limestone lithologies differ from section to section. In Dowber Gill, Wharfedale, the limestone is all medium grey whereas at Greenhow and Brough all beds are dark grey. Elsewhere there are alternations of grey and dark grey beds (Figure 7).

Beds above the Hardrow Scar Limestone These beds form one of the most complex intervals between major limestones. In a belt stretching through Kingsdale, Garsdale, Dentdale and much of Wensleydale and upper Swaledale they consist of four minor cycles, locally interrupted by a sandstone-filled channel which cuts through most of this sequence. Moore (1958) records a wealth of details in Wensleydale and upper Swaledale and defines three minor limestones designated IIIA – C, terms which are here used informally for ease of reference. In the Brough area and in Wharfedale most sections show two limestones, and the detailed correlation with the central area is not everywhere certain. Generally over the north, west and central ground these beds average some 70 ft (21 m) in thickness and vary less than most Yoredale cyclothems (Figure 10, map D). In the south-east the uppermost two-thirds of the sequence is commonly a sandstone called the Dirt Pot Grit.

One of the best sections showing the divisions into four minor cycles comes from Backstone Gill, Kingsdale [SD 7046 8008 to 7055 7995, Figure 7, section 10]. The three minor limestones (IIIA – C), originally mapped by Strahan in 1883, are well exposed and the associated mudstones have yielded goniatites, unfortunately not of zonal significance, at two levels. The calcareous sandstone overlying limestone IIIC is remarkable for its fretted weathering which emphasises sedimentary features with the clarity of a text book illustration. Beds seen are as follows:

	Thickness		
	ft	in	m
SIMONSTONE LIMESTONE	–	–	–
Sandstone, fine-grained; well-bedded	13	0	3.96
Gap	5	6	1.67
Mudstone, shaly	3	3	1.00
Sandstone, calcareous, fine-grained; fretted weathering enhances cross-laminae, vertical borings	5	6	1.67
Limestone (IIIC), dark grey, biosparite; abundant *Linoprotonia dentifer*, *G. okensis* [with some shells approaching the form of *G. crassiventer*]; other forms are *Lithostrotion portlocki*, *Syringothyris sp.*	3	7	1.10
Mudstone; *Rugosochonetes celticus*, *Schellwienella* cf. *crenistria* in basal 1 ft (0.30 m)	6	0	1.82
Limestone, dark grey, nodular, argillaceous		8	0.20
Mudstone, calcareous; abundant ostracods, common *Brachythyris sp.*, *Spiriferellina perplicata*, *Leiopteria thompsoni*; other forms include *?Lithostrotion pauciradiale*, *Fenestella* cf. *plebeia*, *Ichthyorachis sp.*, *Sulcoretepora parallela*, *Echinoconchus elegans*, *Eomarginifera setosa*, *Gigantoproductus edelburgensis*, *Tornquistia polita*, *Retispira sp.*, *Edmondia transversa*	1	0	0.30
Gap about	1	0	0.30
Mudstone; abundant *Rugosochonetes celticus*, common *Fenestella* cf. *parallela*, *Productus productus*, *Rugosochonetes hardrenis*, *Pernopecten sowerbii*, other forms include *Fluctuaria undata*, *Spirifer striatus* group, *Retispira sp.*, *Aviculopecten interstitialis*, *Edmondia laminata*, *Palaeolima simplex*, *Sanguinolites* cf. *abdenensis*, *Streblochondria anisota*, indet. Beyrichoceratid	2	8	0.80
Limestone (IIIB), dark grey, silty, ferruginous weathering; abundant *Zoophycos* at top with *Pleuropugnoides pleurodon*, *Pugilis* cf. *senilis*	3	3	1.00
Sandstone, fine-grained, thickly bedded, with some cross-laminae	12	6	3.81
Gap	6	6	1.98
Mudstone, dark grey, shaly; siderite nodules (only scree seen)	7	3	2.20
Limestone (IIIA), dark grey, argillaceous; common *Gigantoproductus* aff. *okensis*, other forms include *G. dentifer*, *G. edelburgensis*, *Latiproductus latissimus*, *Productus* cf. *productus*, *Zoophycos*	2	10	0.85
Gap (mudstone fragments, siderite nodules)	5	0	1.52
Mudstone, dark grey, shaly; common *Fenestella* cf. *frutex*, *Eomarginifera setosa*, *Aviculopecten interstitialis*; other forms include *Sulcoretepora parallela*, *Chonetipustula carringtoniana*, *E.* cf. *tissingtonensis*, *Latiproductus latissimus*, *Megachonetes dalmanianus*, *Tropidostropha sp.*, *Koninckopecten desquamata*, indet. goniatite	5	0	1.52
HARDROW SCAR LIMESTONE	–	–	–

The first minor cycle consists of fossiliferous mudstones locally overlain by fine-grained sandstone with seatearth at the top. Moore (*op. cit.*) records a coal in some sections. In Dentdale and the River Rawthey Hicks (1957) notes that these beds below limestone IIIA are only some 6 ft (1.80 m) thick.

The second minor cycle commences with limestone IIIA, a dark grey micrite with *Gigantoproductus* in Kingsdale and Garsdale. In Mill Gill, Wensleydale (Figure 7, section 3), the lower part of Moore's section is now better exposed and the true IIIA can be seen 12 ft 6 in (3.80 m) above the top of the Hardrow Scar Limestone; limestone 'IIIA' is now renamed IIIB. Mudstones overlying limestone IIIA are usually unfossiliferous (Moore, *op. cit.*). The overlying fine-grained sandstone is locally capped by a seatearth. In Mere Gill, Ingleborough, it has been identified by Hicks (1959) as the sandstone with stellate bioturbation traces first noted by Garwood and Goodyear (1924).

The third minor cycle starts with limestone IIIB which attains its maximum thickness of 13 ft (3.96 m) in lower Dentdale (Hicks, 1957) and is commonly the thickest of the three limestones, but the least fossiliferous. In the River Clough, Garsdale, it is richly shelly, however, and occurs within calcareous sandstone. IIIB is also exposed in Mere Gill, Ingleborough (Hicks, 1959). The mudstones above limestone IIIB are locally very fossiliferous, as for instance in Kingsdale, and are overlain by sandstones, which are often thickly bedded with a seatearth at the top in places.

Limestone IIIC is not present in all sections in Moore's (*op. cit.*) study area. This dark grey limestone is thickest in Kingsdale and Dentdale, reaching a maximum of 7 ft (2.13 m), where it is locally sandy and commonly is rich in *Gigantoproductus* (Hicks, 1957 and the present work). Beds overlying IIIC are usually mudstones overlain by fine-grained sandstones passing upwards from well-bedded into blocky lithologies, locally with a seatearth.

In Wharfedale, Crook Gill and Buckden Beck (Figure 7, sections 12, 15) show only one limestone. It seems likely that one or more of the minor limestones dies out in the Buckden area and it is not known which one persists. Similarly in lower Wensleydale only two limestones are recorded in Keld Heads Mine (Figure 7, section 7).

In the south-east, minor limestones are rarely recorded. They are absent in the records of Coalgrove Head Shaft in the Grassington Moor Mining Field, for instance, and a single limestone (IIIC?) is seen at the top of the sequence in Ashfold Side Beck No. 1 Borehole near Greenhow (Figure 7, sections 20, 22). Grimwith No.4 Borehole (Figure 7, section 21), midway between Grassington and Greenhow, shows three limestones which may be correlatives of IIIA – C. They do not persist in all boreholes on the Grimwith site, however, and could be strictly local. The mudstones yield a rich fauna at many levels which includes *Goniatites sp.*, bryozoa, brachiopods and bivalves (Wilson, Dunham and Pattison, in preparation). The mudstones overlying the Hardrow Scar Limestone in Bewerley Mines No. 4 Borehole sunk on the Providence Vein near Greenhow include *Goniatites sp.* and *Beyrichoceratoides*? with brachiopods, gastropods and bivalves.

The Greenhow boreholes are chiefly remarkable for the great local thickening of the Dirt Pot Grit, a fine-grained sandstone with mudstone partings. Borings on the Providence Vein show thicknesses of 36 ft (10.97 m) in Bewerley Mines No. 1 [SD 1140 6610], 70 ft (21.34 m) in Bewerley Mines No. 4 [SD 1200 6621] thickening southwards in 1 mile (1.6 km) to a local maximum of 224 ft (68.28 m) close to the Waterhole North Vein at Greenhow in Coldstones No. 3 [SD 1160 6459, Figure 7, section 23].

The Dirt Pot Grit around Grassington and Greenhow is commonly fine-grained and not cross-bedded. Despite its thickness it was not obviously laid down in a distributary channel. In its thickest development in Coldstones No. 3 Borehole it is a laminated sandstone with mudstone clasts at several levels. Bioturbation occurs at two horizons suggesting deposition in shallow-water environments.

A feature of the sequence in the centre of the area is a major distributary channel infilled with cross-bedded coarse- to fine-grained sandstone. This has been traced by Moore (*op. cit.*) southwards from Upper Swaledale across Wensleydale into Bishopdale. A typical section is that in Greenseat Beck, Swaledale (Figure 7, section 2) where the indications are that all three minor limestones are absent due to the effects of the distributary channel. Studies in the west have not revealed any more channel sandstones (Hicks, 1957). The sandstone underlying the probable limestone IIIB on Green Haw Moor (Figure 7, section 9) forms a strong feature and is partly medium-grained with cross-bedded units. Whilst it is not known to wash out limestone IIIA, it may be in a minor distributary channel on a different line from that noted by Moore. Elsewhere in Garsdale and Mallerstang sandstones dominate the sequence, but limestones persist, with some modification to calcareous sandstone. None are known to be washed out.

Simonstone Limestone This limestone was originally called the Simonside Limestone by Phillips (1836) but altered to Simonstone by Dakyns and others (1891) on the grounds that Phillips had apparently intended to name the formation after Simonstone hamlet in Wensleydale. This is one of the most consistent Brigantian limestones in thickness, ranging, normally, from some 20 to 41 ft (6.10–12.50 m). At Greenhow, however, it seems likely that it has thickened to 73 ft (22.25 m) in Ashfold Side Beck No. 1 Borehole (Figure 7, section 22), where it is represented by the coarsely crinoidal lower half of the Toft Gate Limestone of Dunham and Stubblefield (1945). Commonly the Simonstone Limestone forms a prominent feature, as for instance on Penhill in Wensleydale. In central Wensleydale, in Whitfield Gill [SD 9351 9202], the continuation of Mill Gill, the limestone consists of two units, each of grey biosparite passing up into dark grey, with some mudstone partings (Figure 7, section 3). The lower unit in Whitfield Gill has varying solubility and weathers in picturesque 'hour glass' forms. In a number of other sections grey biosparites overlie dark grey beds, but there is no general rule and in several sections all beds are grey in colour. Locally in Dentdale a shaly mudstone parting up to 3 ft 6 in (1.07 m) thick occurs near the middle of the formation (Hicks, 1957).

In most sections fossils are not particularly common, but an important *Orionastraea* horizon occurs at the top (Hudson, 1929a), locally associated with an algal band. In upper Kingsdale the algae in the top 6 in (0.15 m) of the limestone are associated with phosphatic nodules (Hicks, 1957). A pseudobreccia band near the middle of the limestone in Wensleydale carries a locally prolific coral fauna (Moore, *op. cit.*).

Beds above the Simonstone Limestone These thin persistently to the south-east from a maximum of 218 ft (66.45 m) near Brough to 14 ft (4.27 m) at Greenhow Hill (Figure 10, map E). Moore (*op. cit.*) has shown that the succession can be divided in Wensleydale into four minor cycles, and designates three minor limestones IVA – C. The two lower limestones are persistent and can split, whereas IVC may fail southwards or become amalgamated with the base of the Middle Limestone. The present study bears out Moore's findings, but there are local difficulties in recognition of individual limestones, especially where splitting has occurred.

The clastic beds of the first minor cycle, which starts with the Simonstone Limestone, are in many places by far the thickest, with the overlying limestone IVA commonly about midway between the Simonstone and Middle limestones. The lower beds are thick mudstones with siderite nodules, usually with fossils towards the base as on Green Haw Moor, Langstrothdale (p. 40). In Gunnerside G1 Borehole (Figure 7, section 2) fossils occur throughout and include a variety of bryozoa, brachiopods, bivalves and trilobite fragments, as on Green Haw Moor, but the most significant find is the goniatite *Hibernicoceras sp*. The mudstones are overlain by well-bedded fine-grained sandstones with current-ripple lamination and trails, becoming thicker bedded upwards with some cross-bedding. These are superbly exposed in Whitfield Gill Force, Askrigg [SD 9342 9219, Figure 7, section 3] and in a waterfall in Combe Gill, Deepdale, in Dentdale [SD 7267 8247]. The base of the sandstone cuts across 3 ft (0.90 m) of underlying mudstone in Garth Gill, Garsdale [SD 7721 9084]. In Barbondale, Hicks (1957) records 35 ft 6 in (10.82 m) of medium- and coarse-grained cross-bedded sandstone at this level overlain by 1 ft (0.30 m) of coal, mapped by the Geological Survey. In the north, seatearths are seen near the top of this sandstone, as in the River Eden section near Catagill Scar (Figure 7, section 1). In the south-east the sandstone dies out, leaving only fossiliferous mudstones in Coverdale and at Greenhow.

The second minor cycle begins with Limestone IVA with a prolific coral fauna in Wensleydale and Swaledale (Moore, *op. cit.*), in Gunnerside G1 Borehole and in Walden and Coverdale (Wilson, 1960a). It is probably the single limestone in the record of Keld Heads Mine, Wensleydale (Figure 7, section 7). In Swaledale, Moore records a sandstone splitting IVA into two leaves, also a feature of the G1 Borehole record (Figure 7, section 2). It seems likely that the limestone is also split by a sandstone parting in Langstrothdale where a stream on Green Haw Moor [SD 8242 7763 to 8218 7768, Figure 7, section 9] shows the following well-exposed section between the Simonstone and Middle limestones:

	Thickness ft	in	m
MIDDLE LIMESTONE	–	–	–
Gap	4	3	1.30
Sandstone, fine-grained, calcareous (equivalent of Limestone IVC?)	2	0	0.60
Sandstone, fine-grained; blocky, becoming well bedded downwards	10	2	3.10
Mudstone, silty near top	19	8	6.00
Limestone IVB, dark grey, muddy partings, ferruginous weathering; *Lithostrotion junceum*, *Syringopora*?, *Zaphrentites sp.*, *Echinoconchus elegans*, *Latiproductus latissimus*, *Spiriferellina sp.*	1	4	0.40
Mudstone	1	0	0.30
Gap	1	8	0.50
Sandstone, fine-grained, calcareous in top 1 ft 4 in (0.40 m)	9	0	2.74
Mudstone, calcareous		10	0.25
Limestone, dark grey (split off IVB?)		8	0.20
Mudstone, dark grey, siderite nodules; lower half very fossiliferous; *Koninckophyllum echinatum*, *Fenestella* cf. *frutex*, *Levifenestella undecimalis*, *Rhabdomeson gracile*, *Sulcoretepora parallela*, *Antiquatonia* cf. *antiquata*, *Brachythris integricosta*, *Dictyoclostus pinguis*, *Eomarginifera setosa*, *E. tissingtonensis*, *Hustedia radialis*, *Latiproductus latissimus*, *Rhipidomella michelini*, *Rugosochonetes celticus*, *Schizophoria connivens*, *Spirifer triangularis*, *Tornquistia polita*, *Ianthinopsis ventricosa*,			

	Thickness ft	in	m
Aviculopecten interstitialis, *Edmondia maccoyii*, *Phestia attenuata*, *Pterinopectinella dumontiana*, *Streblochondria anisota*, trilobite glabella, ostracods.	14	0	4.27
Limestone IVA, dark grey, medium- grained biosparite	2	8	0.82
Mudstone, grey	1	0	0.30
Gap	3	8	1.12
Sandstone, fine-grained	4	0	1.22
Mudstone, dark grey	1	0	0.30
Limestone, dark grey (split off IVA?); *Heterophyllia*?, trepostome bryozoa, aff. *Antiquatonia insculpta*, *Echinoconchus* cf. *elegans*, *Latiproductus latissimus*, *Schellwienella rotundata* [abundant], *Koninckopecten*?		7	0.17
Mudstone with rare thin sandstone bands	32	0	9.75
Mudstone, dark grey; siderite nodules; abundant fossils including *Emmonsia parasitica*, *Fenestella* cf. *frutex*, *Penniretepora* cf. *pluma*, *Sulcoretepora parallela*, *Brachythyris pinguis*, *Coledium rhomboideum*, *Krotovia spinulosa*, *Orbiculoidea nitida*, *Pleuropugnoides pleurodon*, *Rhipidomella michelini*, *Rugosochonetes celticus*, *Schizophoria connivens*, *Semiplanus semiplanus*, *S. striatus* group, *Aviculopecten interstitialis*, *Limipecten dissimilis*, *Streblochondria anisota*, *Sulcatopinna flabelliformis*, trilobite pygidium	1	8	0.50
Gap	8	6	2.59
SIMONSTONE LIMESTONE, algal nodules at top, with gap	24	6	7.47

In Birkett Cutting and on the River Eden, limestone IVA is rich in corals, as in Wensleydale, but has a rubbly texture and is covered in ferruginous blotches. It resembles the bed at the top of the Single Post Limestone in the Birkett Cutting (p. 41) and is likely also to be a caliche limestone, modified under emergent conditions. Limestone IVA has been traced into Dentdale (Figure 7, sections 5, 6), but is not as fossiliferous as in Wensleydale. In Kingsdale indications are that the limestone is here very thin and locally represented by calcareous sandstone (Figure 7, section 10). Locally, at Garth Gill, Garsdale, it is absent (Figure 7, section 4).

Beds above limestone IVA are mudstones with a rich fauna in Wensleydale (Moore, *op. cit.*) and on Pen-y-ghent. These are usually overlain by thinly bedded sandstones with trace fossils, worked for roofing tiles for Dent Church in Kirk Bank Quarry, Dentdale [SD 728 870] and also in Wensleydale. Hicks (1957) records convoluted bedding near the base of the sandstone in Barbondale, also seen at this level in Gunnerside G1 Borehole.

The third minor cycle begins with limestone IVB which is not as frequently exposed as IVA and has not yet been recorded with certainty in Deepdale, on Ingleborough or on Pen-y-ghent so that it seems likely that it fails to the south. Mudstones with chonetoids may be representative of this bed on Pen-y-ghent. It is not in the record of Keld Heads Mine, Wensleydale, but is known to occur on nearby Penhill. Moore notes that IVB is locally split by shale partings and totals 6 ft (1.83 m) in thickness. In Gunnerside G1 Borehole, however, there are complications which make Moore's numbering scheme difficult to apply. No less than four distinct limestones occur within 59 ft (18 m) of strata (Figure 7, section 2). Each limestone is followed by a miniature cyclothem and each rests on a seatearth. This complex of minor cycles resembles that of the 'Alternating Beds' described at this level by Burgess and Holliday (1979) in the Brough area. The picture is rendered even more varied when the Birkett Cutting section in beds of nearly identical overall thickness is considered. Here no limestone higher than IVA was seen by the Geological Survey when the cutting was freshly dug, nor is one visible now. However a fine-grained sandstone with big crinoid columnals is perhaps equivalent to Limestone IVB. Limestone IVB also appears to split on Green Haw Moor, Langstrothdale, with an upper leaf rich in corals (p. 39), and in Hazle Bank Gill (Figure 7, sections 8, 9).

Moore (*op. cit.*) notes that in Swaledale a thick sandstone locally lies across the horizon of limestone IVB and merges into the overlying and underlying sandstones to make one thick false-bedded formation. Where seen in impressive exposures in Mossdale Gill, Wensleydale [SD 819 914, Moore, personal communication] and in an incomplete section in Greenseat Beck [SD 9004 9746] it is medium- and coarse-grained, but is elsewhere finer according to Moore (1960).

In Garth Gill, Garsdale, (Figure 7, section 4) fine-grained sandstones step down abruptly across 7 ft (2.13 m) of the underlying mudstones. There is no trace of IVB and it is presumed to be washed out. The limestone reappears to the south-west in Scotchergill, Dentdale (Figure 7, section 5).

Mudstones overlying IVB are seldom fossiliferous (Moore, *op. cit.* and the present study). They are overlain by sandstone, with current-ripple lamination and trace fossils, passing up into a more blocky lithology with local cross-bedded units. A coal overlying the sandstone has been worked locally in Wensleydale.

The fourth minor cycle starts with limestone IVC, which dies out locally in Wensleydale (Moore, 1958). On Penhill, it occurs very close to the base of the Middle Limestone and is virtually amalgamated with it in places, a feature also noted by Moore in the south-east of his area. Whilst IVC in the Gunnerside G1 Borehole may be a multiple horizon, it was seen in very few sections during the present survey and is presumed to have either amalgamated with the Middle Limestone or died out south of Wensleydale. In Garth Gill, Garsdale, IVC is a sandy limestone with desiccation cracks in its upper surface.

Beds above the Simonstone Limestone thin to the south-east. In Walden and Coverdale there are up to six separate limestone bands, likely to be IVA and B with many local splits (Wilson 1960a, fig. 2). Associated mudstones are richly fossiliferous and include clisiophyllid corals and latissimoid productoids. A single sandstone beneath the Middle Limestone thins to the south and is absent in the Grassington Mining Field at Coalgrove Head Shaft (Figure 7, section 20) and at Greenhow. Up to two thin limestones, possibly correlatives of limestones IVA – B, were seen in boreholes near Greenhow in a sequence of fossiliferous mudstones as little as 14 ft (4.27 m) thick (Figure 7, sections 22, 23).

Middle Limestone This formation was named by Phillips (1836) though Sedgwick (1836) had referred to it as the Mossdale Moor or Wold Limestone. It persists over the whole area but shows large thickness and lithological variations (Figure 10, map F). It is notable for its several faunal bands and biostromes. Dakyns and others (1891) believed that the Middle Limestone of the Askrigg Block resulted from the union of three separate limestones on the Alston Block, the Scar, Cockle Shell and Single Post. This is now generally accepted. The splitting of the Middle Limestone into three members takes place progressively in a northerly direction. Around Gunnerside in Swaledale, Clough mapped the Middle Limestone in three distinct members split by mudstone partings (Yorkshire six-inch County Sheet 51). Towards Kirkby Stephen the partings thicken and sandstones come in. In the south of the Askrigg Block the tripartite division of the Middle Limestone is, locally, hard to see due to lack of clastic partings and a blurring of lithological and faunal distinctions, but an emergence surface at the top of the lowest, Single Post, member is an aid to detailed correlation. In the south-east, around Greenhow, the Middle is equivalent to the upper part of the Toft Gate Limestone (Figure 8, sections 20, 21).

The Single Post Limestone member is some 6 ft 6 in to 40 ft (2–12.20 m) in thickness and consists of a thickly-bedded medium grey or light grey medium-grained biosparite which commonly forms a line of scars. A blotchy crystallisation has been noted by several workers on this formation. Towards the base distinctive coral bands are present (Wilson, 1960a, p. 290).

The uppermost stratum of the Single Post Limestone has, in several places, an undulating surface, with patchy red staining in the top 0.30 m, with an angular disconformity where seen in Hazle Bank Gill, Langstrothdale [SD 8655 8262]. Dr Ramsbottom has recently been investigating this emergence surface, which he suggests is the boundary between two mesothems making up the Brigantian Stage.

Beds between the Single Post and Cockle Shell members are thickest in the Stainmore area where up to 56 ft (17 m) of sandstones and mudstones lie between the two limestones (Burgess and Holliday, 1979). They thin to 31 ft 6 in (9.60 m) in the Birkett Cutting section [NY 7743 0288, Figure 8] whilst in Dentdale, Garsdale and upper Wensleydale a parting 3 to 8 ft (0.90–2.45 m) thick persists between the Single Post and Cockle Shell members, but dies out entirely to the south and east. Locally the beds at this level include, besides mudstone, a calcareous sandstone with a characteristic fluted weathering well seen in Garth Gill, Garsdale [SD 7711 9099]. Hudson (1929b) recorded nodules with the sponge *Erythrospongia lithodes* from the top part of the parting in mid-Wensleydale, but Hicks (1957) and the present authors have been unable to find the band to the west and east.

The Cockle Shell member typically consists of well-bedded, medium dark grey, fine-grained biosparite, with a foetid smell when fractured, ranging in thickness from some 4 ft (1.20 m) in Garsdale to at least 41 ft (12.50 m) in Walden. In much of Wensleydale and Swaledale, shells of *Gigantoproductus* are common, as are colonies of *Lithostrotion junceum* in the position of growth. *Gigantoproductus* persists southwards into the Kettlewell and Grassington areas of Wharfedale but is commonly absent in other parts of Wharfedale, in Dentdale and Garsdale. Where *Gigantoproductus* fails and mudstone partings are also absent, as in much of Wharfedale and Ribblesdale, the definition of the top of the Cockle Shell member is difficult. North of Grassington the *Gigantoproductus* beds pass southwards into crinoid banks with blastoids in the overlying beds (Black, 1950; Joysey, 1955).

Beds between the Cockle Shell and Scar members consist of up to 77 ft (23.50 m) of mudstones with siderite nodules overlain by fine-grained sandstone and a thin coal seam in the country around Brough and Kirkby Stephen. The maximum thickness is in the Birkett Cutting, at a location some 5 miles (9 km) SW of the area of greatest thickness (near Brough) recorded in the lower clastic parting. Farther south, clastic beds above the Cockle Shell Limestone thin to 7 ft (2 m) in Gunnerside G1 Borehole. In mid-Wensleydale, as in Garsdale and Dentdale a mudstone about 5 ft (1.50 m) thick is typical at this level (Hicks; Moore, personal communications). To the south and east the parting dies out completely, but a porcellanous calcilutite 1 ft (0.30 m) thick seen in Deepdale and in Force Gill, Whernside may be near this horizon.

The Scar Limestone division of the Middle Limestone shows the greatest thickness variation from about 5 ft 6 in (1.70 m) to some 122 ft (37 m). The area of maximum thickness runs in an arc from Penhill through Nidderdale to Grassington (Figure 10, map F). Typically, this member is a thickly-bedded, grey, medium-grained biosparite with a local mudstone parting 6 ft (1.80 m) thick on Penhill. Where the limestone is thickest, on Penhill, portions consist almost exclusively of articulated crinoid stems about ½ in (1 cm) in diameter. Locally at Knarleton Knot, on Penhill [SE 0256 8684], these form well-defined mounds, some with cores of calcite mudstone. At Long Ing Wood, Penhill [SE 023 861], a bryozoan biostrome rich in trepostomes forms the uppermost 15 ft (4.50 m) of the Scar Limestone member. A deduction of 10 ft (3 m) from the overall limestone thickness on Penhill has been made to allow for the likelihood that the Five Yard Limestone is here in direct contact with the Middle Limestone and indistinguishable from it (p. 43). This is not the case at Long Ing Wood, however, where the Five Yard Limestone and underlying clastic beds map out separately above bryozoan beds in the Middle Limestone.

In the Kettlewell – Grassington Moor area and Upper Nidderdale the Middle Limestone is unusually thick, mostly due to thickening of the Scar Limestone member. *Gigantoproductus* commonly occurs at two levels, the lower being broadly equivalent to the Cockle Shell Limestone and the upper one lying within the Scar Limestone. Overall thickness of the Middle Limestone ranges from 103 ft (31.40 m) at Providence Mine, Kettlewell to a probable 195 ft (59.44 m) at Coalgrove Head Shaft (Figure 8, sections 20, 22). Intermediate values are 158 ft (48.16 m) at Mossdale Mine [SE 0128 6966] and 182 ft (55.47 m) at Brunt's Shaft [SE 0299 6670], according to Geological Survey Vertical Section Sheet 28, 1874.

Recent boreholes at Greenhow show that the Middle Limestone as a whole ranges greatly in thickness from 65 to about 180 ft (19.80–55.00 m) in a distance of about 1 mile (1.6 km). Beds 11 ft 3 in (3.43 m) thick with *Gigantoproductus* (Figure 8, section 21) are recorded, but die out to the north where the limestone thins. These are high in the sequence and probably at the level of the band in the Scar Limestone member. Lithology of the Middle Limestone is variable, but includes grainstones with many very large crinoid columnals similar to those recorded in Toft Gate Quarry (Dunham and Stubblefield, 1945, p. 225). These occur both above, below and in the beds containing *Gigantoproductus*.

In Garsdale, Dentdale and Kingsdale, where the Scar Limestone member is only some 5 ft (1.50 m) thick, Hicks (1957) records a widespread algal nodule band near the top of the limestone which extends eastwards into Upper Wensleydale (Hudson, 1929b; Moore, 1958). Turner (1956) records a band, some 1 ft (0.30 m) thick, rich in *Gigantoproductus*, over a wide area between Garsdale and Kirkby Stephen, confirmed in the log of G1 Borehole. A similar band also occurs in Nidderdale, but perhaps at a higher level in the member.

Beds above the Middle Limestone These beds show a marked southward thinning from a maximum of 158 ft (48.16 m) in Gunnerside G1 Borehole to as little as 2 ft (0.61 m) in Wharfedale, a trend which was initially worked out in detail by Moore (1958) in Wensleydale. The rock succession can be divided into two main areas, one broadly north of the River Ure where sandstone is dominant and one south of the Ure where sandstones are absent (Figure 10, map G).

In the northern area the succession is thick and dominated by sandstone, notably in the North Swaledale Mineral Belt. The sequence in G1 Borehole between 336 and 494 ft 1 in (102.41–150.60 m) may be taken as typical:

	Thickness ft	in	m
Five Yard Limestone	–	–	–
Mudstone, dark grey, silty, with cross- laminated sandstone bands	5	3	1.61
Mudstone, dark grey; *Aclisina sp.*, *Promytilus emaciatus*, palaeoniscid scale	2	3	0.68
Sandstone, fine-grained, cross-bedded	7	8	2.33
Coal		2	0.04
Seatearth mudstone and sandstone	1	2	0.35
Sandstone, very fine-grained, on silty mudstone	8	7	2.63
Sandstone, probably seatearth	1	0	0.30
Sandstone, fine-grained, laminated	5	11	1.80
Mudstone, dark grey; *Productus sp.*	5	3	1.60
Sandstone, fine-grained, cross-bedded, trough cross-laminated; interbedded with mudstones	13	3	4.05

	Thickness		
	ft	in	m
Sandstone, fine-grained; micaceous laminae	25	1	7.65
Sandstone, fine-grained, laminated, trough cross-laminated, current-ripple laminated, bioturbated; with some mudstone bands becoming more common downwards; plant debris	70	6	21.49
Mudstone, dark grey; juvenile invertebrates	7	5	2.26
Mudstone, dark grey, scattered siderite nodules; *Cladochonus sp.*, *Fenestella* cf. *parallela*, *Phricodothyris sp.*, *Schizophoria* cf. *resupinata*, *Parallelodon semicostatus*	4	7	1.40
MIDDLE LIMESTONE			

In Reigill near Kirkby Stephen [NY 8030 0668], three seatearths were noted (Figure 8, section 1), one of them overlain by a coal, and two further coals occur in the Greta section near Barnard Castle (Wells, 1957; Mills and Hull, 1976). Sandstones dominate the sequence and include both cross-bedded sandstone in thick units and better laminated beds with cross-laminations, current ripple laminations, trace fossils and abundant drifted plant debris. Throughout the North Swaledale Mineral Belt the sandstones are thick, but from Arkengarthdale eastwards are poorly exposed due, chiefly, to landslip debris; they have been quarried for building stone on Calver and below Fremington Edge, Reeth. Two sandstones figure in the record of the Hurst Mines (Figure 8, section 5). An exposure in Rispey Wood, Holgate Beck [NZ 0624 0438, Figure 8, section 4] is in thickly bedded fine-grained sandstone, with some cross-bedding, a lithology recorded by Wells (*op. cit.* fig. 2) in the Gilling Dome and River Greta sections.

Farther south, sections such as Sar Gill [SD 9072 9196] in Wensleydale and Garth Gill [SD 7707 9102] Garsdale show a fairly equal proportion of mudstone and sandstone in a typical coarsening upwards Yoredale cycle, without the seatearth horizons developed farther north. As in Gunnerside G1 Borehole, the basal mudstones contain siderite nodules, and a rich invertebrate fauna in their lowest beds. Bioturbation and trails are seen in the lower, better laminated parts of the sandstones.

South of the River Ure sandstones die out. The succession consists solely of mudstone with siderite nodules which continue to thin towards the south, apart from some local thickening near Greenhow. Marine fossils, notably brachiopods and trilobites, occur towards the top and base of the sequence (Moore, 1958), but farther south all the beds are fossiliferous. In Wharfedale, where the beds are very thin, they are seldom exposed; but their presence is commonly indicated by the toppling forward of large blocks of Five Yard Limestone, given minimal support by the 2 to 5 ft (0.60–1.50 m) thick band of soft mudstone.

Two thin limestone bands occur in the lower half of the cyclothem at Ivelet Force, Swaledale [SD 9362 9820]. Both appear to rest on seatearth horizons. Elsewhere limestones are recorded locally in Wensleydale, high in the cyclothem (Moore, *op. cit.*), and in Spice Gill, Dentdale [SD 747 873] (Hicks, 1957).

In the country south of the River Dee in Dentdale as far as Ingleborough, the Five Yard Limestone is only recognised in one or two places. This leads to difficulties in definition of the underlying measures. In Gastack Beck, Deepdale [SD 7104 8260], a mere 4 ft 6 in (1.37 m) of mudstone rich in brachiopods and bryozoa lies beneath the probable Five Yard Limestone, but farther east in Force Gill [SD 7572 8220] thickly bedded sandstones make up most of the beds between the Middle Limestone and a calcareous sandstone developed at the horizon of the Five Yard Limestone where seen 3000 ft (900 m) farther north (Figure 8, sections 11, 12). On Fountains Fell (E. W. Johnson, personal communication) and Ingleborough the relevant beds above the Middle Limestone are unexposed (Figure 8, sections 13, 17).

In the Greenhow area the mudstones above the Middle Limestone are probably represented by a 6-ft (1.80 m) thick shale in the uppermost part of the Toft Gate Limestone (Dunham and Stubblefield, 1945, p. 227) which thickens greatly 1 mile (1.6 km) farther north in borings sunk in the Providence Vein. Here, Bewerley Mines No. 3 Borehole yielded important new goniatites (p. 22) between the depths of 717 ft 9 in and 751 ft 3 in (218.77–228.98 m):

	Thickness		
	ft	in	m
Mudstone, grey, slightly silty and calcareous; *Brachythyris sp.*, *Rugosochonetes sp.*, *Spirifer sp.*, *Tornquistia polita*, *Koninckopecten alternata*	11	3	3.43
Limestone, muddy; scattered crinoid debris		10	0.25
Mudstone, slightly silty; a few bands of muddy limestone; *Fasciculophyllum sp.*, *Fenestella* cf. *frutex*, *F.* cf. *plebeia*, *Krotovia spinulosa*, *Schizophoria sp.*, *Spirifer bisculcatus* group, *Edmondia maccoyii*, *Girtyoceras sp.*, *Goniatites granosus*, non-mucronate trilobite pygidia, *Paladin sp.*	18	1	5.51
Limestone, pale grey; abundant small crinoid debris	1	0	0.30
Mudstone grey, slightly silty; *Petalodus acuminatus*	2	4	0.71
TOFT GATE (MIDDLE) LIMESTONE	–	–	–

In upper Nidderdale and the Kettlewell area, as at Greenhow, the beds above the Middle Limestone are fossiliferous mudstones ranging in thickness from 5 ft (1.52 m) at Providence Mine, Kettlewell to 4 ft 6 in to 20 ft (1.37–6.10 m) in Nidderdale. Farther south at Mossdale Mine [SE 0128 6966] 12 ft (3.66 m) of mudstone with thin limestone bands are probably at this level. On Grassington Moor shaft records (Geological Survey Vertical Sections Sheet 28, 1874) show considerable variations in thickness of mudstone partings at or near the top of the Middle Limestone. A mudstone parting selected as immediately overlying the Middle Limestone in Coalgrove Head Shaft (Figure 8, section 22) correlates well with one in the nearby Moss Shaft. However the records of nearby Brunt's and Sarah shafts show no mudstone at this level, making this correlation tentative. It is possible that the thin mudstone 28 ft (8.53 m) lower in the succession marks the top of the Middle Limestone. This is found in all four shafts, but is locally only 3 in (0.08 m) thick.

Five Yard Limestone This formation was variously called the Five Yard, Third Set or Shallop Limestone by Swaledale miners, and the Geological Survey used the first two terms in its maps and memoirs. Phillips (1836) referred to it as the 'Impure Productal Limestone'. Current usage is restricted to the term Five Yard Limestone, which reflects its average thickness in the North Swaledale Mineral Belt.

The Five Yard Limestone is very variable in thickness, with a minimum of some 2 ft 3 in (0.69 m) in upper Wensleydale (Moore, 1958) and a maximum of 60 ft (18.29 m), according to a table of rock thicknesses at Keld Heads Mine, Wensleydale. In Garsdale, Dentdale and in a belt stretching eastwards through the southern tributaries of Wensleydale as far as Coverdale, the Five Yard Limestone is typically about 7 ft (2.13 m) thick and consists of two leaves split by a fossiliferous mudstone parting.

In Dentdale and Garsdale the following is an average section, typical of many others (Hicks, personal communication):

	Thickness		
	ft	in	m
Limestone, dark grey; weathering with a blotchy reddish brown hematite-rich outer skin	1	4	0.40
Mudstone, shaly; crinoid columnals	1	8	0.50
Limestone, dark grey, fine-grained biosparite; rich in *Gigantoproductus*	3	0	0.90

The fossiliferous lower leaf thins northwards (Moore, *op. cit.*) and is locally represented by calcareous sandstone with a thin capping of siliceous limestone. In Swaledale the Five Yard Limestone is usually in one leaf, probably the thickened upper leaf (Moore, *op. cit.*) except perhaps in the Gunnerside G1 Borehole which shows a mudstone parting, fairly low in the limestone (p. 44).

The upper leaf of the limestone is coarsely crinoidal in parts of Swaledale and the north flank of Wensleydale (Moore, *op. cit.*). This facies is particularly well seen in an old quarry [SE 0795 9079] close to the mouth of Keld Heads Level, Wensleydale, as follows:

	Thickness		
	ft	in	m
Limestone, grey to light grey crinoidal biosparrudite (grainstone), with many columnals (½ in wide) and stems (6 in long)	26	0	7.92
Mudstone, dark grey; very fossiliferous	1	0	0.30
Limestone, pinkish grey	1	0	0.30

The thin limestone close to the base of the formation is possibly representative of the lower member. Records from Keld Heads Mine suggest a total thickness of 60 ft (18.29 m) for the whole formation, much of it above the quarry section. In the first surface rise at Bolton Park Mine [SE 0406 9290], 2½ miles (4 km) WNW of Keld Heads Mine, the crinoidal Five Yard Limestone is 57 ft (17.37 m) in thickness. It thins in about 200 yd (180 m) westwards to only 21 ft (6.40 m) (unpublished MS of Richmond memoir)

The Five Yard and Middle limestones come close together on the watershed between the Ure and Wharfe and this situation appears to persist throughout Wharfedale and Littondale. Though exposure is seldom ideal, the thin limestone 1 ft 8 in to 9 ft (0.51–2.74 m) thick is recognised in a number of sections. In most, only one leaf of dark grey limestone was seen. On Fountains Fell its presence has not been verified.

On Penhill and in lower Coverdale, Wilson (1960a) noted that the Five Yard Limestone appeared to be missing with clastics only separating the Middle and Three Yard limestones. Moore (personal communication) has later suggested that the exceptionally thick Middle Limestone here may incorporate the Five Yard Limestone in its upper beds. Mapping by the authors on the slopes of Penhill tends to confirm this suggestion.

The log of Middleton Tyas No. 1 Borehole [NZ 2371 0603] shows that the probable Five Yard Limestone is separated by only 2 ft 6 in (0.75 m) of strata, mostly sandstone, from the Middle Limestone. This situation is reminiscent of that postulated on Penhill, but the Five Yard Limestone at Middleton Tyas is an unusual 60 ft 3 in (18.35 m) in thickness. It is not in a crinoid bank facies as it contains only scattered crinoid debris (Figure 8, section 7).

The Five Yard Limestone presents another problem in the south-western ground from the south side of Dentdale as far as Ingleborough. In this area neither Dakyns and others (1891) nor Hicks (1957) has traced any definite Five Yard Limestone. Available evidence is ambiguous. Strahan noted the Five Yard with its 'horseshoe' shells (*Gigantoproductus*) at many localities on the north side of Dentdale, but only in one small area on the south side, namely on the north-eastern slopes of Whernside [SD 7596 8319]. According to his mapping some 50 ft (15 m) of strata separate this from the (here unexposed) Middle Limestone. In a number of other sections in the southern tributaries of Dentdale, Strahan mapped a red-topped limestone a short distance above the Middle Limestone which he appears (Dakyns and others, 1891, p. 44) to assign to the highest part of the Middle Limestone. This has been examined by us in Gastack Beck [SD 7103 8258, Figure 8, section 11] where it is only 11 in (0.28 m) in thickness, lacks *Gigantoproductus*, but has a red hematitic crust reminiscent of that so commonly seen in the upper leaf of the Five Yard Limestone in Garsdale and northern parts of Dentdale. This bed is thus possibly the Five Yard Limestone. A section in Force Gill, Whernside [SD 7566 8222] shows no Five Yard Limestone despite an almost continuous exposure (Figure 8, section 12). A calcareous sandstone 53 ft (16.15 m) above the top of the Middle Limestone, according with the level of the 'horseshoe' limestone recorded only ⅔ mile (1 km) away by Strahan, has been correlated with the Five Yard Limestone.

The Five Yard Limestone was mapped by Wilson (1960a) in Upper Nidderdale and by comparison of the records at Greenhow with that area it is now possible to assign the highest member of the Toft Gate Limestone recorded in the Cockhill Adit by Dunham and Stubblefield (1945) to this formation. Recently it has been penetrated in several boreholes on Providence Vein where it is seen to be some 23 ft (7 m) in thickness, consisting of grey to dark grey, fine-grained limestone, with a porcellanous band near the top (Figure 8, section 20).

Shafts on Grassington Moor (Coalgrove Head, Brunts, Moss, Sarah) all show mudstone partings at or near the top of the Middle Limestone, though the sections differ in detail. It seems likely that the Five Yard Limestone is present and is at least 10 ft (3 m) thick (Figure 8, section 22).

Beds above the Five Yard Limestone Thickness variations differ in pattern from those of underlying Brigantian cyclothems. The thickest beds lie in the centre and west of the area, rather than the north. As in other cyclothems the greatest thinning is in the south-east. The facies patterns fall into four distinct areas which grade laterally into each other.

1 North of the River Swale Three minor cyclothems occur, each consisting of mudstone with marine fossils overlain by sandstone and usually capped by a seatearth, exemplified by the section in Gunnerside G1 Borehole between the depths of 226 ft 6 in and 336 ft (69.05 and 102.41 m) (Figure 8, section 3) as follows:

	Thickness		
	ft	in	m
THREE YARD LIMESTONE	–	–	–
Mudstone, greyish black; drifted plants	–	1	0.03
Seatearth mudstone	3	6	1.08
Mudstone and sandstone	1	1	0.34
Seatearth sandstone, very fine-grained	2	1	0.64
Sandstone, very fine-grained	2	4	0.72
Mudstone, dark grey, shaly, pyritic; *Fenestella sp.*, *Hemitrypa sp.*, *Fluctuaria sp.*, *Productus sp.*, *Pugnax sp.*, *Microptychis sp.*, *Edmondia sp.*, *Limipecten dissimilis*, *Solenomorpha* cf. *parallela*, *Streblochondria sp.*	8	6	2.59
Sandstone, fine-grained; micaceous laminae	5	8	1.73
Seatearth mudstone	1	1	0.32
Sandstone, fine-grained; siltstone and mudstone partings in upper half; cross-laminated; U-shaped burrow	16	8	5.09
Mudstone, dark grey; fish fragments and small shells	–	8	0.21
Siltstone with sandstone laminae	1	2	0.36
Probable seatearth mudstone, silty	–	6	0.14
Siltstone overlying mudstone; probably bioturbated	3	7	1.10
Sandstone, fine-grained; load casts near base	23	8	7.22
Siltstone interlaminated with very fine- grained sandstone, cross-laminated, load casts; crinoid debris at base	5	4	1.63
Mudstone, dark grey, shaly, siderite nodules; very calcareous near base, fossiliferous throughout; trepostome bryozoa, *Antiquatonia sulcata*, *Brachythyris* cf. *integricosta*, *Schizophoria sp.*, *Spirifer sp.* [cf. *striatus* group], *Aviculopecten sp.*, *Limipecten dissimilis*, *Lithophaga lingualis*, *Coleolus sp.*	10	2	3.10

	Thickness ft	in	m
FIVE YARD LIMESTONE			
Limestone, dark grey, medium-grained biosparite with some argillaceous biomicrite; *Rugosochonetes celticus* group, *Schizophoria* cf. *resupinata*	16	9	5.10
Mudstone, dark grey, very calcareous	–	5	0.13
Limestone, dark grey, fine-grained biosparite, argillaceous in part	6	0	1.83

Shales of the highest cycle with marine fossils were noted also in Greenseat Beck, Swaledale [SD 8989 9689]. Elsewhere there is developed at this level a calcareous sandstone [Reigill, NY 8046 0686] or a limestone 1 ft (0.30 m) thick (River Greta *in* Mills and Hull, 1976, p. 19 and Rispey Wood [NZ 0627 0426] *in* unpublished memoir for one-inch Old Series Sheet 97 NE). Further east in Middleton Tyas No. 1 Borehole [NZ 2371 0603] the limestone comprises 1 ft 9 in (0.54 m) of dark grey wackestone with *Zoophycos*.

2 *Lower Wensleydale* Thick fine-grained, in places cross-bedded sandstones, resting on mudstones and a thin sandstone occur, but the sequence is seldom fully seen. The sandstones are the Preston Grits, finely exposed in old quarries at Preston under Scar [SE 0700 9121] and recorded in the workings of the nearby Keld Heads Mine (Figure 8, section 15). Westwards, a thin sandstone comes in within the marine mudstones, effectively making two minor cycles. This is well seen in Sar Gill, Wensleydale [SD 9036 9213] (Figure 8, section 9). The sandstone persists as far south as Hazle Bank Gill, Wharfedale [SD 8665 8271]. In Sar Gill marine fossils, notably bryozoa and trilobite fragments occur abundantly in mudstone with siderite nodules in both minor cycles.

3 *West-central area, from Garsdale southwards to Whernside* This is the area of greatest thickening and sandstones dominate the sequence, typified by outstandingly good sections in Gastack Beck, Deepdale [SD 7102 8251], Force Gill, Whernside [SD 7560 8224], and Bank Gill, Sleddale [SD 8500 8468] (Figure 8, sections 11, 12, 14). Cross-bedding is generally rare. The upper beds in particular include thinly bedded sandstones with siltstone interbeds usually lacking cross-cutting relationships; bioturbation and trails are very common. The lithologies are consistent with interdistributary deposits laid down as a result of crevasse splays in a deltaic system.

4 *The Southern Belt from Ingleborough eastwards* This area is marked by the dying out of sandstones and is characterised by marine mudstones, often rich in big unfossiliferous siderite nodules up to 1 ft (30 cm) in length. The fauna, concentrated in the lower half of the mudstone, is commonly abundant, with a rich variety of invertebrates. The mudstones have been recognised in Upper Nidderdale where they attain a thickness of 46 ft 6 in (14.17 m) (Wilson, 1960a) and in boreholes in the Providence Vein, Greenhow (Figure 8, section 20). The following section dipping at 40° between 669 ft 3 in and 680 ft 6 in depth (203.99 and 207.42 m) in the inclined Bewerley Mines No. 3 Borehole at Providence Vein yielded a goniatite:

	Thickness ft	in	m
GRASSINGTON GRIT FORMATION; sandstone	–	–	–
Mudstone, dark grey, shaly; *Productus sp.*, *Spirifer bisulcatus*, *S.* cf. *trigonalis*	5	1	1.55
Limestone, grey, muddy; crinoidal	2	2	0.66
Mudstone, dark grey, shaly; *Rhopalolasma rylstonense*, numerous species of bryozoa, *Globosochonetes parseptus*, *Phricodothyris sp.*, *Schizophoria sp.* and *Sudeticeras stolbergi* group	4	0	1.22

The section in Mere Gill, Ingleborough [SD 7456 7515] is at the edge of the fourth facies area since a thin (about 7 ft—2.10 m) median sandstone like that in Sar Gill is present. Mudstones in the upper part of the sequence, formerly referred to the probable middle cyclothem, have yielded *Neoglyphioceras sp.* (Hopkins, 1956 as amended by Ramsbottom, 1974).

Three Yard Limestone This formation was first named by the Swaledale miners who also referred to it as the Riddings or Third limestone. As its name suggests, over much of Swaledale and Wensleydale it is a grey medium-grained biosparite averaging some 9 ft (2.70 m) in thickness. In parts of the Brough area (Burgess and Holliday, 1979) and locally in upper Swaledale, as, for instance, in Gunnerside G1 Borehole, the limestone is split by a thin fine-grained sandstone parting. This split in the Three Yard Limestone is best seen in Coverdale and Langstrothdale (Figure 9, map). In Coverdale the sandstone parting is absent in the lower reaches of the valley, but at Lord's Gill [SE 019 789] it is 1 ft 6 in (0.46 m) thick. Westwards this thickens in 1 mile (1.6 km) to 19 ft (5.79 m) in Downs Gill [SE 007 786], and to 28 ft 6 in (8.69 m) in the nearby valley of Walden [SD 977 791] (Figure 8, section 16). In Langstrothdale the split persists with 21 to 33 ft (6.40–10.06 m) of sandstone between the two leaves (Figure 9). Typically the sandstone is fine or very fine-grained, calcareous in part, with a fretted type of weathering emphasising bioturbated lamination.

The upper member of the Three Yard Limestone is usually a grey biosparite 3 ft to 13 ft 6 in (0.9–4.10 m) in thickness. The lower member is also a single limestone in Coverdale, but is split by several mudstone partings in Walden and Upper Wharfedale. Where the splits occur the limestones, too, become argillaceous, with a fauna of brachiopods and bivalves. Locally in Coverdale a band of algal nodules occurs. The splitting of the lower member into numerous silty limestones and calcareous siltstones interbedded with mudstones is at a maximum in two streams making up a complementary section on Green Haw Moor [SD 8264 7747 and SD 8296 7784, Figure 9, section 2] where they occur through 108 ft (32.80 m) of beds:

	Thickness ft	in	m
Gap (to estimated base of UNDERSET LIMESTONE)	79	0	24.08
Mudstone, silty	23	6	7.16
Gap (probably mudstone)	5	3	1.60
Limestones and intercalated beds of THREE YARD LIMESTONE, medium-grained sparite ?*Amplexizaphrentis curvilinea*	2	0	0.60
Limestone, biomicrite, dark grey	2	1	0.64
Gap	3	3	1.00
Limestone, grey richly bioclastic packstone, *Alitaria panderi*, *Antiquatonia sulcata*, *Echinoconchus elegans*, *Schizophoria resupinata*	1	10	0.55
Gap	2	6	0.75
Limestone, grey, fine-grained sparite	1	6	0.45
Gap (with blocks of shelly calcareous sandstone)	2	0	0.60
Sandstone, fine-grained; well-bedded to thickly-bedded	25	3	7.70
Gap (blocks of shelly calcareous sandstone)	6	6	1.98
Limestone, very silty, tough; common *Buxtonia scabricula*, *Productus carbonarius*, *Myalina pernoides*	1	8	0.50
Gap (mudstone debris)	2	4	0.70
Sandstone, fine-grained tough	2	0	0.60
Mudstone, silty, with two bands of calcareous siltstone; the lower band yields abundant *Productus carbonarius* and *Schellwienella rotundata*; other fossils include *Actinopteria persulcata*, *Sulcatopinna flabelliformis*	6	6	1.98
Mudstone, silty, with eight bands of calcareous siltstone with the following common forms: *Buxtonia scabricula*, *Edmondia* cf. *sulcata*, *Myalina*			

	Thickness		
	ft	in	m
pernoides; other fossils include *Productus concinnus*, *Spirifer* cf. *triangularis*, *Actinopteria persulcata*, *Pernopecten sowerbyi*, *Sanguinolites plicatus*, *Schizodus impressus*, *Wilkingia elliptica*	29	0	8.84
Mudstone, silty	17	0	5.18
Limestone, muddy, fine-grained; rare small brachiopods	3	9	1.14
Siltstone, very calcareous	2	6	0.76
Mudstones of FIVE YARD CYCLOTHEM	–	–	–

In Mere Gill, Ingleborough [SD 7458 7503, Figure 8, section 13] a partial exposure of at least five thin limestones interbedded with mudstone (Figure 9, section 1) distributed through 20 ft (6.10 m) of strata has previously been referred, tentatively, to the Five Yard Limestone (Hicks, 1959). When compared with the newly examined sections in Wharfedale it seems more likely that these beds are the lower member of the Three Yard. A higher limestone, identified as the Three Yard by Hicks, is the upper member of the present account. The ill-exposed beds between the upper and lower members are mostly mudstone here rather than sandstone as in Wharfedale.

On the western Askrigg Block the Three Yard persists as a thin limestone from Mallerstang as far south as Dentdale. In Deepdale, a southern tributary of Dentdale, the attenuated 1-ft (0.30-m) limestone is apparently present in some of its headwater streams and not others, but reappears in Force Gill on Whernside (4 ft, 1.20 m).

Clearly the limestone is locally very thin and has not been found on the south side of Pen-y-ghent, where sandstones occur at the likely horizon, or on Fountains Fell where thick mudstones are seen (E. W. Johnson, personal communication) (Figure 8, section 17). In Waitgate Gill (Wells' 'Rake Beck'), Swaledale, sandstone directly overlies 5 ft (1.50 m) of Three Yard Limestone with unconformity; transgression by the sandstone is presumed, by Wells (1957), to have removed the limestone in the Gilling Dome to the east.

The Three Yard Limestone persists into Nidderdale (Wilson, 1960a) and it consists of 19 ft (5.80 m) of grey medium-grained limestone at the western end of a series of boreholes drilled on the Providence Vein, Greenhow (Figure 8, section 20). Eastwards and southwards it is cut out by the unconformity at the base of the Grassington Grit Formation.

Beds above the Three Yard Limestone These beds are thicker in the north and west where they attain 190 ft (58 m) (Hicks, 1957). In the south-east, where sandstones are virtually absent, there is marked thinning to a minimum of 6 ft (1.83 m) at the head of Nidderdale (Figure 10, map I). In the north and west sandstones in the upper part of the cyclothem, the equivalents of the Natrass Gill Hazle of the Alston Block (Dunham, 1948), are very thick. They form fine cliffs on the River Eden [SD 777 989], in Gunnerside Gill [NY 939 007] and on the River Greta [NZ 0804 1242] (Figure 8, sections 2, 3, 6).

Mudstones, locally with siderite nodules, just above the Three Yard Limestone are usually sparsely fossiliferous, but are, sporadically, rich in brachiopods (e.g. Mills and Hull, 1976, p. 21).

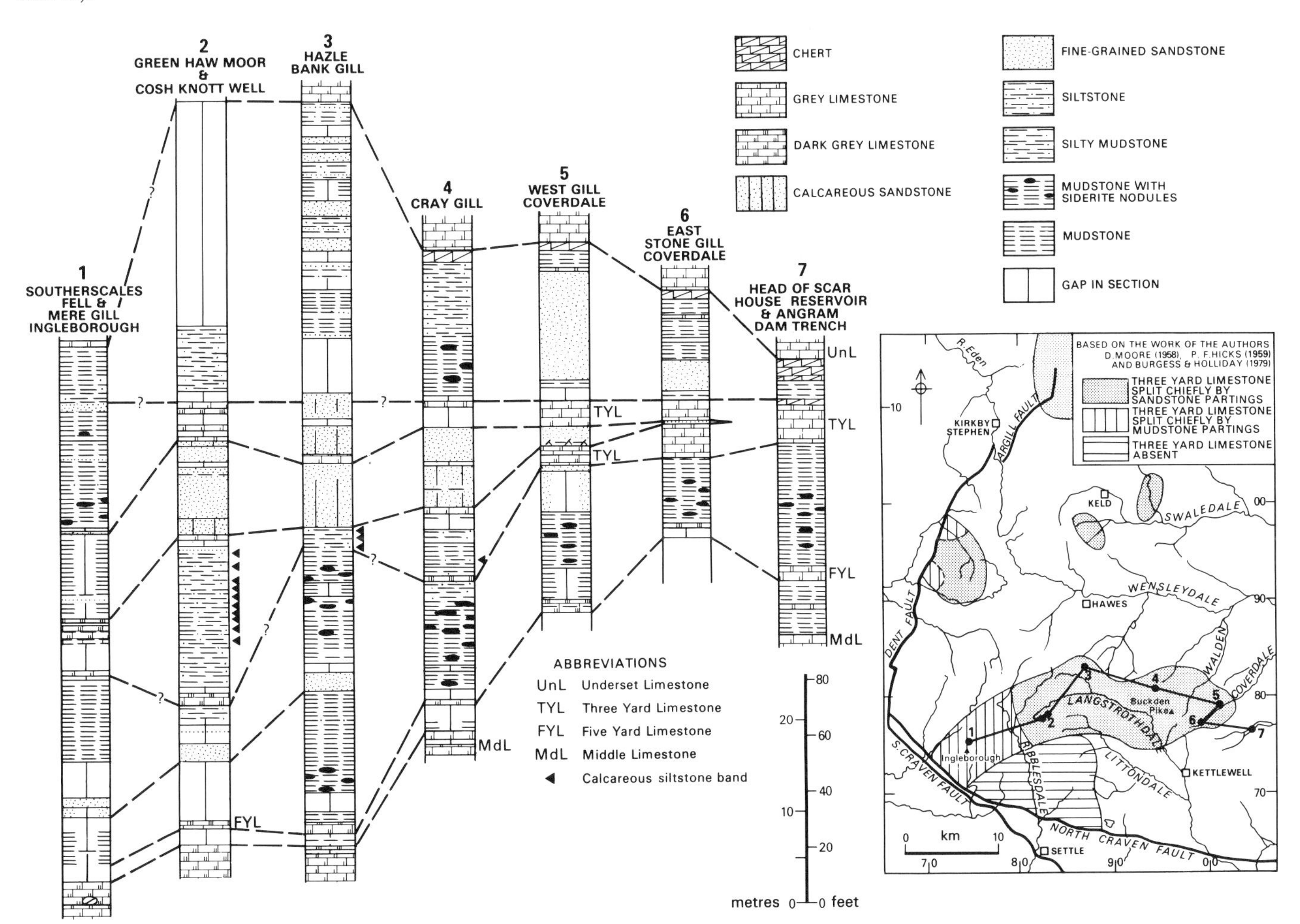

Figure 9 Stratigraphy of the Three Yard Limestone in Langstrothdale and adjacent valleys

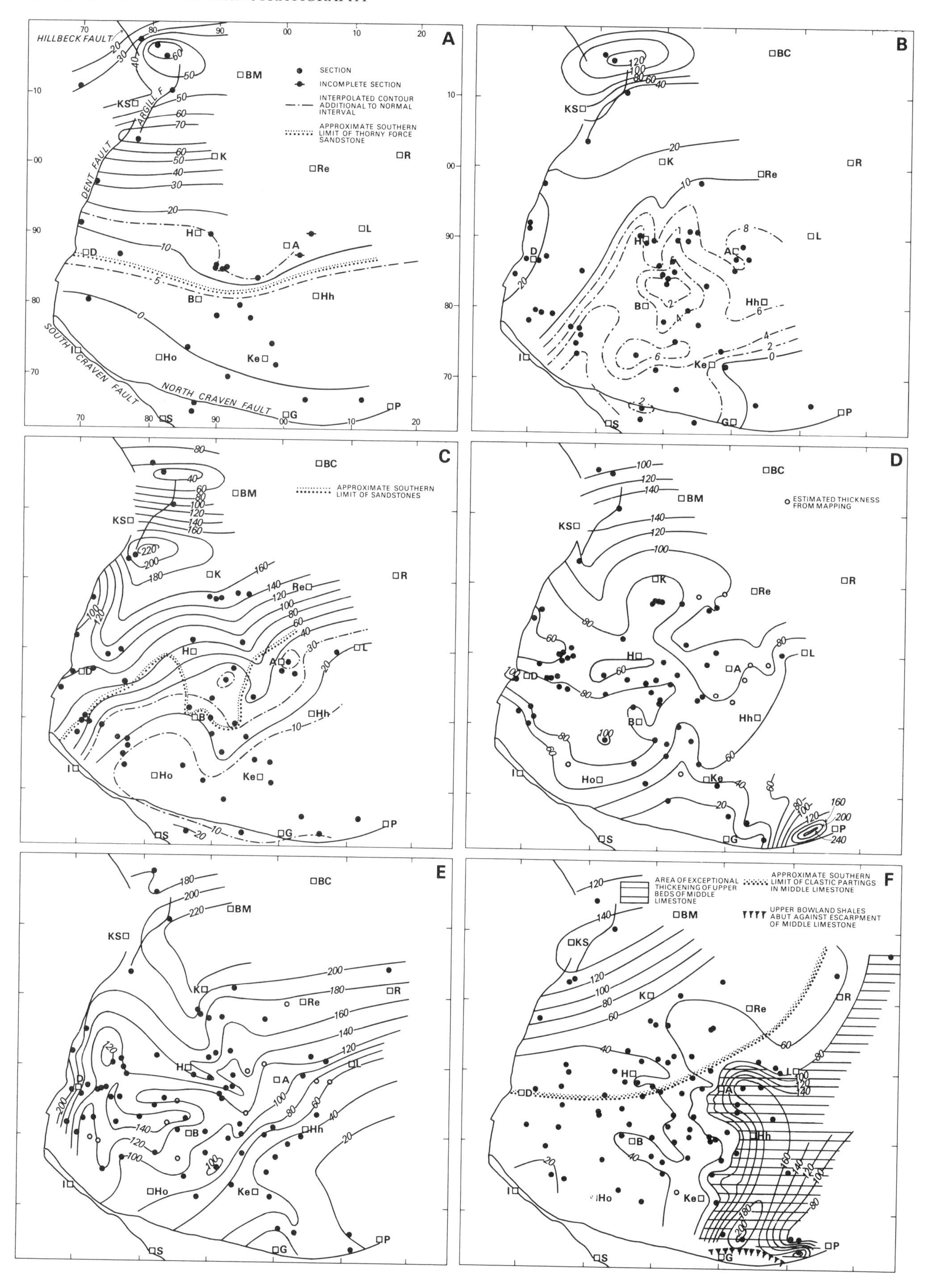
A
HILLBECK FAULT
ARGILL F.
DENT FAULT
SOUTH CRAVEN FAULT
NORTH CRAVEN FAULT
SECTION
INCOMPLETE SECTION
INTERPOLATED CONTOUR ADDITIONAL TO NORMAL INTERVAL
APPROXIMATE SOUTHERN LIMIT OF THORNY FORCE SANDSTONE
B
C
APPROXIMATE SOUTHERN LIMIT OF SANDSTONES
D
ESTIMATED THICKNESS FROM MAPPING
E
F
AREA OF EXCEPTIONAL THICKENING OF UPPER BEDS OF MIDDLE LIMESTONE
APPROXIMATE SOUTHERN LIMIT OF CLASTIC PARTINGS IN MIDDLE LIMESTONE
UPPER BOWLAND SHALES ABUT AGAINST ESCARPMENT OF MIDDLE LIMESTONE
BC
BM
KS
K
R
Re
H
A
L
D
B
Hh
I
Ho
Ke
S
G
P

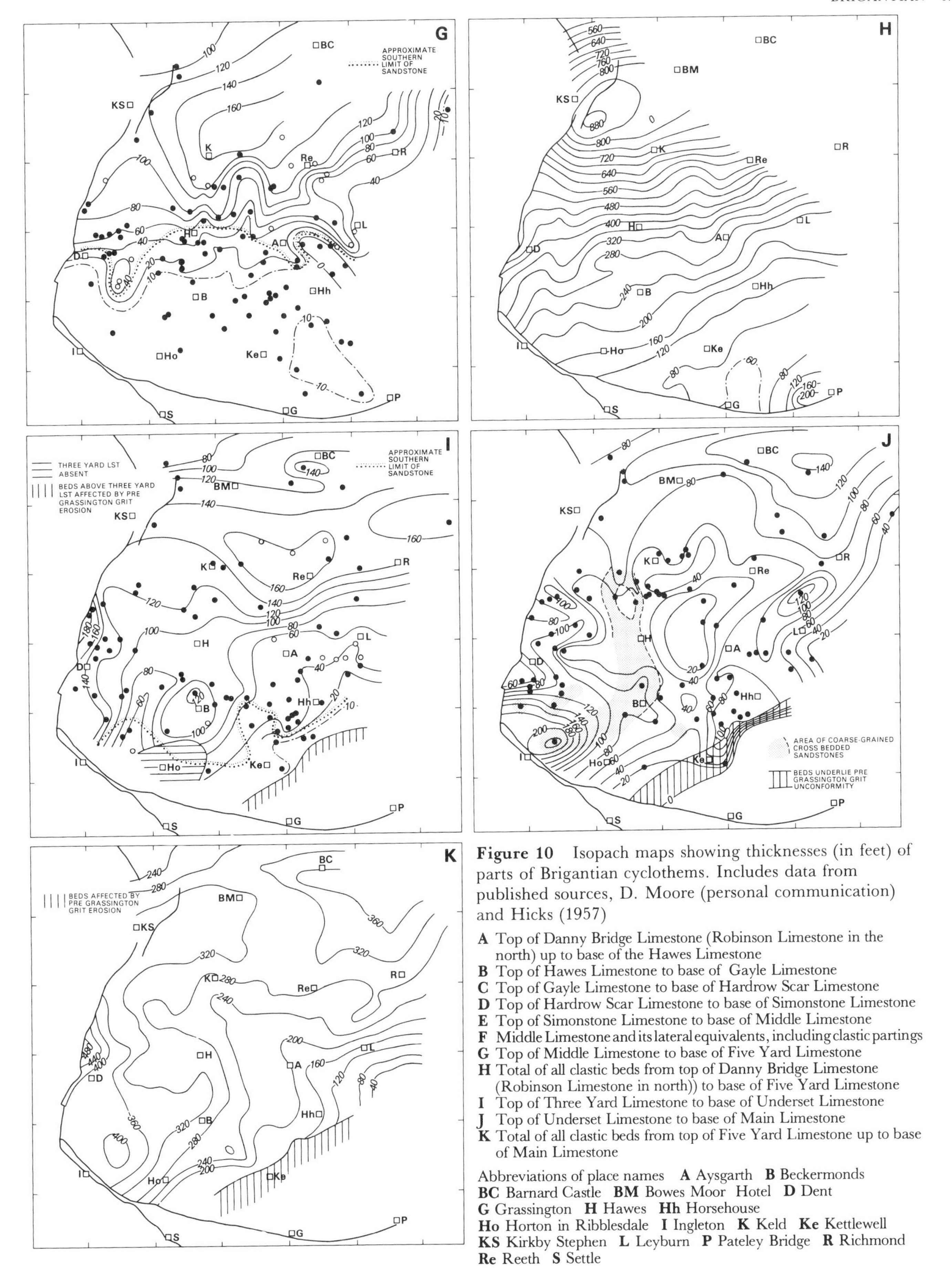

Figure 10 Isopach maps showing thicknesses (in feet) of parts of Brigantian cyclothems. Includes data from published sources, D. Moore (personal communication) and Hicks (1957)

A Top of Danny Bridge Limestone (Robinson Limestone in the north) up to base of the Hawes Limestone
B Top of Hawes Limestone to base of Gayle Limestone
C Top of Gayle Limestone to base of Hardrow Scar Limestone
D Top of Hardrow Scar Limestone to base of Simonstone Limestone
E Top of Simonstone Limestone to base of Middle Limestone
F Middle Limestone and its lateral equivalents, including clastic partings
G Top of Middle Limestone to base of Five Yard Limestone
H Total of all clastic beds from top of Danny Bridge Limestone (Robinson Limestone in north)) to base of Five Yard Limestone
I Top of Three Yard Limestone to base of Underset Limestone
J Top of Underset Limestone to base of Main Limestone
K Total of all clastic beds from top of Five Yard Limestone up to base of Main Limestone

Abbreviations of place names **A** Aysgarth **B** Beckermonds **BC** Barnard Castle **BM** Bowes Moor Hotel **D** Dent **G** Grassington **H** Hawes **Hh** Horsehouse **Ho** Horton in Ribblesdale **I** Ingleton **K** Keld **Ke** Kettlewell **KS** Kirkby Stephen **L** Leyburn **P** Pateley Bridge **R** Richmond **Re** Reeth **S** Settle

The overlying cross-bedded sandstones in Brignall Banks on the River Greta cut across 26 ft (7.92 m) of these mudstones (Wells, 1957, p. 239) and apparently fill a channel. Moore (1958) noted cross-sets with an amplitude of 8 ft (2.44 m) in Ellerkin Scar, north-east of Askrigg, and the sandstones in the headwaters of the Eden are extensively cross-bedded, though chiefly fine-grained. Hicks (1957) reports medium-grained cross-bedded sandstones at Near Gill Laids [SD 704 929] which extend into quarries in nearby Garsdale. All the above occurrences could be channel deposits.

North of the River Ure from three to eight seatearth horizons occur within the sandstone, notably in Middleton Tyas No. 1 Borehole [NZ 2371 0603], Reigill [NY 803 067, Figure 8, section 1, 7], Taythes Gill [SD 712 951, Hicks, 1957] and Swinner Gill [NY 111 009]. One of the most complete sections is in Gunnerside G1 Borehole (Figure 8, section 3) between the depths of 53 ft 4 in and 226 ft 6 in (16.26 and 69.05 m) as follows:

	Thickness ft	in	m
Underset Limestone	–	–	–
Seatearth sandstone, very fine-grained	2	9	0.84
Sandstone, fine-grained; mostly well-laminated; current-ripple and trough cross-laminated; flaser bedding; a few mudstone partings	24	4	7.42
Seatearth sandstone, fine-grained	1	9	0.54
Sandstone, fine-grained; cross- laminated, trough cross-laminated; flaser bedding	15	3	4.65
Mudstone, silty and siltstone; bioturbation near base	13	1	3.99
Sandstone, fine-grained		4	0.11
Coal, shaly		1	0.02
Sandstone, fine-grained; laminated, cross-laminated and cross-bedded	22	4	6.81
Sandstone, fine-grained; cross-bedded near top	10	3	3.12
Sandstone, fine-grained; laminated in part; plant debris	20	1	6.13
Sandstone, fine-grained. thickly bedded	11	10	3.60
Sandstone, fine-grained, laminated; rich in plant debris	12	4	3.75
Mudstone, with some bands of siltstone in upper half; rare spiriferoids, *Pernopecten concentricum*, *Euphemites sp.*	28	0	8.52
Mudstone, with siderite bands; *Martinia sp.*, *Spirifer sp.*	2	7	0.79
Three Yard Limestone			
Limestone, biosparite, chiefly medium- to coarse-grained	4	2	1.26
Sandstone, fine-grained, calcareous; small shells		4	0.10
Limestone, fine-grained biosparite; probable flattened algal nodules, *Buxtonia sp.*	2	4	0.71
Limestone, muddy and mudstone; indet. solitary coral, *Avonia*, *Edmondia sp.*, ostracods	4	2	1.27

The sandstone below the Underset Limestone, called the Twentyseven Fathom Grit, figures in the records of the nearby Lownathwaite Mine and extends eastwards into Arkengarthdale as a very thick formation, much obscured by huge landslips on Fremington Edge, Arkengarthdale and at Winterings above Gunnerside. The record from the Hurst Mine (Figure 8, section 5) shows two thick sandstones separated by a 4-in (0.10-m) coal. In Waitgate Gill [NZ 082 050] thick sandstones are again developed, also with a coal seam (Figure 8, section 4). The lowest sandstone has an erosive base on the Three Yard Limestone (Wells, 1957). The upper sandstone is 90 ft (27.43 m) thick in the Gilling Dome at Aske Moor [NZ 155 040] (Wells, *op. cit.*). In Middleton Tyas No. 1 Borehole [NZ 2371 0603] 139 ft (42.20 m) of fine-grained sandstones underlie the Underset Limestone. Thin mudstone partings occur, two of them with a marine fauna. Seatearth sandstones occur at eight levels (Figure 8, section 7). The underlying mudstone has some marine fossils, especially just above the Three Yard Limestone.

South of the River Ure to the southern limit of sandstones there is usually a coarsening upwards cycle, but a seatearth seldom occurs. Well bedded sandstones are common and have been worked for roofing tiles at High Pike, Deepdale [SD 719 826]. Southwards the sandstones are split by many mudstone partings, as in Hazle Bank Gill, Wharfedale [SD 8680 8288]. On Ingleborough the sandstones have virtually died out and mudstones at this level have yielded *Posidonia corrugata* [SD 7583 7661]. On Fountains Fell also, thick mudstones occur below the Underset Limestone, but the Three Yard Limestone is absent (E. W. Johnson, personal communication, Figure 8, sections 13, 17). On the south side of Pen-y-ghent both the Underset and Three Yard limestones seem to be absent and sandstones overlain by mudstones are apparently at the level of these beds. Eastwards in Wharfedale, south-east of Kettlewell, and in Nidderdale, only mudstones, some 6 to 14 ft (1.83–4.27 m) thick occur between the Three Yard and probable Underset limestones. The basal 4 ft 4 in (1.45 m) of dark grey mudstones with rare fish fragments and bryozoa were encountered in boreholes on the Providence Vein, Greenhow, but are cut out in nearby borings to the south and east by unconformity at the base of the Grassington Grit Formation.

Around Buckden Pike and on Penhill, platy cherts almost directly underlie the Underset Limestone. Some 2 to 4 ft (0.60–1.20 m) in thickness, they contain a distinctive fauna of fenestellid bryozoans, small productoids and bivalves. In Buckden Beck [SD 9541 7806], locally at least 7 ft (2.10 m) of ochreous sandstone directly overlies these cherts. In places thin limestones occur in the chert (Moore, 1958) or a little beneath it.

At four locations limestones were seen some 10 to 20 ft (3–6 m) beneath the Underset Limestone. The most notable is at High Pike [SD 7147 8243] where quarries for tile stone reveal up to 3 ft (0.90 m) of biosparite, rich in *Gigantoproductus*, 20 ft (6.10 m) beneath the Underset Limestone (Figure 8, section 11). A limestone was also seen in Force Gill, Whernside (Figure 8, section 12) and on Pen-y-ghent. In Reigill [NY 8088 1691] a 1-ft thick (0.30-m) ferruginous siltstone with spiriferoids was seen 20 ft (6.10 m) below the Underset Limestone. Six inches (0.15 m) of fine-grained seatearth sandstone, also with abundant spiriferoids, was noted in Rake Beck [NZ 0802 0547] some 10 ft (3 m) below the Underset Limestone (Figure 8, section 4).

Underset Limestone This formation, named by Phillips (1836), is usually about 30 ft (9 m) thick but is prone to abrupt local thickening. Its distribution is patchy in the south of the area, but elsewhere it is continuously developed. Commonly the limestone forms a good feature, but usually a less strong one than the overlying Main Limestone. Examples are Satron Low Walls in Swaledale and Low Dove Scar, Penhill.

Typically the Underset Limestone is a grey biosparite with crinoid debris abundant especially towards the top; bryozoa are also an important faunal component (Moore, 1958). A coral biostrome extends southwards from Middleton in Teesdale to the latitude of Dentdale (Turner, 1956, pl. 34). This is a single band rich in simple corals, notably *Dibunophyllum bipartitum subspp.* and, in the south, *Diphyphyllum*. A similar, higher band occurs locally. Above the persistent lower coral bed, Moore (*op. cit.*) notes the presence of lenticular bioherms of crinoidal limestone, some with cores of dolomitised cavernous limestone. Hicks (1957) notes similar crinoidal thickenings, up to 60 ft (18 m) thick, at Great Wold on Whernside [SD 743 847], in Barbondale, Garsdale and on Great Coum. Farther east, Blue Scar, Wensleydale [SD 9855 9107] shows at least 51 ft (15.50 m) of richly crinoidal limestone (grainstone) which is almost devoid of bedding in the biohermal core.

In the Middleton Tyas area the Underset Limestone, which carries much of the copper mineralisation, is exceptionally thick, attaining 80 ft 3 in (24.47 m) in Middleton Tyas No. 1 Borehole [NZ 2371 0603]. It consists dominantly of light grey crinoid debris grainstones with several layers rich in big crinoid stems. *Gigantoproductus* occurs at several levels (Figure 8, section 7).

Like the Three Yard Limestone, the Underset thins southwards in the south of the area, notably on Whernside (Hicks, 1957). An attenuated 'gingerbread' limestone on Park Fell, Ingleborough (Hicks, 1959), is possibly the Underset (Figure 8, section 13). Though the Underset Limestone is present on Fountains Fell, no trace has been seen of it on Pen-y-ghent. In Wharfedale and Littondale it is present largely on the higher slopes of the valleys. The limestone seems to have been removed by channelling above Litton (Dunham and others, 1953) where, on the slopes of the Crystal Beck valley [SD 910 760], there is a persistent feature in some 60 ft (18 m) of ill-exposed coarse- and medium-grained sandstone midway between the Three Yard and Main limestones and at the level where Underset Limestone should occur. On the east side of Wharfedale at Kettlewell, and again at Angram Reservoir in upper Nidderdale, the Underset Limestone is apparently chiefly represented by cherts which include a blocky, very pale grey, china-like lithology.

Underset Chert This is chiefly seen in Swaledale where its thickness of some 30 ft (9 m) commonly equals that of the underlying Underset Limestone. Typically this is a blocky, very dark grey or greyish black, fine-grained chert. Some bands are almost black, glassy and with a cuboidal fracture and were formerly worked for buhrstones along Fremington Edge, east of Reeth. Rarely, the cherts are pale grey in colour, as for instance near Kexwith [NZ 0648 0401]. Fossils comprise rare *Zoophycos*, with, locally, a variety of invertebrate and fish remains. On Widdale Fell a distinctive highly fossiliferous calcite mudstone directly overlies the attenuated Underset Chert and is likely to be a lateral equivalent of its highest beds. The section in Cote Gill, Garsdale, [SD 7880 9037] is as follows:

	Thickness ft	in	m
Mudstone, shaly; abundant *Emmonsia parasitica*, *Fenestella* cf. *frutex*, *Martinia glabra*, *Pernopecten concentricum*, *Streblochondria anisota*, ostracods	6		1.83
Gap	1	4	0.40
Calcite mudstone, grey, muddy, fissile except near top; abundant sponge spicules, *Lingula mytilloides*, *Pleuropugnoides pleurodon*, *Productus productus* and *Septimyalina redesdalensis*; less			

Table 4 The Underset Cyclothem

Area	Mine or section	National Grid reference	Underset Limestone ft in	m	Underset Chert ft in	m	Sandstone under Main Limestone ft in	m	Underset Cyclothem ft in	m
10	Hazelgill	SD 7793 9979	20 0	(6.10)	21 0	(6.40)	37 9	(11.50)	108 3	(33.00)
11	Great Sleddale Beck	SD 8387 9926	29 6	(9.00)	—		86 4	(26.30)	115 10	(35.30)
	East Gill	NY 8971 0152	34 0	(10.35)	30 10	(9.40)	38 9	(11.80)	119 5	(36.40)
	Swinner Gill	NY 9105 0081	27 7	(8.40)	33 8	(10.25)	28 8	(8.75)	123 8	(37.70)
	North Hush, Lownathwaite	NY 9369 0140	27 3	(8.30)	30 6	(9.30)	27 7	(8.40)	126 8	(38.60)
	Lownathwaite Mine	NY 937 015	25 5	(7.75)	20 6	(6.25)	26 3	(8.00)	100 1	(30.50)
	Gunnerside Gill	NY 9378 0201	20 4	(6.20)	32 0	(9.75)	39 4	(12.00)	111 6	(34.00)
	Gunnerside G1 Borehole	NY 9385 0125	25 3	(7.70)	23 7+	(7.20+)	*not drilled*		*not drilled*	
	Friarfold Hush	NY 9406 0136	29 6	(9.00)	31 6	(9.60)	19 0	(5.80)	119 5	(36.40)
	Old Gang Mine, Bell's Shaft, Brandy Bottle Incline, Pedley's Cross Cut	NY 960 020	25 5	(7.75)	22 4	(6.80)	24 11	(7.60)	110 5	(30.60)
	Little Punchard Gill	NY 9661 0439	22 0	(6.70)	33 2	(10.10)	26 3	(8.00)	96 2	(29.30)
	Surrender Mine	NY 9796 0257	23 8	(7.20)	18 9	(5.70)	29 6	(9.00)	86 7	(26.40)
	Wetshaw Fourth Whim	NY 9803 0243	19 6	(5.94)	36 0	(10.97)	40 6	(12.34)	111 0	(33.83)
	Blackside Vein at Danby under Level	NY 9867 0255	18 10	(5.75)	36 1	(11.00)	21 0	(6.40)	105 0	(32.00)
	Hurst Mines	NZ 046 024	18 0	(5.49)	24 0	(7.32)	16 0	(4.88)	96 0	(29.26)
	Waitgate Gill	NZ 0859 0420	26 3	(8.00)	30 6	(9.30)	12 9	(3.90)	139 1	(42.40)
12	Chantry Borehole	NZ 2468 0706	80 3	(24.47)	—		26 3	(8.00)	120 5	(36.70)
13	Grainy Gill	SD 8726 9719	40 8	(12.40)	1 8	(0.50)	5 7	(1.70)	76 9	(23.40)
	Greenseat Beck	SD 8980 9658	55 9	(17.00)	—		19 0	(5.80)	104 4	(31.80)
	Summer Lodge Beck	SD 9640 9532	31 6	(9.60)	8 3	(2.50)	1 8	(0.50)	49 2	(15.00)
	Preston Moor Shafts (composite)	SE 063 929	36 1	(11.00)	22 0	(6.70)	—		76 9	(23.40)
	Keld Heads Mine and Barney Beck	SE 075 919 SE 0510 9245	15 5	(4.70)	10 10	(3.30)	24 0	(7.32)	102 0	(31.09)
14	Coal Gill, Whernside	SD 7508 8240	8 6	(2.60)	—		—		121 9	(37.10)
	Cote Gill, Widdale Fell	SD 7876 9038	15 5	(4.70)	3 3	(1.00)	26 3	(8.00)	103 4	(31.50)
	Fountains Fell	SD 8648 7229	32 10	(10.00)	13 2	(4.00)	—		65 7	(20.00)

	Thickness		
	ft	in	m
common fossils include *Bellerophon sp.*, *Retispira sp.*, *Aviculopecten knockonniensis*, *Lithophaga lingualis*, *Palaeoneilo laevirostrum*	4	0	1.22
Gap (with mudstone fragments)	1	8	0.50
UNDERSET CHERT			
Gap (with blocks of brown-weathering chert)	3	6	1.07
UNDERSET LIMESTONE			
Limestone, grey, medium-grained biosparite	5	6	1.68
Limestone, medium-grained, dark grey biosparite	8	6	2.59

In the northern fringes of the present area the Underset Chert is separated from the underlying limestone by up to 37 ft 9 in (11.51 m) of fossiliferous mudstones. In the Mount Pleasant Borehole these yielded an important P_{2c} goniatite fauna (Figure 3, horizon 15).

Beds above the Underset Chert Strata in this division range in thickness from 10 to 237 ft (3.0–72.20 m) and do not follow regional trends established in lower cyclothems. The thickest beds are on Ingleborough, but the exact figure is dependent on the accurate recognition of the supposed Underset Limestone (Figure 8, section 13). Thick medium- and coarse-grained, commonly cross-bedded, feldspathic sandstones dominate the sequence. Smaller thicknesses of lithologically similar sandstone are seen in sections on Widdale Fell and Whernside (Figure 8, sections 10, 12) and Moore traces this broad belt of sandstones northwards into two major channels on Shunner Fell (Figure 10, map J). Great Sleddale stands at the northern end of the branching channels where the sequence between the Underset and Main limestones consists of 86 ft (26.20 m) of fine- and medium-grained superbly cross-bedded sandstone. Above the cross-bedded sandstone, a worked coal occurs at several scattered locations (Moore, 1958, p. 120). These are Coal Sike, Cotterdale, 1 ft 6 in (0.45 m) [SD 836 944], Coal Sike, Whernside, 10 in (0.25 m) [SD 7490 8235] and on Widdale Fell, 8½ in (0.21 m) [SD 784 888], Crag Fell and in the Rawthey valley. Coarse-grained sandstone with an erosive base has also been noted in Littondale (Dunham and others, 1953).

In the eastern part of the area, the beds above the Underset Limestone are a more conventional cycle, coarsening upwards, and averaging some 60 ft (18 m) in thickness. The Main Limestone is usually underlain by some 13 to 40 ft (4–12 m) of sandstone, the equivalent of the Tuft of the Alston Block. Current-ripple lamination and trace fossils are well seen in Swinner Gill and Gunnerside Gill in the North Swaledale Mineral Field. Underlying strata are mudstones with rare marine fossils which are locally abundant in the basal beds, notably in Cote Gill, Widdale Fell (p. 49) and elsewhere in Garsdale (Hicks, 1957).

In the country between Reeth and Hawes, the beds of the Underset cyclothem are usually thin, with a minimum of 10 ft (3 m) in Summer Lodge Beck (Table 4). They are still only 30 ft thick (9.14 m) at Browna Gill [SE 010 971] near Harker Mine. Similarly in the east in Lower Coverdale and Chantry Borehole values from 11 to 40 ft (3.35–12.19 m) are recorded (Figure 10, map J).

Over the greater part of the Alston Block, a marine horizon, the Iron Post Limestone, locally represented by a fossiliferous shelly mudstone, occurs about midway between the Underset and Main limestones (Dunham, 1948, p. 23). This band is also known in the Brough area (Burgess and Holliday, 1979) and in the Cotherstone Syncline (Reading, 1957). Near Hazelgill, Mallerstang, a section beside an underpass to the railway line [SD 7785 9959] shows limestone and cherty limestone 15 ft 6 in (4.70 m) in thickness which is apparently midway between the Underset and Main limestones; it rests on at least 5 ft (1.50 m) of tough fine-grained sandstone and is clearly separate from the Underset Chert (Figure 8, section 2). Dakyns and others (1891) note a similar occurrence in Grisedale, and state that on Garsdale Common this same bed is a little below a pebbly feldspathic sandstone. It appears therefore that this thin limestone predates the feldspathic channel sandstones which are such a major feature of the western half of the present area. Farther south, limestones are generally absent except for a local thin band in Melbecks Gill, Coverdale [SE 0616 8497] and on Pen-y-ghent [SD 8422 7340].

CHAPTER 5
Country and cover rocks: Namurian and Westphalian stratigraphy

Except in the extreme south-eastern part of the region, at Grassington and along Ashfold Side Beck, mineral deposits are found only in beds lying beneath the intra-Pendleian unconformity. In Wharfedale, the Main Limestone, the lowest bed of the Pendleian Stage of the Namurian, appears beneath this unconformity, and additional members of the early Pendleian succession progressively appear towards the north.

On the northern side of Swaledale the full succession including the Stonesdale Limestones and the Mirk Fell Ganister is present, and much of the early Pendleian constitutes potential mineral-bearing ground. The thick shales in the Namurian above the unconformity appear to have prevented mineralising fluids from reaching the sandstones and grits, and this part of the succession may be regarded, except at Grassington and Ashfold Side, as providing cover rocks to the mineralisation.

EARLY PENDLEIAN

Main Limestone This formation, named by Phillips (1836), was also known to the miners of Swaledale as the Twelve Fathom Limestone because of its fairly constant thickness in the mineral field. Typically the limestone is some 70 to 80 ft (21.30–24 m) thick, attaining a local maximum on Kisdon, Swaledale of 130 ft (40 m) (Rowell and Scanlon, 1957a). It usually forms a feature with a line of scars, as on Satron High Walls, Swaledale and High Clint, Wensleydale. In Swaledale, in particular, picturesque ravines like Gunnerside Gill and Swinner Gill are in this limestone. Several old hushes have steep cliffs in this formation, the most notable being the Stodart and North Rake Hushes in Arkengarthdale and Friarfold Hush in Swaledale (Plates 5, 6, 7).

The Main Limestone is typically a medium grey or medium dark grey, medium-grained biosparite, with corals or gigantoproductoids concentrated in one or two bands, excepting the uppermost 30 ft (9 m) which is largely unfossiliferous. In the south-eastern part of the Askrigg Block in particular, as for instance on Pen-y-ghent, faunal bands are rare. At a number of localities in Swaledale some 3 to 10 ft (0.90–3.00 m) of beds with gigantoproductoids occur near the middle of the Main Limestone, notably in their thickest development of Greenseat Beck [SD 8975 9643]. A widespread coral biostrome rich in *Chaetetes* occurs on the Alston Block (Johnson, 1958) and throughout most of Mallerstang in the present area (Turner, 1962).

In the middle part of the Main Limestone, crinoid banks occur locally, with a great abundance of crinoid stems. The best exposure is at High Clint, Stags Fell [SD 880 922] where several crinoid banks up to 20 ft (6 m) thick are overlain by limestones rich in gigantoproductoids at several levels (Moore, 1958). In scars above Parpin Gill Moss, Sleddale [SD 8521 8437] crinoid banks extend through a vertical height of 27 ft (8.20 m). Hughes (1909) and Hicks (1959) record a highly crinoidal facies of the Main Limestone near the summit of Ingleborough and at Greensett Craggs, Whernside [SD 748 817].

Bioturbation is a feature of the upper beds of the Main Limestone in parts of Swaledale. This is particularly so in Little Punchard Gill [NY 9597 0364] where three bands of vertical borings and additional horizons of vermiform bioturbation occur in the uppermost 14 ft (4.27 m) of the limestone.

A limited area around Swarth Fell (Rowell and Scanlon, *op. cit.*) shows two leaves of Main Limestone. An upper leaf 5 ft (1.50 m) thick is separated from the main body of the limestone by up to 19 ft (5.80 m) of mudstone. Oxnop No. 2 Borehole [SD 9311 9470] shows 52 ft (15.85 m) of fine- and medium-grained sparite with a coral band in the middle (Figure 11, section 2).

The Main Limestone shows marked thinning in the west where thicknesses under 25 ft (7.60 m) are common (Rowell and Scanlon, *op. cit*; Hicks, 1957) and in the east in Lower Coverdale it is absent, locally, at Swineside [SE 0626 8437]. It is tempting to ascribe this thinning to erosion at the base of the Little Limestone, as suggested by Wells (1957) for the Leyburn area, but, near Coverham, the Main Chert is present above only 16 ft (4.88 m) of Main Limestone, suggesting that much of the thinning is depositional. A borehole at Chantry [NZ 2469 0705] near Middleton Tyas showed only 29 ft 2 in (8.88 m) of medium grey, fine-grained sparite, with several bioturbated horizons near the top and base of the Main Limestone.

Main Chert This formation was first named by Swaledale miners and appears on old cross sections and the logs of the mine shafts, including Fourth Whim, Surrender Mine, Lownathwaite Mines, Blackside Vein, and Prys Mine workings (Table 5). It directly overlies the Main Limestone and, throughout the Swaledale mining field, is a more or less well bedded dark grey, blocky or platy chert, with some very fissile bands. Wells (1955) points out that the upper beds of the chert are locally flinty near Bowes and the lower beds in places become a cherty limestone with some 20 per cent silica.

Throughout the North Swaledale Mineral Belt the chert averages some 15 ft (4.60 m) in thickness, with local maxima of 40 ft (12.20 m) in West Stonesdale and of 30 ft (9.14 m) in Moresdale Gill (Wells, *op. cit.*, pl. 19). A prolongation of the area with cherts extends north-westwards towards Bowes and could mark a source of silica supply. South of the River Ure up to 9 ft (2.70 m) of the Main Chert have been recorded at three isolated outcrops.

In the North Swaledale Mineral Belt the Main Chert is spectacularly exposed above the Main Limestone in many hushes, the best of which are probably Friarfold [NY 9417 0135], Stodart [NY 983 030]. and Fell End [NZ 021 024] (Plates 5, 7, 8; Figure 11). These three sections, and many more (Figure 11), are remarkable for the abundance of *Latiproductus latissimus* group, previously noted at this level by Rowell and Scanlon (1957a) on Winton Fell. Slabs covered with specimens showing great variation in form were collected in Little Punchard Gill [NY 9592 0355] from which the following list was compiled: solitary coral, *Fenestella* cf. *parallela*, *Antiquatonia sulcata*, *Brachythyris sp.*, *Buxtonia sp.* [common], *Dictyoclostus?*, *Dielasma sp.*, *Eomarginifera longispina* [common], *Latiproductus latissimus* group [abundant], *Orbiculoidea sp.*, *Pleuropugnoides sp.*, *Productus carbonarius*, *Pustula?*, *Rugosochonetes celticus* group [large], *Spirifer trigonalis*, *Palaeostylus (Stephanozygus)? subplanatus*, *Aviculopecten interstitialis*, *Conocardium* cf. *rostratum*, *Edmondia expansa*, *Limipecten dissimilis*, *Parallelodon sp.*, *Pernopecten sp.*, *Poterioceras sp.*, *Coleolus sp.*

A locality at Sun Hush [NZ 0523 0219], Hurst shows 8 ft (2.44 m) of blocky and thinly bedded Main Chert (not the full thickness) with an apron of debris. Besides abundant *Latiproductus latissimus* there are no less than 22 bivalve species present, see Pattison and Wilson (in preparation).

Table 5 The Main Cyclothem

Mine or section	National Grid reference	Main Limestone ft in	m	Main Chert ft in	m	Main Cyclothem ft in	m
Area 10							
Deep Gill	NY 7778 0031	80 0	(24.38)	—		130 0	(39.62)
Area 11							
Great Sleddale Beck	NY 8320 9910	94 0	(28.65)	—		106 0	(32.31)
Starting Gill Shaft	NY 8856 0362	72 0	(21.95)	28 0	(8.53)	100 0	(30.48)
East Gill	NY 8975 0202	88 7	(27.00)	5 3	(1.60)	109 11	(33.50)
Swinner Gill	NY 9118 0118	79 5	(24.20)	12 6	(3.80)	107 3	(32.74)
Lownathwaite Mine	NY 937 015	72 2	(22.00)	17 9	(5.40)	91 10	(28.00)
Blakethwaite Mine	NY 940 030	72 2	(22.00)	18 4	(5.60)	96 6	(29.40)
Friarfold Hush	NY 9420 0142	69 0	(21.00)	14 5	(4.40)	96 2	(29.30)
Little Punchard Gill	NY 9597 0364	64 0	(19.50)	20 0	(6.10)	83 8+	(25.50+)
Brandy Bottle Incline, Pedley's Cross Cut	NY 960 020	79 7	(24.25)	16 9	(5.10)	104 6	(31.85)
Surrender Mine	NY 9796 0257	71 4	(21.75)	15 1	(4.60)	89 7	(27.30)
Wetshaw Fourth Whim	NY 9803 0243	72 0	(21.95)	18 0	(5.49)	93 0	(28.35)
Stodart Hush	NY 9823 0300	64+	(19.50+)	17 5	(5.30)	90 3+	(27.50+)
Blackside Vein at Danby Under Level	NY 9867 0255	72 0	(21.95)	18 0	(5.49)	95 0	(28.96)
Slei (Furn) Gill	NZ 0216 0323	60 4	(18.40)	20 4	(6.20)	90 3	(27.50)
Hurst Mines	NZ 046 024	72 0	(21.95)	18 0	(5.49)	95 6	(29.11)
Area 12							
Church Gill	NZ 1429 0055	46 11	(14.30)	12 2	(3.70)	59 1	(18.00)
Chantry Borehole	NZ 2467 0705	29 2	(8.88)	—		47 7	(14.50)
Area 13							
Fossdale Gill	SD 8638 9551	68 0	(20.73)	5 11	(1.80)	81 9	(24.93)
Grainy Gill	SD 8706 9708	75 6	(23.00)	2 0	(0.60)	91 3	(27.80)
Greenseat Beck	SD 8975 9643	76 2	(23.20)	4 11	(1.50)	84 0	(25.60)
Oxnop No. 2 Borehole	SD 9311 9470	51 6	(15.70)	8 10	(2.70)	60 4	(18.40)
Summer Lodge Beck	SD 9643 9516	66 3	(20.20)	7 5	(2.25)	76 9	(23.40)
Browna Gill	SE 0098 9702	59 1	(18.00)	7 3	(2.20)	70 10	(21.60)
Redmire Quarry	SE 0470 9300	75 6	(23.00)	4 7	(1.40)	80 1	(24.40)
Preston Moor shafts (composite)	SE 063 929	73 2	(22.30)	—		80 8	(24.60)
Keld Heads Mine	SE 075 919	63 0	(19.20)	—		63 6	(19.35)
Blea Grin Gill	SD 7862 8771	76 5	(23.30)	—		110 11	(32.80)
Pen-y-ghent	SD 8428 7346	96 2	(29.30)	—		96 2?	(29.30?)
Fountains Fell	SD 8744 7124	31 2	(9.50)	—			

Wells (1955) believed the Main Chert to be largely a primary deposit, a co-precipitate of calcite and cryptocrystalline silica. Only the upper part, rich in chalcedony, in the Bowes area, was thought by him (*op. cit.*, p. 191) to be secondary in origin.

Beds between the Main Chert and Little Limestone South of Leyburn, the Little and Main limestones are in juxtaposition, but as these formations are traced to the north and north-west, intervening beds up to 150 ft (45.70 m) in thickness come in (Figure 11, section 3 and Figure 12, map A). These are mudstones with between one and three fine- to medium-grained sandstones, locally with cross-bedding or ripple marks, referred to as the Coal Sills, a term derived from the Alston area (Forster, 1809). Locally coal has been worked around Baugh Fell, the Rawthey valley and Grisdale. In the north-east, in Sleightholme Beck, a particularly good section, with two major sandstone members, has been studied by Elliott (1975) as part of a larger sedimentological study of the Coal Sills (Figure 11, section 3). The lower sandstone member, probably the High Coal Sill of the Alston area according to Burgess and Holliday (1979), is interpreted by Elliott as a prograding deltaic deposit in a coarsening-upwards sequence. The upper sandstone member (probably the White Hazle of the Alston area) is interpreted as a barrier island – lagoon complex on top of the abandoned deltaic lobe.

On the Alston Block, the Coal Sills, which there comprise both the High and the Low Coal Sill, include two major channels filled with thick sandstones (Dunham, 1948, p. 28). Work on the Askrigg Block by Rowell and Scanlon (1957a) and Hicks (1957) indicates the presence of very coarse-grained, commonly pebbly, sandstones in the lower sandstone in the Coal Sills around Baugh Fell and Swarth Fell (Figure 12, map A). These are perhaps channel fillings.

The Coal Sills Marine Band which is locally present above the Upper Coal Sill consists of a thin limestone overlying fossiliferous mudstone as, for instance, in Black Sike, Arkengarthdale (Figure 11, section 3). In the Mallerstang area the basal sandstone of the Coal Sills is locally calcareous with a fauna of brachiopods. Rowell and Scanlon (*op. cit.*) record *Brachythyris* cf. *aemula*, *Eomarginifera derbiensis* and other brachiopods, some preserved with attached spines.

In Black Sike (Figure 11, section 3) the White Hazle directly underlying the Little Limestone is remarkable for the abundance of annelid trails [NY 9952 0739].

Most of the North Swaledale – Arkengarthdale Mineral Field lies outside the limit of the Coal Sills. A thin mudstone (Figure 11, section 1) separates the Main Chert from the Little Limestone over much of this field. In Shaw Beck, Hurst, the upper part of this mudstone yielded an extensive fauna (p.55). In Chantry Borehole, Middleton Tyas, 18 ft 5 in (5.62 m) of mudstone at this level also are fossiliferous. In East Grain [NY 9120 0119] the section which includes the marginal remnant of the Coal Sills is as follows:

	Thickness		
	ft	in	m
Limestone, grey (LITTLE LIMESTONE)	4	0	1.22
Mudstone		8	0.20
Sandstone, fine-grained, calcareous (?White Hazle)	6 in to 1	8	0.15 to 0.50
Mudstone (presumed Coal Sills Marine Band); *Pleuropugnoides sp.*, *Productus* cf. *carbonarius* [common], *Euphemites sp.*, *Stegocoelia sp.*, *Worthenia* [common], *Aviculopinna mutica*, *Edmondia transversa*, *Streblochondria anisota*, *Cravenoceras?*	3	6	1.07
Sandstone, fine-grained, calcareous (?High Coal Sill)	3	4	1.02
Mudstone, shaly, ferruginous; basal 3 ft 3 in (1 m) fossiliferous with pyrite blebs; *Fenestella* cf. *frutex*, trepostome bryozoa, *Buxtonia* cf. *scabricula*, *Derbyia?* [common], *Eomarginifera longispina*, *Linoproductus sp.*, *Productus carbonarius* [common], *Spirifer bisulcatus* group [common], *Girtyspira sp.*, *Naticopsis variata*, *Palaeostylus* [*Stephanozygus*] *rugiferus*, *Retispira sp.*, *Limipecten dissimilis*, *Pernopecten sowerbii* [common], *Coleolus sp.*, Ostracods	7	6	2.30
MAIN CHERT with latissimoid productoids			

Sections on Ingleborough, Pen-y-ghent and Fountains Fell show mudstones with local sandstones between the probable basal bed of the Grassington Grit Formation—a coarse-grained feldspathic sandstone—and the top of the Main Limestone (Figure 13, section 10). In the absence of any marker bands it is not possible at present to correlate them with the Coal Sills.

Little Limestone This formation, also called Upper Little (also 'Lime Bed' or 'Thin Lime' on several Swaledale sections), was first named by the Alston miners. It is a medium grey, crinoidal biosparite averaging 4 ft (1.20 m) in thickness. Typically it shows a bored upper surface with infillings, some 1 in (3 cm) in width and 1 ft 4 in (40 cm) in depth, consisting of chert with scattered grains of glauconite (Wells, 1955). In places the borings make a double row spaced about 1 ft 8 in (50 cm) apart (Figure 11, section 3, Redmire Quarry). Despite local failure of this bored horizon the 'pipe bed' has been a most valuable marker band, especially in Wensleydale where the Little Limestone is commonly in direct contact with the lithologically similar Main Limestone. Locally in Lower Swaledale the Little Limestone is unusually thick, as at Church Gill and Ellerton Scar (Figure 11, section 2). At the latter locality, much of the overlying Richmond Cherts is crinoidal grainstone and the Little Limestone could form part or all of a continuous run of limestone 34 ft (10.36 m) thick.

To the west and south, the Little Limestone ceases to be a distinctive bed. On Widdale Fell [SD 7850 8641] the attenuated remnant of the combined Little Limestone and overlying Richmond Cherts is only 2 ft 6 in (0.75 m) of silty chert. Farther south Hicks (1957) reports 6 ft (1.80 m) of crinoidal sandstone at the north end of Great Coum [SD 7028 8391] and 1 ft (0.30 m) of black limestone with *Zoophycos* at the south end of the hill [SD 7029 8294], at the horizon where the Little Limestone is expected. On Whernside and Ingleborough the Litte Limestone has not been recorded and is presumed to have died out (Figure 12, map B).

Richmond Cherts The 'Richmond Chert Series' (now called Richmond Cherts) was first defined by Wells (1955) as the siliceous limestones and cherts overlying the Little Limestone. They consist of cherts, cherty limestones and limestones, with mudstone partings, attaining a thickness of about 110 ft (33.50 m) W of Richmond and 120 ft (36.6 m) 4 km SW of Leyburn (Figure 12, map B).

Wells (1955) contributed greatly to an understanding of regional variations in thickness and lithology of the Richmond Cherts. The present study has focussed attention on the hushes and adds detail in the North Swaledale – Arkengarthdale Mineral Belt, an area not much studied by Wells. It is in this ground that old sections of the Blakethwaite, Surrender, Hurst, Tanner Rake North Level and other mines, and sections quoted by Phillips (1836, p. 66) showed a broadly uniform stratigraphical usage within the Richmond Cherts exemplified by the section at Fourth Whim, Arkengarthdale (Dakyns and others, 1891, p. 116):

	Thickness	
	ft	m
Chert or Iron Bed, rusty siliceous mudstone	12	3.66
'Red Beds'	13	3.96
Shale	7	2.13
'Black Beds', generally black siliceous mudstones	14	4.27

Wells (1957) considered the most prevalent mining terms, the Red Beds and Black Beds, to be of little stratigraphical value but the present work confirms their usefulness over a wide area (area A, see below). The Iron Beds are less easy to recognise in the field, but this term was only used on a few of the old sections and they are here included in the Red Beds.

South of the River Swale, the Black Beds were only locally recognised and the whole of the Richmond Cherts was referred to in old mine records and Geological Survey six-inch maps as 'Red Beds' or alternatively 'Main Chert'. A separate 'black chert' is noted on some shaft records just above the Red Beds.

The area of the Richmond Cherts outcrop has been divided into six parts (Figure 11) as follows:

AREA A The Richmond Cherts here exhibit an orderly stratigraphy in three main members in upwards sequence as follows:

1 Black Beds These are medium dark grey, silty, platy cherts up to 36 ft (11 m) in thickness with few macrofossils. Typically the beds split to reveal an abundance of arcuate trace fossils, with a sheet-like cross-section ornamented by chevrons corresponding to arcs on the bedding planes. Formerly called 'caudi-galli' these markings are referred to *Zoophycos*, now generally thought to be the burrow systems of deposit-feeding infaunal animals (Ekdale, 1977, p. 175).

To the east and south, the Black Beds lose their characteristic lithology and are indistinguishable from the rest of the Richmond Cherts. Westwards in Upper Swaledale, above Keld, the whole of the Richmond Cherts resembles the Black Beds in lithology, and is apparently their lateral equivalent (Figure 11, section 1). A mine record at Starting Gill Shaft [NY 8856 0362], West Stonesdale, separates Red Beds and Black Beds, but field work in nearby East Gill does not confirm the presence of the upper, Red Beds, member.

2 Mudstone member This consists of up to 13 ft (4 m) of ill-exposed shaly mudstones, locally with beds of fissile chert (Figure 11, section

1). These beds tend to form clayey slopes separating lines of cliffs in the Black Beds and Red Beds.

3 Red Beds Typically these are blocky cherts with abundant crinoid debris, often weathering with a highly irregular surface due to patchy distribution of silica. Some bands are 'fampy' and weather to a ferruginous reddish brown rottenstone, hence the term 'Red Beds'. The Red Beds average some 21 ft (6.40 m) in thickness around Gunnerside Gill and thicken to up to 58 ft (17.6 m) on the east side of Arkengarthdale where they are split by a mudstone parting.

The present work shows that the above development of the Richmond Cherts extends to substantial parts of the Swale – Ure watershed (Figure 11, section 2). Indeed, the lithological differences which set apart the Red and Black Beds are nowhere better seen than at Harker Mine [SE 0155 9723] and Browna Gill [SE 0103 9696]. East of Harker Mine, however, the Black Beds cease to be a distinct entity and probably pass laterally into a litho- and biofacies indistinguishable from Red Beds.

In Hurr Gill and Black Sike, Arkengarthdale (Figure 11, section 3), two blocky chert horizons lie at higher levels than Richmond Cherts elsewhere. Only the lower and thicker horizon is here included in the Richmond Cherts since the thinner one is separated from the main body of cherts by thick mudstones. Both these high chert horizons resemble the Black Beds in the presence of *Zoophycos*, but small brachiopods are more common.

Mr Pattison notes that the fauna of the 'Black Beds' is limited to small brachiopods including *Echinoconchus elegans*, *Eomarginifera longispina* and *Pleuropugnoides greenleightonensis*. It is noticeable that shelly fossils are rare or absent where *Zoophycos* (see above) is most abundant.

The fauna of the 'Red Beds' and its lateral equivalents is more varied and includes corals, bryozoa, molluscs and bradyodont fish teeth in addition to numerous brachiopods. The most common brachiopods are rhynchonelloids and the small productoids *Avonia youngiana*, *Buxtonia sp.* and *Eomarginifera spp.* as well as the form called *Semicostella sp. nov.* (Pattison and Wilson, in preparation). The Red Beds and their lateral equivalents are especially noteworthy for the abundance of sponge remains comprising body forms, isolated small spicules and the long anchoring basalia rods referred here and in most earlier literature to species of *Hyalostelia*, although Reid (1968, p. 1247) states that the genus should not be identified from basalia alone as similar rods can occur in other sponges. The sponge bodies seen include both amorphous masses with reticulate surface patterns and various, generally circular, shapes. None has been identified generically but etching reveals them to be mostly siliceous hexactins and monaxons, which, together with the apparent absence of triradiate spicules suggests that the sponges were originally siliceous forms.

Fasciculate colonies of *Diphyphyllum* have been noted, extending for several metres, in outcrops in the middle of the Red Beds at four localities in Arkengarthdale and on the Swale – Ure interfluve (Figure 11, sections 1, 3) and amongst the dump of an old adit on Crackpot Moor [SD 9573 9599]. Rare solitary corals also occur in the Red Beds.

Authors who have described collections of Richmond Chert fossils include Davis, 1883 (fish), de Koninck and Wood, 1858 (crinoids), Vine, 1881 (bryozoa) and Brady, 1876 (foraminifera).

AREA B The Red Beds appear to thin out on the west side of area A leaving only Black Beds. Thus exposures such as those in the headwaters of the River Swale (Figure 11, section 1) are exclusively in blocky grey cherts with *Zoophycos*, but no crinoid debris.

AREA C West of the Hurst Mines, where old records show Black Beds and Red Beds (Figure 11, section 1), the distinction between the two formations appears to break down, probably due to lateral passage of Black Beds eastwards into a Red Beds facies, largely lacking *Zoophycos*. Area C is distinctive dissected table-land country north of Marske and includes the spectacular Whitcliffe Scar in Richmond Cherts. The entire formation here consists of blocky cherts and cherty limestones with some bands of crinoidal grainstone.

The beds of area C are characterised by a fauna similar to that of the Red Beds. Surfaces rich in *Hyalostelia* anchoring rods are numerous, as, for instance, in Chantry Borehole and at Whitcliffe Scar (Figure 11, section 1). Crinoid debris is in many places abundant and from an old quarry 600 yd (549 m) NE of Helwith, de Koninck and Wood (1858) described new species of *Woodocrinus* preserved as complete calices. During the present investigation, nowhere was the fauna more abundant than in the extensive section in White Scar, Shaw Beck, Hurst [NZ 0695 0278 to 0636 0236] (see also Figure 11, section 1) as follows:

	Thickness		
	ft	in	m
Beds of LITTLE LIMESTONE cyclothem			
Mudstone, very cherty	1	8	0.50
Clay, probably weathered mudstone	2	4	0.70
Chert, platy; 'gingerbread' weathering	1	0	0.30
Mudstone, cherty	1	4	0.40
Gap (probably mudstone)	2	0	0.60
RICHMOND CHERTS			
Chert, very platy	4	0	1.22
Gap	1	7	0.50
Limestone, very light grey, cherty, biosparite, chert nodules; crinoid debris abundant	9	2	2.80
Chert, light grey, with irregular and porous weathering: *Hyalostelia parallella*, *H. smithii*, sponge spicules, sponge body, *Dibunophyllum sp.*, zaphrentoid, *Derbyia sp.* [common], *Dictyoclostus pinguis* [common], *Eomarginifera sp.* [common], *Productus* cf. *carbonarius*, *Spirifer* cf. *trigonalis*	11	6	3.50
Chert, light grey, flinty (Marske Chert Band)	1	2	0.30
Chert, light grey, with irregular and porous weathering	13	9	4.20
Limestone, light grey, slightly cherty, crinoidal biosparite	4	3	1.30
Chert, grey, blocky, locally with large scale cross-bedded units up to 16 ft 6 in (5 m) in thickness; sparse *Zoophycos*. Fauna collected in these beds and strata below Marske Chert Band, are: *Hyalostelia parallela* [common], *H. smithii*, sponge bodies, *Fenestella* cf. *plebeia*, *Polypora sp.*, *Rhombopora sp.*, *Alitaria panderi*, *Avonia youngiana* [common], *Brachythyris integricosta*, *Buxtonia sp.* [abundant], *Dictyoclostus pinguis* [common], *Echinoconchus elegans* [common], *E. punctatus*, *Eomarginifera lobata*, *E.* cf. *longispina*, *Georgethyris profecta*, *Krotovia spinulosa*, *Martinia sp.*, *Overtonia fimbriata*, *Pleuropugnoides greenleightonensis* [common], *Pliocochonetes crassistria*, *Productus* cf. *carbonarius*, *Rhipidomella sp.*, *Schizophoria sp.*, *Semicostella sp. nov.* [as form in Faraday House Shell Bed in Brough district], *Spirifer* cf. *trigonalis, Bellerophon sp.*, *Edmondia* cf. *senilis, Sanguinolites sp.*, *Catastroboceras sp.*, ostracods, *Petalodus* tooth	16	6	5.00
Chert, grey, very platy	2	0	0.60
Limestone, light grey, cherty, biosparite, crinoidal; *Actinoconchus sp.*, *Avonia youngiana*, *Dielasma sp.*, *Echinoconchus* cf. *elegans*, *Phricodothyris sp.*, *Pleuropugnoides greenleightonensis*, *Pugilis pugilis*, *Schizophoria sp.*, *Spirifer sp.*	7	6	2.30

	Thickness		
	ft	in	m
Chert, grey, silty, blocky, sparsely fossiliferous	6	6	2.00
Limestone, light grey, cherty, crinoidal, biosparite; *Avonia youngiana*, *Brachythyris* cf. *ovalis*, *Eomarginifera* cf. *longispina*, *Latiproductus latissimus* group [abundant], *Overtonia fimbriata*, *Schizophoria sp.*, cf. *Semicostella sp. nov.*	5	0	1.52
Chert, platy, with *Fenestella* cf. *frutex*, *Eomarginifera* cf. *longispina*, *Rugosochonetes celticus* group, *Schizophoria resupinata* [common], *Naticopsis* sp.	1	8	0.50
LITTLE LIMESTONE			
Limestone, medium grey, slightly cherty sparite; 'Pipe Bed' seen in section farther upstream [NZ 0606 0258]	4	0	1.22
Mudstones below RICHMOND CHERTS			
Mudstone, shaly; *Fenestella* cf. *parallela*, *Arthyris lamellosa*, *Brachythyris sp.* [common], *Eomarginifera longispina*, cf. *Semicostella sp. nov.* [? as form in Faraday House Shell Bed in Brough district], *Spirifer bisulcatus* group, *Naticopsis* aff. *planispira*, *Porcellia puzo*, *Aviculopecten sp.*, *Cardiomorpha sp.*, *Conocardium alaeforme* [common], *C.* aff. *inflatum*, *Cypricardella* aff. *annae*, *Edmondia* aff. *lyelli*, *Limipecten sowerbii*, *Catastroboceras* aff. *quadratum*, trilobite fragments, ostracods [common], bradyodont tooth	3	3	1.00
Gap	14		4.27
MAIN CHERT			
Chert, grey, platy; *Latiproductus latissimus* group [common]	5	0	1.52
Chert, grey, blocky	3	3	0.99
Chert, medium dark grey, platy, very fissile near top	6	6	1.98

The irregular cavernous weathering below the Marske Chert Band, and the big cross-bedded units indicative of high current activity during deposition are notable.

AREA D This comprises a boomerang-shaped area in which crinoid debris grainstones are interbedded with subordinate blocky cherts. The finest exposure, not previously noted, is in Ellerton Scar (Figure 11, section 2). The crinoid grainstone belt extends from Downholme Quarry [SE 1134 9815] through quarries on Leyburn Moor to the north-western slopes of Penhill (Wilson, 1960a).

Plate 3 Richmond Cherts in Tanner Rake Hush, Arkengarthdale

The observer is standing on the glassy Marske Chert Band, which caps the cliff and its apron of debris in the Red Beds

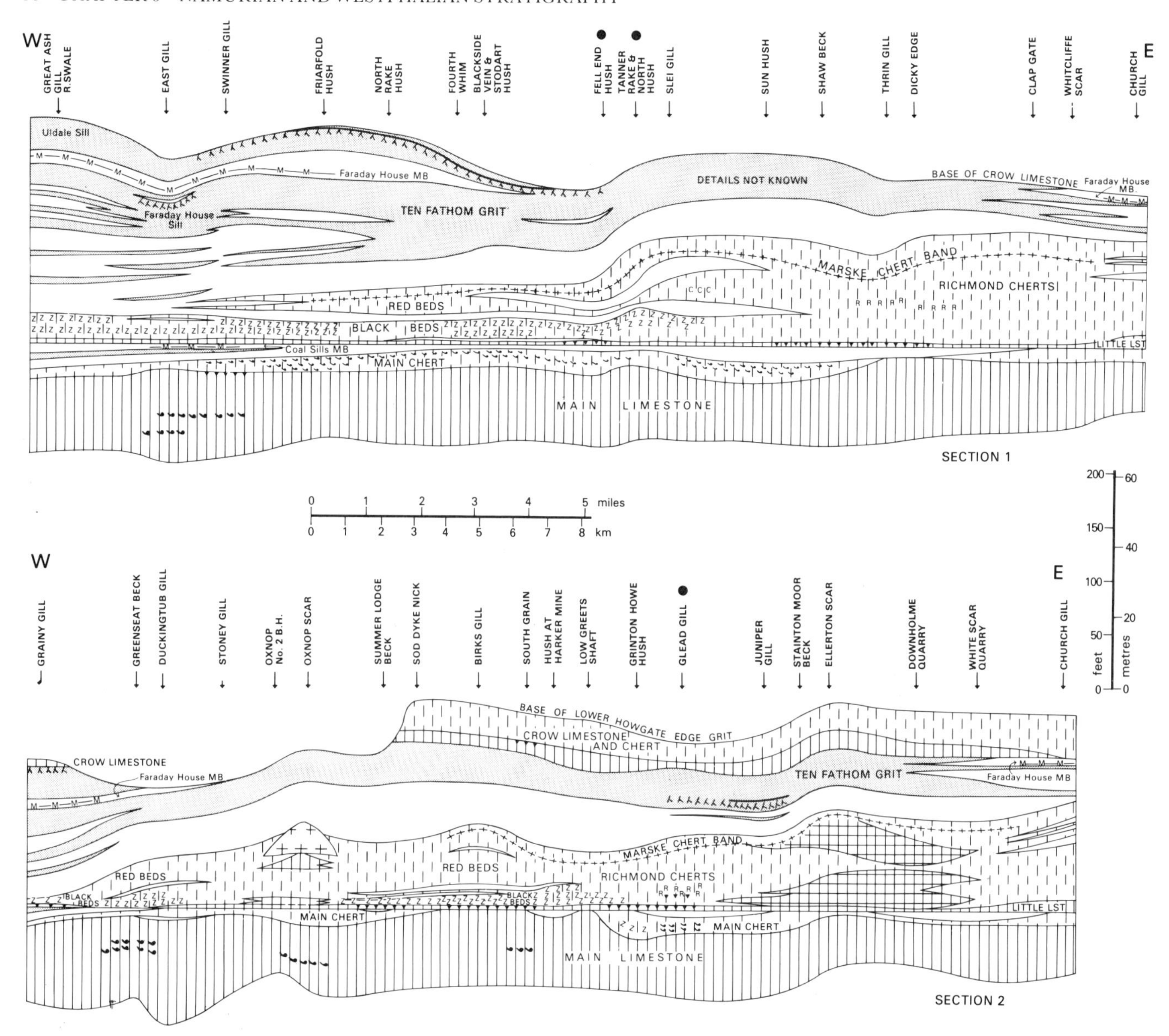

Figure 11 Map and sections to show the lithological variations in the Richmond Cherts and associated strata

AREA E A small area of the Swale – Ure watershed around Oxnop shows an anomalous development of the Richmond Cherts. The best section is in Oxnop No. 2 Borehole [SD 9311 9470, Figure 11, section 2], with a rapid alternation of limestones and cherts containing some *Hyalostelia*. Limestone beds equivalent to the highest strata in the borehole pass southwards into cherts on a prominent feature east of Oxnop Scar. At yet higher levels in the Richmond Cherts, three ovoid hills some 100 m in length are seen on Satron Moor [SD 9425 9597]. Slopes 13 to 40 ft (4–12 m) in height are almost lacking in exposure, but are strewn with chert debris rich in crinoid columnals, spiriferoids and orthotetoids. These are possibly original mounds of accumulation.

AREA F In the Leyburn area and Coverdale, the Richmond Cherts are at their thickest, but are planed off progressively to the south by pre-Grassington Grit erosion (Wilson, 1960a). These beds are dominantly blocky cherts, with local gentle cross-cutting relationships indicative of current activity during deposition, and, in places, with *Zoophycos*. Locally, lenses of crinoid debris grainstone occur, in places with marked depositional dips. On Penhill and west of Leyburn, the blocky cherts pass rapidly westwards into thick grainstones of area C. A possible equivalent of the Marske Chert Band has been recognised in some sections.

In areas A, C and D in Swaledale, a common feature of the upper part of the Richmond Cherts is a light grey flinty chert up to 4 ft (1.20 m) in thickness (Plate 3). This is the Marske Chert Band of Wells (1955). During the present work it has been found as far south as Glead Gill [SE 0533 9589] on Grinton Moor and westwards to North Hush [NY 9339 0139] and Friarfold Hush [NY 9427 0144], Gunnerside Gill. The limits of this band are shown on Figure 12, map B. Cherts and cherty limestones of the Red Beds occurring in close association with the Marske Chert Band are often rich in crinoid columnals. A second area of thin flinty chert in Coverdale is a possible correlative of the Marske Chert Band.

Beds between Richmond Cherts and Crow Limestone These beds are mudstones overlain by thick sandstones, the Ten Fathom Grit of the Swaledale miners. From a maximum of 218 ft (66.45 m)

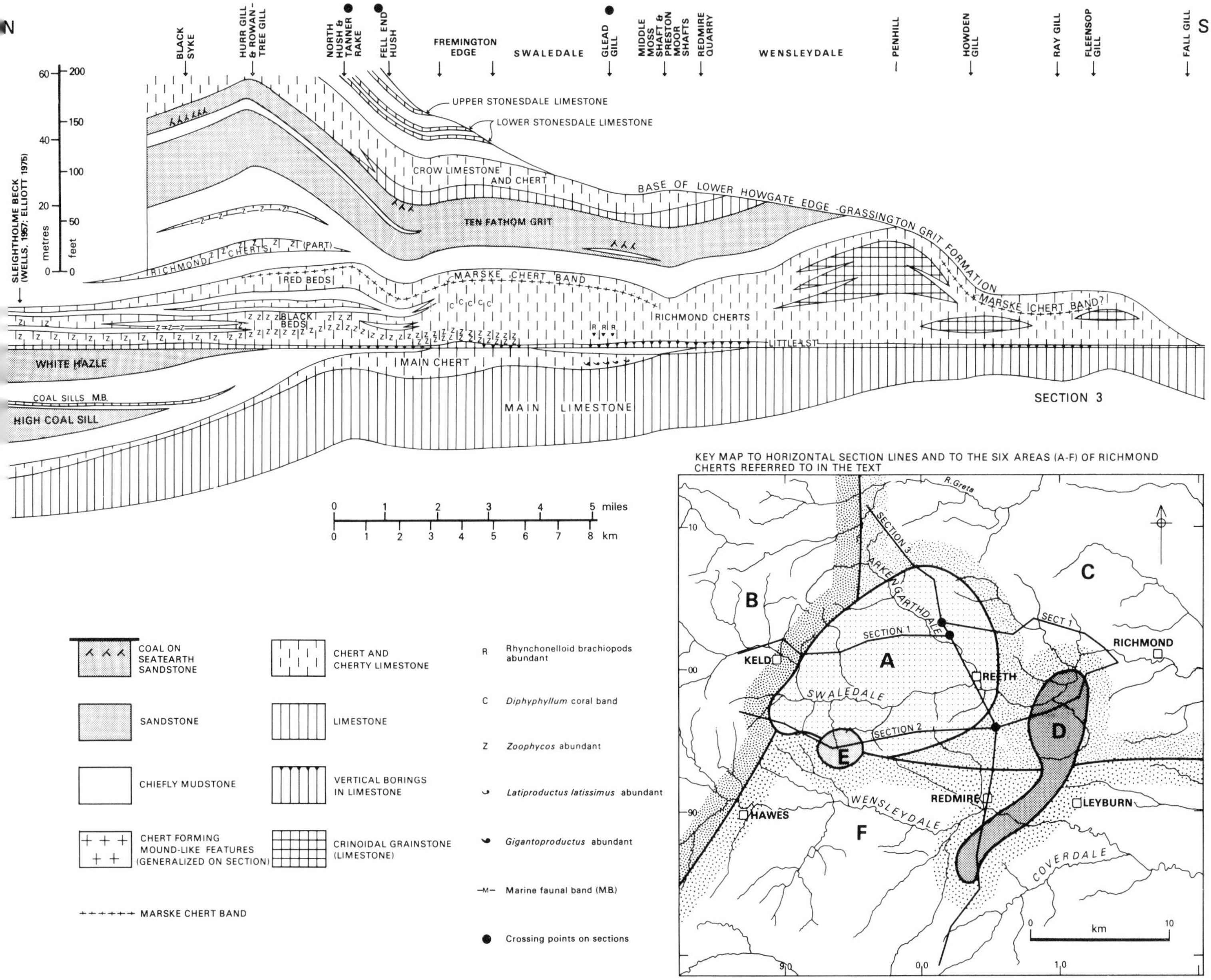

in Whitsun Dale these beds thin to the south-east (Figure 12, map C). South of Great Shunner Fell they are affected by intra-E_1 erosion and are therefore absent in much of the country south of the River Ure (Figure 13). Dakyns and others (1891) noted a marine band in a mudstone parting within the Ten Fathom Grit of Faraday Gill [NY 817 066], named the Faraday House Marine Band by Turner (1955). This mudstone splits the grit into two leaves in upper Swaledale, called the Faraday House Sill (Rowell and Scanlon, 1957a) and the Uldale Sill (Turner, 1955), both of which vary greatly in thickness (Rowell and Scanlon, *op. cit.*, figs. 5, 6).

Beds just above the Richmond Cherts consist of mudstones with siderite nodules. These are thickest around Garsdale (30 ft—9 m Hicks, 1957). Mr J. Pattison found *Modiolus sp.* 23 ft (7 m) above the top of the Richmond Cherts of Swinner Gill [NY 9131 0121]. These beds pass up into silty mudstones with thin bands of fine-grained sandstone. Locally in Blea Grin Gill, Widdale Fell [SD 7863 8768], a seatearth sandstone occurs just above the attenuated Little Limestone (Figure 13, section 9).

The Faraday House Sill is typically a fine-grained sandstone, usually thinly bedded with ripple-marked surfaces and annelid trails. In the Brough area a coarse-grained cross-bedded facies also occurs (Burgess and Holliday, 1979), probably indicative of channel-fill in a delta complex. The maximum thickness of 140 ft (42.67 m) was recorded by Rowell and Scanlon in Whitsun Dale [NY 866 026], with complete disappearance by lateral passage into mudstone around High Seat. The section in Great Ash Gill (Figure 11, section 1), with many sandstone bands in mudstone, illustrates a stage in this lateral passage. This formation has died out south of Reeth, where only a single sandstone underlies the Crow Limestone. Hicks (1957) has mapped the Faraday House Sill as far south as Garsdale where it is 30 ft (9 m) thick. On Widdale Fell the sandstone is only 3 ft (0.9 m) thick in Blea Grin Gill and is presumed to die out southwards.

The Faraday House Marine Band occur in a mudstone of variable thickness between the Faraday House and Uldale sills. It is commonly a limestone, locally cherty, or a fossiliferous mudstone. Rowell and Scanlon (1957a) record a varied fauna of brachiopods, some bivalves, *Hyalostelia sp.*, *Chaetetes sp.*, *Zaphrentites sp.*, and *Tylonautilus nodiferus* early form. Hicks notes that the band is usually a sparsely fossiliferous, calcareous, micaceous sandstone in the

Garsdale area. A thin limestone, first noted during the primary survey mapping of Widdale Fell [SD 7865 8763, SD 7803 8756], is a 6 to 8 in (0.15–0.20 m) thick and grey calcilutite, weathering to an ochrous rottenstone, with orthotetoid brachiopods. This is the farthest south that the marine band has been identified. In the Swaledale mineral belt the band has been traced as far east as Gunnerside Gill [NY 9365 0289] where silty mudstones 6 ft (1.8 m) above the Faraday House Sill yield *Euphemites* cf. *ardenensis*, *Edmondia?* and *Sanguinolites* aff. *variabilis*. To the east of Arkengarthdale the marine band has been located by us in Church Gill [NZ 1416 0042] (Figure 11, section 1). It here contains abundant well preserved *Lingula sp.*

The Uldale Sill is a fine-grained, commonly thinly bedded sandstone, with ripple marks and annelid trails, becoming thickly-bedded towards the top. Rowell and Scanlon (*op. cit.*) believed this to be the more persistent leaf of the Ten Fathom Grit and it seems likely that south of Reeth, where only one leaf is developed, this is equivalent to the Uldale Sill. Thickening of the formation occurs in the west, notably at Nor Gill, West Baugh Fell [SD 714 941] where Hicks (1957) records 75 ft (23 m). He traces the formation southwards through Garsdale as far as Great Coum [SD 702 831], where it is over 50 ft (15 m) thick and Whernside [SD 735 803], over 30 ft (9 m) thick. He notes that the top of the sandstone is usually a ganister up to 25 ft (7.60 m) thick. A number of mine and shaft sections in the North Swaledale Mineral Belt (Wet Shaw Fourth Whim, Surrender Mine, Hurst Mines) show a coal seam up to 2 ft 6 in (0.76 m) in thickness close to the top of the Uldale Sill (Figure 11, section 1). This seam is exposed in Gunnerside Gill [NY 9324 0310] where it is 4 in (0.10 m) thick. It is virtually at the same stratigraphical level as the Crag Coal of the Alston area. At

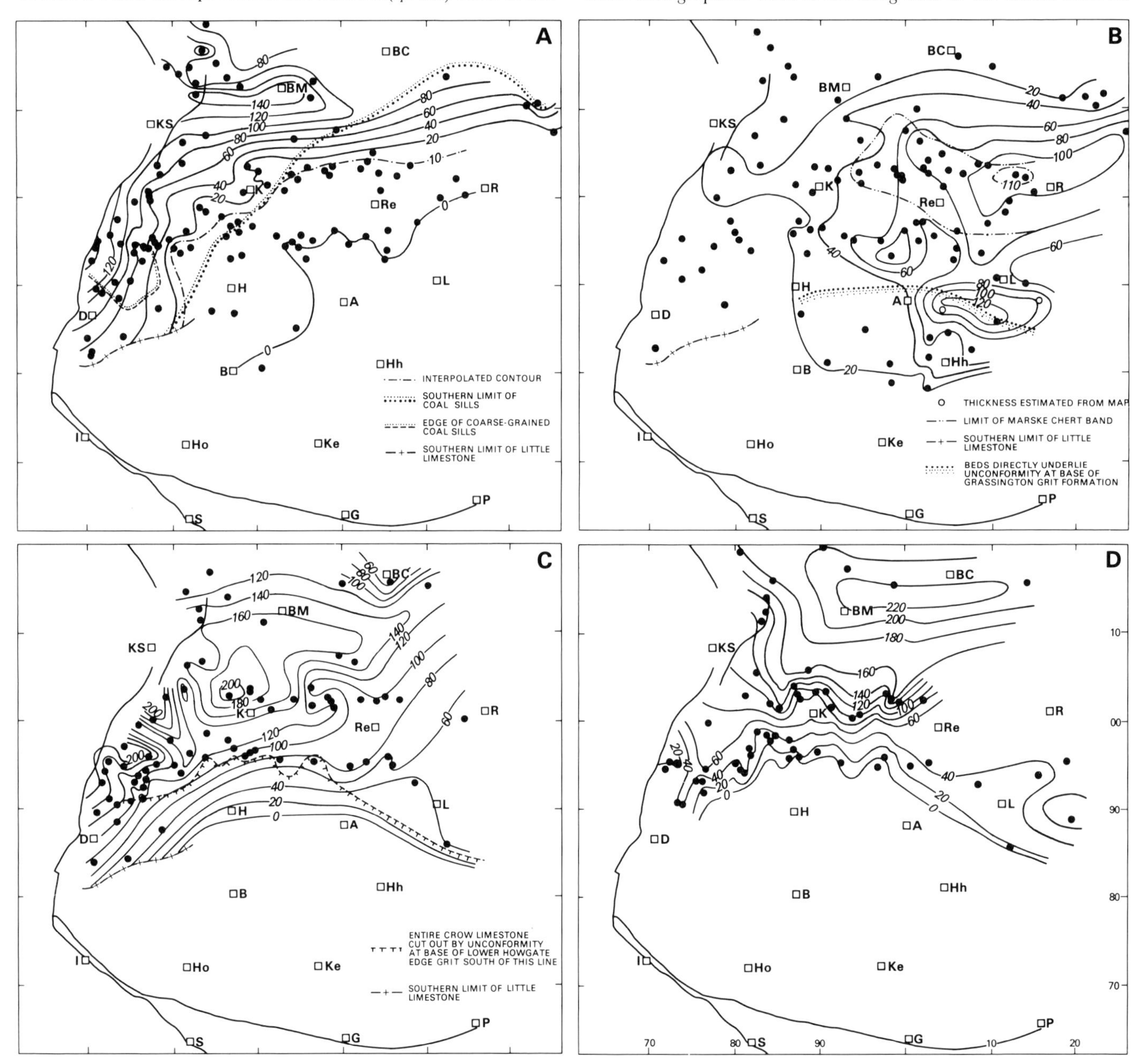

Figure 12 Isopach maps showing thicknesses (in feet) of Namurian strata. Includes data from published sources

A Top of Main Limestone to base of Little Limestone; **B** Little Limestone and Richmond Cherts; **C** Top of Richmond Cherts to base of Crow Limestone; **D** Base of Crow Limestone to base of Mirk Fell Ganister–Lower Howgate Edge Grit.

several localities in upper Swaledale, Rowell and Scanlon (1957a) recorded a fauna of brachiopods and bivalves in a calcareous sandstone developed at the top of the Uldale Sill. Locally in Birkdale there is a silty parting with *Lingula* in the Uldale Sill.

South of Reeth, the Ten Fathom Grit is recorded in Middle Moss [SE 0521 9408] and Low Greets [SE 0253 9549] shafts (Figure 13, section 5, Figure 11, section 2), and a shaft in Herontree Allotment [near SE 083 931]. It ranges from 10 to 38 ft (3.00–11.60 m) in thickness and consists of a single leaf directly under the Crow Limestone, probably equivalent to the Uldale Sill. Where exposed, in Summer Lodge Beck [SD 9653 9502] and Devis Hole Gill [SD 0514 9566], south of Reeth, the Ten Fathom Grit is a thinly bedded sandstone with annelid trails, current-ripple lamination and some cross-bedding, becoming blocky upwards. A coal seam 4 in (0.10 m) thick was seen in Juniper Gill [SE 0764 9588]. East of Arkengarthdale the grit thins progressively to the east, but is still at least 58 ft (17.60 m) thick at Fell End Hush. Eastwards there is further thinning and splitting of the grit towards Church Gill (Figure 11, section 1).

Crow Limestone First named by the Swaledale miners, this is the highest mineralised formation. It consists of up to 77 ft (23.50 m) of limestone, cherty limestone and chert of variable lithology, with associated mudstone partings which are especially thick in the Stainmore Syncline (Burgess and Holliday, 1979, fig. 31). In some sections the more pure limestones are concentrated towards the bottom of the formation, and in the Kirkby Stephen district a lower Crow Limestone and upper Crow Cherts have been differentiated, but this distinction is in fact hard to make at many exposures (Rowell and Scanlon, 1957a, p. 13 and the present work).

Old mining sections of Wetshaw Fourth Whim (Dakyns and others, 1891), Lownathwaite, Surrender (High Shaft and adjacent workings), Old Moulds (Phillips, 1836, p. 66) and Hurst Mines all show the Crow Chert in two or three leaves overlying a thin Crow Limestone (Figure 13, section 2). The highest leaf, commonly labelled 'flinty chert', is 12 to 15 ft (3.66–4.57 m) in thickness. This is particularly well seen at Hungry Hushes, Arkengarthdale, in tilted beds lying between the Folly and New Rake veins [NY 9857 0282], where the Crow Chert makes a double cliff. A 10 ft (3 m) gap between the features covers the mudstone parting in the record of the nearby Wetshaw Fourth Whim. The upper cliff includes 6 ft 6 in (2 m) of grey glassy chert overlying a distinctive banded chert, a sequence also seen at Tanner Rake [NZ 0137 0323] on the opposite side of Arkengarthdale. As far away as South Grain, Apedale [SE 0055 9491], a glassy chert has been noted at the top of the Crow Limestone, but it is unlikely that it is as continuous a horizon as the Marske Chert Band of the Richmond Cherts. On the northern fringes of the North Swaledale Mineral Belt the upper part of the Crow Limestone lacks any glassy chert, nor was any seen in large exposures in Gunnerside Gill below Blakethwaite Dam (see below), this despite records of 'flinty chert' at the top of the Crow Limestone in the Blakethwaite mine record.

In the North Swaledale Mineral Belt a common lithology is a soft orange coloured rottenstone, a ferruginous decalcified chert. This is particularly well seen 100 yd (90 m) downstream from Barras End High Level [NY 9887 0116] and in Botcher Gill [NY 9312 0068]. The lithology resembles that in the Lower Felltop Limestone of the Alston area which has been worked for iron ore.

Rowell and Scanlon (1957a) show that glauconite becomes a major constituent, up to 48 per cent of the whole rock, in the area between Baugh Fell, Brownber Edge and Nine Standards Rigg. In addition, phosphatic nodules, perhaps coprolites, about ½ in (1 cm) in diameter, with spines and spicules, occur. On the northern slopes of Great Shunner Fell, the Crow Limestone appears to be absent in a zone half a mile (800 m) wide on the south side of the Stockdale Vein, conceivably related to a contemporaneous movement on the vein (Rowell and Scanlon, *op. cit.*, p. 15). In the south-west of the Askrigg Block, in common with the Richmond Cherts the Crow Limestone becomes very thin. Hicks (1957) reported thicknesses of about 3 ft (0.90 m) of limestone on Baugh Fell [SD 720 943] and on Rise Hill [SD 731 885]. South of Dentdale in Great Coum, he saw 1 ft 6 in (0.46 m) of unfossiliferous calcareous sandstone at this horizon. South of a wavy line (Figure 13) following the watershed between Swaledale and Wensleydale, the Crow Limestone is cut out by intra-Pendleian erosion.

The Crow Limestone is 68 ft (20.73 m) thick in the Richmond area where it was studied by R. W. Hey (1956). He considered its cherty nature to be due to penecontemporaneous silicification. The occurrence of limestone in Swaledale in varying degrees of silicification seems to fit in with this interpretation.

The excellent natural exposures in Gunnerside Gill below Blakethwaite Dam and in the tributary stream of Benty Gutter [NY 9360 0298] are notable for the ease with which a fauna can be extracted from the Crow Limestone:

	Thickness		
	ft	in	m
Mudstone, shaly, ferruginous			
Gap about	13	0	4
Chert, grey, blocky	1	4	0.40
Gap (probably mudstone)	2	0	0.60
Chert, grey, blocky	1	9	0.53
Gap (probably mudstone)	1	9	0.53
Chert, grey, blocky	1	2	0.35
Mudstone, fissile, soapy	2	0	0.61
Chert, grey, blocky; *Zoophycos* in fallen blocks. Basal 3 ft (0.90 m) yields sponge spicules [abundant], *Fenestella* cf. *hemispherica*, *Antiquatonia sulcata*, *Buxtonia sp.*, *Echinoconchus punctatus*, *Eomarginifera longispina*, *Lingula mytilloides*, *Orbiculoidea sp.*, *Productus sp.*, *Spirifer bisulcatus* group, *Catastroboceras kilbridense*, *C. quadratum*, ostracods	8	6	2.60
Mudstone, cherty, very platy; *Hyalostelia smithii*, *Rhombopora sp.*, *Antiquatonia sulcata*, *Buxtonia sp.*, *Schizophoria sp.*, *Cypricardella sp.*	1	6	0.46
Mudstone, dark grey; a few brachiopods	1	9	0.53
Gap	4		1.22
Mudstone, weathered	2	8	0.81
Limestone, grey, slightly cherty	1	8	0.50
Mudstone, grey		9	0.23
Gap	1	8	0.50
Limestone, grey, biosparite, becoming slightly cherty downwards	18		5.49
Limestone, cherty; irregular weathering and spongy texture common. Blocks from this and the overlying unit yielded *Hyalostelia smithii*, *Dibunophyllum sp.*, *Fenestella spp.*, *Thamniscus?*, *Buxtonia spp.*, *Echinoconchus punctatus*, *Eomarginifera sp.*, *Productus sp.*, *Rugosochonetes sp.*, *Schellwienella sp.*, *Spiriferellina sp.*	10	0	3.05
Limestone, grey, cherty	6	6	1.98
Mudstone, grey, cherty; stick bryozoa, trepostome bryozoa, *Avonia youngiana*, *Brachythyris integricosta*, *Composita ambigua*, *Fluctuaria* cf. *undata*, orthotetoids, *Overtonia fimbriata*, *Productus concinnus*, *Pustula* cf. *rugata*, *Spirifer bisulcatus* group, *S.* cf. *triangularis*, *S. trigonalis*, *Porcellia sp.*, *Streblochondria anisota*, ostracods	3	6	1.07
Gap	1	6	0.46
Limestone, grey, with cherty patches; passing up into ferruginous blocky chert	2	8	0.81
Chert, grey, platy	3	3	1.00
Siltstone		10	0.25
Coal		4	0.10

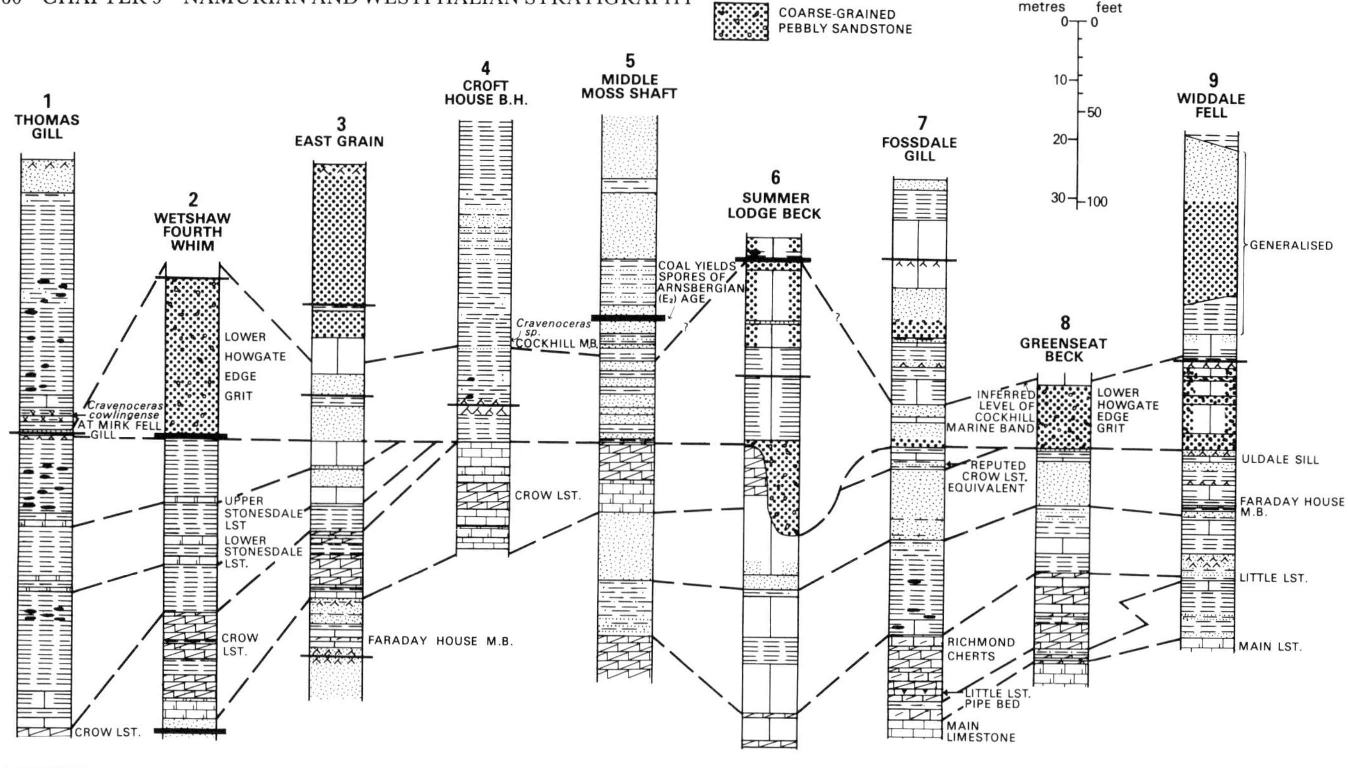

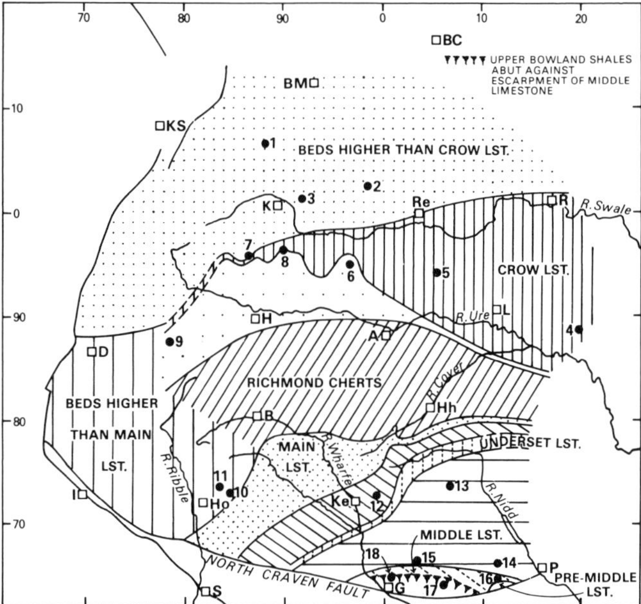

Figure 13 Sections in the Grassington Grit Formation with a map to show the formations beneath the plane of the intra-E_1 unconformity (base of Grassington Grit)

As in the Richmond Cherts, corals are very rare but *Hyalostelia* spicules are locally abundant (Rowell and Scanlon, 1957a, p. 14 and the present work). *Zoophycos* occurs at numerous localities, notably in Swinner Gill [NY 9144 0123], but not as abundantly as in the Black Beds of the Richmond Cherts. At four widely separated localities in Swaledale, a 'pipe bed' (*Scolithos*) was seen in the basal 5 ft (1.50 m) of the Crow Limestone. At two places the pipes were oblique rather than vertical. It is not as persistent a bioturbated bed as the 'pipe bed' in the Little Limestone.

Beds between Crow Limestone and Mirk Fell Ganister/Lower Howgate Edge Grit Strata in this division are virtually restricted to the country north of Wensleydale due to the effect of the intra-Pendleian unconformity which cuts out these beds to the south. Beds are thickest in the north and there is great southerly thinning independent of the unconformity (Figure 13, sections 1–3). In the north Swaledale mining field the sequence is excellently seen in Thomas Gill, West Stonesdale where 162 ft (49.40 m) of mudstone, with siderite nodules at several levels, overlies the Crow Chert (Figure 13, section 1). Limestones, first named in this area by Hudson (1941), occur at two levels, namely the Lower Stonesdale Limestone (2 ft—0.60 m thick) in three leaves and the Upper Stonesdale Limestone (1 ft—0.30 m thick) in two leaves. The Lower Stonesdale Limestone and the overlying mudstones have yielded a variety of brachiopods, bivalves and nautiloids (Rowell and Scanlon, 1957a) at a number of localities in Upper Swaledale. Three genera of trilobites occur in the overlying mudstone. The Upper Stonesdale Limestone, a well-jointed limestone with a ferruginous top, yields a more distinctive fauna due to the abundance of *Schellwienella rotundata* and *Pustula rugata* traced over country between Mallerstang and Arkengarthdale by Turner (1955, fig. 2).

The Stonesdale limestones have not been recorded west of Mallerstang, and Hicks (1957) has seen continuous sequences of shaly mudstone on Baugh Fell. Locally, the Lower Stonesdale Limestone is absent in the eastern part of the Cotherstone Syncline. Eastwards, the Upper Stonesdale Limestone has been newly noted high up Gunnerside Gill [NY 9331 0308] where it yields *Productus* cf. *carbonarius*, *Pustula rugata* [abundant], *Schellwienella rotundata* [abundant], *Aviculopecten* aff. *losseni*, *Myalina mitchelli*, *Schizodus* cf. *portlocki*, *Sulcatopinna flabelliformis* and ostracods. In Arkengarthdale, Dakyns and others (1891) noted a cherty facies of the Upper Stonesdale Limestone (or Fell Top Limestone). The chert was referred to as the 'Top Crow Chert', immediately overlain by 'Fell Top Limestone'. A typical exposure, in Great Punchard Gill [NY 9475 0420] shows:

	Thickness		
	ft	in	m
Limestone, dark grey, muddy; fossiliferous	0	2½	0.07
Seatearth sandstone	0	5½	0.14
Sandstone, very cherty, spongy weathering; very fossiliferous with *Schellwienella sp.* at top	4	3	1.30

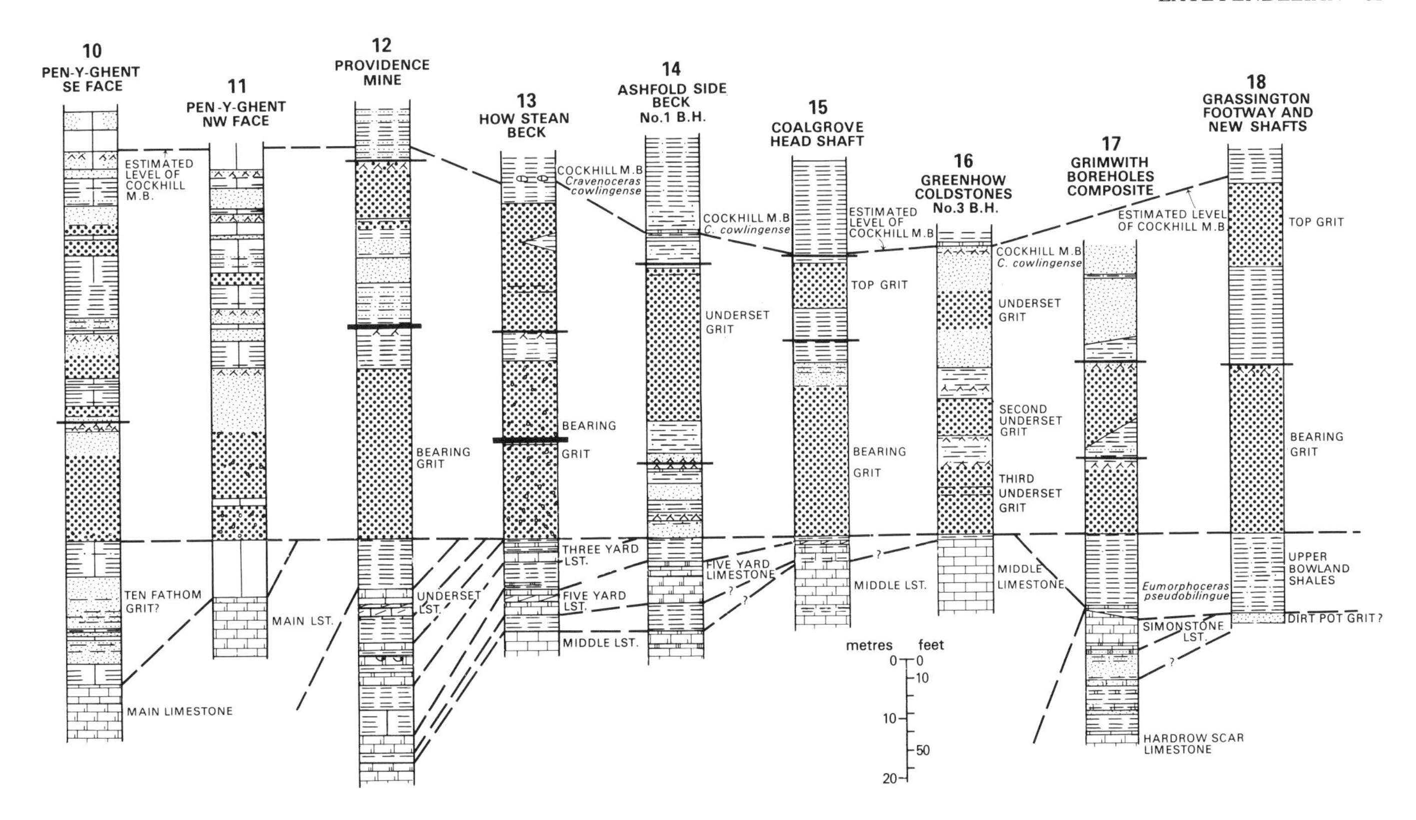

Mine records between Gunnerside Gill and Hurst show up to three separate thin limestones in mudstones overlying the Crow Limestone (Figure 13, section 2, Figure 11, section 3). It is likely that these are the Stonesdale limestones. The Upper Stonesdale Limestone has been noted in Bleaberry Gill [NY 9877 0124] (Turner, 1955) and in Hungry Hushes, a scar [NY 9877 0279] in 26 ft 6 in (8.08 m) of mudstone shows 6 in (15 cm) of weathered limestone with *Schellwienella* cf. *crenistria*. The interval above the Crow Limestone is obscured by faulting.

Reading (1957) and Burgess and Holliday (1979) show that the measures above the Crow Limestone thicken greatly eastwards within the Cotherstone Syncline. They include a third limestone, the Hunder Beck Limestone, formerly thought to be the same as the Lower Stonesdale Limestone by Reading (1957). Up to three sandstones are recorded, the thickest being a coarse-grained sandstone beneath the Hunder Beck Limestone. A few feet below the Upper Stonesdale Limestone a coal is sporadically developed. Dakyns and others (1891) record a coal at this level near Annaside Head, Arkengarthdale and a seatearth occurs in Gunnerside Gill.

Upper Bowland Shales These are dark grey platy mudstones, with thin calcareous bands, attaining 400 ft (122 m) in the Cracoe area and averaging 250 ft (76 m) near Clitheroe. They thin northwards towards the Askrigg Block (Black, 1958, fig. 3) and extend about 1 mile (1.6 km) northwards on to the block at Grassington (Black, 1950) and Malham (Hudson, 1944). At Malham and Bordley there was a pre-Bowland Shale fault scarp on the Middle Craven Fault over which part of the Bowland Shale transgressed (Garwood and Goodyear, 1924, pl. 21; Hudson, 1944, pl. 1) but at Grassington there was a shallower feature with no evidence of an active fault (Black, 1950). At the latter locality a thin sandstone is developed at the feather edge of the Bowland Shale (Black, 1950, pl. 5). The Upper Bowland Shales appear to have been penetrated in New Shaft [SE 0157 6480] with 44 ft (13.41 m) of 'shales and girdles with iron pyrites' (Figure 13, section 18). Within the shale succession at Grassington is the Scale Haw Limestone, a possible correlative of the Main Limestone (Black, 1950). New boreholes at Grimwith Reservoir between Grassington and Greenhow (Figure 13, section 17) show the presence of 47 ft (14.33 m) of Upper Bowland Shales over half a mile (0.8 km) N of the North Craven Fault in an area where these beds were not hitherto known to occur. At Grimwith, as in the Grassington and Malham areas, *Eumorphoceras pseudobilingue*, indicating Pendleian (E_1) beds, has been recorded. Garwood and Goodyear (1924) also record a higher fauna in the Moor Close Gill section of the Bordley area with *Cravenoceras malhamense*. The sequence above the *E. pseudobilingue* band is not here known.

In the Cracoe knoll country, the Bowland Shales were banked against the reef knolls and finally engulfed them just prior to the onset of deposition of the Skipton Moor Grits (Black, 1958, fig. 2). It is likely that the Upper Bowland Shales and the Skipton Moor Grits–Pendle Grits of the Craven lowlands were being deposited contemporaneously with strata north of the Craven Faults ranging from the Main Limestone up to the Stonesdale limestones.

LATE PENDLEIAN

Unconformity beneath the Grassington Grit The presence of an unconformity at this level was probably first suspected by Dakyns (*vide* Chubb and Hudson, 1925, p. 238). At Greenhow, in particular, where pre-Grassington Grit earth movements were greatest, a detailed picture has been built up of domal flexuring and faulting followed by erosion (Dunham and Stubblefield, 1945). Beds under the Grit here range from the Asbian Greenhow Limestone up to the Toft Gate Limestone (Middle Limestone Member). Dunham and Stubblefield mapped a WNW–ESE-trending fault, apparently a branch of the North Craven Fault, which intersects the Waterhole North Vein at a small angle. This dislocation juxtaposes the Coldstones and Toft Gate limestones at surface and the Toft Gate and Greenhow limestones in the Cockhill Adit (Dunham and Stubblefield, *op. cit.*, fig. 2). Beds between the Toft Gate and Coldstones limestones are thus faulted out at surface and were not seen underground; details of intervening beds were unknown. A recent inclined borehole, Coldstones No. 3, drilled close to the line of Gillfield Adit (660 ft—200 m E of Cockhill Adit) has confirmed much of the succession between the two abovementioned limestones. This proves to be remarkably thick, notably in the 223 ft (68 m) thick Dirt Pot Grit which is nearly four times its already inflated thickness at Providence Vein 1 mile (1.6 km) to the north (Figures 7, 17). The estimated pre-Grassington Grit throw on this fault and a small branching fault in the borehole is 760 ft (232 m) down north, a throw comparable in value with that of the North Craven Fault hereabouts, but in the opposite direction. In post-Grassington Grit times the fault, at its eastern end, apparently moved the Grassington Grit down south, making a second small movement in an opposite direction to the first. The unusual thickness of the Dirt Pot Grit may be related to early movements on the fault during Brigantian sedimentation.

The unconformity at Greenhow has been traced northwards as far as Swaledale (Chubb and Hudson, 1925; Rowell and Scanlon, 1957a; Wilson, 1960b). At its northern limit, in the Stainmore area, there is erosive channelling at the base of the grit in Lower Baldersdale and Deepdale, but this dies out to the west (Burgess and Holliday, 1979, fig. 32). Southwards the Grassington Grit and its lateral equivalents rest on progressively lower horizons in the early Pendleian – Brigantian (Figure 13). There is rarely any sign of any angular discordance and the erosion surface is generally almost flat except for local channels as in Coverdale (Wilson, 1960a) and near Summer Lodge Beck, Swaledale. Also the Crow and Main limestones tend to come in abruptly and probably their outcrop under the grit is locally marked by a gentle escarpment, as for instance at Caseker Crag, Great Whernside [SD 992 743]. The progressive nature of the unconformity has been ascribed to tilting of the Askrigg Block with greater erosion in the south-east corner. Since this was a local tectonic effect related to tilting of the block Ramsbottom (1977) treated it as an event within Mesothem N1. However, work in progress in the Sykes Anticline, Slaidburn (R. S. Arthurton, personal communication) suggest a marked unconformity beneath the Brennand Grit, the local correlative of the Grassington Grit. Thus tectonic factors besides tilting of the Askrigg Block may be involved.

Grassington Grit Formation and equivalents Dakyns (1892) first named the Grassington Grit. Later, Hudson (1939) referred to it as a group, but this usage is here discontinued owing to the modern tendency to use this term in a wider sense. Typically this formation is some 160 ft (49 m) thick but locally attains 218 ft (66.45 m). In the north-west there is drastic thinning to a mere 4 ft (1.20 m) where its equivalent is called the Mirk Fell Ganister.

In the southern half of the area and in Arkengarthdale in the north-east, the Grassington Grit and its equivalents include much cross-bedded, pebbly, coarse-grained, feldspathic sandstone. There are up to five seatearths, some with local coal seams. Besides fine-grained sandstones, interbedded mudstones occur, locally with *Lingula* and fish remains. There are great lateral variations in thickness and lithology, as, for instance, on Pen-y-ghent (Figure 13, sections 10, 11 situated only 600 yd (549 m) apart). These beds were evidently laid down in a deltaic complex, with periodic marine incursions indicated by *Lingula* faunas.

Grassington Moor is of prime interest since there the lower half of the formation is a thick coarse-grained feldspathic sandstone termed the Bearing Grit (Phillips, 1836), because of the presence in it of rich lead oreshoots. It is thickest in Ringleton Shaft [SE 0284 6734] where it attains 103 ft (31.40 m), thinning eastwards to 59 ft (18 m) in Sarah Shaft [SE 0357 6731] in 800 yds (731 m). It is overlain by an impersistent coal seam up to 20 in (0.51 m) in thickness. A further two sandstones occur, (only one in some sections), the higher being termed the Top Grit (Geological Survey Vertical Sections, Sheet 28) (Figure 13, sections 15, 18).

To the east of Grassington Moor, the Bearing Grit splits. Here, new reservoir boreholes at Grimwith add detail to an old section in the Kelshaw Level (Dakyns, *in* MS) (Figure 13, section 17). Farther east, at Greenhow, the Grassington Grit was formerly divided into the Underset, Second Underset and Third Underset grits (Newbold, *in* MS). Sections were measured in these beds by Dunham and Stubblefield (1945) in Cockhill Adit and Bell's Shaft. More recently, borings have been made through the grit in the valley of Ashfold Side Beck near Providence Vein and also in Coldstones No. 3 Borehole, alongside the line of Gillfield Adit, Greenhow [SD 1160 6459] (Figure 13, sections 14, 16). These records confirm that the thickest sandstone in the Greenhow area is at the top, not the base of the formation as at Grassington. This, the Underset Grit of Newbold, ranges from coarse- to fine-grained and attains 85 ft (25.90 m) in thickness. The much thinner Second Underset and Third Underset grits are both coarse-grained in Coldstones No. 3 Borehole, the lower or Third Grit being pebbly at several levels. One mile (1.6 km) farther north at Ashfold Side Beck No. 1 Borehole the thick medium- and coarse-grained Underset Grit is underlain by three thin fine-grained sandstones, the attenuated remnants of the Second and Third grits.

North-west of Grassington, the Grassington Grit forms the cappings of the high fells. Exposure is particularly good on Pen-y-ghent where a thick, cross-bedded, coarse-grained, basal sandstone shows great variations in thickness. Higher beds include fine-grained, current-ripple laminated sandstone with mudstones. Seatearths are seen at four levels. Southwards on Fountains Fell two coals occur, one of which was extensively worked. A recent survey (E. W. Johnson, personal communication) has failed to confirm the presence of seven seams as suggested by Dakyns and others (1890). On Ingleborough, Whernside, Dodd Fell, Birks Fells and Widdale Fell (Figure 13, section 9), the Grassington Grit and its equivalent, the Lower Howgate Edge Grit, include much coarse-grained sandstone. The top of the formation is hard to fix, however, due to the non-exposure of the Cockhill Marine Band, if, indeed, it exists so far west.

North of Grassington Moor, the coarse-grained Bearing Grit is thick in Providence Mine, Kettlewell, and persists into highest Coverdale and How Stean, Nidderdale where cross-bedded units are finely exposed (Figure 13, sections 12, 13). Thinning and splitting of the Bearing Grit takes place northwards down Coverdale (Wilson, 1960a, pl. 8) and also in highest Nidderdale (Tonks, 1925, pl. 17) where worked coal seams occur (Figure 13, section 13). Thus, in much of lower Wensleydale and lower Coverdale, the Grassington Grit Formation is thin and consists chiefly of mudstone, though thinly-bedded sandstones were worked for tilestone on Penhill. In Croft House Borehole north of the River Ure [SE 1982 8883] the whole formation, here overlain by the Cockhill Marine Band, totals only 52 ft 6 in (16 m) in thickness and consists entirely of mudstones with two seatearths and a thin coal seam (Figure 13, section 4). North-westwards from Croft House, the mudstone facies, with thin sandstone bands, is found in incomplete exposures on Stainton Moor [SE 089 951], but the Cockhill Marine Band was not seen. The upper limit of the formation is uncertain in Middle Moss Shaft [SE 0521 9408] for the same reasons, though spores from coal chips from an old pit [SE 0654 9360] near the shaft suggest an Arnsbergian age for the worked coal, and hence for the same coal in the Middle Moss Shaft section, above the top of the Grassington Grit Formation (Figure 13, section 5).

West of Middle Moss Shaft, the pebbly, cross-bedded Lower Howgate Edge Grit, the local equivalent of the Grassington Grit (Rowell and Scanlon, 1957a), is present all along the Ure – Swale watershed between Apedale Head and Great Shunner Fell (Figure 13, sections 6, 7, 8). Generally there is one sandstone, but locally, around Summer Lodge Moor, the grit appears to be in two leaves, the lower of which dies out abruptly eastwards against a palaeoslope in Crow Chert.

In upper Swaledale, in the latitude of Muker, the Lower Howgate Edge Grit passes northwards into the Mirk Fell Ganister, a fine-grained sandstone as little as 4 ft (1.2 m) thick. Locally in Little Sleddale it thickens to 50 ft (15.20 m) and contains two thin coals (Rowell and Scanlon, 1957a). Eastwards in Swaledale (north of the river) there is a great thickening and coarsening of the Mirk Fell Ganister to a pebbly, cross-bedded, feldspathic sandstone, here referred to the Lower Howgate Edge Grit. This is at least 86 ft (26.20 m) thick in Wetshaw Fourth Whim [NY 9803 0243] (Figures 20, 13, section 2) and well exposed in crags on Fell End Moor [NZ 0274 0300] on the east side of Arkengarthdale (Figure 27). A similar eastward thickening was observed farther north in the Cotherstone Syncline by Burgess and Holliday (1979, fig. 32). It seems likely that a major channel filled by coarse-grained, cross-bedded sandstone extended southwards from Baldersdale into Arkengarthdale.

ARNSBERGIAN AND CHOKIERIAN

Cockhill Limestone and its equivalents This important marker band has been known at Greenhow since Newbould (*in* MS, 1805) called it the Top Limestone. It was examined in Cockhill and Gillfield adits by Dunham and Stubblefield (1945), who first noted the abundant specimens of *Cravenoceras cowlingense* (see also p. 23). In numerous borings on Providence Vein it is 2 to 3 ft (0.60–0.90 m) thick and in some of them it is overlain by mudstones with limestone bullions also crowded with goniatites. There is southward thickening to 5 ft (1.52 m) in Cockhill Adit and 8 ft (2.44 m) in Bell's Shaft, Greenhow (Dunham and Stubblefield, 1945).

In the Grassington Moor mining field numerous old shaft sections (e.g. Vertical Section Sheet 28) do not note a limestone above the Grassington Grit Formation. In all likelihood there are only some limestone bullions in mudstone, such as have been observed by the authors on the tip of Red Scar Shaft [SE 0558 6661] and by Tonks at Low Peru Shaft [SE 0348 6751]. Locally these bullions have yielded fossils on the dump of Cook's Shaft [SE 0314 6538] at the head of Bolton Gill, with *C. cowlingense* and the type specimen of *Eumorphoceras bisulcatum grassingtonense*.

Northwards in upper Nidderdale, the Cockhill horizon is typically a band with *C. cowlingense* in large limestone bullions up to 3 ft 6 in (1.07 m) diameter within fossiliferous mudstone, here called the Cockhill Marine Band (Wilson and Thompson, 1959). Overlying the limestone is some 3 ft (0.90 m) of fossiliferous, dark grey mudstone with irregularly shaped phosphatic nodules, which can be traced northwards into Coverdale. Croft House Borehole near Newton-le-Willows [SE 1982 8883] shows a similar mudstone with *Cravenoceras sp.* and *Posidoniella* aff. *vetusta* (this latter bivalve is also common in the Cockhill Marine Band of Nidderdale). West of Leyburn, no marine band has been found at this level.

North of the River Swale *Cravenoceras cowlingense* has been recorded by Hudson (1941) in shales between the Mirk Fell Ironstones of Mirk Fell Gill and by Wilson and Thompson (1959) in Brigstone Gill, Birk Dale [NY 839 019]. Recently Burgess and Ramsbottom (1970) reported *C. sp. nov.* aff. *cowlingense* in the Lad Gill Limestone of Lovely Seat, a substantially higher horizon than the Mirk Fell Ironstones according to Rowell and Scanlon (1957a). The question arises as to whether the Lad Gill Limestone could be the correlative of the Cockhill Marine Band as the underlying beds are of a facies broadly similar to the Grassington Grit Formation (i.e. coarse-grained sandstones and mudstones which include a coal seam). Despite the faunal contrast in the presence of zaphrentoids in the Mirk Fell Ironstones and their virtual absence in the Cockhill Marine Band of Nidderdale, there is an interesting lithological resemblance between the two marine formations. Both of them contain irregularly shaped phosphatic nodules within the mudstones and these also occur in the geographically intermediate Croft House Borehole. The existing correlation of Cockhill Marine Band with Mirk Fell Ironstones is here followed though it is conceivable that it might be incorrect.

The Cockhill Marine Band has not been proved on Buckden Pike, Dodd Fell Hill, Great Coum, Whernside, Pen-y-ghent or Widdale Fell.

Nidderdale Shales and their equivalents This is a distinctive formation 234 to 304 ft (71–93 m) thick in the main 'Millstone Grit' outcrop between Greenhow and the River Ure and on Buckden Pike. In the Ashfold Side Beck No. 1 Borehole on Providence Vein [SE 1160 6615] the shales are 292 ft (89.00 m) thick. They consist dominantly of mudstone and silty mudstone with thin, fine-grained sandstones, lacking in seatearth horizons; thicker medium-grained sandstones occur, notably on Penhill (Wilson, 1960b). Within the lower 100 ft (30 m) of the Nidderdale Shales at Greenhow there are eight thin bands of ferruginous silty limestone (Dunham and Stubblefield, *op. cit.*), which have been subsequently verified in several borings on the Providence Vein.

Ashfold Side Beck No. 1 Borehole on Providence Vein yielded *Naticopsis?* and *Sanguinolites* cf. *striatus* (Wilson, 1977). At Croft House Borehole, north of the River Ure [SE 1982 8883] the same band contains *Cravenoceras?* and *Sanguinolites*. This marine band is a likely correlative of the main *E. bisulcatum* band to the south and of the Lad Gill Limestone with *C. sp. nov.* aff. *cowlingense* to the north-west. The Nidderdale Shales of the Croft House Borehole differ from those at Greenhow in the presence of scattered bivalves throughout the lower half of the shales.

In the area between Baugh Fell and Tan Hill, a workable seam, the Tan Hill Coal, up to 3 ft 6 in (1.07 m) in thickness, occurs, whereas in the south-east, seatearth horizons are totally absent. The Upper Howgate Edge Grit is largely coarse-grained and prone to great variations in thickness (Rowell and Scanlon, 1957a, figs. 10, 11). In places it evidently forms a washout channel, also present in the east of the Brough area (Burgess and Holliday, 1979, fig. 32). Outliers on Widdale Fell and Whernside include suggested coarse-grained correlatives of the Upper Howgate Edge Grit (Figure 14, sections 4, 5) but the underlying Cockhill Marine Band has not been found. Higher beds in the north-west include the Lad Gill Limestone and a sandstone with sporadic brachiopod and crinoid debris, the Fossil Sandstone (Figure 14).

Red Scar Grit This is a coarse-grained, cross-bedded, feldspathic sandstone which extends from Stainmore southwards to the Lancaster Fells. Its most spectacular exposure is in a 2 mile (3.2 km) long cliff 100 ft (30 m) in height along Mallerstang Edge [NY 798 014] (Rowell and Scanlon, 1957a). Here called Pickersett Edge Grit, it also forms the higher slopes of Great Shunner Fell, Wildboar Fell and Rogan's Seat, and correlates with the High Wood Grit of Stainmore (Figure 14).

In boreholes sunk on Providence Vein, apart from the more southerly section of Cockhill Adit where Dunham and Stubblefield (1945, fig. 2) record a single coarse-grained sandstone, in Colsterdale and parts of Upper Nidderdale, the Red Scar Grit consists of two members split by a mudstone parting with *Lingula* and other brachiopods and bivalves (Figure 14, sections 7, 8). Locally the parting carries a coal seam, the Woogill Coal of the Colsterdale area. The lower member is a coarse-grained, feldspathic sandstone. The upper member is quartzitic, prone to great thickness variation coupled with local 20 ft (6 m) high cross-sets; it is shelly in places in the north and is locally a sandy limestone in lower Wensleydale (Phillips, 1836; Wilson, 1960b).

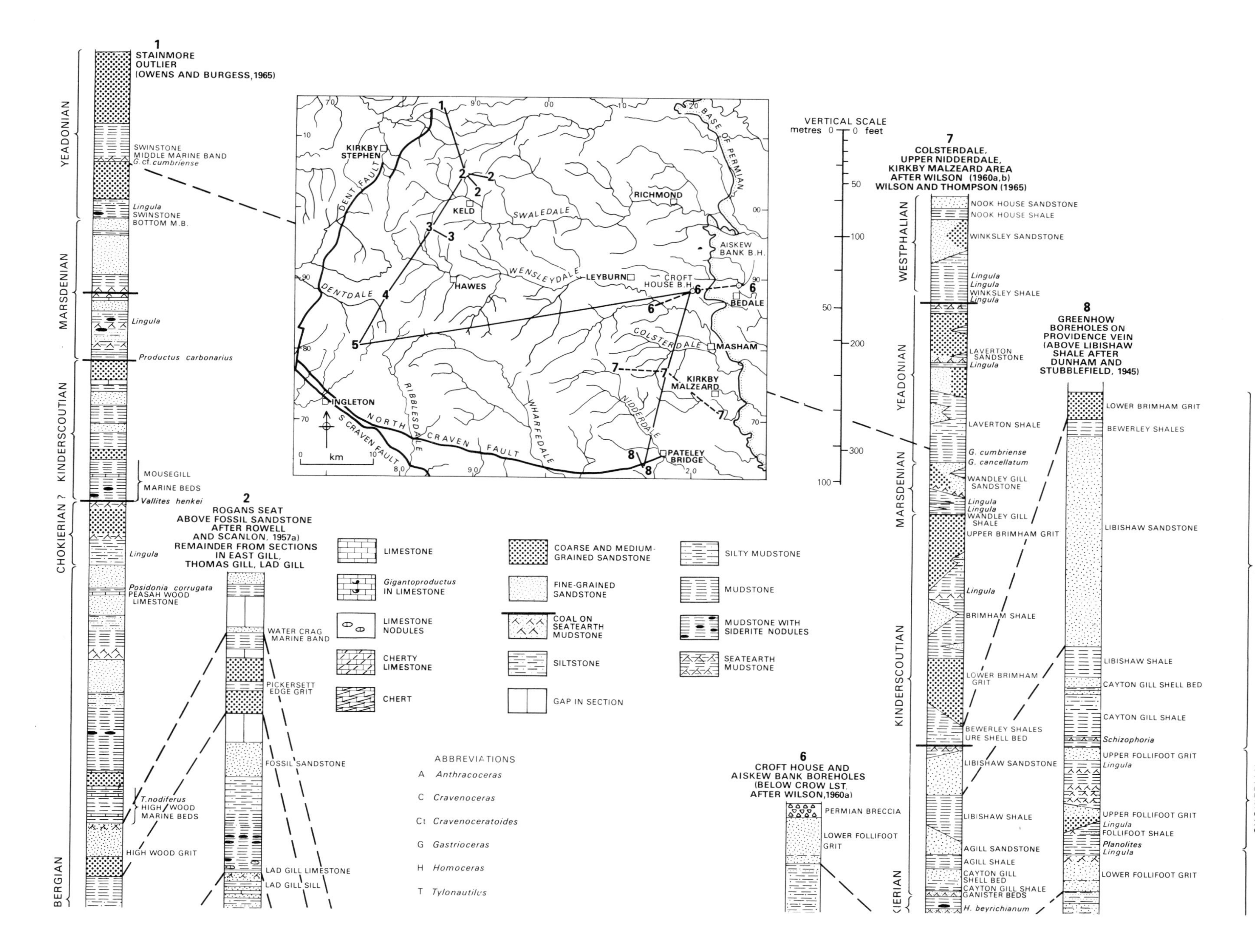
1
STAINMORE OUTLIER (OWENS AND BURGESS, 1965)
YEADONIAN
MARSDENIAN
KINDERSCOUTIAN
CHOKIERIAN ?
BERGIAN
SWINSTONE MIDDLE MARINE BAND
G. cf. cumbriense
Lingula
SWINSTONE BOTTOM M.B.
Lingula
Productus carbonarius
MOUSEGILL MARINE BEDS
Vallites henkei
Lingula
Posidonia corrugata
PEASAH WOOD LIMESTONE
T. nodiferus
HIGH WOOD MARINE BEDS
HIGH WOOD GRIT
2
ROGANS SEAT ABOVE FOSSIL SANDSTONE AFTER ROWELL AND SCANLON, 1957a) REMAINDER FROM SECTIONS IN EAST GILL, THOMAS GILL, LAD GILL
WATER CRAG MARINE BAND
PICKERSETT EDGE GRIT
FOSSIL SANDSTONE
LAD GILL LIMESTONE
LAD GILL SILL
KIRKBY STEPHEN
KELD
SWALEDALE
RICHMOND
BASE OF PERMIAN
AISKEW BANK B.H.
WENSLEYDALE
LEYBURN
CROFT HOUSE B.H.
BEDALE
HAWES
DENTDALE
DENT FAULT
COLSTERDALE
MASHAM
KIRKBY MALZEARD
INGLETON
RIBBLESDALE
WHARFEDALE
NIDDERDALE
NORTH CRAVEN FAULT
S CRAVEN FAULT
PATELEY BRIDGE
km
LIMESTONE
Gigantoproductus IN LIMESTONE
LIMESTONE NODULES
CHERTY LIMESTONE
CHERT
COARSE AND MEDIUM-GRAINED SANDSTONE
FINE-GRAINED SANDSTONE
COAL ON SEATEARTH MUDSTONE
SILTSTONE
GAP IN SECTION
SILTY MUDSTONE
MUDSTONE
MUDSTONE WITH SIDERITE NODULES
SEATEARTH MUDSTONE
ABBREVIATIONS
A Anthracoceras
C Cravenoceras
Ct Cravenoceratoides
G Gastrioceras
H Homoceras
T Tylonautilus
VERTICAL SCALE
metres
feet
6
CROFT HOUSE AND AISKEW BANK BOREHOLES (BELOW CROW LST. AFTER WILSON, 1960a)
PERMIAN BRECCIA
LOWER FOLLIFOOT GRIT
7
COLSTERDALE, UPPER NIDDERDALE, KIRKBY MALZEARD AREA AFTER WILSON (1960a, b) WILSON AND THOMPSON (1965)
WESTPHALIAN
NOOK HOUSE SANDSTONE
NOOK HOUSE SHALE
WINKSLEY SANDSTONE
Lingula
Lingula
WINKSLEY SHALE
Lingula
LAVERTON SANDSTONE
Lingula
LAVERTON SHALE
G. cumbriense
G. cancellatum
WANDLEY GILL SANDSTONE
Lingula
Lingula
WANDLEY GILL SHALE
UPPER BRIMHAM GRIT
Lingula
BRIMHAM SHALE
LOWER BRIMHAM GRIT
BEWERLEY SHALES
URE SHELL BED
LIBISHAW SANDSTONE
LIBISHAW SHALE
AGILL SANDSTONE
AGILL SHALE
CAYTON GILL SHELL BED
CAYTON GILL SHALE
GANISTER BEDS
H. beyrichianum
8
GREENHOW BOREHOLES ON PROVIDENCE VEIN (ABOVE LIBISHAW SHALE AFTER DUNHAM AND STUBBLEFIELD, 1945)
LOWER BRIMHAM GRIT
BEWERLEY SHALES
LIBISHAW SANDSTONE
LIBISHAW SHALE
CAYTON GILL SHELL BED
CAYTON GILL SHALE
Schizophoria
UPPER FOLLIFOOT GRIT
Lingula
UPPER FOLLIFOOT GRIT
Lingula
FOLLIFOOT SHALE
Planolites
Lingula
LOWER FOLLIFOOT GRIT
CHOKIERIAN

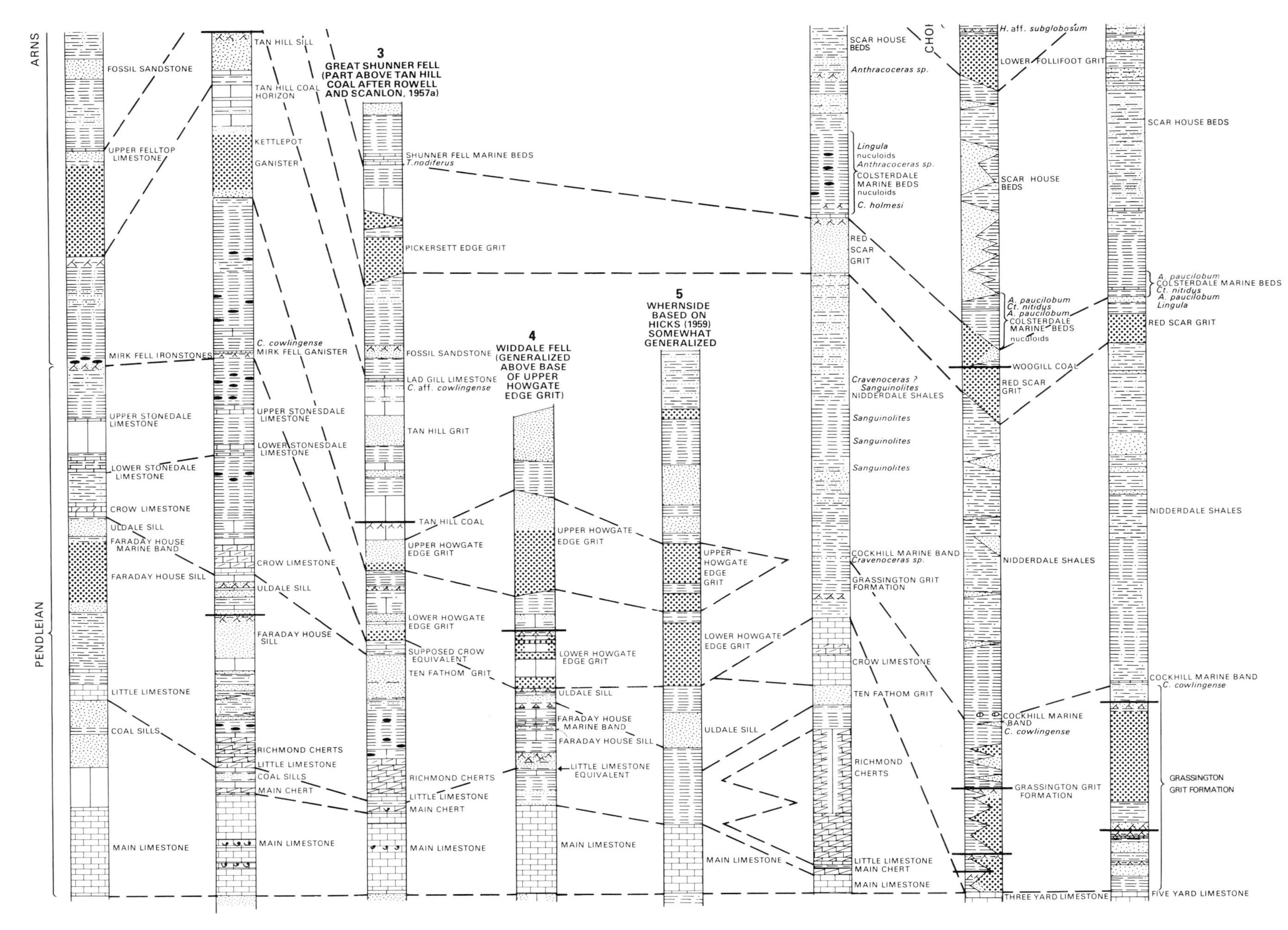

Figure 14 Comparative sections of Namurian strata

Colsterdale Marine Beds These are shaly mudstones with *Anthracoceras paucilobum*, very rare *Tylonautilus nodiferus* and abundant bivalves, which contain the distinctive, well-jointed Colsterdale Limestone, about 1 ft (0.30 m) in thickness, with abundant *Cravenoceratoides nitidus* usually preserved as flattened impressions (see also p. 23). The Marine Beds extend from Greenhow northwards into Colsterdale (Bisat, 1914; Dunham and Stubblefield, 1945; Wilson and Thompson, 1959) and recent boreholes north of the River Ure show some 70 ft (21 m), triple the normal thickness south of the river. In the northern reaches of Colsterdale, Wilson and Thompson (1959) linked this thickening to the incoming of mudstones containing nuculoids and gastropods near the base of the formation. North of the Ure all the marine beds are of this biofacies and the Colsterdale Limestone is absent, possibly due to lateral passage into the upper beds with nuculoids. New finds, in boreholes, of *Cravenoceras holmesi* low down in the Marine Beds north of the Ure are probably from beds lower than the *Ct. nitidus* equivalent (Figure 14, section 6). Marine mudstones, with limestones containing *Tylonautilus nodiferus*, located close to the summits of Shunner Fell and High Seat (Rowell and Scanlon, 1957a) and in the High Wood Marine Beds of the Stainmore outlier (Owens and Burgess 1965) are probable lateral equivalents of the Colsterdale Marine Beds, as is the Water Crag Marine Band, a fossiliferous sandstone on Rogan's Seat (Rowell, personal communication, Figure 14, sections 1–3).

Beds between the Colsterdale Marine Beds and the Lower Follifoot Grit At Greenhow these beds, estimated by Newbould as being 150 ft (45.70 m) thick, were originally called by him the Ellenscar Plate (Dunham and Stubblefield, 1945); they were renamed the Scar House Beds by Wilson (1960b). These are silty mudstones with variable interbeds of fine-grained sandstone, best seen in boreholes on the Providence Vein. Here the thickness varies from 192 to 210 ft (58.50–64.00 m). Northwards these beds are very variable, with sandstone dominating the sequence in a belt running northwards from upper Nidderdale into central Colsterdale. Cross-bedding is not seen, but sole markings are common, consistent with an origin as proximal turbidites.

To the north and east, the equivalent strata include bands of abundant marine fossils, as for instance in the Stainmore outlier (Owens and Burgess, 1965) and in the Aiskew Bank Borehole near Bedale [SE 2667 8888] (Figure 14, sections 1, 6).

Lower Follifoot Grit and overlying shales The Lower Follifoot Grit is a persistent formation forming extensive dip slopes in Colsterdale. It is commonly a feldspathic, medium- or coarse-grained sandstone, becoming quartzitic to the south where it has been worked for silica bricks at Smelthouses in Nidderdale [SE 196 645] (A. T. Thompson, personal communication). The overlying shales include two *Lingula* bands in boreholes near Greenhow (Figure 14, section 8), the probable lateral equivalents of the top *Nuculoceras nuculum* band and *Homoceras beyrichianum* band (the latter is of Chokierian age) of the Simonseat – Fewston area to the south (Hudson, 1939; Wilson, 1977).

Upper Follifoot Grit The Upper Follifoot Grit has also been recently penetrated in boreholes along the Providence Vein, Greenhow. Two fine-grained sandstone members vary greatly in thickness and are split by three or four thick seatearth mudstones at the western end of the series of boreholes. Seatearths are also a notable feature of the equivalent strata in Colsterdale (Figure 14, sections 7, 8).

ALPORTIAN

No marine bands of Alportian age have been recorded in the area, and Alportian strata may well be absent, as suggested by Ramsbottom (1977).

KINDERSCOUTIAN

Beds between the Upper Follifoot and Lower Brimham grits These are best known at Greenhow (Dunham and Stubblefield, 1945) and in the Colsterdale and Kirkby Malzeard areas (Wilson, 1960b; Wilson and Thompson, 1965). Shelly sandstones and siltstones occur at numerous levels in the sequence in the area from the River Ure southwards to Otley and Harrogate (Wilson *in* Ramsbottom, 1974, fig. 30). At Greenhow the chief of these, the Cayton Gill Shell Bed, is preserved on the southern, downthrow side, of the Providence Vein (Dunham and Stubblefield, *op. cit.*) and also forms an outlier, not hitherto recorded, 1 mile (1.6 km) N of the vein on the upthrow side on Gouthwaite Moor [SE 070 690], (A. T. Thompson, personal communication). New boreholes at Greenhow show that, beneath the sandstone of the Cayton Gill Shell Bed, 38 ft (11.58 m) of silty mudstone with *Productus carbonarius* rest on a distinctive thin limestone packed with *Schizophoria* (Wilson, Dunham and Pattison, in preparation).

The above group of beds includes the Mousegill Marine Beds with *Vallites henkei* (R_{1a} *R. circumplicatile* Zone) in the Stainmore outlier (Owens and Burgess, 1965). These are at about the level of the Cayton Gill Shale of the south-eastern ground where Wilson (1977) records *R. circumplicatile* group 5 miles (8 km) S of the North Craven Fault. Thompson (*in* Godwin, 1973) records *R.* aff. *pulchellum* in Libishaw Shale just south of the line of the North Craven Fault east of Pateley Bridge, at the level of the *R.* cf. *paucicrenulatum R. umbilicatum* (R_{1a}) fauna recorded farther south near Fewston by Wilson (1977). This marine band is not known north of the Craven faults.

Brimham Grits The Lower Brimham Grit of the south-eastern ground is a massive, coarse-grained, cross-bedded, feldspathic sandstone which forms spectacular tors and lines of cliffs, as at Brimham Rocks. A. T. Thompson (personal communication) notes that there is evidence that the Lower Brimham Grit thickens abruptly southwards across the North Craven Fault at Pateley Bridge, from 30 to 400 ft (9–122 m) indicating a contemporaneous effect of the fault on sedimentation. The Upper Brimham Grit is usually less massive than the Lower and is separated from it by shales with a *Lingula* band (Figure 14, section 7). The probable equivalent beds in the Stainmore outlier contain finer grained sandstones than the Brimham grits in the south-east.

MARSDENIAN

Wandley Gill Shale and Sandstone In the south-eastern ground these total some 48 ft (14.60 m) in thickness and are apparently the sole representatives of the 600 ft (183 m) thick Marsdenian beds of the Bradford area. The several *Reticuloceras* bands near Bradford are represented by only two *Lingula* bands. The Wandley Gill Sandstone, commonly about 27 ft (8.20 m) thick on the southern fringes of Colsterdale has thickened southwards to 92 ft (28.10 m) in the Winksley Borehole [SE 2507 7151]. The Wandley Gill Sandstone in the borehole consists of four small cycles, probably equivalent to thicker cycles with named sandstones in the 262 ft (80 m) thick sequence in the Farnham Borehole near Knaresborough [SE 3469 5996] (Burgess and Cooper, 1980). There is one *Lingula* band at about the level where *R. superbilingue* might be expected. An earlier record by Godwin (1973) of the *R. superbilingue* band at a nearby outcrop has not been confirmed in the borehole and it seems likely that this is a record within the *G. cancellatum* band which is known to yield *R. superbilingue* elsewhere in northern England.

YEADONIAN

Laverton Shale and Sandstone The Laverton Shale is a valuable datum horizon in the south-east since it contains two goniatite bands extensively developed in the north of England, the *Gastrioceras cancellatum* and *G. cumbriense* bands. The former is a new discovery by C. G. Godwin, later confirmed in the Winksley Borehole (Figure 14, section 7). The latter was first found by A. T. Thompson (Wilson and Thompson, 1965; Godwin, 1973). Farther north in the Stainmore outlier the Swinstone Middle Marine Band yielded *G.* cf. *cumbriense* (Owens and Burgess, 1965).

The Laverton Sandstone of the Kirkby Malzeard area and its equivalent in the Stainmore outlier are coarse-grained sandstones like their probable correlative to the south, the Rough Rock. Locally the Laverton Sandstone contains mudstone partings, one of them with a *Lingula* band (Figure 14, section 7).

WESTPHALIAN

In the Stainmore outlier Ford (1955) and Owens and Burgess (1965) record, above Namurian strata, sandstone and mudstones, with coal seams, of Westphalian A and B age, the highest strata belonging to the Lower *A. similis – A. pulchra* Zone. In the Kirkby Malzeard area there are several outliers, with up to 107 ft (32.60 m) of probable Westphalian strata overlying the Laverton Sandstone. The drilling at Winksley confirms the data from exposure that the Winksley Shale, as in equivalent beds at Stainmore, contains only a *Lingula* band at the level where *G. subcrenatum* would be expected to occur (Figure 14, section 7).

REFERENCES (Chapters 3, 4 and 5)

Aisenverg, D. E., Brazhnikova, N. E., Vassilyuk, N. P., Vdovenko, M. V., Gorak, S. V., Dunaeva, N. N., Zernetskaya, N. V., Poletaev, V. I., Potievskaya, P. D., Rotai, A. P. and Sergeeva, M. T. 1979. The Carboniferous sequence of the Donetz Basin: a standard section for the Carboniferous System. In *The Carboniferous of the USSR*. Wagner, R. H., Higgins, A. C. and Meyen, S. V. (Editors) (Leeds: Yorkshire Geological Society.) 247 pp.

Bisat, W. S. 1914. The Millstone Grit sequence between Masham and Great Whernside. *Proc. Yorkshire Geol. Soc.*, Vol. 19, pp. 20–24.

— 1924. The Carboniferous goniatites of the north of England and their zones. *Proc. Yorkshire Geol. Soc.*, Vol. 20, pp. 40–124.

— 1928. The Carboniferous goniatite zones of England and their continental equivalents. *C.R. Congr. Int. Stratigr. Géol. Carbonif.* (Heerlen, 1927), pp. 117–133.

— 1934. The goniatites of the *Beyrichoceras* Zone in the north of England. *Proc. Yorkshire Geol. Soc.*, Vol. 22, pp. 280–309.

— 1957. In discussion of Shirley, J. The Carboniferous Limestone of the Monyash – Wirksworth area, Derbyshire. *Q. J. geol. Soc. London*, Vol. 114, pp. 411–429.

Black, W. W. 1940. The Bowland Shales from Thorlby to Burnsall, Yorkshire. *Trans. Leeds Geol. Assoc.*, Vol. 5, pp. 308–321.

— 1950. The Carboniferous geology of the Grassington area, Yorkshire. *Proc. Yorkshire Geol. Soc.*, Vol. 28, pp. 29–42.

— 1952. A zaphrentoid fauna from the shales above the Simonstone Limestone, Fountains Fell, Yorkshire. *Trans. Leeds Geol. Assoc.*, Vol. 6, pp. 182–187.

— 1957. A boulder-bed in the Bowland Shales near Burnsall, Yorkshire. *Trans. Leeds Geol. Assoc.*, Vol. 7, pp. 24–33.

— 1958. The structure of the Burnsall – Clitheroe district and its bearing on the origin of the Cracoe Knoll-Reefs. *Proc. Yorkshire Geol. Soc.*, Vol. 31, pp. 391–414.

Bond, G. 1950. The Lower Carboniferous reef limestones of Cracoe, Yorkshire. *Q. J. Geol. Soc. London.*, Vol. 105, pp. 157–188.

Bott, M. H. P. 1961. A gravity survey off the coast of north-east England. *Proc. Yorkshire Geol. Soc.*, Vol. 33, pp. 1–20.

— 1967. Geophysical invetigations of the northern Pennine basement rocks. *Proc. Yorkshire Geol. Soc.*, Vol. 36, pp. 139–168.

— and Johnson, G. A. L. 1967. The controlling mechanism of Carboniferous cyclic sedimentation. *Q. J. Geol. Soc. London.*, Vol. 122, pp. 421–441.

Brady, H. B. 1876. Carboniferous and Permian foraminifera. *Palaeontogr. Soc.* [Monogr.]. 166 pp.

Burgess, I. C. a (in preparation). Lower Carboniferous sections in the Sedbergh district, Cumbria. *Trans. Leeds Geol. Assoc.*

— b (in preparation). The Raydale (BGS) Borehole near Askrigg, North Yorkshire.

— and Cooper, A. H. 1980. The Farnham (IGS) Borehole near Knaresborough, north Yorkshire. *Rep. Inst. Geol. Sci.*, No. 80/1, pp. 12–17.

— and Harrison, R. K. 1967. Carboniferous Basement Beds in the Roman Fell district, Westmorland. *Proc. Yorkshire Geol. Soc.*, Vol. 36, pp. 203–225.

— and Holliday, D. W. 1979. Geology of the country around Brough-under-Stainmore. *Mem. Geol. Surv. G.B.*, 170 pp.

— and Mitchell, M. 1976. Viséan lower Yoredale limestones on the Alston and Askrigg blocks and the base of the D_2 Zone in northern England. *Proc. Yorkshire Geol. Soc.*, Vol. 40, pp. 613–630.

— and Ramsbottom, W. H. C. 1970. A new goniatite horizon in the Hearne Beck Limestone (Namurian, E_2) near Lovely Seat, upper Wensleydale. *J. Earth Sci.*, Vol. 8, pp. 143–147

Butterfield, J. A. 1920. The conglomerates underlying the Carboniferous Limestone in the N. W. of England. *Naturalist*, pp. 249–252, 281–284.

— 1939a. Petrological study of some Yoredale sandstones. *Trans. Leeds Geol. Assoc.*, Vol. 5, pp. 264–284.

— 1939b. Brookite in the Millstone Grit of Yorkshire. *Geol. Mag.*, Vol. 76, pp. 220–228.

Chubb, L. J. and Hudson, R. G. S. 1925. The nature of the junction between the Lower Carboniferous and the Millstone Grit of north-west Yorkshire. *Proc. Yorkshire Geol. Soc.*, Vol. 20, pp. 257–291.

Cope, F. W. 1940. *Daviesiella llangollensis* (Davidson) and related forms: morphology, biology and distribution. *J. Manchester Geol. Assoc.*, Vol. 1, pp. 199–231.

Cousins, J. 1977. The sedimentology of the Middle, Underset and Crow limestones of the mid-Carboniferous Yoredale Group of the Askrigg Block. (Unpublished Ph.D. thesis, University of Southampton.) 482 pp.

Dakyns, J. R. 1892. On the geology of the country between Grassington and Wensleydale. *Proc. Yorkshire Geol. Soc.*, Vol. 12, pp. 133–144.

— 1894. A sketch of the geology of Nidderdale and the Washburn, north of Blubberhouses. *Proc. Yorkshire Geol. Soc.*, Vol. 12, pp. 294–299.

— Tiddeman, R. H., Gunn, W. and Strahan, A. 1890. The geology of the country around Ingleborough with parts of Wensleydale and Wharfedale. *Mem. Geol. Surv. G.B.*, 103 pp.

— Russell, R., Clough, C. T. and Strahan, A. 1891. The geology of the country around Mallerstang with parts of Wensleydale, Swaledale and Arkendale. *Mem. Geol. Surv. G.B.*, 213 pp.

DAVIES, T. A. and SUPKO, 1973. Oceanic sediments and their diagenesis: some examples from deep-sea drilling. *J. Sediment. Petrol.*, Vol. 43, pp. 381–390.
DAVIS, J. W. 1883. On the fossil fishes of the Carboniferous Limestone Series of Great Britain. *Trans. R. Dublin Soc.*, 2nd ser., Vol. 1, pp. 327–548.
DIXON, E. E. L. and HUDSON, R. G. S. 1931. A mid-Carboniferous boulder-bed near Settle. *Geol. Mag.*, Vol. 68, pp. 81–82.
DOUGHTY, P. S. 1968. Joint densities and their relation to lithology in the Great Scar Limestone. *Proc. Yorkshire Geol. Soc.*, Vol. 36, pp. 479–512.
— 1974. *Davidsonina* (*Cyrtina*) *septosa* (Phillips) and the structure of the Viséan Great Scar Limestone north of Settle, Yorkshire. *Proc. Yorkshire Geol. Soc.*, Vol. 40, pp. 41–47.
DUNHAM, K. C. 1948. Geology of the Northern Pennine Orefield: Vol. 1, Tyne to Stainmore. *Mem. Geol. Surv. G.B.*, 367 pp.
— 1950. Lower Carboniferous sedimentation in the Northern Pennines (England). *Int. Geol. Congr.*, Part 4, pp. 46–63.
— HEMINGWAY, J. E., VERSEY, H. C. and WILCOCKSON, W. H. 1953. A guide to the geology of the district round Ingleborough. *Proc. Yorkshire Geol. Soc.*, Vol. 29, pp. 77–115.
— and ROSE, W. C. C. 1974. Geology of the iron-ore field of south Cumberland and Furness. *Wartime Pamphlet Geol. Surv. G. B.,* No. 16, 26 pp.
— and STUBBLEFIELD, C. J. 1945. The stratigraphy, structure and mineralisation of the Greenhow mining area, Yorkshire. *Q. J. Geol. Soc. London*, Vol. 100, pp. 209–268.
EARP, J. R., MAGRAW, D., POOLE, E. G., LAND, D. H. and WHITEMAN, A. J. 1961. Geology of the country around Clitheroe and Nelson. *Mem. Geol. Surv. G.B.,* Sheet 68, 346 pp.
EKDALE, A. A. 1977. Abyssal trace fossils in worldwide Deep Sea Drilling Project cores. In *Trace Fossils 2.* CRIMES, T. P. and HARPER, J. C. (Editors). (Liverpool: Seel House Press.) 351 pp.
ELLIOTT, T. 1975. The sedimentary history of a delta lobe from a Yoredale (Carboniferous) cyclothem. *Proc. Yorkshire Geol. Soc.*, Vol. 40, pp. 505–536.
EVANS, W. B., WILSON, A. A., TAYLOR, B. J. and PRICE, D. 1968. Geology of the country around Macclesfield, Congleton, Crewe and Middlewich. *Mem. Geol. Surv. G. B.*, 328 pp.
FALCON, N. L. and KENT, P. E. 1960. Geological results of petroleum exploration in Britain 1945–1957. *Mem. Geol. Soc. London.*, No. 2.
FOLK, R. L. 1959. Practical petrographic classification of limestones. *Bull. Am. Assoc. Pet. Geol.*, Vol. 43, pp. 1–38.
FORD, T. D. 1955. The Upper Carboniferous rocks of the Stainmore Coalfield. *Geol. Mag.*, Vol. 92, pp. 218–230.
FORSTER, W. 1809. *A treatise on a section of the strata from Newcastle-on-Tyne to the mountain of Cross Fell, in Cumberland; with remarks on mineral veins in general.* (Alston.)
FOWLER, A. 1944. A deep bore in the Cleveland Hills. *Geol. Mag.*, Vol. 81, pp. 193–206, 254–265.
FRIEDMAN, G. 1964. Early diagenesis and lithification in carbonate sediments. *J. Sediment. Petrol.*,Vol. 34, pp. 777–813.
GARWOOD, E. J. 1913. The Lower Carboniferous succession in the north-west of England. *Q. J. Geol. Soc. London.*, Vol. 68, pp. 449–586.
— 1922. On a freshwater shale with *Viviparus* and associated beds from the base of the Carboniferous rocks in Ribblesdale, Yorkshire. *Geol. Mag.*, Vol. 59, pp. 289–293.
— and GOODYEAR, E. 1924. The Lower Carboniferous succession in the Settle district. *Q. J. Geol. Soc. London.*, Vol. 80, pp. 184–273.
GEORGE, T. N. 1958. Lower Carboniferous palaeogeography of the British Isles. *Proc. Yorkshire Geol. Soc.*, Vol. 31, pp. 227–318.
— 1978. Eustasy and tectonics: sedimentary rhythms and stratigraphical units in British Dinantian correlation. *Proc. Yorkshire Geol. Soc.*, Vol. 42, pp. 229–262.
— JOHNSON, G. A. L., MITCHELL, M., PRENTICE, J. E., RAMSBOTTOM, W. H. C., SEVASTOPULO, G. D. and WILSON, R. B. 1976. A correlation of Dinantian rocks in the British Isles. *Spec. Rep. Geol. Soc. London.*, No. 7, 87 pp.
GILLIGAN, A. 1920. The petrography of the Millstone Grit of Yorkshire. *Q. J. Geol. Soc. London*, Vol. 74, pp. 251–294.
GODWIN, C. G. 1973. The geology of the Ure valley water main trenches 1969–71. *Proc. Yorkshire Geol. Soc.*, Vol. 39, pp. 537–546.
GREENSMITH, J. T. 1957. Lithology, with particular reference to cementation, of Upper Carboniferous sandstones in northern Derbyshire, England. *J. Sediment. Petrol.*, Vol. 5, pp. 405–416.
HALLETT, D. 1970. Foraminifera and algae from the Yoredale 'Series' (Viséan – Namurian) of northern England. *C. R. Congr. Int. Stratigr. Géol. Carbonif.* (Sheffield, 1967), Vol. 2, pp. 323–330.
HEY, R. W. 1956. Cherts and limestones from the Crow Series near Richmond, Yorkshire. *Proc. Yorkshire Geol. Soc.*, Vol. 30, pp. 289–299.
HICKS, P. F. 1957. *The Yoredale Series and Millstone Grit of the south west corner of the Askrigg Block.* (Unpublished Ph.D. thesis, University of Leeds.)
— 1959. The Yoredale Rocks of Ingleborough, Yorkshire. *Proc. Yorkshire Geol. Soc.*, Vol. 32, pp. 31–43.
HILL, D. 1938–41. Carboniferous rugose corals of Scotland. *Palaeontogr. Soc.* [Monogr.]. 213 pp.
HOLLIDAY, D. W., NEVES, R. and OWENS, B. 1979. Stratigraphy and palynology of early Dinantian (Carboniferous) strata in shallow boreholes near Ravenstonedale, Cumbria. *Proc. Yorkshire Geol. Soc.*, Vol. 42, pp. 343–356.
HOPKINS, W. 1956. *Neoglyphioceras* cf. *spirale* (Phillips) from Ingleborough. *Geol. Mag.*, Vol. 93, pp. 173–174.
HUDSON, R. G. S. 1924. On the rhythmic succession of the Yoredale Series in Wensleydale. *Proc. Yorkshire Geol. Soc.*, Vol. 20, pp. 125–135.
— 1925. Faunal horizons in the Lower Carboniferous of north-west Yorkshire. *Geol. Mag.*, Vol. 62, pp. 181–186.
— 1926. Lower Carboniferous zonal nomenclature. *Rep. Br. Assoc.*, (Southampton, 1925), pp. 1–9.
— 1929a. On the Lower Carboniferous coral *Orionastraea* and its distribution in the north of England. *Proc. Leeds Philos. Lit. Soc.*, Vol. 1, pp. 440–457.
— 1929b. A Carboniferous lagoon deposit with sponges. *Proc. Yorkshire Geol. Soc.*, Vol. 21, pp. 181–196.
— 1930. The Carboniferous of the Craven reef Belt; the Namurian unconformity at Scaleber near Settle. *Proc. Geol. Assoc.*, Vol. 41, pp. 290–322.
— 1932. The pre-Namurian knoll topography of Derbyshire and Yorkshire. *Trans. Leeds Geol. Assoc.*, Vol. 5, pp. 49–64.
— 1938. Pp. 295–330 *in* 'The geology of the country around Harrogate'. *Proc. Geol. Assoc.*, Vol. 49, pp. 293–352.
— 1939. The Millstone Grit succession of the Simonseat Anticline, Yorkshire. *Proc. Yorkshire Geol. Soc.*, Vol. 23, pp. 319–349.
— 1941. The Mirk Fell Beds (Namurian E_2) of Tan Hill, Yorkshire. *Proc. Yorkshire Geol. Soc.*, Vol. 24, pp. 259–289.
— 1944. A pre-Namurian fault-scarp at Malham. *Proc. Leeds Philos. Lit. Soc. Sci. Sect.*, Vol. 4, pp. 226–232.
— and COTTON, G. 1945. The Lower Carboniferous in a boring at Alport, Derbyshire. *Proc. Yorkshire Geol. Soc.*, Vol. 25, pp. 254–330.

— and Fox, T. 1943. An upper Viséan zaphrentoid fauna from the Yoredale beds of north-west Yorkshire. *Proc. Yorkshire Geol. Soc.*, Vol. 25, pp. 101–126.

Hughes, T. M. 1909. Ingleborough, Part VI. The Carboniferous Rocks. *Proc. Yorkshire Geol. Soc.*, Vol. 16, pp. 253–320.

Jefferson, D. P. 1980. Cyclic sedimentation in the Holkerian (Middle Viséan) north of Settle, Yorkshire. *Proc. Yorkshire Geol. Soc.*, Vol. 42, pp. 483–503.

Johnson, G. A. L. 1958. Biostromes in the Namurian Great Limestone of northern England. *Palaeontology*, Vol. 1, pp. 147–157.

— 1959. The Carboniferous stratigraphy of the Roman Wall district in western Northumberland. *Proc. Yorkshire Geol. Soc.*, Vol. 32, pp. 83–130.

— 1960. Palaeogeography of the northern Pennines and part of north eastern England during the deposition of Carboniferous cyclothemic deposits. *Int. Geol. Congr.*, Part 12, pp. 118–128.

— 1961. Lateral variation of marine and deltaic sediments in cyclothemic deposits with particular reference to the Viséan and Namurian of Northern England. *C. R. Congr. Int. Stratigr. Géol. Carbonif.* (Heerlen, 1958), Vol. 2, pp. 323–330.

— 1967. Basement control of Carboniferous sedimentation in northern England. *Proc. Yorkshire Geol. Soc.*, Vol. 36, pp. 175–194.

— and Dunham, K. C. 1963. The geology of Moor House. *Monogr. Nat. Conserv.*, No. 2, 182 pp.

— Hodge, B. L. and Fairbairn, R. A. 1962. The base of the Namurian and of the Millstone Grit in north-eastern England. *Proc. Yorkshire Geol. Soc.*, Vol. 33, pp. 341–362.

— and Marshall, A. E. 1971. Tournaisian beds in Ravenstonedale, Westmorland. *Proc. Yorkshire Geol. Soc.*, Vol. 38, pp. 261–280.

Joysey, K. A. 1955. On the geological distribution of Carboniferous blastoids in the Craven area, based on a study of their occurrence in the Yoredale Series of Grassington, Yorkshire. *Q. J. Geol. Soc. London*, Vol. 111, pp. 209–224.

Kendall, P. F. and Wroot, H. E. 1924. *Geology of Yorkshire.* (Vienna: published privately.) 995 pp.

Kent, P. E. 1966. The structure of the concealed Carboniferous rocks of north-eastern England. *Proc. Yorkshire Geol. Soc.*, Vol. 35, pp. 323–352.

Koninck, L. G. De and Wood, E. 1858. On the genus *Woodocrinus*. *Rep. Br. Assoc.*, (Dublin, 1857) Sect. C., pp. 76–78.

Krynine, P. D. 1948. The megascopic study and field classification of sedimentary rocks. *J. Geol.*, Vol. 56, pp. 130–165.

Marr, J. E. 1929. The rigidity of north-west Yorkshire. *Naturalist*, 1921, pp. 63–72.

McCabe, P. J. 1975. The structure of the Viséan limestones between the Craven Faults, Settle, Yorkshire: a discussion. *Proc. Yorkshire Geol. Soc.*, Vol. 40, pp. 563–564.

Miller, A. A. and Turner, J. S. 1931. The Lower Carboniferous succession along the Dent Fault and the Yoredale beds of the Shap district. *Proc. Geol. Assoc.*, Vol. 42, pp. 1–28.

Mills, D. A. C. and Hull, J. H. 1976. Geology of the country around Barnard Castle. *Mem. Geol. Surv. G. B.*, 385 pp.

Mitchell, M. 1978. Dinantian. Pp. 168–177 in *The Geology of the Lake District.* Moseley, F. (Editor). (Leeds: Yorkshire Geological Society.) 284 pp.

Moore, D. 1958. The Yoredale Series of Upper Wensleydale and adjacent parts of north-west Yorkshire. *Proc. Yorkshire Geol. Soc.*, Vol. 31, pp. 91–148.

— 1959. Role of deltas in the formation of some British Lower Carboniferous cyclothems. *J. Geol.*, Vol. 67, pp. 522–539.

— 1960. Sedimentation units in sandstones of the Yoredale Series (Lower Carboniferous) of Yorkshire, England. *J. Sediment. Petrol.*, Vol. 30, pp. 218–227.

Moore, E. W. J. 1936. The Bowland Shales from Pendle to Dinckley. *J. Manchester Geol. Assoc.*, Vol. 1, pp. 167–192.

— 1941. Sections in the Bowland Shales west of Barnoldswick. *Proc. Yorkshire Geol. Soc.*, Vol. 24, pp. 252–258.

Mundy, D. J. C. 1978. Reef Communities. Pp. 157–167 in *The Ecology of Fossils.* McKerrow, W. S. (Editor). (London: Duckworth.) 384 pp.

O'Connor, J. 1964. The geology of the area around Malham Tarn, Yorkshire. *Field Stud.*, Vol. 2, pp. 53–82.

Owens, B. and Burgess, I. C. 1965. The stratigraphy and palynology of the Upper Carboniferous outlier of Stainmore, Westmorland. *Bull. Geol. Surv. G. B.*, No. 23, pp. 17–44.

Pattison, J. 1981. The stratigraphical distribution of gigantoproductoid brachiopods in the Viséan and Namurian rocks of some areas in northern England. *Rep. Inst. Geol. Sci.*, No. 81/9.

— Wilson, A. A. (in preparation). The macropalaeontology and stratigraphy of the late Brigantian and Pendleian cherts of North Yorkshire.

Phillips, J. 1836. *Illustrations of the geology of Yorkshire,* Part II. *The Mountain Limestone District.* (London: John Murray.) 253 pp.

Ramsbottom, W. H. C. 1973. Transgressions and regressions in the Dinantian: a new synthesis of British Dinantian stratigraphy. *Proc. Yorkshire Geol. Soc.*, Vol. 39, pp. 567–607.

— 1974. Dinantian. Pp. 47–73 in *The geology and mineral resources of Yorkshire.* Rayner, D. H. and Hemingway, J. E. (Editors). (Leeds: Yorkshire Geological Society.) 405 pp.

— 1977. Major cycles of transgression and regression (mesothems) in the Namurian. *Proc. Yorkshire Geol. Soc.*, Vol. 41, pp. 261–291.

— Calver, M. A., Eagar, R. M. C., Hodson, F., Holliday, D. W., Stubblefield, C. J. and Wilson, R. B. 1978. A correlation of Silesian rocks in the British Isles. *Spec. Rep. Geol. Soc. London*, No. 10, 81 pp.

Rayner, D. H. 1953. The Lower Carboniferous rocks in the north of England: a review. *Proc. Yorkshire Geol. Soc.*, Vol. 28, pp. 231–315.

Reading, H. G. 1957. The stratigraphy and structure of the Cotherstone Syncline. *Q. J. Geol. Soc. London*, Vol. 113, pp. 27–56.

Reid, R. E. H. 1968. *Hyalostelia smithii* (Young and Young) and the sponge genus *Hyalostelia* Zittel (Class Hexactinellida). *J. Palaeontol.*, Vol. 42, pp. 1243–1248.

Rotai, A. P. 1979. Carboniferous stratigraphy of the USSR: Proposal for an international classification. In *The Carboniferous of the USSR.* Wagner, R. H., Higgins, A. C. and Meyen, S. V. (Editors). (Leeds: Yorkshire Geological Society.) 247 pp.

Rowell, A. J. and Scanlon, J. E. 1957a. The Namurian of the north-west quarter of the Askrigg Block. *Proc. Yorkshire Geol. Soc.*, Vol. 31, pp. 1–38.

— — 1957b. The relation between the Yoredale Series and the Millstone Grit on the Askrigg Block. *Proc. Yorkshire Geol. Soc.*, Vol. 31, pp. 79–90.

Schwarzacher, W. 1958. The stratification of the Great Scar Limestone in the Settle district of Yorkshire. *Liverpool Manchester Geol. J.*, Vol. 2, pp. 124–142.

Sedgwick, A. 1836. Description of a series of longitudinal and transverse sections through a portion of the Carboniferous chain between Penigent and Kirkby Stephen. *Trans. Geol. Soc. London*, Vol. 4, pp. 69–101.

Somerville, I. D. 1979. A sedimentary cyclicity in early Asbian (lower D_1) limestones in the Llangollen district of

North Wales. *Proc. Yorkshire Geol. Soc.*, Vol. 42, pp. 397–404.

Stevenson, I. P. and Gaunt, G. D. 1971. Geology of the country around Chapel en le Frith. *Mem. Geol. Surv. G. B.*, 444 pp.

Taylor, B. J., Burgess, I. C., Land, D. H., Mills, D. A. C., Smith, D. B. and Warren, P. T. 1971. Northern England. *Br. Reg. Geol. Inst. Geol. Sci.*, Fourth Edition.

Tonks, L. H. 1925. The Millstone Grit and Yoredale rocks of Nidderdale. *Proc. Yorkshire Geol. Soc.*, Vol. 20, pp. 226–256.

Turner, J. S. 1927. The Lower Carboniferous succession in the Westmorland Pennines and the relations of the Pennine and Dent faults. *Proc. Geol. Assoc.*, Vol. 38, pp. 339–374;

— 1950. Notes on the Carboniferous Limestone of Ravenstonedale, Westmorland. *Trans. Leeds Geol. Assoc.*, Vol. 6, pp. 124–134.

— 1955. Upper Yoredales and Millstone Grit relations in the Stainmore Coalfield. *Geol. Mag.*, Vol. 92, p. 350.

— 1956. Some faunal bands in the upper Viséan and early Namurian of the Askrigg Block. *Liverpool Manchester Geol. J.*, Vol. 1, pp. 410–419.

— 1959. Pinskey Gill Beds in the Lune Valley, Westmorland, and Ashfell Sandstone in Garsdale, Yorkshire. *Trans. Leeds Geol. Assoc.*, Vol. 7, pp. 78–87.

— 1962. A note on biostromes in the Middle and Main Limestones (Yoredale Beds) of the north-western part of the Askrigg Block. *Proc. Leeds Philos. Lit. Soc. Sci. Sect.*, Vol. 8, pp. 243–246.

— 1963. Some reflections on the medium-scale cartography of the British Lower Carboniferous strata. *Trans. Leeds Geol. Assoc.*, Vol. 7, pp. 151–174.

Varker, W. J. and Higgins, A. C. 1979. Conodont evidence for the age of the Pinskey Gill Beds of Ravenstonedale, north-west England. *Proc. Yorkshire Geol. Soc.*, Vol. 42, pp. 357–369.

Vine, G. R. 1881. Notes on the Carboniferous polyzoa of north Yorkshire. *Proc. Yorkshire Geol. Soc.*, Vol. 7, pp. 329–341.

Walker, C. T. 1955. Current-bedding directions in sandstones of lower *Reticuloceras* age in the Millstone Grit of Wharfedale, Yorkshire. *Proc. Yorkshire Geol. Soc.*, Vol. 30, pp. 115–132.

— 1967. Relation of Upper Viséan sedimentology to the Bowland Shale overlap in Yorkshire, England. *Sediment. Geol.*, Vol. 1, pp. 117–136.

Waltham, A. C. 1971. Shale units in the Great Scar Limestone of the southern Askrigg Block. *Proc. Yorkshire Geol. Soc.*, Vol. 38, pp. 285–292.

Weller, J. M. 1930. Cyclic sedimentation of the Pennsylvanian Period and its significance. *J. Geol.*, Vol. 38, pp. 97–135.

— 1956. Argument for diastrophic control of Late Palaeozoic cyclothems. *Bull. Am. Assoc. Pet. Geol.*, Vol. 40, pp. 17–50.

Wells, A. J. 1955. The development of chert between the Main and Crow Limestones in North Yorkshire. *Proc. Yorkshire Geol. Soc.*, Vol. 30, pp. 177–196.

— 1957. The stratigraphy and structure of the Middleton Tyas - Sleightholme Anticline, North Yorkshire. *Proc. Geol. Assoc.*, Vol. 68, pp. 231–254.

— 1960. Cyclic sedimentation: A review. *Geol. Mag.*, Vol. 97, pp. 389–403.

Wilson, A. A. 1960a. The Carboniferous rocks of Coverdale and adjacent valleys in the Yorkshire Pennines. *Proc. Yorkshire Geol. Soc.*, Vol. 32, pp. 285–316.

— 1960b. The Millstone Grit Series of Colsterdale and neighbourhood, Yorkshire. *Proc. Yorkshire Geol. Soc.*, Vol. 32, pp. 429–452.

— 1974. Developments in limestone geology in the Ingleton - Settle area. *Trans. Br. Cave Res. Assoc.*, Vol. 1, pp. 61–64.

— 1977. The Namurian Rocks of the Fewston area. *Trans. Leeds Geol. Assoc.*, Vol. 9, pp. 1–42.

— and Cornwell, J. D. 1982. The IGS borehole at Beckermonds Scar, north Yorkshire. *Proc. Yorkshire Geol. Soc.*, Vol. 44, pp. 59–88.

— Dunham, K. C. and Pattison, J. (in preparation). The stratigraphy of recent boreholes in the Greenhow and Grimwith areas, Yorkshire.

— and Thompson, A. T. 1959. Marine bands of Arnsbergian age (Namurian) in the south-eastern portion of the Askrigg Block, Yorkshire. *Proc. Yorkshire Geol. Soc.*, Vol. 32, pp. 45–67.

— — 1965. The Carboniferous succession in the Kirkby Malzeard area, Yorkshire. *Proc. Yorkshire Geol. Soc.*, Vol. 35, pp. 203–227.

Yates, P. J. 1962. The palaeontology of the Namurian rocks of Slieve Anierin, Co. Leitrim, Eire. *Palaeontology*, Vol. 5, pp. 355–443.

CHAPTER 6

Structure

INTRODUCTION

The orefield described in this volume lies within the southern half of the Rigid Block of North-West Yorkshire (Marr, 1921); to this half the name Askrigg Block was applied by Hudson (1938). Its structural boundaries are the Cotherstone Syncline on the north, the Stainmore Monocline and Dent Fault-system on the west and the Craven faults to the south. Towards the east the block is covered by Westphalian, Permian and later formations and the exact position of the eastern margin is uncertain, though it may be assumed to lie west of Cleveland Hills No. 1 Borehole (Kent, 1966). The concept of the rigid block stems from the uniformity in style of Carboniferous sedimentation in the area, which evidently subsided more slowly during Dinantian and Namurian times than the surrounding basins. The Cotherstone Syncline, previously regarded as a minor but convenient sag separating the Askrigg from the Alston Block, has assumed greater importance as the probable site of a Carboniferous gulf or trough as a result of recent research; the size of the Askrigg fault-block has been correspondingly reduced. Nevertheless, by the end of Pendleian times, the Cotherstone sag had ceased to be effective and throughout the later Namurian, (save near Barnard Castle, Mills and Hull, 1976) the sediment thicknesses total much the same whether they occur within the syncline or over the Askrigg Block.

It has clearly emerged as a result of geophysical investigation followed by boring that the Alston and Askrigg blocks owe their identity to tectonic and intrusive events during Lower Palaeozoic time, particularly during the closing stages of the Caledonian Orogeny. Two phases of movement, probably in late Arenig and in pre-Caradoc times, may have affected the basement rocks; the final phase in late Silurian – early Devonian time certainly did so, and this was accompanied or closely followed by the intrusion of the granites which provide the cores of the blocks.

During the Carboniferous epoch, the sea may have penetrated into the Cotherstone – Stainmore gulf early in the Tournaisian, at the Ivorian Stage, but it did not submerge the Askrigg Block until the Holkerian or early Asbian stages. From the later part of the Asbian, through Brigantian to the Pendleian Stage, it can be argued that subsidence had a certain rhythmic character while the Yoredale sedimentary facies was laid down both over the Askrigg Block and the Cotherstone sag. After each clear-water phase, during which limestone accumulated, deltaic sedimentation followed until the sedimentary pile built up to sea level (when coal swamps flourished) or almost to sea level. Subsidence then restored clear-water conditions and the limestone of the next cyclothem began to form. It is not proposed to review here the numerous explanations that have been offered for this process, but merely to note that a total of 12 or 13 reasonably complete cyclothem units were laid down over the Askrigg Block, where this style of sedimentation is just as characteristic as it is over the Alston Block. A rhythmic quality can also be seen in the succession of grits, marine bands and shales that make up the late Pendleian to Yeadonian strata. Dinantian and later Carboniferous structural history is thus one of gentle epeirogenic or perhaps eustatic movements. The rhythmic sequence was, however, interrupted by one important period of non-sequence or emergence, when uplift centred upon Greenhow tilted the Askrigg Block towards the north and west, and sediments were stripped off, or not deposited—probably mainly the former—in progressively greater thicknesses towards the centre of disturbance. This event can be precisely dated as coming between the Upper Stonesdale Cyclothem of the Pendleian and the arrival of the Grassington Lower Howgate Edge Grit. It gave rise to the intra-Pendleian unconformity, first described by Chubb and Hudson (1925), though Hudson (1938) continued to postulate regression northwards of the Yoredale facies. At the end of the Carboniferous the Hercynian movements, still epeirogenic in character, were expressed by gentle folding of the block accompanied by fracturing particularly along the southern and eastern sides. The Stainmore Monocline (the Dent Line of Turner, 1935), defining the western margin of the block, and the first movements of the Inner Pennine Fault* belong to this period. The Middleton Tyas – Sleightholme Anticline (Wells, 1957) and the extensive region of complex fracturing lying south-west of it, here called the North Swaledale Mineral Belt, are of the same age for the anticline is clearly truncated by the Rotliegendes unconformity beneath the Permian. The Hercynian fracturing prepared the way for the ingress of the mineralising fluids that emplaced the primary ores.

The chief later movements were those that elevated the Pennine blocks. Bott (1974, 1978) has postulated that notwithstanding the reverse movement on the Inner Pennine Fault, the Carboniferous sediments on the Alston Block stood at a higher level than those of the Vale of Eden as a result of the Hercynian movements, and that they were further elevated relative to the Vale during Permo-Triassic time. Critical evidence is lacking to decide whether this was also the case for the Stainmore (Dent Line) Monocline. All observers are, however, agreed that both blocks were elevated to their present positions by post-Triassic movements on the Outer Pennine, Dent and Craven faults. Formerly it was usual to see these major uplifts as strain effects in the foreland of the Alpine Orogeny of Tertiary date; but the geological investigation of the North Sea has revealed that another period of movement, the Cimmerian, reaching the maximum of its effect towards the end of the Jurassic (see, for example, Ziegler, 1975) may be of equal or greater

*See Shotton, F. W., 1935, p. 676

importance. Critical evidence is lacking for the present area but arguments will be advanced for regarding one or both of these periods as post-dating the mineralisation.

STRUCTURE OF THE LOWER PALAEOZOIC BASEMENT

Direct observation of the basement rocks of the Askrigg Block is possible only in the small inlier at Capon Hall and the Ribblesdale and Austwick inliers (north of the North Craven Fault and east of Ingleborough) and Chapel-le-Dale (west of Ingleborough). Only two borings have penetrated these rocks within the block, Raydale [9026 8474] (Dunham, 1974b) and Beckermonds Scar [8635 8016] (Wilson and Cornwell, 1982) (see pp. 8, 9). The Ingleton Group rocks, seen in Chapel-le-Dale and around Horton-in-Ribblesdale, exhibit a markedly different structural style from that of the later Ordovician and Silurian rocks of the Craven inliers. Leedal and Walker (1950) have demonstrated by means of 'way up' studies that they are in tight isoclinal folds, overturned towards the south and striking about N40°W. These rocks are now believed to be Arenig in age (p. 8) and are thus coeval with part of the Skiddaw Group of Cumbria. There is growing agreement that the Skiddaw Group was folded by late Arenig movements, prior to the Borrowdale volcanic episode (for example, Soper and Moseley, 1978) but post the late-Arenig Eycott volcanism (Wadge, 1978). The pre-Borrowdale folds in the Skiddaw Group trend and plunge to the north in the Lake District, and it is not therefore obvious that the folding in the Ingleton Group belongs to this phase, though it may do so. A second phase of Caledonian movement, preceding the deposition of the Caradocian/Ashgillian Coniston Limestone Group, is well attested in Cumbria and this, too, may have affected the Ingleton Group. Non-sequences in the Coniston Group of the Craven inliers (Ingham and Rickards, 1974) may indicate the continuation here of movements later than the so-called pre-Caradoc movements. The conspicuous axial-plane cleavage of the Ingleton Group would normally be regarded as an accompaniment of the tight folding, and Soper and Moseley's conclusion (*op. cit.*, p. 53) which implies that all the cleavage is of late Silurian – early Devonian age may not be justified in this area. The late Caledonian folding of the Coniston Limestone Group and the overlying Silurian rocks produced, as already noted, a much more open style of folding, striking N70–80°W and plunging gently to the east; cleavage here is much less conspicuous. The fact that the fold trends (north-west to west-north-west) in the Craven inliers are not at all consistent with the normal directions of Caledonian folds in the north of England (east – west to north-east) has been commented upon by Moseley (1972), who also notes a tendency of the trends to swing round the Shap granite.

The Rb:Sr dating of the Wensleydale granite at 400 ± 10 Ma makes it reasonable to assume that the intrusion is of post-tectonic type, like the exposed Caledonian granites of the Lake District. The latest outline for the batholith, deduced from gravity measurements (Bott, *op. cit.*) shows it elongated in a west-north-west direction, with its southern margin more or less paralleling the strike of the folding and cleavage in the Craven inliers. The conspicuous gravity low beneath middle Wensleydale is taken to represent a cupola above the general level of the batholith. The belt of magnetic anomaly already mentioned (p. 8), described by Bott (1967), parallels the southern margin of the batholith, but extends for a long distance both to the west-north-west and east-south-east. Although this may be explicable in Langstrothdale by the presence of layers carrying sedimentary magnetite in the Ingleton Group, perhaps reinforced by contact metamorphism related to the granite, this explanation would seem difficult to sustain over the distances involved. At the same time it may be used as local evidence of the strike of the Ingleton Group beneath the Carboniferous cover and above the main roof of the batholith. Whether the east – west striking Silurian rocks of the Howgill Fells, exposed west of the Dent Fault, also continue over the Askrigg Block is entirely unknown, but, from analogy with the Pennine Fault, it can be argued that the Dent Fault had a strong downthrow west prior to the deposition of the ?Devonian and Carboniferous rocks, and thus that the Lower Palaeozoics east of the Dent Fault may be older than those of the Howgill Fells. The question of the extension of Skiddaw Slates from the Cross Fell and Teesdale inliers southward beneath the Cotherstone Syncline and the northern part of the block is also unsolved. It would not, however, be surprising if the grain of the Lower Palaeozoic foundation of the Askrigg Block had a predominantly west-north-west trend, in view of the much greater prominence of this direction among the post-Carboniferous fractures when compared with the Alston Block, where an east-north-east strike predominates among the veins, reflecting basement trends. Some indication has, however, been obtained of the depth to basement beneath the Cotherstone Syncline, and Stainmore Trough. P. M. Swinburn (1976), interpreting seismic refraction data, finds depths of 9600 and 8200 ft (2.93 and 2.5 km) in the centre of the syncline, and 11 500 ft (3.5 km) under the Stainmore Trough. These figures may be compared with 2000 ft (0.6 km) in the centre of the Alston Block and approximately twice this figure on the north side of the Askrigg Block. There is some uncertainty about these figures because the Skiddaw Group may not provide a strong contrast with the basal Carboniferous and perhaps some deeper interface is involved.

STRUCTURES MARGINAL TO THE ASKRIGG BLOCK

Cotherstone Syncline and Stainmore Trough

The general trend of the syncline is nearly east – west, and there is a gentle eastward plunge at 125 ft per mile (24 m/km) until a nearly flat bottom is reached near sea level. Namurian strata occupy the centre of the syncline, which has weathered to a broad topographic depression, loosely known as the Pass of Stainmore and followed by route A66 from Bowes to Brough under Stainmore. The first detailed account of the syncline is that of Reading (1957), whose work includes a structure-contour map. Some revision in detail has resulted from the resurvey of 1958–1967 leading to the publication of the Brough-under-Stainmore 1:50 000 map (Sheet 31) and its description (Burgess and Holliday, 1979). The contours used in Figure 15 are derived from the later publication.

If the Brigantian and Asbian succession established for the Alston Block continued unchanged beneath the syncline, the base of the Carboniferous would have been at about 1600 ft (488 m) below sea level north-east of Bowes, or less than 3300 ft (1 km) below surface, figures that may be compared with those of Swinburn cited above. The investigation of the Swindale Beck Fault (linked to the Closehouse – Lunedale faults, which form the northern margin of the syncline) had shown an abrupt change, across the fault, in the thickness of the Asbian Melmerby Scar Limestone from 130 ft (40 m) on the north side to over 350 ft (107 m) on the south, suggesting thickening into a trough associated with the syncline (Dunham, 1948, p. 63) but this gave little indication of the magnitude of the depression. Kent's (1966) reconstruction of Carboniferous palaeogeography suggested a linkage from a gulf containing basinal facies (as indicated by the Cleveland Hills No. 1 Borehole) through Stainmore with Ravenstonedale. The gravity measurements by Bott (1967) showed Stainmore, including the Cotherstone Syncline and the country for several miles to the south, as a 'quiet' area compared with the granite-cored blocks to the north and south, suggesting a deep trough of thick Carboniferous between the two blocks. Johnson and Marshall (1971) have implied that early Dinantian (Tournaisian) sediments, not present on the blocks but exposed in the Basement Group of Ravenstonedale, could continue through the gulf. The concept of a deep trough between the blocks is a new and important development since the first volume of this memoir was written and it is now usual for structure maps to indicate this (for example, Johnson and Marshall, *op. cit.*, fig. 1 and Bott, 1978, fig. 18). The southern margin of the gulf, as distinct from that of the syncline, has not yet been established by drilling, but Bott (1967) has indicated that the line of the Stockdale Disturbance (p. 128) would satisfy the gravity data. The section (Figure 4) shows a transition from trough to block rather than a break, but in either case the highly mineralised North Swaledale Belt (Chapter 9) must overlie basinal rather than block sediments, though this is not apparent in the exposed Brigantian sediments there, which range stratigraphically down to the Middle Limestone. However, as Figure 8 shows, there is a marked thickening of the late Brigantian beds in the area between Gunnerside Gill and Kirkby Stephen. Accepting the existence of the trough, there is in the present state of knowledge a question about where shelf facies as seen in Ravenstonedale gives place, at each stage, to true basin sediments.

Stainmore (Dent Line) Monocline and Dent Faults

The major structural zone that marks the western boundary of the present area trends south-south-west from Stainmore Common, near Kirkby Stephen to Leck near Ingleton, a distance of 25 miles (40 km). To the west of it, the thick Carboniferous of the Howgills – Ravenstonedale area has a dip of about 12° to the north, steeper than the average inclination of the Carboniferous of the orefield. The displacement across the Dent Fault zone therefore increases to the south, and reverses in direction. Opposite Brough Sowerby there is a net westerly downthrow; the base of the Great Limestone (Main Limestone) reaches 1500 ft (451 m) OD east of the zone, whereas beneath the Permian of the Belah basin the same horizon is below 200 ft OD. However, a series of swells such as the asymmetric Hartley dome, and the smaller Limes Head dome of Turner (1935, figs. 12, 13) reverse this position before Kirkby Stephen is reached, (see Figure 15), and thereafter the net displacement is down to the east. These folds are consistent with the origin of the zone as an anticline overturning to the east, as advocated by Bott (1974) for the Inner Pennine Zone bounding the Alston Block; and as Turner suggested, the Dent disturbances are properly regarded as the southward continuation of Shotton's (1935) reverse faults. In an analysis of the Dent structure, the belt of domal swells along the western margin may be regarded as the first element, but they do not appear to persist along the full length of the disturbance. The second element is the Dent Line Monocline of Turner (*op. cit.*) (= the Stainmore Monocline of Burgess and Holliday *op. cit.*), the course of which is shown on Figures 15 and 24. Except where affected by minor strike faulting, the beds involved appear in normal stratigraphical order; they are upturned to the west as shown by Strahan (*in* Dakyns and others, 1891, figs. 7, 8 and 10) and the structure faces, and in effect downthrows, east. The third element is the Dent Fault *sensu stricto*, a high angle reverse fault downthrowing east, from which the monocline diverges north of Kirkby Stephen Common, the fault meanwhile continuing, but diminishing in throw, until the base of the Permian is reached, beyond which it is not recognised. The evidence is unfortunately not clear enough to prove that the movement on the fault was entirely pre-Permian, indeed it seems very likely that uplift of the Askrigg Block must have taken place along this line, as well as along the Craven faults that join it, in Cimmerian or later time. It is certain that faults such as the Argill and Augill faults of the Brough (31) Sheet, which bring down an outlier of Westphalian within the monocline, reverse its eastern displacement; these faults probably post-date the New Red Sandstone. Nevertheless, around Dent the net displacement must be at least 3600 ft (1405 m) down east, and may be greater. The elements in the structure may be summarised by a west – east traverse in the neighbourhood of Kirkby Stephen which would show (a) the thick Carboniferous sequence of Ravenstonedale dipping steadily north; (b) a belt of domes and basins (one of which contains Permian rocks), occupying a width of up to 1¼ miles (2 km) immediately west of the disturbance giving place farther south to the Dent Fault; (c) the monocline, facing east but cut by normal faults downthrowing west; (d) the thinner Brigantian – Pendleian sequence of the high fells, cut by a series of west-north-west normal faults. Around Hartley Birkett mineralisation occurs both within the disturbance and, on a small scale, in the domal belt.

The Craven faults

These well-known faults, first mapped and described by Dakyns and others (1890), have given rise to an extensive literature, important contributions to which have come from Garwood and Goodyear (1924), Wager (1931), Hudson (1944) and Black (1950). Subsurface evidence for the eastern stretch of the northern fault is given in Dunham and Stubblefield (1945) and Wilson, Dunham and Pattison (in preparation). Three main faults make up the system with many minor links (Figure 15). The South Craven Fault affects Westphalian strata and overlying red conglomerates that some have regarded as equivalents of the Permian Brockrams (most recently, Moseley, 1972, fig. 11), though others disagree (Dunham and others, 1952, p. 108). The fresh scarp made by this fault at Giggleswick, near Settle, has been regarded as evidence of geologically recent movement. This fault runs east-south-east into the Craven Basin and is of no further concern here. The Middle and North Craven faults are associated with, although they do not entirely coincide with, the hinge belt of Dinantian time separating the Askrigg Block and Craven Basin. Across the hinge, thick Asbian limestone overlain by Brigantian cyclothems give place to basin facies black shales, shaly limestones and occasional sandstones. The tectonic conditions along the hinge promoted the accumulation of the knoll and patch reefs of Cracoe and Malham, which coincide geographically with it. Ramsbottom (1974, fig. 22) has illustrated stages in the structural history of the reefs as they formed and were buried by Bowland Shales. Though they lie mainly south of the faults, the small mineral deposits found in them are described in the present account. Further evidence of contemporaneous movement during the Dinantian comes from the Middle Craven Fault, which Hudson (1944) and D. J. C. Mundy and R. S. Arthurton (personal communication) show to have formed a scarp against which Bowland Shales were banked, transgressing it in places. As already noted, elevation was taking place around Greenhow, at the east end of the North Craven Fault, prior to the deposition of the Grassington Grit in the late Pendleian and at least one fault, the South Waterhole Vein, with a throw of as much as 700 ft (213 m) down to the north, is covered by the grit. The North Craven Fault, with a similar throw in the opposite direction, may have begun to move at this time, but no proof of this has been found. Mapping suggests that Upper Bowland Shales are substantially thicker to the south of the fault than to the north at Greenhow, but they cross the line of the fault, in strength, (Black, 1950) at Hebden and Grassington.

The North Craven Fault, like most large faults, is, in detail, a complex of subparallel fractures. In the western one-third of its course the fault exactly parallels the fold-axes and cleavage of the Ingletonian rocks of the inliers from its junction with the Dent Fault as far east-south-east as Austwick. Beyond this, the fault changes to a more easterly course, apparently in sympathy with the strike of the Caradocian and later Lower Palaeozoic rocks. A further change to slightly north of east occurs east of Grassington, but here no information is available about a possible basement control of the trend. Splitting of the fault into a more widely-spaced series of fractures occurs after it has crossed the Wharfe valley between Grassington and Linton (Black, 1958) and one of the these can be shown to downthrow north (p. 201). An underground exposure of the main fault-zone was seen in Hebden Trial Level (see Figure 34), and evidence is given later that the zone is at least 400 ft (122 m) wide at Greenhow, where parts of it were sufficiently mineralised to be worth working. The Middle Craven Fault and several diverging fractures carry ore near Malham.

The net post-Westphalian displacement on the combined North and South Craven faults at the west end of the zone can hardly be less than 2000 ft (610 m) and may be considerably more. In the vicinity of the Craven inliers, the throw of the North Craven Fault, calculated from the position of the base of the Carboniferous north of the fault, and that of the Girvanella Band to the south is not less than 850 ft (260 m). If the interpretation of the geology of the Bradford Aqueduct Tunnel at Greenhow, given by Dunham and Stubblefield (*op. cit.*), is correct, this fault throws 750 ft (228 m) down to the south but the displacement diminishes eastward to less than 200 ft (61 m) SE of Coldstones, and near Pateley Bridge the structure dies out. These estimates assume normal dip-slip movement, but Wager (*op. cit.*) has proposed an elegant model to explain the change in joint directions near the fault that implies dextral transcurrent movement. This movement, if accepted, fits in better with northerly-directed stress, as proposed by Anderson (1951) than with the east – west stress system required to produce the Inner Pennine and Dent Line movements and the other northerly strain effects discussed by Moseley (1972) in his general synthesis. An initial northward stress field is more consistent with the accepted plate tectonic situation of Variscan time, but it may be suggested that this could generate rotation of the Lake District and Northern Pennines rigid blocks, in turn re-orienting the whole stress field, a concept implied in Dunham (1933).

Of the several folds in the Craven Basin, one continues east-north-east to become the Skyreholme Anticline, the axis of which has been considered by some authors to converge upon the North Craven Fault. In fact, the evidence of the Bradford Aqueduct Tunnel suggests that the axis turns and runs eastward parallel to the fault, and that the Greenhow Anticline is in effect a parallel structure, and not the continuation of the Skyreholme structure (Dunham and Stubblefield, *op. cit.*, fig. 5). On a simple model, the positions of the Ribblesdale folds could be consistent with dextral slip on the Craven faults. The Greenhow Anticline, comprising three domes and two basins, also raises interesting questions about the nature of the Hercynian stress field; the long axis is east – west; the basins may be regarded as cross folds on north – south axes.

INTERNAL STRUCTURE OF THE OREFIELD

The Askrigg Half-dome

The broad structure of the field has been investigated by means of structure-contours on the base of the Main (= Great) Limestone. In the northern half of the area this is reasonably satisfactory because there are many exposures of this limestone and there is useful subsurface data. South of Wensleydale, however, the Main Limestone forms only disconnected outliers, save in the Masham (Sheet 51) area, and a second reference datum, the Simonstone Limestone has been used. The first attempt (Dunham, 1958) was based upon a standard stratigraphical separation of the two limestones of 550 ft (168 m), but recent investigations have provided the information summarised in the isopachyte maps (Figure 10), and Figure 15 thus gives a more accurate picture than the 1958 version (also used in Dunham, 1959; 1974a). The general form of the gentle structure remains unchanged.

Although investigators of strong folding in adjacent areas may regard the cover rocks of the Pennine blocks as flat-lying (e.g. Moseley, 1972), both blocks are in fact half-domes, the distortion

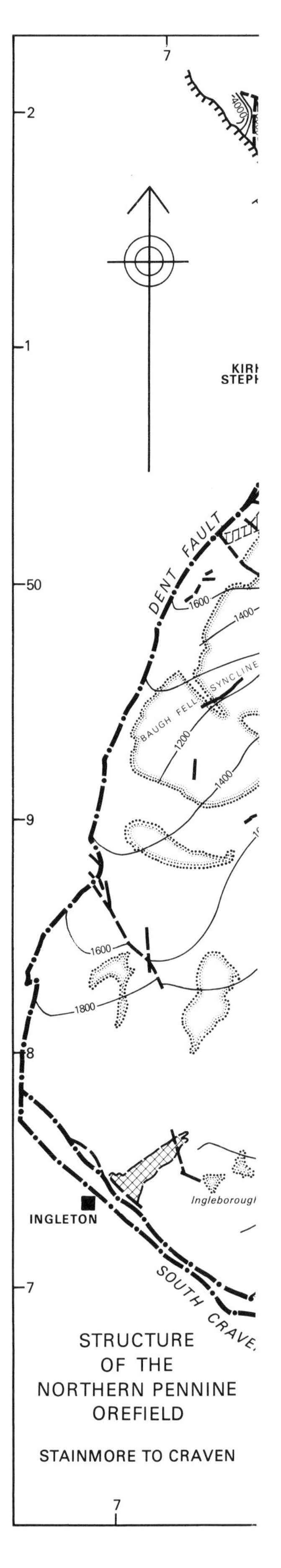

Figure 15 Structure of the North Pennine Orefield (Stainmore to Craven). Contours (in feet above Ordnance Datum) on the base of the Main Limestone, with indirect data from the Crow, Little, Underset, Middle and Simonstone limestones. Data are incorporated from Reading (1957), Wells (1957), Rowell and Scanlon (1957), Dunham (1959) and Burgess and Holliday (1979).

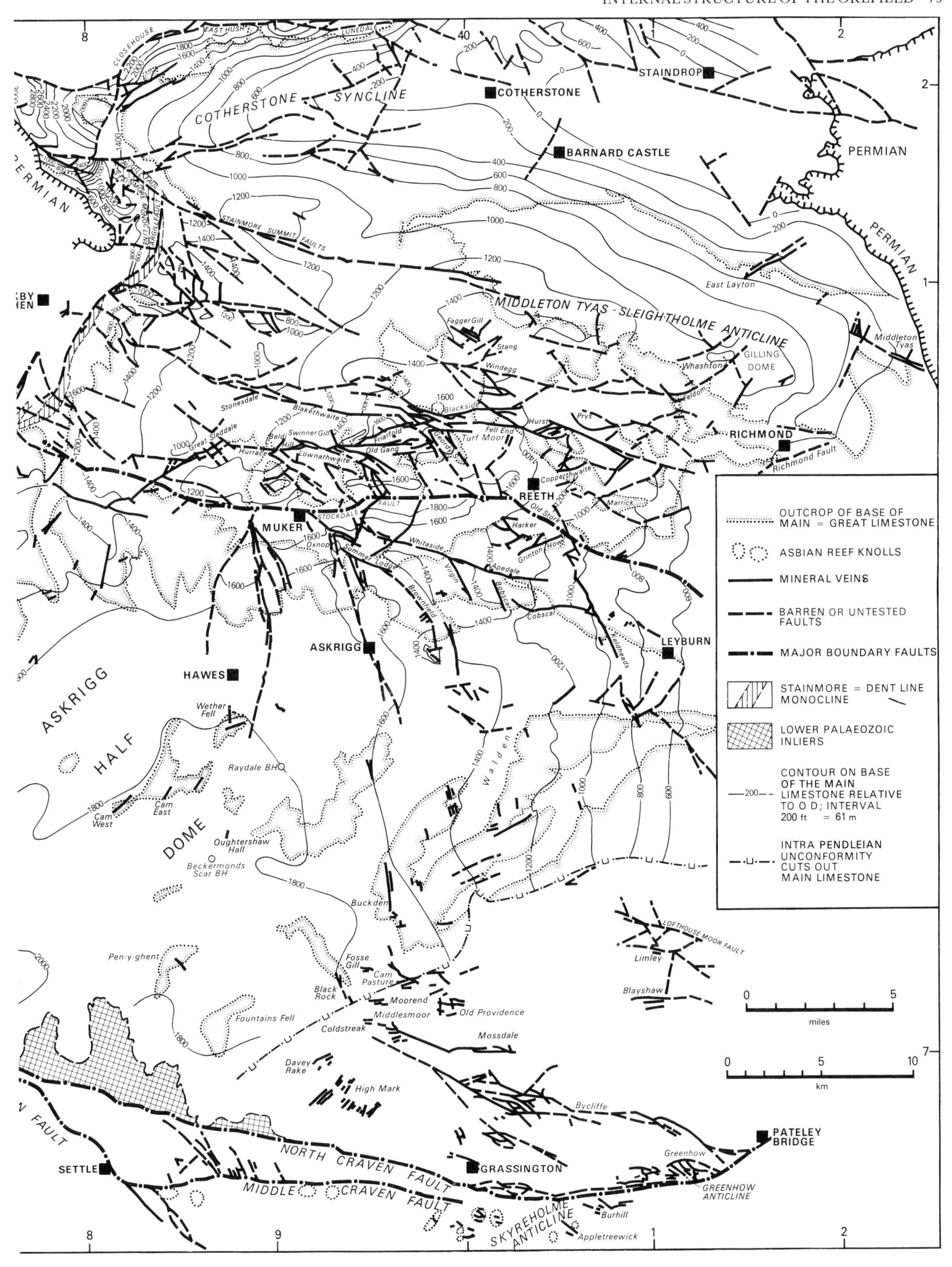

COTHERSTONE SYNCLINE
COTHERSTONE
STAINDROP
BARNARD CASTLE
PERMIAN
STAINMORE SUMMIT FAULTS
MIDDLETON TYAS - SLEIGHTHOLME ANTICLINE
GILLING DOME
East Layton
Middleton Tyas
Whashton
Feldom
Faggergill
Stang
Windegg
Blackside
Hurst
Prys
RICHMOND
Richmond Fault
Stonesdale
Blakethwaite
Great Sleddale
Beldi Hill
Swinner Gill
Friarfold
Turf Moor
Fell End
Hurrace
Lownathwaite
Old Gang
Copperthwaite
REETH
Marrick
MUKER
STOCKDALE FAULT
Harker
Old Stork
Oxnop
Summer Lodge
Whitaside
Grinton How
Apedale
Virgin
Bobscar
Brownfield
Cobscar
Keldheads
ASKRIGG
HAWES
LEYBURN
ASKRIGG HALF DOME
Wether Fell
Raydale BH
Cam East
Cam West
Oughtershaw Hall
Beckermonds Scar BH
Walden
Buckden
Pen-y-ghent
Fosse Gill
Cam Pasture
Black Rock
Moorend
Old Providence
Middlesmoor
Coldstreak
Fountains Fell
Mossdale
LOFTHOUSE MOOR FAULT
Limley
Blayshaw
Davey Rake
High Mark
Bycliffe
PATELEY BRIDGE
Greenhow
GREENHOW ANTICLINE
NORTH CRAVEN FAULT
MIDDLE CRAVEN FAULT
SETTLE
GRASSINGTON
SKYREHOLME ANTICLINE
Burhill
Appletreewick
OUTCROP OF BASE OF MAIN = GREAT LIMESTONE
ASBIAN REEF KNOLLS
MINERAL VEINS
BARREN OR UNTESTED FAULTS
MAJOR BOUNDARY FAULTS
STAINMORE = DENT LINE MONOCLINE
LOWER PALAEOZOIC INLIERS
CONTOUR ON BASE OF THE MAIN LIMESTONE RELATIVE TO O D; INTERVAL 200 ft = 61 m
INTRA PENDLEIAN UNCONFORMITY CUTS OUT MAIN LIMESTONE
miles
km

being of significance as a necessary background to the emplacement of the ores.

The Askrigg half-dome is open towards the Craven faults, and has a gentle northward plunge. The beds dip north-west, north and east away from the maximum elevations of the Main Limestone which lie above 2000 ft (610 m) OD on Ingleborough and Fountains Fell. The central part of the half-dome, covering High Mark, Littondale and Langstrothdale and thus a considerable part of the Hawes (Sheet 50) area, has a structural slope to the north at only 40 ft per mile (7.6 m/km) and is thus nearly a structural plateau. This is reflected in the fact that it is almost devoid of faults and of fractures other than joints; it is correspondingly very poorly mineralised. After the 1600 ft structure-contour is passed, however, the inclination increases in all outward directions. To the north-west towards the Dent Fault, the average dip increases to 160 ft per mile (30.3 m/km) but the structure is increasingly disturbed by faults, dominantly west-north-west to north-west in direction. To the north the 'nose' of the half-dome along the Ure/Swale interfluve is cut by many north – south and west-north-west veins and faults. On the east side of Wharfedale, the dip increases to 120 ft per mile (22.7 m/km) and continues through Walden and Coverdale, and though the Main Limestone to the south-east is cut out by the intra-E_1 unconformity (Figure 13), Wilson and Thompson's (1965) observations in the Kirkby Malzeard area indicate that a comparable dip continues above the unconformity. The appearance of substantially productive veins on the east side of Wharfedale coinciding with the increase of dip begins in the north at Buckden Gavel, and extends through the Kettlewell and Conistone fields to the major deposits of Grassington High and Low moors; the anticlinal structural ridge of Greenhow and the Skyreholme Anticline are also extensively mineralised.

Two remarkable features of the northern part of the half-dome are the small but deep basins apparently associated with the Brownfield – Summer Lodge Fault (Figures 31, 32) at Carperby and Oxnop. It is not easy to understand how such features could arise in the sedimentary cover of the granite-cored block; perhaps bodily removal by solution of limestone or of a small lens of evaporites could generate the neccessary subsidence.

Stockdale Fault (Disturbance)

This occupies a position in some respects comparable with the Burtreeford Disturbance that bisects the Alston Block, though its direction is 90° different. It has variously been called the Stockdale Vein (Dakyns and others, 1891) and Monocline (Dunham, 1959). It nowhere carries workable mineralisation and is clearly in part a compression feature. As already noted, Bott (1967) considers that it coincides with the north margin of the Askrigg Block *sensu stricto*. The main course is from Great Sleddale Head through Thwaite and along the north slope of Swaledale to near Healaugh, where it is lost under drift. It is clear that its eastward course cannot continue much farther; it seems likely that it is linked to, and dissipates its throw along, the south-east-trending Old Stork Vein, which itself continues as a fault downthrowing north-east for a further 11 miles (17.6 km) to Constable Burton [SE 144 908]. West of Great Sleddale, two faults link it with the Dent Disturbance so that the orefield is in effect bisected by this continuous structural line. The elements of the structure include a fault or faults throwing 450 ft (137 m) near Gunnerside Gill and a monoclinal dip, particularly from the downthrow side, which brings the Main Limestone down as much as 300 ft (91 m) on Gunnerside Pasture. It appears to be another illustration of the old Durham Coalfield saying that 'beds rise to a dipper, and dip to a riser' when approaching a fault, an anomalous situation which would not be expected from purely tensional faulting. Clough's section (*in* Dakyns and others, 1891, fig. 11) shows the monocline and fault but gives a much simplified view of the disturbance. For part of the course of the disturbance a small anticline rather than a monocline lies on the hanging wall side of the fault, as shown on Figure 25. KCD

Folding in association with the Stockdale Fault is seen in almost continuous exposure in Ivelet Beck [SD 9367 9796 to 9358 9824]. Beds between the Gayle and Hardrow Scar limestones are thrown into a series of folds in a 690-ft (210-m) wide belt which becomes progressively more gentle away from the fault (Figure 16). There are 111 ft (33.8 m) of fine-grained sandstones exposed with a few bands of siltstone and mudstone. These beds show dips ranging up to 58°, and a very local dip of 80° close to the fault where gaps in exposure could conceal additional fault planes.

The fault plane itself is seen in a small waterfall, hading 22° in a direction N23°W. A 30-ft (9-m) gap in exposure could conceal any associated crush zone. North of the fault, which is estimated to throw down north 462 ft (141 m), is a continuous section in Ivelet Force in almost flat lying strata. AAW

This section shows the detail of the folding on the footwall side of the fault; the inferred fold to the north lies beyond the end of the drawn section. At Thwaite the shift of the base of the Main Limestone across the dale, resulting mainly from the disturbance, is 600 ft (183 m) down north. As in the case of the Burtreeford Monocline, there is no correlation in the line of vein-fractures lying on opposite sides of the disturbance. From the north side, for example, the Watersykes, Barbara and Kining veins (Figure 25) of the North Swaledale Belt strike south-east towards the disturbance but do not pass through it; while Friar Intake Vein (Figure 30) worked almost up to it from the south side was never found to the north. Nearing Healaugh, the disturbance is shown on the Richmond (41) Sheet as terminating against a small north-west fault near Kearton. However, we consider that it most probably continues to join up with the west-north-west fault coming down from Feetham Pasture, linking across via Harkerside Place [SE 027 988], north of the old flagstone levels near Maiden Castle, with the Old Stork Vein. This vein, at Redway Head Mine, throws 280 ft (85 m) NE (see Figure 30).

Richmond anticlinal belt

The term follows Kent (1974) and is taken to include the Middleton Tyas – Sleightholme Anticline (Gilling Dome) and the collapsed anticlinal or domal region cut through by the North Swaledale Mineral Belt, lying south-west of it. North of the Stockdale Disturbance, the 'nose' of the Askrigg half-dome can still be discerned with the beds dipping off it to the north-west, north and east, but in a much broken-up condition resulting from the fracturing in the mineral belt. This belt contains the most extensive mineralisation in the area herein described; the vein-fissures, some of them faults

Figure 16 Section through the Stockdale Disturbance in Ivelet Beck, Swaledale. Stratigraphic symbols as on Figure 8.

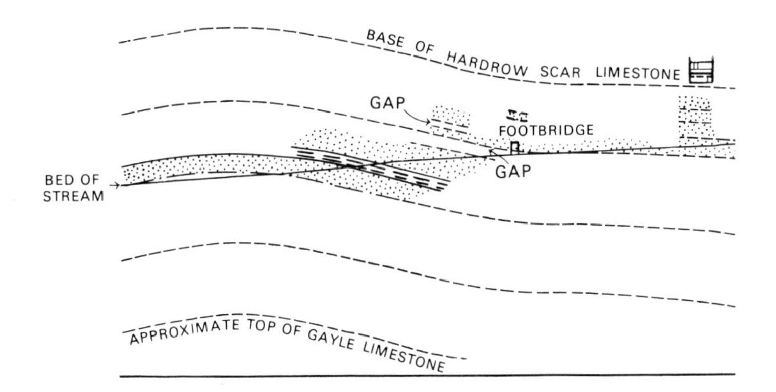

with displacements into hundreds of feet, run west-north-west, east – west, east-north-east and north-north-west and will be discussed later. The fragmentation of the area comes out most clearly when the ground with the base of the Main Limestone between 1600 and 1400 ft (488 and 427 m) is emphasised as on the inset to Figure 25. North and east of this fragmented area lies the Middleton Tyas Anticline, trending west-north-west, which itself provides the south-west flank of the Cotherstone Syncline. Dakyns and Gunn first mapped the anticline, but nothing was published about it before Marr (1910) and Kendall (1911); the latter author suggested that it was continuous with the Stowgill Anticline (cf. Dakyns and others, 1891; Turner, 1935), but Versey (1927, 1942) disagreed and the Stowgill structure, as a result of recent mapping in the Brough (Sheet 31) district, is no longer recognised as a continuous structure. The first detailed study of the Middleton Tyas – Sleightholme Anticline was that of Wells (1957), whose structure-contour map provided some of the data for Figure 25. He demonstrated the continuity of this structure from Sleightholme [NY 959 101 NZ 250 040] where it disappears beneath the unconformable base of the Permian; the axial trend is N.75°E and axial plunge 100 ft per mile (19 m/km) to the east; the inclination of the north limb averages 4° or 350 ft per mile (66.6 m/km); there are three crestal regions, respectively at Stang Forest, Whashton and at Gilling (Figure 15). The remarkable feature is the almost total absence of fracturing, and therefore mineralisation, except at the south and east margins of the structure. The age of this fold is, no doubt, Hercynian but speculation continues that both it and the Cleveland Hills Dome, affecting the north-east Yorkshire Jurassic rocks, overlie a common belt of crustal instability. Dr G. A. L. Johnson (personal communication) suggests that the Middleton Tyas Anticline might be due to halokinesis by Lower Carboniferous salt or other evaporites but no direct evidence of what underlies the anticline is at present available.

The vein system of the North Swaledale Mineral Belt is included in the discussion below, and Chapter 9, with a structural map (Figure 25), gives details. The most noteworthy general point is that there are fragmented structural highs in the central region, flanked by relative lows to the west and east. The general trend parallels that of the Stockdale Disturbance as far as the entrance to Great Sleddale, but the Belt then turns west-south-west and links with the disturbance at Sleddale Head.

West and north of the Mineral Belt, still more complex faulting characterises the north-west part of the present area, though very little mineralisation has been found, so far, in association with it. The detailed stratigraphical and structural study of Rowell and Scanlon (1957) has provided data for the structure contours on Figure 15 here. It will be noted that structurally high ground lies adjacent to parts of the Dent Line, cut by many north-west faults.

Joints

The brittle Carboniferous rocks throughout the area, even over the structural 'plateau' in the Hawes (50) Sheet, are everywhere cut by joints almost perpendicular to bedding, spaced at intervals between a few inches and as much as 10 ft (3 m). They include shear joints produced in conjugate sets by tear movement on the micro-scale; extension joints where the walls have been forcibly torn apart; and torsional joints arranged, usually on a small scale only, *en échelon*. Phillips (1836) description of the joints of this area, with a statistical diagram, must stand as one of the first in the history of the subject. His measurements show a strong maximum in a north-north-west direction, one that all subsequent work in the Pennines has substantiated. He also found east – west and east-north-east directions represented. A study of the limestone pavements of Kilnsey Moor by Wager (1931) brought out the fact that the dominant north-north-west joints are modified to a more westerly point as they come near the North Craven Fault, (except at Skirethorns near Grassington, where the modification is in the opposite sense). This fault he interpreted as a tear, with sinistral movement between Leck and Malham, but dextral around Grassington. Wager noted that a second joint-set, more or less at right angles to the north-north-west set, is usually present, and also a weaker east – west, north – south conjugate set. In detail, his careful observations on the dominant set show that variation in trend, even away from the Craven faults, is quite wide, the range being north to north-west with the average perhaps N20°W. This is consistent with observations made by Mr Tonks when he measured joints near the clusters of weak but numerous mineral veins in the Great Scar Limestone at Middlesmoor and Cam Pasture, Wharfedale; he noted a weaker east-north-east set, and a very weak conjugate set bisecting roughly the angles between the other two. The relation between joints and mineral veins is discussed in the next section. The predominance of the north-north-west joints established in the Alston Block (Dunham, 1933; 1948) by statistical methods also applies widely over the Askrigg Block; this direction is one of three (the others being east-north-east and west-north-west) in the Middleton Tyas Anticline (Wells, *op. cit.*) and it is represented, though not dominant, in the Cotherstone Syncline (Reading, *op. cit.*). The extensive study of jointing in north-west England by Moseley and Ahmed (1967) brings out the fact that outside the blocks, and as far south as Derbyshire, more westerly directions, perhaps averaging N45 – 50°W. are dominant. The cleat of coals in the Yorkshire and

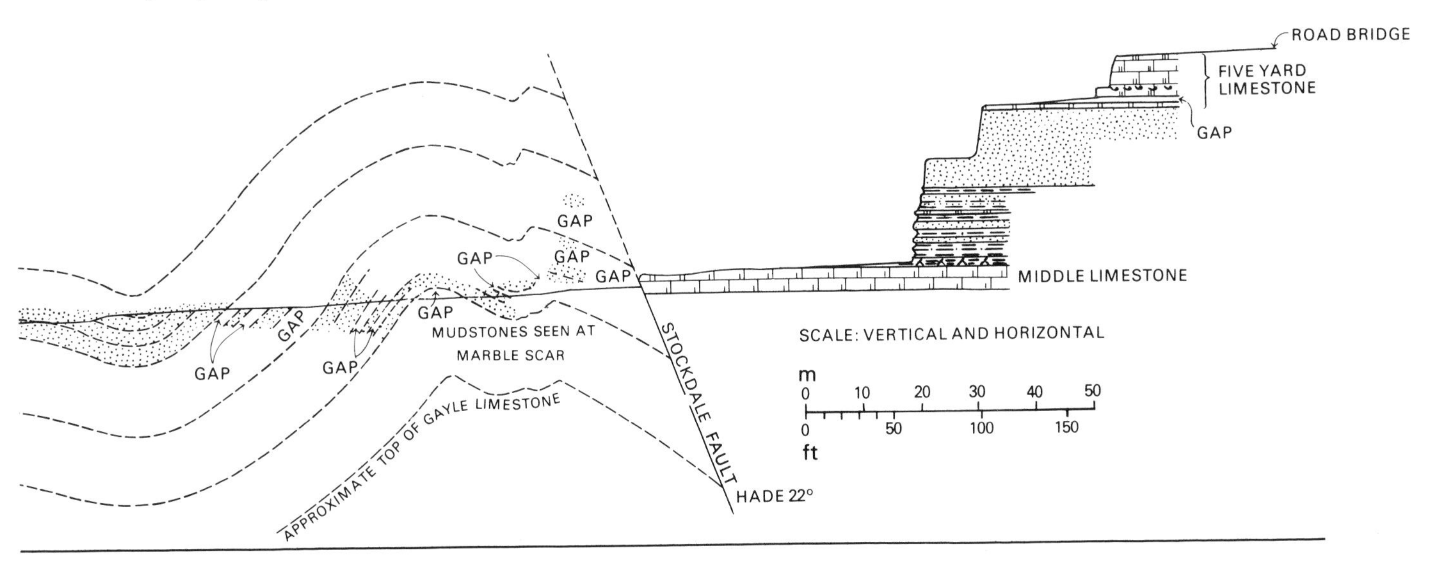

Lancashire coalfields corresponds with this direction (Kendall and Wroot, 1924, pp. 130, 131). Although the hypothesis advocated by many authors from Wager on, that the north-west or north-north-west direction and the set more or less perpendicular to it are shear joints, is an attractive one, this fails to explain the great predominance of the north – west or north-north-west set over the other direction, or the virtual absence of a north-east set in the coals, and it appears that there must be a regional rather than local cause for this. The inconsistency of the near 90° angle with rock mechanics experimental results indicating an acute angle substantially less than 90° between the shears, facing the direction of maximum compressive stress even in confined systems, also requires explanation. Wells (1957) takes the view that the east-north-east – west-north-west pair are true shear joints, but here the difficulty is that the north-north-west set must lie nearly at right angles to the maximum stress direction, where normally only the axial plane cleavage of folds is produced. Doughty's (1968) interesting observation of plumose structures on the faces of conjugate joints in the Great Scar Limestone, has led him to point out that experiments by Bagsar (1948) on mild steel under tension produced feather structures on cleavage fractures perpendicular to the principal tensile direction; and that Davis (1948), experimenting on steel cylinders filled with oil to act as energy reservoirs found they would shear at 45° to the direction of compression. Doughty therefore considers that inception of two sets of joints nearly at right angles is acceptable if the fractures, initiated as shears, are propagated as fracture cleavage, and if the rock acts as a strain energy reservoir to be released upon uplift into an environment where the confining pressure is less. Price (1966) has demonstrated the importance of residual strain energy in the structural evolution of brittle rocks, but the obstacle to applying this concept to the present problem appears to be that it requires jointing to be a late, not an early effect, whereas our experience strongly suggests that in the orefield, jointing preceded faulting and introduction of mineral veins; and that the latter were certainly in place before elevation of the blocks to their present positions. If Bott's (1974) proposal for elevation of the Alston Block prior to the deposition of the Penrith Sandstone but after the injection of the Whin Sill is also applied to the Askrigg Block it would mean that strain energy accumulated at this early stage, rather than at the time of the Cimmerian movements or in the Tertiary. As Doughty recognises, a later period of further uplift would be required, but this presents no difficulty in the light of the evidence provided by the mineral deposits, which were repeatedly reopened after emplacement and which in many cases exhibit post-mineralisation slickensides. All this happened after the jointing of the country rocks, for although minerals associated with the veins commonly extended into joints, the deposits are not traversed by the joint systems. It will be evident that several unsolved problems remain with regard to this phase of the tectonic history of the orefield.

Vein-fractures

The ore deposits are epigenetic veins, emplaced in fault-controlled fractures traversing stratigraphical levels from the Asbian up to the Pendleian, with associated replacement deposits in certain carbonate rocks where joints or minor fractures have acted as feeders. The courses of the veins are shown as solid lines on Figure 15 and 25, and the faults not known to be mineralised as broken lines. Compared with the Alston Block, the west-north-west direction here is very much more prominent, the east – west direction slightly more important, and the east-north-east decidely less so; in fact it may be claimed that the direction of maximum tensional opening in the Askrigg field was about N70 – 75°W, whereas in the Alston field it was N75°E (Dunham, 1959, p. 8). It may be noted that these two directions are respectively perpendicular to the Dent Line and the Inner Pennine Line. The North Swaledale fracture belt (Figure 25) shows a series of at least three nodes from which a fan of fractures radiated between north-east and south-east. These could be interpreted in terms of the experiments of W. Riedel (1929) as indicating that the North Swaledale fracturing was propagated from west to east, further emphasising the probable connexion with the Dent Line movements. Even where pronounced folds are being traversed, as at Greenhow Hill and Skyreholme (Chapter 13), the direction mentioned remains the axis of maximum opening. One vein-direction that may be distinguished from those already mentioned is the set lying between N5°W and 25°W, particularly on the south side of the Beldi Hill – Old Gang complex in the North Swaledale Belt, and in the Swale/Ure watershed. This corresponds with the 'cross vein' direction of the Alston field, where most examples are barren. Here, however, the direction is 30 – 35° nearer to the axis of maximum tension. Several useful oreshoots follow this trend. Very little evidence of post-mineralisation shift of other veins along this direction has been found here.

Fractures with small displacements provided the most effective channels for mineralisation. The figures given by Lonsdale Bradley (1862) for the veins in Swaledale, collected from all the mines while they were still in full operation, bring this out, as the histogram (Figure 19) based on Bradley's data shows. A similar situation, with faults of minor displacement as the ore carriers, holds for the field as a whole. The reason is to be found in the behaviour of minor fractures, perhaps unique to the northern Pennines in having small hades, generally not over 10° in the brittle strata (chert, limestone and quartzitic sandstone), with much higher inclinations as they pass through weaker beds such as mudstone, shaly sandstone and some sandstones. The effect is to give rise to laterally extensive, ribbon-like openings within the brittle beds, while the fissures remain closed in the softer beds. These characteristics have been discussed in detail (in Dunham, 1948, pp. 71, 72 and figs. 9, 10) and can be seen in the section at Surrender Mine (Figure 18) as well as in a cross-section from Grassington (Dunham, 1959, fig. 2). The particular conditions giving rise to this kind of structure are, it is suggested, a combination of cyclothemic sedimentary rocks with gentle doming, the tensional stress accompanying which is accommodated by this fracture strain. The width of the openings is related to the throw and there must be an upward limit beyond which such openings cannot be maintained. Empirically, this is probably around a maximum throw of 20 ft (6.1 m). Larger faults plane out the irregularities caused by the varying physical response of the wallrocks and this is accompanied by the development of gouge to which the shales largely contribute (the Whamp, Vamp, Douk, Donk or Platy Sample of the mining districts—Bradley, *op. cit.*, p. 11). The presence of this material greatly impedes the passage of fluids, and hence few of the larger faults are good ore-carriers.

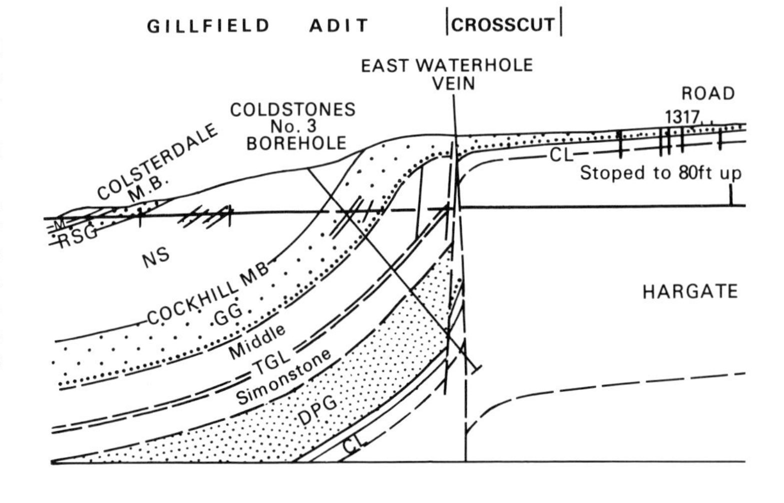

Figure 17 Cross-section through the Greenhow Anticline on the line of Gillfield Adit.

A further indication that the vein fractures formed under tensile stress is provided by the way in which, when numerous veins are present in a restricted belt, like the Surrender and Arkengarthdale part of the North Swaledale area, the downthrows are not in any consistent direction. The small blocks defined by the fractures moved freely up or down to accommodate the strain, a situation that would not be expected under conditions of compression (Figures 18 and 25).

For the numerous small veins around Proctor High Mark on Kilnsey Moor both Wager (*op. cit.*) and Raistrick (1938) have concluded that these are merely mineralised east-north-east joints. It is true that the displacement along many of them is small but some can be shown to be faults; and Wager's mapping of adjacent well-exposed joints gives them directions not coincident with the fractures. Mr Tonks' mapping of joints in relation to the many small veins on Cam Pasture, Kettlewell, as well as at Coldstreak, in Littondale, and at Middlesmoor Pasture (analysed in detail on pp. 187–189) shows that the principal joint directions are very poorly represented among the trends of the veins. The scrin complexes near West Feldom and Preston (pp. 152, 180) are now so poorly exposed that no conclusions can be drawn about these east-north-east, closely-spaced fractures. It may, however, be said that in general, joints only become mineralised near through-going channels of ancient fluid movement, provided by minor faulting which, as Moseley and Ahmed (1967) have concluded, has an origin different from that of the joints. Our conclusion is that jointing demonstrably preceded the vein-fracturing.

Other faults

Outside the main area of mineralisation, faults are most numerous in the high fell country north of the Stockdale Disturbance and its two western branches, extending up to the Cotherstone Syncline. Most of this ground, in the Kirkby Stephen (Sheet 40) district, was examined for structure by Rowell and Scanlon (*op. cit.*). Almost all the faults trend between west-north-west and north-west and there is no consistent direction of downthrow. The two links from the Stockdale Disturbance to the Dent Fault throw down north and both displace the Dent Line Monocline; thus it could be argued that after the compressive movements from the west that produced, at the Hercynian epoch, the Dent and Inner Pennine overfold and faults, there was a stage of north – south tension, as also indicated for the mineral veins; and that, in the centre of the Stockdale Disturbance, this followed a stage of relative upfolding to the north, the later fault more than cancelling the earlier gentle monocline. The depressed area north of Baugh and Wild Boar fells is succeeded by the Bleakham and Kitchen Gill faults (Figure 24) throwing in the opposite direction, and the base of the Main Limestone climbs up to 2000 ft (610 m) OD adjacent to the Bastifell Fault, only to descend to below 900 ft (274 m) as it approaches the Wrenside Fault. The Bastifell Fault is linked by a ramifying series of fractures with the productive Blakethwaite Vein of the North Swaledale Belt (p. 139). Two of these fractures have been tested, beneath the Pendleian and higher cover rocks, from West Stonesdale, but for most of their length the Blakethwaite – Bastifell faults do not lend themselves to easy exploration at promising levels, the Main Limestone being deeply buried. The structure of the ground of the Brough (Sheet 31) area has recently been described by Burgess and Holliday (*op. cit.*). The Stainmore Summit faults can be followed east-south-east for over 6 miles (9.6 km) from their junction with the Augill Fault (which has some mineralisation associated with it) (Dunham, 1948) to the old Spanham lead mines in the Barnard Castle (Sheet 32) area (Mills and Hull, 1976), but possible productive measures are concealed except at the two ends of this stretch. The Closehouse faults, part of the Lunedale belt forming the north side of the Cotherstone Syncline, and the Lunehead faults to the south, are well mineralised but lie in the area described by Dunham (1948). Generally, the faults cutting through the west-facing slopes of Edendale seem to be barren of mineralisation.

Apart from the five northward-trending mineral veins cutting through the Swale/Ure interfluve (Chapter 9), mapping has suggested the presence of at least six others in the Kirkby Stephen (Sheet 40) district with displacements of 20 to 40 ft (6–12 m), some to the west, others to the east. Only a few of these cross Wensleydale to the south side. A few examples occur in Upper Wharfedale and the adjacent valleys to the east; at Old Providence, Kettlewell, a long tunnel followed one such fracture, but it was barren. The Grassington field is remarkably free of 'cross veins', but at Skyreholme and the Craven Cross part of the Greenhow mining area they become numerous; almost without vertical displacement, they effect both dextral and sinistral shifts of the productive veins varying from a few feet up to as much as 100 ft (30 m) (Figure 36).

As might be expected, a number of faults which contain oreshoots in the North Swaledale Mineral Belt continue outside the mineralised area. Examples include the west-north-west fault crossing Cocker Hill [NZ 047 068], feebly mineralised at its western end but continuing to bring Main Limestone and Chert to the north into contact with Richmond Chert to the south for nearly 2 miles (3 km). The Alcock North Vein of Arkengarthdale continues both west-north-west towards Tan Hill and east-north-east along Moresdale Ridge for 3 miles (4.8 km) further, the eastern mile being along the crest of an anticline with 200 ft (61 m) amplitude. In Rispey Wood [NZ 064 043] where the south limb of the fold

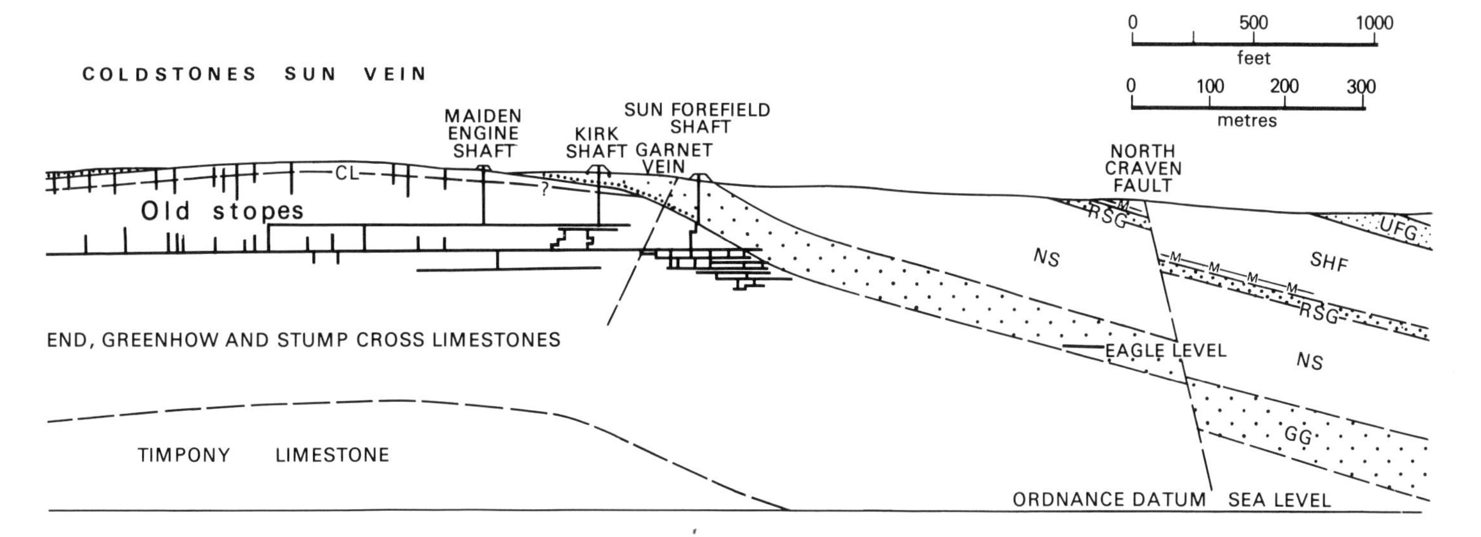

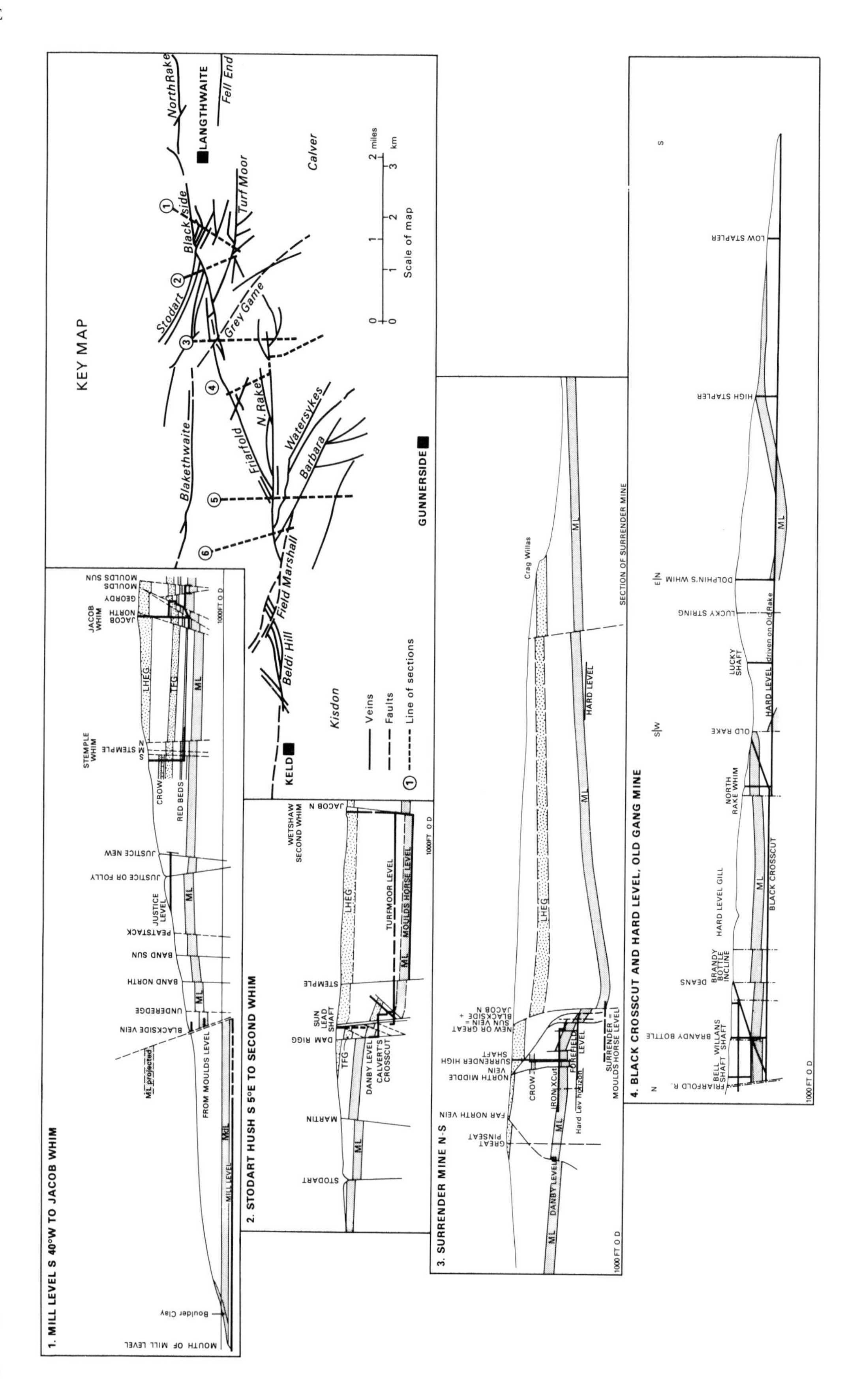

Figure 18 Cross-sections to show veins in the North Swaledale Mineral Belt. Data from the records of the A.D and C.B. companies and the primary geological survey.

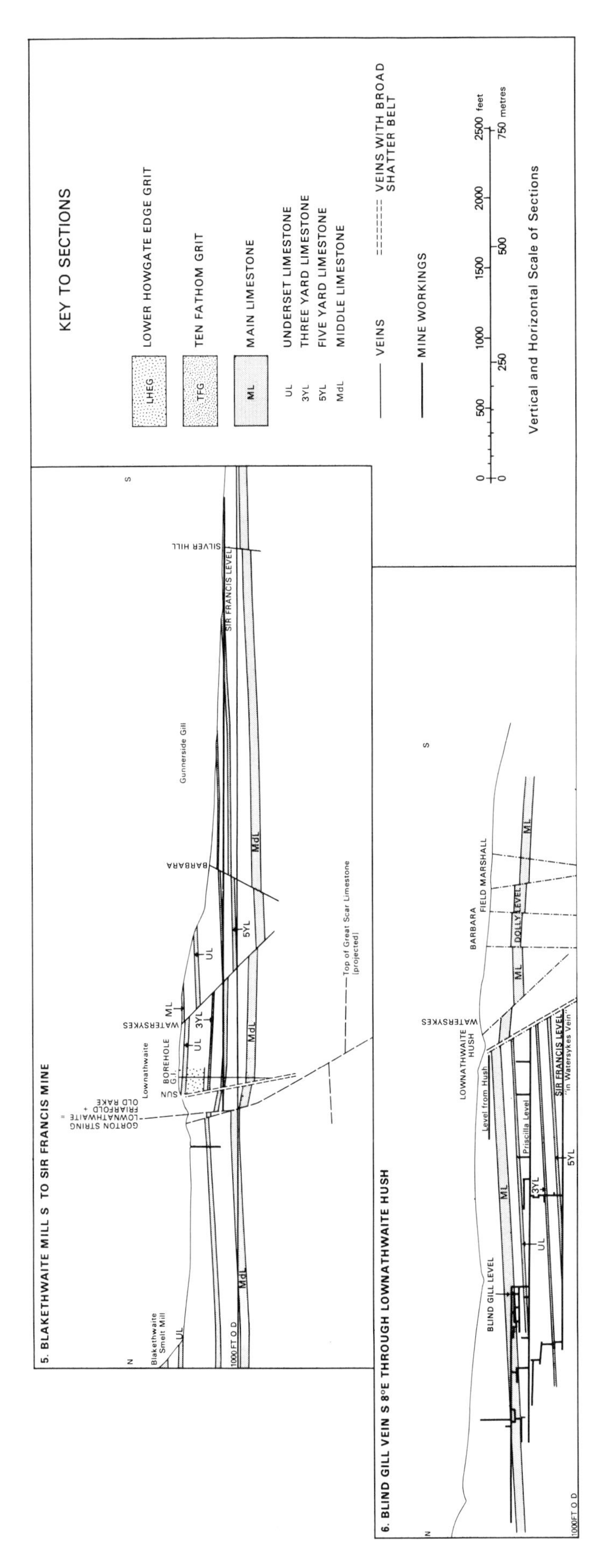

makes impressive features, branches are given off but the main fault changes direction abruptly to east-north-east for a further 1¼ miles (2 km) before following an irregular easterly course across Rake Beck to join up with the west-north-west High Waitgate Vein (Figure 25), linking it with the Feldom Vein and Fault of the Richmond Copper Area. The whole linked system, 13 miles (21 km) long, downthrows south. Almost 2 miles (3 km) farther south, the Wallnook North Vein, which forms the north side of the remarkable linking vein-complex at Hurst (p. 149 and Figure 27), continues south-east for 5 miles (8 km) beyond its known productive limit as the Marske Fault, also downthrowing south. Two linking barren faults in this area are noteworthy: the powerful Friarfold – Surrender – Blackside – North Rake runs of veins, trending north of east deteriorate beyond Slei Gill [NZ 020 031] into a north-east fault (shown as a mineral vein on One-inch Richmond Sheet (41) but in fact worthless) linking with the Moresdale Ridge Fault already mentioned. Similarly, the small Marrick vein-complex is linked by an equally barren fault to the Marske Fault. Though the density of faulting in the North Swaledale Mineral Belt (including the vein-fissures) is much greater than in the central 'plateau' of the Askrigg Block, it is hardly as great as the density in either the Durham or Yorkshire coalfield. A transition towards the latter is perhaps represented by the south-east part of the area where the Lofthouse Moor Fault brings in an inlier of Brigantian in generally Namurian country (Wilson, 1960b). A zone of inter-connected faults continues east-south-east north of Lofthouse; the small faults mostly belong to the 'cross vein' direction (Wilson, 1960a, b; Wilson and Thompson, 1965).

TECTONIC HISTORY

A number of summaries have been attempted since those of Turner (1927), Dunham (1933) and Shotton (1935), the latest, for north-west England as a whole, being that of Moseley (1972). The following lists the stages recognisable in the history of the Askrigg Block and Cotherstone – Stainmore Trough; the interpretation of some of these may be regarded as controversial:

1 The Pre-Cambrian crystalline basement—if such a basement exists here—probably lies many kilometres beneath this area, and unless it is revealed by the seismic refraction experiment of Swinburn (1975), which seems improbable, no knowledge of its position is available.

2 The earliest sediments (Ingleton Group; equivalent to part of the Skiddaw Group) form a thick series, dominated by turbidites, deposited in the proto-Atlantic or Iapetus Ocean. Two periods of earth movement or of sliding on the slope—late Arenig and early Caradoc—may have affected these rocks prior to the next stage. Cleavage is more prominent than in later rocks.

3 In Ashgill time, the Coniston Limestone and associated sediments were laid down, in part under shelf conditions, but in the Silurian deeper water sedimentation was resumed, until the late Silurian - early Devonian movements gave rise to moderate folding, with associated cleavage, not quite coaxial with the Ingleton Group folds.

4 The Wensleydale granite batholith was intruded at about 400 Ma, perhaps during the later stages of the folding (as with the Skiddaw Granite) or perhaps post-tectonically. Minor intrusions of diorite, lamprophyre and porphyrite accompanied it, but on the existing evidence, not extensively.

5 The Caledonian Mountains, of which this area may have formed a small part, were undergoing erosion throughout

the Devonian and no proved sediments of this age have been preserved in the area. The Roman Fell and Mell Fell conglomerates to the north and north-west perhaps represent molasse of this stage.

6 Early in Carboniferous time a gulf or graben trending east – west to east-south-east opened between the granite-cored stable blocks of Alston and Askrigg, permitting sedimentation to begin in possibly the Ivorian Stage. Sedimentation did not spread generally over the blocks until the Holkerian Stage, and their subsidence continued to be less rapid than that of the gulf or graben, probably until the end of the Dinantian.

7 During the Dinantian the western margins of the two blocks were not defined as at present, but the south margin of the Askrigg Block was a zone of near emergence separating the block area of shelf sedimentation from the basin sedimentation to the south. Part of the Craven zone emerged above sea level during Pendleian time, but up to that stage the Askrigg Block had subsided intermittently as the Yoredale deltas spread slowly and retreated suddenly over it.

8 Epeirogenic events during the Westphalian – Rotliegendes interval produced strain effects in the Carboniferous rocks some of which acted as structural controls for later mineralisation:

(a) North – south to north-north-east maximum horizontal stress was transmitted, as Anderson (1951) has proposed, from the Armorican Orogeny 300 miles (480 km) to the south. The strain effects include the Cotherstone Syncline, Middleton Tyas – Sleightholme Anticline and minor monoclinal or anticlinal folds (or perhaps horsts) at Wrenside, Bastifell, Moresdale Ridge – Rispey Wood and along the Stockdale Line. We consider that the strong north-north-west and weaker east-north-east joints may have been initiated at this stage and are attracted by Doughty's (1968) proposed mechanism of shear followed by fracture cleavage; but an explanation of the enhanced importance of the north-north-west set is required.

(b) The granite-cored, Borrowdale-stiffened mass which is now the Lake District began to move east. The west sides of the Alston and Askrigg blocks now became defined as steep anticlines giving place to reverse faulting (Dakyns and others, 1891; Bott, 1974) along the Inner Pennine and Dent lines. The trend of the latter is structurally anomalous (NNE) and may indicate clockwise rotation of the Askrigg Block. An elaborate system of tension fractures opened in brittle beds of the Carboniferous cover of the Askrigg Block related to a west-north-west axis of maximum tensile strain.

(c) Meanwhile according to Bott's (1967) hypothesis, the granite core of the block showed a tendency to rise. This gently domed the cover sediments about a centre in Wensleydale. The tensions set up controlled the form of openings filled by the mineralising fluids, commencing at this time.

9 Strong upward movements along the Craven and Dent zones tilted the block to the north, converting the gentle dome into a half dome; these movements are assumed to be post-New Red Sandstone, either Cimmerian or Tertiary in age. Mineralisation now effectively ceased.

10 During Pleistocene time the whole area was depressed by ice-loading and in the Holocene recovery released strain energy causing joints within 660 ft (200 m) of surface to open; karstification and oxidation of ores ensued. KCD

REFERENCES

ANDERSON, E. M. 1951 *The dynamics of faulting and dyke formation with applications to Britain.* 2nd Edit. (Edinburgh.) 206 pp.

BAGSAR, A. B. 1948. Development of cleavage fracture in mild steel. *J. Am. Soc. Mech. Engrs.*, Vol. 70, pp. 751–809.

BLACK, W. W. 1950. The Carboniferous geology of the Grassington area, Yorkshire. *Proc. Yorkshire Geol. Soc.*, Vol. 28, pp. 29–42.

— 1958. The structure of the Burnsall – Cracoe district and its bearing on the origin of the Cracoe Knoll – Reefs. *Proc. Yorkshire Geol. Soc.*, Vol. 31, pp. 391–414.

BOTT, M. H. P. 1967. Geophysical investigations of the northern Pennine basement rocks. *Proc. Yorkshire Geol. Soc.*, Vol. 36, pp. 139–168.

— 1974. The geological interpretation of a gravity survey of the English Lake District. *J. Geol. Soc. London*, Vol. 130, pp. 309–331.

— 1978. Deep Structure. Pp. 25–40 in *The Geology of the Lake District.* MOSELEY, F. (Editor). (Leeds: Yorkshire Geological Society.) 284 pp.

BRADLEY, L. 1862. *Inquiry into the deposition of lead ore, in the mineral veins of Swaledale.* (London: Edward Stanford.) 40 pp.

BURGESS, I. C. and HOLLIDAY, D. W. 1979. Geology of the country around Brough-under-Stainmore. *Mem. Geol. Surv. G.B.*

CHUBB, L. J. and HUDSON, R. G. S. 1925. The nature of the junction between the Lower Carboniferous and the Millstone Grit of North-West Yorkshire. *Proc. Yorkshire Geol. Soc.*, Vol. 20, pp. 257–291.

DAVIS, E. A. 1948. The effect of size and stored energy on the fracture of tubular specimens. *J. App. Mech.*, Vol. 15, pp. 216–221.

DAKYNS, J. R., TIDDEMAN, R. H., GUNN, W. and STRAHAN, A. 1890. The geology of the country around Ingleborough, with parts of Wensleydale and Wharfedale. *Mem. Geol. Surv. G. B.*

— TIDDEMAN, R. H., RUSSELL, R., CLOUGH, C. T. and STRAHAN, A. 1891. The geology of the country around Mallerstang, with parts of Wensleydale, Swaledale and Arkendale. *Mem. Geol. Surv. G. B.*

DOUGHTY, P. S. 1968. Joint densities and their relation to lithology in the Great Scar Limestone. *Proc. Yorkshire Geol. Soc.*, Vol. 36, pp. 479–508.

DUNHAM, K. C. 1933. Structural features of the Alston Block. *Geol Mag.*, Vol. 70, pp. 241–254.

— 1948. Geology of the Northern Pennine Orefield: Vol. I, Tyne to Stainmore. *Mem. Geol. Surv. G.B.*, 1st Edit.

— 1958. Non-ferrous mining potential of the Northern Pennines. *Proc. Symposium on the future of non-ferrous mining in Great Britain.* London, September 1958, pp. 115–145.

— 1959. Epigenetic mineralisation in Yorkshire. *Proc. Yorkshire Geol. Soc.*, Vol. 32, pp. 1–29.

— 1974a. Epigenetic Minerals. Pp. 293–298 in *The geology and mineral resources of Yorkshire.* RAYNER, D. H. and HEMINGWAY, J. E. (Editors). (Leeds: Yorkshire Geological Society.) 405 pp.

— 1974b. Granite beneath the Pennines in North Yorkshire. *Proc. Yorkshire Geol. Soc.*, Vol. 40, pp. 191–194.

— HEMINGWAY, J. E., VERSEY, H. C. and WILCOCKSON, W. H. 1952. A guide to the geology of the district around Ingleborough. *Proc. Yorkshire Geol. Soc.*, Vol. 29, pp. 77–115.

— and STUBBLEFIELD, C. J. 1945. The stratigraphy, structure and mineralisation of the Greenhow Mining Area. *Q.J. Geol. Soc. London*, Vol. 100, pp. 209–268.

GARWOOD, E. J. and GOODYEAR, E. 1924. The Lower Carboniferous succession in the Settle district, and along the line of the Craven Faults. *Q.J. Geol. Soc. London*, Vol. 180, pp. 184–273.

HUDSON, R. G. S. 1938. The Carboniferous rocks. *In* The geology of the country around Harrogate. *Proc. Geol. Assoc.*, Vol. 49, pp. 306–330.

— 1944. A pre-Namurian fault scarp at Malham. *Proc. Leeds Philos. Lit. Soc. (Sci Sect.)*, Vol. 4, pp. 226–232.

INGHAM, J. K. and RICKARDS, R. B. 1974. The Lower Palaeozoic rocks. Pp. 29–44 in *The geology and mineral resources of Yorkshire.* RAYNER, D. H. and HEMINGWAY, J. E. (Editors). (Leeds: Yorkshire Geological Society.) 405 pp.

JOHNSON, G. A. L. and MARSHALL, A. E. 1971. Tournaisian beds in Ravenstonedale, Westmorland. *Proc. Yorkshire Geol. Soc.*, Vol. 38, pp. 261–280.

KENDALL, P. F. 1911. The geology of the districts around Settle and Harrogate. *Proc. Geol. Assoc.*, Vol. 22, pp. 27–60.

KENDALL, P. F. and WROOT, H. E. 1924. *Geology of Yorkshire.* (Leeds.) 995 pp.

KENT, P. E. 1966. The structure of the concealed Carboniferous rocks of north-eastern England. *Proc. Yorkshire Geol. Soc.*, Vol. 35, pp. 323–352.

— 1974. Structural history. Pp. 13–28 in *The geology and mineral resources of Yorkshire.* RAYNER, D. H. and HEMINGWAY, J. E. (Editors). (Leeds: Yorkshire Geological Society.) 405 pp.

LEEDAL, G. P. and WALKER, G. P. L. 1950. A re-study of the Ingletonian Series of Yorkshire. *Geol. Mag.*, Vol. 87, pp. 57–66.

MARR, J. E. 1910. The Lake District and neighbourhood—Upper Palaeozoic and Mesozoic times. Pp. 642–666 in *Geology in the field.* MONCKTON, H. W. and HERRIES, R. (Editors). (London: Geologists' Association.) 916 pp.

— 1921. The rigidity of North-West Yorkshire. *Naturalist* (for 1921), pp. 63–72.

MILLS, D. A. C. and HULL, J. H. 1976. Geology of the country around Barnard Castle. *Mem. Geol. Surv. G.B.*

MOSELEY, F. 1972. A tectonic history of northwest England. *J. Geol. Soc. London*, Vol. 128, pp. 561–598.

— and AHMED, S. M. 1967. Carboniferous joints in the north of England and their relation to earlier and later structures. *Proc. Yorkshire Geol. Soc.*, Vol. 36, pp. 61–90.

PHILLIPS, J. 1836. *Illustrations of the geology of Yorkshire. Part II. The Mountain Limestone.* (London: John Murray.) 253 pp.

PRICE, N. J. 1966. *Fault and joint development in brittle and semi-brittle rock.* (Oxford.) 176 pp.

RAISTRICK, A. 1938. Mineral deposits in the Settle - Malham district, Yorkshire. *Naturalist.*, pp. 119–125.

RAMSBOTTOM, W. H. C. 1974. Dinantian. Pp. 47–73 in *The Geology and mineral resources of Yorkshire.* Rayner, D. H. and HEMINGWAY, J. E. (Editors). (Leeds: Yorkshire Geological Society.) 405 pp.

READING, H. G. 1957. The stratigraphy and structure of the Cotherstone Syncline. *Q.J. Geol. Soc. London*, Vol. 113, pp. 27–56.

RIEDEL, W. 1929. Zur mechanik geologische brucherserscheinungen. *Centralbl. Min.*, Abt. B., pp. 354–368.

ROWELL, A. J. and SCANLON, J. E. 1957. The Namurian of the north-west quarter of the Askrigg Block. *Proc. Yorkshire Geol. Soc.*, Vol. 31, pp. 1–38.

SOPER, N. J. and MOSELEY, F. 1978. Structure. Pp. 45–67 in *The Geology of the Lake District.* MOSELEY, F. (Editor). (Leeds: Yorkshire Geological Society.) 284 pp.

SHOTTON, F. W. 1935. The stratigraphy and tectonics of the Cross Fell Inlier. *Q.J. Geol. Soc. London*, Vol. 91, pp. 639–704.

SWINBURN, P. M. 1975. The crustal structure of northern England. Unpublished Ph.D. Thesis, University of Durham.

TURNER, J. S. 1927. The Lower Carboniferous succession in the Westmorland Pennines and the relations of the Pennine and Dent faults. *Proc. Geol. Assoc.*, Vol. 38, pp. 339–374.

— 1935. Structural geology of Stainmore, Westmorland, and notes on the late Palaeozoic (late Variscan) tectonics of the north of England. *Proc. Geol. Assoc.*, Vol. 46, pp. 121–151.

VERSEY, H. C. 1927. Post-Carboniferous movements in the Northumbrian fault-block. *Proc. Yorkshire Geol. (Polyt.) Soc.*, Vol. 21, pp. 1–16.

— 1942. The build of Yorkshire. *Naturalist*, pp. 27–37.

WADGE, A. J. 1978. Classification and stratigraphical relationships of the Lower Palaeozoic rocks. Pp. 68–78 in *The Geology of the Lake District.* MOSELEY, F. (Editor). (Leeds: Yorkshire Geological Society.) 284 pp.

WAGER, L. R. 1931. Jointing in the Great Scar Limestone of Craven and its relation to the tectonics of the area. *Q.J. Geol. Soc. London*, Vol. 87, pp. 425–458.

WELLS, A. J. 1957. The stratigraphy and structure of the Middleton Tyas - Sleightholme Anticline, North Yorkshire. *Proc. Geol. Assoc.*, Vol. 68, pp. 231–254.

WILSON, A. A. 1960a. The Carboniferous rocks of Coverdale and adjacent valleys in the Yorkshire Pennines. *Proc. Yorkshire Geol. Soc.*, Vol. 32, pp. 285–316.

— 1960b. The Millstone Grit Series of Colsterdale and neighbourhood, Yorkshire. *Proc. Yorkshire Geol. Soc.*, Vol. 32, pp. 429–452.

— and CORNWELL, J. D. 1982. The IGS borehole at Beckermonds Scar, North Yorkshire. *Proc. Yorkshire Geol. Soc.*, Vol. 44, pp. 59–88.

— and THOMPSON, A. T. 1965. The Carboniferous succession in the Kirkby Malzeard area, Yorkshire. *Proc. Yorkshire Geol. Soc.*, Vol. 35, pp. 203–228.

ZIEGLER, W. H. 1975. Outline of the geological history of the North Sea. Pp. 165–187 in *Petroleum and the continental shelf of North-West Europe.* Vol. 1. Geology. WOODLAND, A. W. (Editor). (Barking: Inst. Petroleum, G.B.) 501 pp.

CHAPTER 7

Mineral deposits: General description

ORESHOOTS

The term is defined as 'a continuous body of ore which may be worked with profit or with hope of profit'. It must be admitted, however, that scarcely a single orebody is known within the present area which is now able to meet this definition. The orefield is almost completely inactive, and the concern of this chapter is to summarise the evidence of extensive mining extending back over nearly two millennia, with the twofold object of contributing to knowledge of metalliferous deposits and of providing a background for the assessment of the future potential of the district (Chapter 14). In the field the evidence comes from large-scale mapping, coupled with a study of the multitudinous spoilheaps from bell-pits, shafts and adits. Extensive hushes (hydraulic opencuts), particularly in and around Swaledale, give a striking impression of the size of some of the oreshoots, although only a few small exposures of the mineralised zones remain. Only a few of the hundreds of kilometres of adit levels remain accessible, and these should only be entered under expert guidance. The written records, though less satisfactory than those for the Alston Block, are fairly extensive and include, in addition to the plans in the Health and Safety Executive Collection at Baynard's House, London W2, a collection of over 180 plans and many other documents from the A.D. Company, the principal mining company in Swaledale, now at the County Records Office, Northallerton. Plans salvaged from the former C.B. Mining Company in Arkengarthdale are now in the possession of the Earby Mines Research Group, while those of the Duke of Devonshire's Grassington mines are held by Cavendish Estates. Photographic copies of these last two sets are held by the Geological Survey. Finally the large collection of plans of underground workings at Greenhow made by the late Mr W. W. Varvill are now in the possession of the Department of Mining Engineering of the University of Leeds, which also has on lease one of the few remaining accessible mines.

Four types of oreshoot have been worked: (1) Vein oreshoots, emplaced in fault fissures mainly of small displacement but in a few cases reaching as much as 420 ft (128 m); the hade is generally between vertical and 30°; (2) Scrins (the term being employed in the same sense as in the Derbyshire orefield); narrow deposits in vertical fractures with no more than micro-displacements, tending to occur in subparallel clusters; (3) Pipes, presumably formed along the junction of two intersecting fractures; and (4) Flots, floats or flats, replacement deposits following the bedding at certain specific horizons in the Brigantian and early Pendleian limestones. In the case of all oreshoots other than the rare pipes, the lateral dimension very much exceeds the height or width.

STRATIGRAPHICAL RELATIONS

Vein Oreshoots

The structural controls have already been discussed (pp. 74–80); these imply that oreshoots were able to form only opposite brittle, competent beds in the wallrocks. This is particularly the case where the displacement is less than about 20 ft (6 m) in which case steep, clean fractures affording open space to be filled against competent beds (chiefly limestone and chert) are found to contrast markedly with the lower dips and closed nature of the same fractures as they pass through softer beds (shale, shaly sandstone and most sandstones in this area). Throughout the northern part of the present region, including Stainmore, Swaledale, Arkengarthdale, Stonesdale and Wensleydale, although there are competent grits and sandstones above the intra-Pendleian unconformity, no mineralisation has ever been found to penetrate this part of the Namurian succession. This seems to be due, not so much to the unconformity as to the presence of thick shale beneath it which acted as an effective cover-rock through which the rising fluids were unable to penetrate. Where, in northern Wharfedale, the unconformity has cut out the thick shales and descended on to the Main Limestone, oreshoots begin to appear in the Grassington Grit above the unconformity; and as it cuts out successively lower beds in the Brigantian, this formation becomes an important bearing bed, the major ore-carrier of the Grassington fields. In Swaledale and Arkengarthdale, the highest mineralised level is the Crow Chert and Limestone, but this is far less important than the group, mainly of hard brittle strata, extending from the top of the Richmond Cherts down to the base of the Main Limestone. It may be safely estimated that more than half the lead output of the southern half of the northern Pennines came from this group, and within it, as Table 6 shows, the Main Limestone and Main Chert were of preeminent importance. This emerges very clearly from the detailed evidence assembled by Bradley (1862) from all the mine-operators in the North Swaledale Mineral Belt from Keldside and Littlemoor in the west to Hurst in the east. Figure 19 summarises this information in relation to the displacement on the fractures, and the horizons worked.

The sandstones of the Brigantian hardly ever behaved as brittle beds, but a few oreshoots were worked in North Swaledale in the Underset Chert and Limestone. The lack of successful workings at lower horizons in the Belt led Carruthers and Strahan (1923) to suggest that the local drainage system has cut right through the mineralised zone. This is borne out by a number of deeper trial levels (Parkes Level, [NY 9107 0089]; Mill Level, [NY 9964 0368]; Barras End Low Level, [NY 9932 0069]; Moulds New Level, [NY 9999 0226]; Storthwaite Hall Level, [NY 0178 0217]; Haggs Level, [NZ 0254 0204]; see Figure 25) all of which were driven beneath major oreshoots at Main Limestone and higher horizons, but failed to find ore beneath the Underset Limestone. There is, however, one important exception. Sir Francis Level (Raistrick, 1975) [NY 9399 0001] driven into the main axis of the North Swaledale Belt in Gunnerside Gill, proved the Lownathwaite and Old Gang oreshoots extend down at least as far as the Middle Limestone. This place, where several veins converge, may be regarded as one of the roots of the very extensive mineralised system extending away from it in all directions at higher levels. Boring has shown that mineralisation is still present at the Simonstone Limestone horizon, but it might be expected to go considerably deeper.

In the South Swaledale – Wensleydale area, the principal emphasis is again on the Main Limestone, but here, replacements in Red Beds Limestone are perhaps even more important (p. 179). Oreshoots (Figure 30) have been found at lower levels at Friar's Intake (Simonstone Limestone), Lover Gill (Five Yard Limestone, Figure 30, loc. 4), Carperby (Simonstone and Middle limestones) and Keldheads (Five Yard Limestone and above loc. 18). Small, doubtfully payable shoots were found at Warton (loc. 6) and Aysgarth in strata as low as the Hawes and Hardrow Scar limestones respectively, while the former is mineralised at West Burton and the latter at Seata (Figure 30).

Table 6 Statistical analysis of horizon of vein oreshoots

Horizon	Number of oreshoots by areas (Figure 1) 10	11	12	13	14	15	Total
Late Pendleian to Kinderscoutian							
Red Scar Grit to Libishaw Sandstone	—	—	—	—	—	4	4
Grassington Grit	—	—	—	—	39 (20)	6	45
Early Pendleian							
Crow Chert and Limestone	—	22	—	3	—	—	25
Ten Fathom Grit	—	9	—	1	—	—	10
Red Beds	—	38	—	23	—	—	61
Black Beds	—	45	—	5	—	—	50
Main Chert	—	105	—	6 (39)	—	—	111
Main Limestone	4	112 (30)	3	23	4	—	146
Brigantian							
Underset Chert	—	18	?7	2	—	—	27
Underset Limestone	—	18	—	7	—	—	25
Five Yard Limestone	—	2	1	3	—	—	6
Middle Limestone	—	2	—	1	8	—	11
Simonstone Limestone	—	—	—	3	7	—	10
Hardrow Scar Limestone	—	—	—	1	9	—	10
Gayle, Upper Hawes, Coldstones limestones	4	—	—	4	(70)	7	15
Asbian							
Hargate End, Upper Kingsdale Limestone	6	—	—	—	(25)	26	32
Greenhow, Lower Kingsdale Limestone	—	—	—	—	—	18	18
Stump Cross Limestone	—	—	—	—	—	5	5
Timpony Limestone	—	—	—	—	—	4	4

Numbers in brackets refer to scrins and related deposits

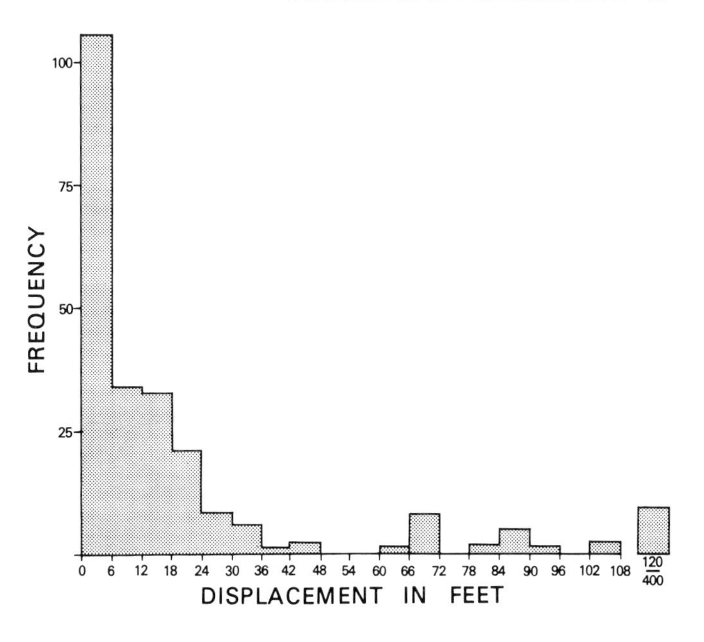

Figure 19 Graph to illustrate the control of oreshoots by displacement on veins in Swaledale. Based on data given by Bradley (1862) for the productive veins.

Northern Wharfedale exposes oreshoots (Figure 15) in the Main Limestone only at Buckden Gavel, where the Grassington Grit, now immediately above the limestone, is also penetrated. Farther south, the unconformity successively cuts out the Main and Underset limestones, but oreshoots are now found in the Middle Limestone (Old Providence, Grassington), the Simonstone and Hardrow Scar limestones (Starbotton, Moor End), and a large number of very small veins west of Wharfedale in beds equivalent to the Gayle and Hawes limestone, extending short distances down into the Kingsdale (Great Scar) Limestone below. However, by far the greater part of the production of lead ore from Wharfedale came from the Grassington mines, from numerous oreshoots in the 'Bearing' and 'Top' grits of the Grassington Grit Formation, details of which are given on Figures 12 and 33. Only a few of these oreshoots were found to extend down into the Middle Limestone, beneath the unconformity.

Finally, at Greenhow (Dunham and Stubblefield, 1945), the intra-Pendleian unconformity has cut down to rest, in places, on the Greenhow Limestone, equivalent to the lower part of the Kingsdale Limestone. Two opposite trends now appear in the disposition of the oreshoots. Along the supposed eastward extension of the Bycliffe Vein (Figure 15) from Grassington into Ashfold Side Beck, substantial orebodies reach up through Red Scar Grit, Follifoot grits and Cayton Gill Beds to the Libishaw Sandstone, stratigraphically the highest level attained in the whole region, and here, even the Grassington Grit has not yet been tested. On the other hand, within the Greenhow Anticline, most oreshoots are in the Hargate End Limestone, some in Greenhow Limestone and a few in the still lower Stump Cross and Timpony formations. On Greenhow Rake a continuous oreshoot extends down 500 ft (152 m) beneath the unconformity. Such a vertical dimension is rivalled only at Lownathwaite in Area 11. The more normal situation is for the height to be determined by the thickness of the controlling hard, brittle bed, less the throw of the vein. As Figure 11 shows, there is a great deal of lithological variation within the Richmond Cherts, some of which has produced conditions unfavourable to the reception of mineralising fluids. Further, it appears from the examination of dump material at mines such as Hurst where there has been extensive working in Main Chert and Richmond Cherts, that the chert beds may fracture differently from the limestones, producing, instead of a single channel, a large number of subparallel small fractures, each of which may receive a thin layer of galena. The same effect probably occurs in the Crow Chert. In wider channels, rock inclusions were probably always present, in some places in such quantity as to make the filling a breccia cemented with introduced minerals. Normally the fragments are angular; a good example can be seen at the head of Fell End Hush [NZ 0252 0241]. Vein-widths in this region are poorly documented, but the impression gained from accessible underground workings and from surface cuts is that many were in the range 3 to 6 ft (roughly 1–2 m). Friarfold Rake, examined in 1952 from Brandy Bottle Incline, appeared to be 6 to 8 ft (1.8–2.4 m) wide, filled with crudely-banded baryte, witherite, colourless fluorite and calcite; a narrow slit for galena had been removed by former miners. From the very extensive tailings dumps in the region (Dunham and Dines, 1945, p. 143), this was one of the largest oreshoots and was continuously mineralised for 1.9 miles (3.1 km). Substantial widths, over 10 ft (3 m), were also found in Waterhole North Vein and Gill Heads Vein at Skyreholme (Figure 35) (both mainly fluorite). Many of the larger veins in this region proved to be in several parallel fractures with 'horses' of rock between. Of this the most spectacular case is Turf Moor in Arkengarthdale, where three parallel main oreshoots and two subsidiary ones were worked, with only narrow limestone screens between them. On many plans a sudden but short change in the direction of a drive on vein, followed by the resumption of the original direction, indicates that a parallel fissure has become the main channel of mineralisation. In Arkengarthdale one of the largest oreshoots was on Dam Rigg North Vein, 2600 ft (792 m) long, in Main Limestone and Chert, and over 80 ft (24 m) high, but the width is not known (Figure 25). At Grassington, two of the largest oreshoots were on Coalgrove Beck Vein, 1360 × 185 ft (415 × 56 m) (Dunham, 1959, fig. 4), extending a short distance below Grassington Grit into Middle Limestone; and the Cavendish oreshoot at Sarah Shaft, 200 ft (61 m) high (half in Middle Limestone) and with a subsidiary oreshoot 30 to 40 ft (9–12 m) high above it, following the Top Grit (Figure 33). These examples give an impression of the largest orebodies, but a majority of the 573 (other than scrins) that have been worked were smaller. It is also generally recognised that many were subject to swells and pinches along their courses.

Scrins

At least six zones of scrins have been recognised, at three different horizons (Table 7). The ancient shallow workings on these have left a ridge and furrow topography readily recognisable on air photographs but unrewarding on the ground because they are largely grassed over. These seem only to have been workable by surface trenching and, except at Grassington, they have not been exploited underground. It is logical to regard the swarms of very narrow veins (Figure 15) mainly in Kingsdale and Hawes limestones around High Mark (Wager, 1931; Raistrick, 1938, 1973) and in Hawes and Gayle limestones at Coldstreak and Middlesmoor, west of Wharfedale, as of similar character to the scrins, though they are not so closely-spaced. A few, where a displacement can be recognised, may be regarded as transitional to vein-oreshoots proper. These clusters are analysed in more detail on pp. 187–188.

Table 7 Scrins and clusters of minor veins

Locality	NGR	Horizon	Direction
South of West Feldom	NZ 100 043	Main Limestone	E–W
East of West Feldom	NZ 107 043	Main Limestone	NE
Cobscar Rake, east	SE 064 923	Limestone in Richmond Cherts	NE, ENE
Lang Scar, Preston	SE 078 926	Richmond Cherts	NE
Heron Tree Allotment, Preston	SE 089 923	Richmond Cherts	NE
Stool Mine, Grassington	SE 015 676	Middle Limestone	WNW
Chatsworth Mine, Grassington	SE 026 672	Grassington Grit	E–W
High Mark	SD 935 673	Kingsdale and Hawes limestones	NE
Coldstreak, Littondale	SD 950 716	Hawes and Gayle limestones	E–W, NW
Middlesmoor	SD 963 713	Hawes and Gayle limestones	E–W, ENE, WNW

The cited map references mark the centre of the cluster for each locality.

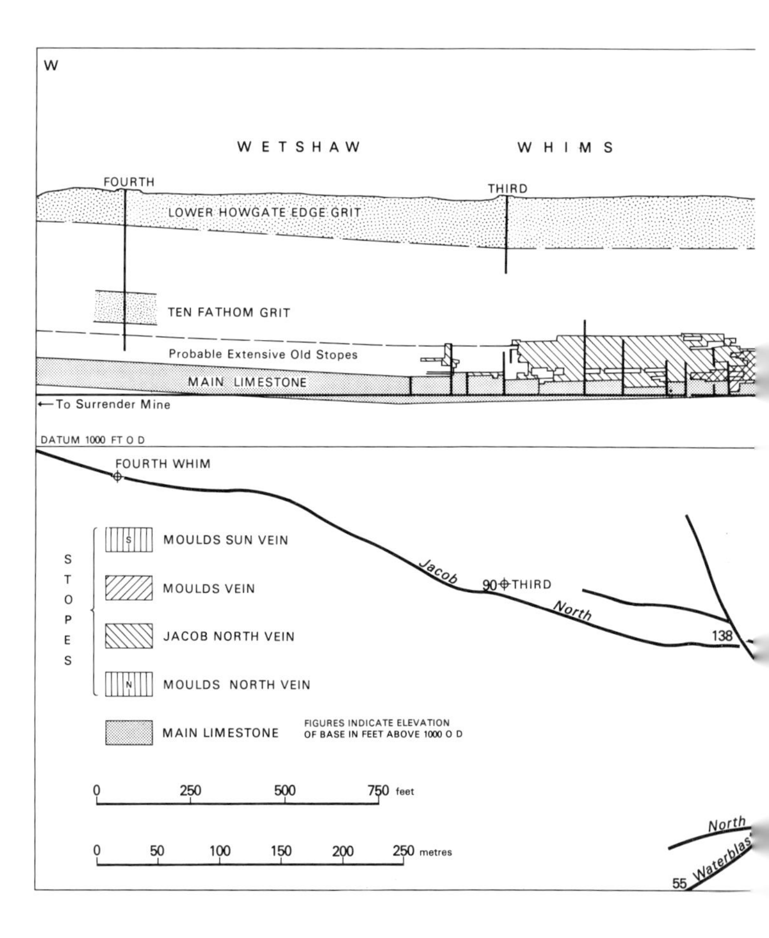

Figure 20 Longitudinal section of Turf Moor Mine, Arkengarthdale, to illustrate the stratigraphical control of oreshoots. Based on reductions by J. R. Earp from plans of the former C. B. Company.

Pipes

These are replacement deposits in limestone, formed either at the crossing of two vertical fissures, in which case they are nearly vertical, or along the crossing of a fissure and a bedding plane, when they are near horizontal. The Gill Shaft pipe on Craven Moor (Varvill, 1937, fig. 25; Dickinson, 1970, pl. 3) is a good example, bottle-like in shape, 130 ft (40 m) high, with a maximum diameter of 60 ft (18 m). In the massive Hargate End and Greenhow limestones, other examples of oreshoots influenced by cross-fractures intersecting productive veins have the characteristic of greater vertical than lateral dimensions. The Pendleton Pipe, related to a minor fracture crossing the Garnet Vein at Greenhow, is an example of a horizontal pipe, also in Hargate End Limestone which contains strong bedding planes. With increasing width, this type of deposit becomes a flat and would belong in the next category, described below.

Flots, floats, flats

All three terms for replacement deposits more or less parallel to the bedding of limestone occur in the written data of this region. Always these deposits are associated with a feeding vein and often they are linked to it by means of narrow leaders, minor fractures which acted as channels for the mineralising fluids. No deposits exactly comparable with the great flats of Weardale, Allendale and Alston Moor are, however, found here; there are not only differences in scale, but mineralogically the wide spreads of ferriferous ankerite – siderite rock are lacking, and fluorite is only rarely present in greater than trace amounts. Nevertheless an appreciable part of the lead production of the region has come from deposits of this sort.

The presence of abundant brown-weathering carbonate (the composition of which is discussed on p. 93) is the most characteristic sign that such deposits have been worked, coupled with the scatter of bell-pits by which they were reached. The two most striking areas are at Windegg, and the east side of Arkengarthdale (Figure 25, loc. 23), and along the Whitaside – Apedale Vein, especially on the west side of the boundary between the two mines (Figure 30). The galena occurred scattered through the carbonate, or forming thin streaks along bedding planes. Other minerals, except calcite, were sparse in most of the deposits. Fluorite was found in good crystals in cavities at West Arn Gill, but hardly anywhere else. Baryte and witherite both occurred as minor constituents of the flats, but only at Nussey Knot near Greenhow was baryte the principal mineral worked (Varvill, *op. cit.*; Dunham and Stubblefield, 1945). Here three definite levels favourable to replacement can be identified, respectively one at the base of the Greenhow Limestone, a second within the Stump Cross Limestone and a third near the top of the Timpony Limestone. In the Main Limestone the 'flat horizons' so widely recognised in the Great Limestone of the Alston Block cannot be identified with certainty in the present region, though the positions of the adits on the scarp face at Windegg suggest at least two persistent levels of metasomatism. No satisfactory underground plans of the flot ground have been kept, and the trends of the deposits at Windegg and elsewhere can only be inferred from the position of the very numerous bell-pits. Even this evidence fails to the east in the area marked 'Brass Pump Floats' on the six-inch map, for here there are no bell-pits and the deposits must have been reached underground from the long Washy Green Adit [NZ 0215 0369]. The primary mineralisation probably increased the permeability of the ground, for everywhere within the present zone of moving groundwaters, extensive cavernisation and other karst action, post-dating the mineralisation, has affected the deposits. In this respect they are similar to the flat pipe deposits of the North Derbyshire Orefield. Where caverns have cut through replaced ground, galena, protected by its insoluble coating of cerussite (p. 106) has persisted in the residual red-brown clay on the

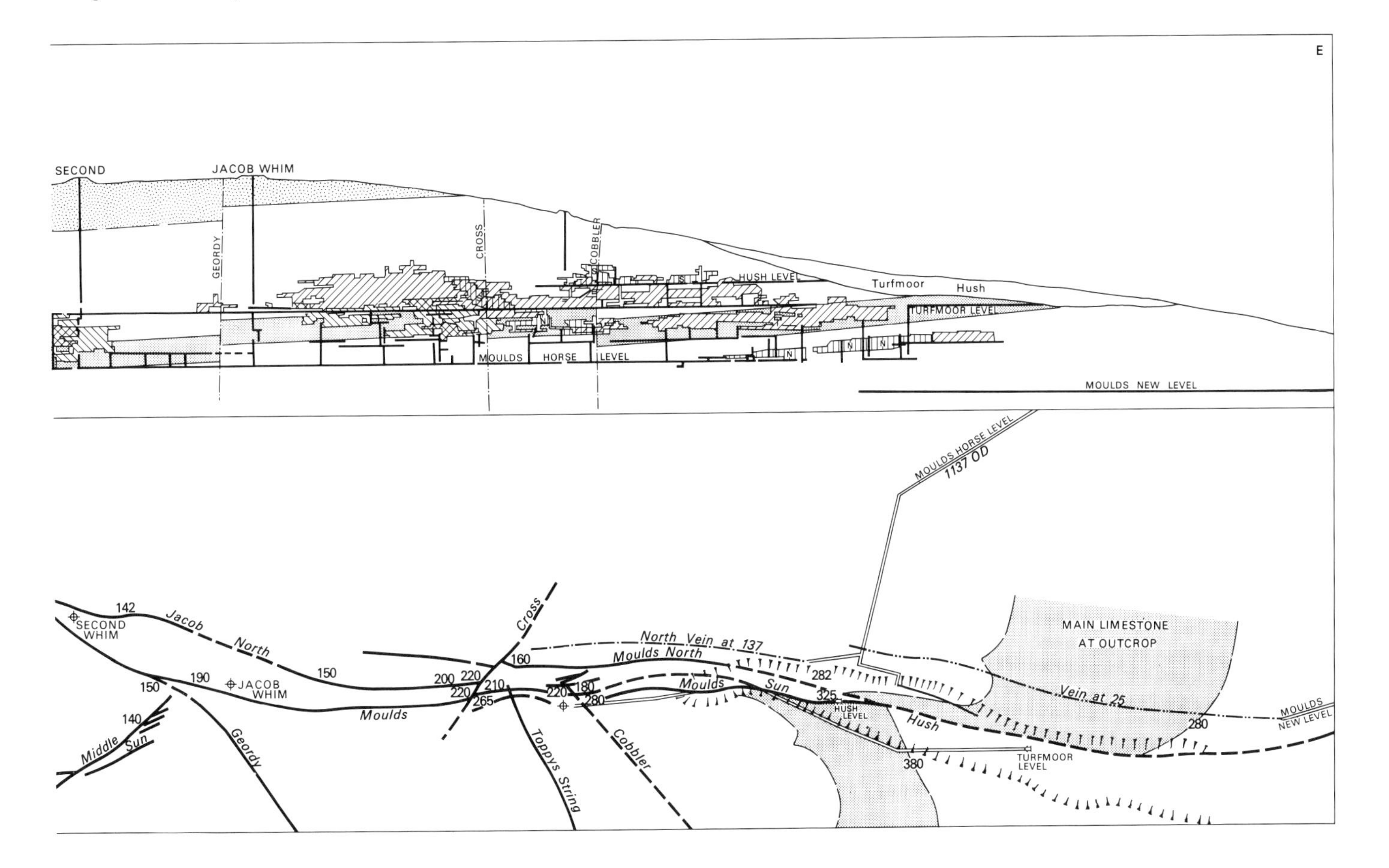

cave floor, where it formed an easily-mined source of ore at localities such as Faggergill, Windegg and, probably, extensively along the Swale/Ure watershed. The secondary zinc mineral smithsonite (calamine) was deposited in caverns of the present cycle at Malham (p. 227) in workable quantity, and both this mineral and the zinc silicate hemimorphite are known from other flot deposits. Table 8 brings out the fact that beds above the Main Limestone assume considerable importance as carriers of flots in the Swale/Ure watershed area. They are not well exposed, but it appears that from the Oxnop valley eastwards, beds of limestone occur within the Richmond Cherts sequence. To these the primary surveyors applied the term Red Beds Limestone, transferring this term from the standard succession recognised by the North Swaledale lead miners. It is likely, however, that these beds are not the strict equivalents of the upper part of the Richmond Cherts in the ground east of Harker Mine (Figure 11, section 2) (p. 53), but come in at horizons equivalent to the lower part also. They, rather than the Main Limestone, may constitute the most important bearing horizon of the watershed in Area 13. It is noticeable that many of the adit levels driven to drain the workings start at the top and not the bottom of the Main Limestone. The records collected by Bradley (*op. cit.*) and the nature of the spoil combine to confirm that flot ground rather than vein stopes provided much of the lead ore mined in this part of the area.

Table 8 Horizons of replacement deposits in limestone

Locality	NGR	Horizon	Minerals*
Area 10			
Hartley Birkett	NY 802 082	Robinson Limestone, Smiddy-Peghorn	F, Ba
Area 11			
Faggergill, Sloat Hole	NY 985 071	Main Limestone	
Windegg–Brass Pump	NZ 015 042	Main Limestone	
Hurst	NZ 040 020	Main Limestone	
Greengates–Cuddy's Grove	NZ 067 021	Main Limestone	
East Arn Gill	SD 915 994	Underset Limestone	F
Area 13			
Spout Gill	SD 936 965	Main Limestone	
Summer Lodge	SD 962 951	'Red Bed' and Main limestones	Ba
Simon Groove–Beezy	SD 948 955	do	
Brownfield	SD 977 924	do	
Whitaside east	SD 985 955	do	W
Virgin Moss	SD 998 943	do	Ba, W
Harkerside	SE 013 974	Main and Underset,? also 'Red Bed' Limestone	
Grinton How	SE 041 962	Main Limestone	
Ellerton Moor	SE 057 960	Main and 'Red Bed' limestones	
Braithwaite	SE120 858	Main Limestone	F
Area 15			
Nussey Knot	SE 080 635	Stump Cross and Timpony limestones	Ba

* Other than carbonates and galena; F = fluorite, Ba = baryte, W = witherite

PRIMARY MINERALS

The stratigraphical relations of the deposits make it perfectly clear that the mineral deposits were, without exception, introduced into their present positions after the consolidation of the wallrocks. The primary minerals, those that persist below the water-table, marking the downward limit of oxidising conditions characteristic of the zone of vadose groundwater, include sulphides, carbonates, fluoride and sulphate, as follows.

Galena, PbS Until the present century, this has been almost the sole mineral sought in the deposits. The sulphide normally occurs in macrocrystalline form, in aggregates of interfering crystals forming bands, continuous or discontinuous, paralleling the walls of veins; as thin layers along the bedding of replaced limestones; or scattered through the matrix of veins and flots. It proved to be recoverable from the matrix or gangue by very simple methods of beneficiation, breaking on stone floors or by simple crushers, followed by handpicking and 'hotching', agitation in water in a sieve so that the heavy particles moved to the bottom. Fines were recovered by arrangements of riffles and eventually by buddles. There were very few mechanised plants, other than waterwheel-driven crushers, even in the heyday of lead mining in this region. The greatest spreads of tailings are found high on the fells, where they could only have been the product of hand-dressing. That this was not an efficient process is evident from the amount of lead left in every tailings heap sample (Tables 15 to 17) and from the fact that galena is still easy to find on the heaps. The sulphide concentrates were all smelted within the area mainly in small mills near the villages. This part of the process was also inefficient, or the concentrates were of poor grade, for even in the 19th century, Hunt's statistics show recoveries of only 60 to 70 per cent lead metal, with the latter figure rarely exceeded (pure PbS contains 86.6 per cent lead). Under the microscope the galena is often found to contain minor amounts of copper as chalcopyrite, zinc as sphalerite and iron as pyrite or marcasite in the form of tiny inclusions but these rarely exceed 1 to 2 per cent of the total sulphide. Among trace elements present in solid solution in the galena, antimony and silver are probably the most important, but bismuth and arsenic may also be present as in the northern region (Dunham, 1948, p.84). No doubt tests were made at least as early as the 18th century to determine the recoverability of the silver, which by that time was an important product in the Alston Pennines, but evidently no concentrate was found to justify the cost of the process; this means that less than 4 oz Ag per ton Pb (112 ppm in Pb, or 129 ppm in PbS) were available, this being approximately the cut-off figure. In the later part of the 19th century, as the industry was expiring, silver was produced from the mines listed in Table 9, but this is likely to have been worth while only at Braithwaite and West Burton, and at Buckden Gavel.

In 1972–75 Dr A. T. Small collected 177 samples of galena from localities in areas 10 to 13, and determined silver and antimony in these by X-ray fluorescence spectrometry (XRF) at the Department of Geological Sciences, University of Durham. His results have appeared in graphic form (Small, 1978, fig. 3) and these in general reiterate and extend, for silver, those given in Table 9. The ranges between individual samples from particular veins and groups of veins are more considerable than can be shown in diagrams giving only mean values, and an abridged version of the figures intended to bring this out is accordingly included as Table 10. Four samples from the Lover Gill Vein, high up on the Swale/Ure watershed, gave very high values, one exceeding 2000 ppm, and Small records that these specimens were searched for free silver minerals such as argentite or silver-bearing sulphosalts, but without finding them. The remaining determinations confirm that the galena of the present region very rarely reached the minimum figure necessary to justify separation of the silver, at least 4 oz Ag/ton Pb, or about 100 ppm Ag in pure PbS. These results make it clear that neither silver nor antimony are uniformly distributed through the galena at any locality, perhaps the most striking example in the case of silver being that of Braithwaite, credited in the official statistics with the highest silver recovery in the region, but yielding a sample in which the metal was below the detection limit. There are, nevertheless, clear regional trends, and these will be considered further in connexion with the zonal arrangement of minerals (p. 95).

The antimony figures, ranging from 0 to 0.4 per cent may be compared with the few figures available for the northern region which range from 0.0035 to 0.1 per cent. So far there is no record of a separate antimony phase such as jamesonite ($4PbS.FeS.3Sb_2S_3$) and it must be assumed that the galena is able to accommodate the maximum amounts recorded within its lattice.

Information on the galena-content of the oreshoots of the present region is sadly lacking. Some idea of the grade necessary for economic working in the 18th and early 19th centuries in the northern region can be gained from the colour coding of the London Lead Company's stope sections, which show the poorest stoped ground as containing 3.3 to 5.5 per cent PbS, intermediate grades of 5.5 to 9.0 per cent, and the best parts of the oreshoots running over 9 per cent (Dunham, 1944), all these figures being on the basis of a running fathom of 6 × 6 × 3 ft. No such sections have been preserved for that company's Swaledale and Wensleydale mines, but it is not unreasonable to think that similar economic conditions applied. The only available sampling of ore still in place, at the Craven Cross Mine, Greenhow, shows 6.8 per cent PbS over a length of 350 ft (107 m), and 7.7 per cent over 400 ft (122 m), both figures assuming a 3 ft working width. The portion sampled is pro-

Table 9 Silver recovered from galena concentrates

Mine	Years	Silver	
		Ounces per long ton lead	ppm in galena
Area 10			
Hartley Birkett	1877	5.0	120
Long Rigg	1877	5.0	120
Mallerstang	1871–1872	6.9	166
Area 11			
Beldi Hill	1882–1883	2.79	68
Old Gang	1876–1882	2.47	59
Arkengarthdale C.B.	1874–1883	2.13	48
Hurst	1875–1880	2.93	71
Area 12			
Merrybent	1866	3.0	72
Area 13			
Apedale	1876–1881	1.79	43
Ellerton Moor	1874–1875	3.8	92
Keld Heads	1875–1881	2.9	70
Braithwaite and West Burton	1854–1862	5.7	137
Worton	1875	3.0	72
Area 14			
Buckden Gavel and Bishopdale	1876–1878	4.9	120
Grassington Moor	1877–1881	3.7	90
Devonshire Royalties, Wharfedale	1863–1864	2.3	55
Area 15			
Greenhow	1875–1883	2.0	48
West Craven Moor	1875–1881	2.9	70
Appletreewick (with Cracoe, Hebden Moor, Old Providence, Starbotton Cam, Moor End, Silver Rake, Stony Grooves)	1864	3.15	76

National Grid references to these mines are given in Chapters 8–13

Table 10 Silver and antimony in galena specimens, areas 10 to 13 (parts per million)

Locality	Samples	Silver		Antimony	
		Range	Mean	Range	Mean
Area 10					
Clouds	7	79–202	120	906–2809	1521
Hartley Birkett	5	57–123	80	1526–2265	1826
Long Rigg	3	70–104	88	1255–1662	1430
Area 11					
Prys Hill	2	41–109	75	2750, 2751	2751
Keld	12	16–86	43	187–1182	603
Arn Gill	1	26		2868	
Gunnerside	10	0–57	29	17–2459	477
Friarfold	6	7–80	32	192–1945	1173
Arkengarthdale	18	0–113	21	81–818	459
Windegg	7	0–158	49	128–1495	553
Hurst	9	0–164	42	0–1721	488
Copperthwaite	9	2–50	27	83–991	557
Cleaburn Pasture	14	2–39	11	86–437	229
Helwith	2	0, 22	11	113, 226	170
Faggergill	9	0–17	5	0–71	27
Stang	3	0–11	5	0–16	5
Downholme	4	0–62	16	81–95	88
Area 12					
Feldom Fault	4	0–1	¼	0–26	6
Area 13					
Lover Gill	7	25–2052	650	1968–3966	2560
Providence (Thwaite)	2	36, 249	125	811–1926	506
Sargill	1	127		1258	
Cottriggs (Raygill Farm)	2	2, 1407	704	1695, 1954	1824
Satronside	5	0–178	57	485–1080	766
Harkerside	8	7–25	17	41–181	204
Apedale	2	9, 38	24	252, 389	321
Grinton	4	9–27	19	21–278	202
Ellerton	7	0–59	20	178–641	367
Cobscar	5	19–60	45	432–991	698
Keld Heads	1	23		1466	
Wet Grooves	5	35–183	71	807–2832	1584
Seata	1	33		1738	
West Burton	1	21		1280	
Braithwaite	1	0		720	
Dodd Fell	3	31–40	35	135–507	264
Wasset Fell	2	0, 21	11	668, 791	729

XRF determinations by A. T. Small, University of Durham, 1972–75 (Small, 1977)

bably near the bottom of the oreshoot on Hargate End Vein, and may underestimate it. At the Old Gang and Friarfold mines in the North Swaledale Belt, the tonnage of tailings from hand dressing has been estimated at 275 000 minimum (Dunham and Dines, 1945); the full tonnage remaining before recent re-treatment for barytes began may have been as great as 500 000 tons. These two mines contributed at least half the estimated pre-1800 production of the A.D. Group of about 260 000 tons lead concentrates. It is probable therefore that the ore brought out was not less than 25 per cent PbS, but this cannot be taken as an estimate of the grade of these wide veins, for they were selectively slit, and probably the crude ore was hand-picked in the stopes before hoisting. As an example of a flat working, the Sloat Hole Level at Faggergill gave access to a network of natural passages along joints in the Main Limestone cutting through a flat. The passages total over 4000 ft (1.2 km) in length and assuming a 6 ft (1.8 m) height and 3 ft (0.9 m) width, they represent the removal in solution of some 5100 tons of limestone. The lead ore production, largely or entirely got by collecting the residual galena in the clay of the passages, amounted to 240 tons, suggesting that the grade of the flat may have been of the order of 4.6 per cent PbS.

Sphalerite, ZnS This was probably the second most abundant sulphide in the veins and flots. In the tables that follow, assays of samples from 79 tailings dumps representing most important parts of the region are given; in these zinc is never absent and the quantity ranges from 0.03 to 40 per cent. Seven values from 4.9 per cent must be discounted as due to secondary concentrations of smithsonite or hemimorphite having been penetrated in the mines, but the remainder averaging 1.1 per cent Zn probably give a fair indication that the deposits on average contained 2 per cent sphalerite. The colour ranges from pale greyish brown to deep brown, probably reflecting the iron-content but this has not been investigated. The mineral is conspicuous along the Whitaside – Apedale Head Vein. Whitaside produced a few tons of zinc concentrates, but otherwise the zinc production from this region has come from one or perhaps two localities yielding calamine (smithsonite). Small (1977) determined cadmium, silver, cobalt and nickel on nine samples, finding ranges of 0.38 to 1.4 per cent Cd, 1 to 261 ppm Ag, cobalt above detection limit in only three samples and nickel in only two. Vaughan and Ixer (1980) analysed sphalerites from Greenhow and Worton, finding only 0.05 per cent iron in them.

Copper Sulphides The commonest of these is chalcopyrite $CuFeS_2$ and it is assumed that most or all the occurrences are primary in origin. It is widespread as minute crystals enclosed in other sulphides, and in fluorite, baryte and witherite; its presence is often betrayed in dump material by green stains of malachite, showing especially clearly on the light-coloured minerals. The average tenor is nevertheless only a fraction of 0.1 per cent in general, and it appears in concentrated form in Area 12 only, where the Feldom, Middleton Tyas, Merrybent and Billy Bank mines produced modest quantities of copper ore. Primary djurleite ($Cu_{1.96}S$) has recently been recognised in this area (p. 157). Chalcopyrite can also be obtained in macroscopic amount at the copper prospect in Great Sleddale (p. 119), and at Clouds, Hartley Birkett and Long Rigg (Figure 24) in Area 10, and near Malham in Area 15, where some copper ore has been produced. About 1950 a small vein carrying coarse chalcopyrite in a witherite matrix was found in the sandstone below the Main Limestone at Forcett Quarry [NZ 155 106], (Figure 28 and Wells, 1957). Dr T. Deans (1951) first recognised the remarkable nodules, like small eggs, of copper sulphides on a bedding plane at the base of the Underset Limestone at Black Scar Quarry, south of Middleton Tyas [NZ 232 052] (Figure 28). The flats worked at Middleton Tyas for copper, but long since inaccessible, may be nodular deposits of similar character. Here, as well as Forcett Quarry, where nodules have also been found, the suite of copper sulphides includes not only chalcopyrite, but bornite Cu_5FeS_4, covelline CuS and chalcocite Cu_2S in addition. To this list, Small (1977) has added digenite $Cu_{1.97}S$. This series of sulphides, all richer in copper than chalcopyrite, are characteristic of deposits that have been secondarily enriched, but both bornite and chalcocite are also well known as primary minerals. The textural evidence supports some secondary enrichment, and covelline, chalcocite and digenite may be related either to a former land-surface (the sub-Rotliegendes surface) or to the present surface, or both. It is, however, quite clear that even if secondary sulphide enrichment has occurred, there must have been primary deposits with a strong concentration of copper at or near the same position before it took place.

Trace-element studies of chalcopyrite by A. T. Small (1977) from East Layton, Kneeton Hall, Feldom, Richmond (Chapter 10), Great Sleddale (Chapter 9) and Clouds and Hartley Birkett (Chapter 8) showed tin below the detection limit in all cases; silver only where galena is associated; nickel, East Layton 0–10 ppm, Kneeton Hall 0–15; Great Sleddale 15–140; Feldom 15–30; but otherwise not detected; and cobalt, East Layton 0–20; Kneeton Hall 0–500; Feldom 15–30; Great Sleddale 18–300 and Hartley Birkett, 100 ppm. The nickel and cobalt figures may signify the presence of bravoite, discussed below.

Bravoite (Fe, Ni, Co)S_2; Pyrite and Marcasite, FeS_2 Springer, Schachner-Korn and Long (1964) produced evidence that the bisulphides of iron, nickel and cobalt are metastably isomorphous, probably in all proportions. Vaughan (1969) has applied Hillebrand's (1907) name bravoite to the resulting solid solutions, where no end member is present to an extent greater than 80 per cent. Superficially bravoite resembles pyrite, but under the reflected-light microscope it shows pale violet to brown shades, as compared with the brassy yellow of pyrite, and composition zoning is common though the mineral generally remains isotropic. The first description from the present region was by Rogers (1974) who found it among the inclusions of sulphides in fluorites from Keldheads Mine (p. 176) and the Wharfedale Mine, Kettlewell (p. 187). With samples from Derbyshire and Skeleron Mine near Clitheroe, as well as those from the Askrigg Orefield he obtained the following average of electron microprobe analyses: iron 37.09; nickel 4.80; sulphur 49.28; copper 0.33; and cobalt 0.21 per cent; but he notes that nickel ranged up to 9.9 per cent. Ixer (1978) recorded bravoite from eight localities in the Askrigg Block, including Greenhow, Swaledale and Arkengarthdale. Vaughan and Ixer (1980) list, in addition to Greenhow and Kettlewell, Trollers Gill, Worton Hall, East Bolton, Summer Lodge Moor, Redmire Scar, Copperthwaite, Hurst, Fell End, Langthwaite and Faggergill; in short, the mineral is widespread throughout the Askrigg Orefield. In samples investigated by these authors it always proved to be the earliest-deposited sulphide, but having regard to the banded nature of the more important veins repetitions must have occurred within the whole paragenesis.

The common iron sulphides are also widespread throughout the orefield in both veins and flots, always in minor amounts, but apparently, less abundant here than in the Alston orefield. Marcasite oxidises more readily than pyrite, leaving a goethite boxwork in which the platy structure of the mineral can sometimes be recognised. Pyrite leaves its characteristic cubes. In ores in limestone, particularly in carbonaceous limestones, Vaughan and Ixer's reflected light work has established the presence of framboids of pyrite of up to 30 μm in diameter that are most probably of synsedimentary origin. They have also identified both nickeliferous pyrite and nickeliferous marcasite among the epigenetic minerals.

Tennantite-Tetrahedrite The presence of fahlerz was first recognised from the grey colour of the sulphide at Clouds by Clough (*in* Dakyns and others, 1891, p. 167); the locality is not among those mentioned by Greg and Lettsom (1858). The mineral, associated with fluorite, chalcopyrite and galena has been confirmed by Small (in press) and found to be tennantite. His average of two electron microprobe analyses yields the following formula:

$$(Cu_{10\cdot06}Ag_{0\cdot02})(Fe_{1\cdot19}Zn_{0\cdot71})Pb_{0\cdot11}(As_{3\cdot71}Sb_{0\cdot31})S_{12\cdot68}$$

Small also discovered a new locality, at Cumpston Hill [Hanging Lund Scar, SD 783 975] which proved to contain the tetrahedrite member of the pair, with the following composition:

$$(Cu_{10\cdot21}Ag_{0\cdot02})(Fe_{0\cdot69}Zn_{1\cdot06})Pb_{0\cdot03}(As_{1\cdot60}Sb_{3\cdot16})S_{12\cdot74}$$

In this case the two microprobe analyses were identical save that the As/Sb ratio varied from 0.28 to 0.79. The Cumpston Hill veins carry quartz, chalcedony and chalcopyrite; an assay of a sample from the most favourable site is given on p. 117, but these low-temperature veins, though large, are very lean in sulphides (p. 117).

It might have been expected that the appreciable amounts of antimony and the smaller amounts of silver in the galena from the orefield might have been explicable, in part, by inclusions of tetrahedrite, but neither Small nor Vaughan and Ixer found any, though the latter authors found one example of a sulphantimonide in galena which they identified as *stibioluzonite* $3Cu_2S(As,Sb)_2S_5$. On the present evidence it must therefore be concluded that, for the most part, antimony and silver do not form separate phases, but substitute for lead in the galena.

Cinnabar, HgS, and other mercury minerals Following the discovery of cinnabar at Masson Hill, Matlock by Kingsbury in 1964 and at Rutland Cavern by Braithwaite and Greenland (1963), Kingsbury found this mineral at the following sites in Areas 14 and 15 (1968): Moss and Turf Pits mines, Grassington; Cockhill, Peat Pasture and Craven Cross mines, Greenhow, and in Greenhow Rake at Greenhow Quarry.* At Peat Pasture the cinnabar pseudomorphs the rare isometric form metacinnabarite, and this form is preserved unaltered in a specimen from Cockhill. In material from Chatsworth Shaft, Grassington, Kingsbury discovered a small amount of native mercury.

Gold, Au Two assays from the records of the Greenhaugh Mining Company of galena from Jamie Vein, Greenhow, indicate respectively 3 and 7.42 dwt/ton gold (4.2 and 10.3 ppm) but this mine also is now inaccessible, and it has not been possible to confirm this record, which is unique in the northern Pennines.

Quartz, SiO_2 Occurrences of quartz in the mineral veins are confined to the western part of Area 10, to the westernmost end of the North Swaledale Belt (Area 11) in Great Sleddale, and to the veins and replacements between Malham and Settle, in the western part of Area 15. The restricted occurrence of this mineral in this region contrasts markedly with the northern region, but follows a pattern similar to north Derbyshire and Flint – Denbigh. The highly mineralised parts of the region contain virtually no quartz contemporaneous with the other introduced minerals, in spite of the extensive occurrence of primary and diagenetic cherts, especially north of the River Ure. In Area 10, there are scattered small quartz veins carrying a small amount of chalcopyrite and pyrite, while in the Settle – Malham area the paragenesis is with lead, zinc and minor copper, and some baryte; here there are patches of silicification in the Kingsdale Limestone associated with the mineral deposits, but elsewhere this type of alteration is rare or absent. Kettlewell 'diamonds' are small idiomorphic quartz crystals weathered out of partly silicified limestone.

* We are grateful to the Keeper of Minerals at the British Museum (Natural History) for allowing us to examine systematically the Main Collection and the Russell Collection for minerals from the present area. As most of the workings are no longer accessible, it was particularly valuable to be able to see good specimens of sulphides, fluorite, carbonates and oxidation-zone minerals preserved from the finest localities in the area.

Fluorite CaF_2† This mineral is one of the five principal gangue minerals of the lead deposits of the region, the others being baryte, witherite, calcite and dolomite. In macroscopic quantity it is restricted to a number of well-defined geographical areas, details of which are given in Figure 21 and Tables 15, 17; but assays of tailings collected outside the fluorine zones indicate its widespread presence in trace amounts in the deposits. In a few veins, restricted to the Greenhow and Appletreewick districts, it was the major constituent of the oreshoots. At Grassington it was generally second in abundance to baryte, while at the Old Gang and Friarfold mines in the North Swaledale Belt it represented only 2 to 8 per cent of the gangue material brought out. The colour is almost everywhere lacking in the strong shades characteristic of the northern region; pale amber yellow and very pale mauve crystals are not uncommon but much of the fluorite is colourless, or white in the mass. These pale colours are consistent with the fact that here the mineral, with only rare exceptions, occurs closely associated with barium minerals, whereas in Weardale and Alston Moor, the strongly coloured green and purple varieties occur in barium-free deposits (Dunham, 1937). A few dark purple cubes were, however, noticed at Burhill and at Carden reef-knoll, Cracoe. The largest crystals were found in Greenhow Rake, reaching 3 in (7.6 cm) on the sides, colourless save for a faint green tinge near the edges. The Adelaide Level at East Arn Gill, Swaledale, yielded smaller but good greenish yellow penetration twins from a flot in the Underset Limestone, while Wet Grooves, Lolly Scar, Old Providence and Sir Francis Level are all localities where pale yellow crystals up to 1 in (25 mm) on the side have been found. Usually the fluorite presents sharp faces towards baryte, witherite or calcite, even where several repetitions occur. At Gill Heads colourless crystals up to 1 in (25 mm) cube were occasionally found, but the wide deposit is mainly massive fluorite. The Warton Hall Level, Wensleydale has yielded crystals in which purple and amber shades are interbanded.

The fluorite from the present region is generally non-fluorescent suggesting on the basis of McClelland's determinations (Dunham, 1952, table II) that the rare-earth elements are present in only very small quantities if at all. A sample from Greenhow Rake examined by him showed 0.02 per cent Y, but La, Eu, Ce and Yb were all below the limit of detection (0.001 per cent in the cases of Eu and Ce). The question of rare-earth content has in recent years been taken up by F. W. Smith, with industrially valuable results in the Weardale area (1974; and Greenwood and Smith, 1977). He has extended his work into the present area, and Small (1977, 1978) has also made many determinations in connection with his study of cryptic zoning. Yttrium is the principal element determined by them, but Smith has also obtained figures for lanthanum and cerium. No further work on europium or other rare-earths has yet been done. Table 11 gives ranges of Y-content supplied by F. W. Smith and A. T. Small.

Baryte, $BaSO_4$† The deposits contain baryte both as a primary mineral and as a secondary mineral arising from the local alteration of barium carbonate minerals by sulphate-bearing groundwater in the oxidation zone. This section deals only with primary baryte. The mineral is present throughout the region, except in the small quartz veins of Area 10, in some carbonate-dominated flots and in the outer fringe where the gangue is calcite (p. 90). Unlike the northern region it is present throughout the fluorite-bearing districts, though at Gill Heads, Skyreholme the amount is very small. The primary occurrences include a few instances where large white platy crystals are developed, the area where Cobscar and Chaytor rakes come together being the most striking, but most commonly closely intergrown material, varying from white to pink in colour, is found.

†The terms fluorite and baryte refer to the pure minerals; fluorspar and barytes to commercial concentrates containing them.

Table 11 Rare-earth content of fluorite

Locality	Analyst	Samples	Yttrium (ppm) Range	Yttrium (ppm) Mean	Others (ppm)
Area 10					
Hartley Birkett, Ladthwaite, Long Rigg	FWS	10	0–28	15	La 0–34 Ce 0–50
Hartley Birkett, Long Rigg	ATS	7	10–75	35	
Bells	FWS	1	21		
	ATS	1	30		
Clouds	FWS	7	12–28	26	
	ATS	7	30–80	59	
Area 11					
Great Sleddale	ATS	2	270, 280	275	
Old Rake, North Rake, Friarfold	FWS	11	46–72	53	La, Ce n.d.
Gunnerside	ATS	2	35, 65	50	
Surrender	FWS	2	36, 37	37	Ce 0.26, La n.d.
Arkengarthdale	FWS	13	17–50	44	
	ATS	4	65–70	67	
Copperthwaite	FWS, ATS	3	58–80	71	
East Arn Gill	ATS	2	65, 75	70	
Area 13					
Thwaite Beck	ATS	1	120	120	
Lover Gill	ATS	1	150	150	
Stags Fell	ATS	1	210	210	
Mossdale Beck	ATS	1	220	220	
Cottriggs (Raydale Farm)	ATS	2	110	110	
Warton	ATS	1	35	58	
	FWS	1	80		La 5, Ce 41
Wet Grooves	FWS	2	28, 37	32	La n.d., Ce 17
	ATS	2	45, 80	63	
Seata	FWS	2	28, 34	31	La n.d., Ce 13
	ATS	1	40	40	
Keld Heads	FWS	2	34	34	La, Ce n.d.
West Burton	ATS	1	45	45	
Braithwaite	ATS	1	70	70	
Area 14					
Starbotton	FWS	2	21, 23	22	La 5, Ce 17
Middlesmoor	FWS	1	26	26	La, Ce n.d.
Bycliffe	FWS	3	27–81	46	La n.d., Ce 5
Palfrey, Low Peru	FWS	1	23	23	
Grassington Middle Vein	FWS	1	26	26	La 14, Ce 6
Yarnbury	FWS	2	11, 31	21	La, Ce n.d.
Cavendish	FWS	1	27	27	La n.d., Ce 4
Area 15					
Lolly	FWS	1	30	30	La 2, Ce 21
Waterhole	FWS	1	29	29	La 15, Ce 26
Greenhow Rake	FWS	3	3–27	19	La n.d., Ce 12
Galloway	FWS	3	15–26	21	La 2, Ce 9
Forest	FWS	1	19	19	La n.d., Ce 18
Burhill	FWS	1	12	12	La 4, Ce 13
Inman	FWS	1	16	16	La 3, Ce 20
Gill Heads	FWS	8	15–25	21	La 1, Ce 12

XRF analyses by F. W. Smith and A. T. Small, University of Durham, 1973–79 (Smith, 1974a, 1974b; Small, 1977, 1978) n.d. = not detected

The very fine-grained variety known as cawk in Derbyshire is rarely found here. Good crystals are rare, and the national collections contain very few from this region; the British Museum has some water-clear crystals collected by Sir Arthur Russell in 1932 from Wet Grooves, but other localities including Friarfold, Victoria Level (Old Gang mines), Danby Level (Arkengarthdale) and Nussey Knot show white to dirty white crystals. The presence of bands and inclusions of unoxidised galena in the platy and finely intergrown material leave no doubt as to the contemporaneity of these minerals. Tiny inclusions of yellow sulphides, probably chiefly chalcopyrite are also common in all parts of the field.

Substitution of a small part of the barium in primary baryte by strontium is quite general. A picked sample of primary baryte from New Whim Shaft, Friarfold Vein, showed on analysis $BaSO_4$, 94 per cent; $BaCO_3$, 0.5; $CaCO_3$, 0.75, $SrSO_4$, 0.9 (C. O. Harvey, Geological Survey No. 1225 *in* Dunham and Dines, 1945). Samples from the northern region (Dunham, 1948, p. 94) ranged from 0.29 to 3.6 per cent $SrSO_4$ and determinations by Small (1977) lie within the same range. No free celestine occurs.

In addition to the normal platy and massive primary baryte, a sharp pointed crystalline variety dominated by (110) faces in combination with (001) is characteristically grown on, and probably produced by alteration of, witherite. Some physical or chemical property of the witherite must stimulate this unique type of crystal growth, and since it was abundantly found in cavities far below the oxidation zone at Settlingstones in the northern region, it is regarded as a late-stage but primary mineral.

Witherite, $BaCO_3$ Though there has been only a very small production of witherite from the present region, the mineral is widespread particularly in the mines of Areas 11 and 13, which share with those on the fringes of the Alston Block the unique position of containing more of this mineral than has been found

Table 12 Analysis of witherite, Virgin Moss Mine

Major elements (%)		Trace elements (ppm)	
BaO	75.79	F	150
SrO	1.18	Ca	170
CO_2	22.41	Fe_2O_3	160
moisture	0.06	SiO_2	420
		MnO	8
	99.44	CuO	2
trace elements	0.16	PbO	27
		ZnO	8
	99.60	SO_3	450
		P_2O_5	160
		Al_2O_3	30
		Y	5

Spectrographic examination showed that the following elements, if present, were in amounts less than the detection limit indicated (in ppm)

Sb	50	Co	10	Nb	20	W	200
As	50	Ga	5	Ag	1	V	5
Be	1	Ge	3	Ta	200	Zr	5
Bi	5	Li	50	Tl	5	Au	10
B	5	Mg	100	Th	200	In	10
Cd	50	Mo	1	Sn	5	Ce	100
Cr	10	Ni	5	Ti	50	La	10

Analysts P. T. Sandon, M. E. Stuart, B. A. R. Tait, 1978: IGS Analytical and Ceramics Unit Report 105.

anywhere else in the world. It is almost certain that before oxidation damaged the deposits, there were major concentrations in the Old Rake, Friarfold and Surrender veins. Witherite was the principal gangue mineral in the deep workings of Sir Francis Level at the A.D. Mines, and some was produced from Barras High Level nearby. At Virgin Moss Mine, flots carrying massive witherite were found; the analysed specimen (Table 12) came from a small heap of picked material near Virgin Level mouth. Other localities include Whitaside, Victoria, Cobscar Rake and, in Area 15, Lolly Mine (enclosing yellow fluorite crystals) and Providence Prosperous. The mineral is normally a distinct yellow colour on a fresh face, but also occurs white and massive. In cavities mamillary growths are found to be incipient pseudo-hexagonal crystals; fully developed crystals in this form are occasionally found, a beautiful example from Danby Level being preserved in the Lady Anne Cox Hippesley Collection (part of the Russell Collection).

Strontianite, $SrCO_3$ This mineral has been found in small quantities at the following localities: Merryfield Mine, near Greenhow (in addition to a specimen from this specific locality, the Natural History Museum Main Collection contains one marked Pateley Bridge which could refer to the same); and Victoria Level, Hard Level Gill.

Barytocalcite, $BaCO_3.CaCO_3$ This rare mineral was first reported from Swaledale by Bradley (1862) but unfortunately he gave no specific locality. At the time he was operating the Blakethwaite and Lownathwaite mines, both of which contain witherite. At the Brandy Bottle Incline on Friarfold Vein, reopened during explorations for baryte in the early 1950s, the wide vein was found to consist mainly of white spar minerals including baryte, witherite and fluorite, but it was found to be difficult to beneficiate this to a density above 4.2 for use in drilling mud. The possibility was canvassed that a substantial part of the orebody might be barytocalcite, but microscopical and XRD examination respectively by Mr Harrison and Mr Young failed to reveal this mineral, though calcite was found to be common. Two new localities were, however, found during the field survey in 1941 where barytocalcite was confirmed by optics and XRD‡ : Beldi Hill, and the Lane Head Mine, Great Sleddale (p. 122). In the collection of the British Museum (Natural History) there is one from the Lolly Mine, Nidderdale (p. 205). During a visit by the Yorkshire Geological Society in 1977, a specimen showing the oblique rhombic prisms characteristic of barytocalcite from the Blaygill, Alston locality (Dunham, 1948, p. 95) was picked up at Sir Francis Level, Gunnerside Gill. The mineral remains, for reasons not understood, a great rarity, and is not a main constituent of veins, even where barium and calcium carbonates have crystallised together.

‡The specimens, numbered MI 27482/3 and MI 27473 respectively, are preserved in the Mineral Inventory at the Geological Museum.

Dolomite $MgCO_3.CaCO_3$ and Ankerite $(Mg,Fe)CO_3.CaCO_3$ Brownish dolomite is the commonest mineral in the limestones adjacent to the veins in all areas, its refractive index ω varying little from 1.685. This indicates an iron-content near zero apart from the minor amount in FeO.OH assumed to be the cause of the colour. It tends to assert its rhombic form during the replacement process and the crystals are usually coarser than those of the adjacent calcite undergoing metasomatism. It is the predominant mineral in the flots at Windegg, Bolton Park, Whitaside – Apedale and Summer Lodge (Table 13) and no carbonate (other than smithsonite and cerussite) has been found in these with ω greater than 1.685. All the worked flots, however, lay within the oxidation zone and goethite-limonite is commonly present in quantity; the possibility that some ankerite was formerly present cannot therefore be ruled out. The presence of sulphides makes it difficult or impossible to establish this chemically, and optical evidence that any remains has not been found.

A number of scattered occurrences of ankerite with appreciable contents of ferrodolomite have nevertheless come to light, though it is clear that the mineral is very much less abundant than in the Alston and Weardale fields. The most interesting occurrence is a replacement of a wide slice of Main Limestone lying between the Middle and Sun veins in Lownathwaite Hush (p. 125), the only place in the region where the rock resembles the protore of the Carricks Mine, Weardale (Dunham, 1948, p. 237). Professor A. C. Dunham made 16 analyses of the unoxidised but highly zoned carbonate from this locality on the electron microprobe at Manchester University. The results (Table 13) show a wide range of compositions within the ankerite series, wider than Smythe and Dunham (1947) were able to establish in any single specimen by optical methods. The average tenor of ferro-dolomite here is about 22 per cent. From Racca High Shaft at Hurst (p. 146), finer-grained vein ankerite was proved by 12 microprobe analyses to vary little around a mean of 9.4 per cent ferro-dolomite. Similar ankerite is associated with Geordy Vein where it crosses Foregill Beck in the Main Limestone (p. 135). Ankerites richer in iron have been collected at Ladthwaite Level (p. 115) and at Stony Grooves Mine near Greenhow (p. 206), both showing $\omega = 1.710$, indicating 42 per cent ferrodolomite. From Hartley Birkett opencut ankerite has $\omega = 1.735$, suggesting 75 per cent ferro-dolomite, a higher content than any found in the northern region (Dunham, 1948, p.92).

Siderite (chalybite)§, Brownspar The $FeCO_3$-$MnCO_3$-$MgCO_3$ minerals are very rare in the veins and flots of this half of the orefield, though common as diagenetic nodules and bands in Brigantian and later shales. The brownspar from Old Providence was found by Mr Harrison to have $\omega = 1.780$. Dr Phemister found siderite with $\omega = 1.830$ in cavities in Copperthwaite veinstuff which also contained hemimorphite. Siderite is also present at Hartley Birkett opencast, accompanying the ankerite.

§ Siderite (of Haidinger) is now the internationally accepted name for chalybite (*Mineral. Mag.*, 1980, Vol. 43, p. 1053).

Table 13 Bulk composition of metasomatic carbonates in altered limestones

Analysis number and locality		$CaCO_3$	$MgCO_3$	$MnCO_3$	$FeCO_3$	Minerals
28 Windegg Flots		71.9	28.1		0.0	Calcite, dolomite
AA1 Lowna-	*range*	54.6–61.0	26.8–31.3	0.0–1.22	0.4–14.6	Ankerite
thwaite Hush	*mean*	56.9	30.5	0.7	10.9	
AA3 Racca Shaft, Hurst		53.7	41.4		4.9	Ankerite
40 Bolton Park		60.3	39.7		0.0	Dolomite, calcite
45 Worton Hall		56.7	43.3		0.0	Dolomite, calcite
47 Old Providence		70.6	17.6		11.8	Brownspar, calcite

AA1, AA3 microprobe analyses by A. C. Dunham, 1977 (University of Manchester); the rest recalculated from tables 17–19

Calcite, Aragonite, $CaCO_3$ In describing Areas 1 to 9 (Dunham, 1948) little is said about calcite because, though often present in small amounts, it is never a principal gangue mineral. In the present region the situation is different; calcite is present in almost every deposit and within a large fringing area it is the dominant gangue mineral of the lead veins, in some deposits the sole accompaniment of galena. Interbanding with the sulphides, fluoride, sulphate and other carbonates leaves no room for doubt that it is as much a primary mineral as any of them; indeed in many deposits, even in sandstone, it forms the first coating on the vein walls. In massive form, calcite is generally white where not stained by iron, zinc or copper hydroxides; the interlocking crystals in some cases assert their rhombic form, but in others do not do so. They vary up to coarsely macrocrystalline, the coarsest material coming from Greenhow Rake, Fielding and other veins at Greenhow, where a columnar form with strongly emphasised prism faces is found. The ordinary-ray refractive index ω varies little from 1.658, suggesting that any substitution in the lattice is of very minor significance. The common crystal forms in cavities are the 'nail head' (rhomb faces dominant over prisms) and the scalenohedron.

Aragonite more rarely forms a massive veinstone, particularly in the wide sparry veins of Old Gang and Arkengarthdale, its place in the sequence usually being a late one. An interesting example from Wasset Fell above Walden reported on by Mr Harrison and Mr Young consists of finely acicular masses giving rise, on broken surfaces, to a silk-like sheen; lines of dark grey staining mark junctions between individual radial growths. The mineral is biaxial negative with low 2V (2E = 16°), refractive indices $\gamma = 1.680$, $\alpha = 1.530$, both ± 0.002. A subordinate pale grey fine-grained carbonate proved to be witherite, both determinations being confirmed by XRD. Massive mamillary banded carbonate from the Stool Mine, Grassington, grown on inner pale fluorite consists of aragonite in radially fibrous form (analysis, Table 14) with optics nearly uniaxial negative, $\gamma = 1.680$. The concentric growth laminae in a specimen about 6 cm wide number about 6 per mm. From the same locality, botryoidal, radially fibrous coarser massive carbonate, with translucent inner and opaque outer zones proved to be calcite ($\omega = 1.658 \pm 0.002$). Compared by energy dispersive analysis carried out on polished thin sections with the aragonite, this proved to have appreciably less $SrCO_3$ in solid solution, but more $MgCO_3$. The stability of aragonite has been attributed by Siegel (1958) to the presence of strontium in the crystal lattice. The two minerals clearly grew in the same geological environment in the Stool Mine case, and it might be argued that the temporary availability of Sr caused the aragonite dimorph to form rather than calcite. However, alteration of aragonite to calcite, with loss of part of the Sr cannot be ruled out.

Table 14 Analyses of aragonite and calcite, Stool Mine, Grassington

	Aragonite 4 fields	Calcite 4 fields	Calcite 2 fields
CaO	56.78	55.48	55.95
MgO	n.d.	0.99	0.82
SrO	0.84	0.13	0.13
ZnO	0.26	1.89	n.d.
SO_3	0.10	0.10	n.d.
CO_2+	42.02	41.41	43.10

Microprobe energy-dispersive analyses by Dr M. T. Styles, 1979; CO_2+ by difference; n.d. = not detected

The massive form is not, however, the best-known mode in the present region, for it has yielded spectacular museum specimens of the 'flos ferri' variety, consisting of curviform and ramifying white tubes. It seems fairly clear that this material comes from open cavities, and that it probably post-dates the oxidation of the ore deposits and the initiation of the present cycle caverns. Specimens in the Natural History Museum main collection came from the Arkengarthdale mines, especially Faggergill (but possibly also including Danby Level), Windegg and Whitaside; the Russell Collection adds Old Gang to this list and the M. I. Collection of the Geological Museum, Providence Mine near Thwaite, Swaledale.

DISTRIBUTION OF PRIMARY MINERALS

The existence of considerable variation in the contents of the mineral veins was first recorded by J. G. Goodchild and C. T. Clough in the Mallerstang memoir (Dakyns and others, 1891, pp. 166–167) as the following extracts show:

> A little copper pyrites, with its usual accompaniments of blue and green carbonates ... occurs in some of the veins traversing the Carboniferous Rocks; but these occurrences are, almost without exception, confined to the part of the district outside the lead-bearing area.
>
> The minerals accompanying the galena vary in different parts of the district. As a rule, calcite and barytes, with more or less iron-pyrites and zinc-blende preponderate, acccompanied in exceptional cases by quartz or fluorspar ... Calcite is, in fact, the commonest mineral in the veins ... Barytes is common along the course of Friarfold Vein ... It also occurs in the Old Rake Vein, the Blakethwaite Vein etc. The fluorspar does not show the rich colours found in some districts.

This refers only to the north-west part of the field, in the Kirkby Stephen (Sheet 40) area, and fluorite has proved to be more important in the wider field than this suggests.

Bradley (1862) defined the miner's term *rider* as any mineral matter existing between the true or perfect sides, cheeks or walls of a vein and distinguished three groups: (i) rock fragments, principally limestone, chert and grit; (ii) crystalline minerals including Calcareous Spar, Fluor Spar, Caulk or Cawk (carbonate and sulphate of baryta), Barytocalcite, Quartz, Iron Pyrites, Black Jack and Grey Jack (according to him, oxides of zinc); (iii) friable and tenacious earthy substances named Whamp or Vamp, Douk or Donk, or Platy Sample. Group (iii) would now be recognised as including fault gouge and limonitic oxidation products. His account distinguished between veins well supplied with rider and those that are not, but he gives no indication of regional distribution of Group (ii) fillings.

Following the publication (Dunham, 1934) of a regional zonal pattern for the minerals of the Alston Block part of the orefield, Hay (1939) applied similar methods to the Greenhow – Grassington area, finding strong fluorite mineralisation at Greenhow giving place both east and west to baryte, on the basis of which he postulated a feeding centre under Greenhow. He also suggested a dextral shift of the zones by the North Craven Fault, but produced a confusing picture of zones in Wharfedale. At Greenhow, Dunham and Stubblefield (1945) recognised three crude zones grading into one another: (1) fluorite with little galena or baryte; (2) fluorite – galena – baryte; (3) baryte – galena with little or no fluorite. To these, Dickinson (1970) added a calcite – galena zone, and noted that calcite may be present in substantial amounts in the fluorite zone.

Most of the data necessary for a complete picture, as far as the evidence goes, of the whole field covered in this volume was collected during the mineral survey by Tonks, Earp and Dunham early in World War II but a number of minor localities, not expected to be of interest at that time, were left unvisited. This work defined three major areas where virtually every deposit carried fluorite: (a) Lownathwaite to Arkengarthdale in the North Swaledale Mineral Belt (b) Wharfedale from Grassington to Buckden (c) Skyreholme – Greenhow, each surrounded by deposits without this mineral; and this information was published in the Fluorspar memoir (Dunham, 1952) and reproduced in general discussions of the orefield (1959; 1974). A number of minor 'inliers' of the fluorite zone were recognised outside the main areas, particularly at Wet Grooves and Keld Heads (Wensleydale), Lolly (Nidderdale) and in the foothills near or in the Dent Disturbance. It was noted that baryte, with some witherite, and calcite are widespread gangue minerals for galena outside the fluorite-bearing areas, and the three principal fluoritic areas were postulated as overlying feeding centres, evidence being presented of a downward increase in fluorite in the case of the North Swaledale centre. The small copper occurrences around Middleton Tyas in the east and the Mallerstang fells in the west were regarded as marginal, similar to some noted in the northern region near the Vale of Eden.

The detailed investigation of 63 collecting areas in the Stainmore Depression and northern part of the Askrigg Block by Small (1977; 1978) has led him to recognise the following overt zonal mineralogy:

1 Q Zone: essential quartz
(a) with chalcopyrite (inner)
(b) also with fluorite (outer)
(c) with baryte, galena, chalcopyrite (Stainmore)
2 F Zone: essential fluorite
(a) with little chalcopyrite and baryte
(b) with galena, little chalcopyrite, ± baryte, witherite, sphalerite
(c) as (b) but with no chalcopyrite
3 G Zone: essential galena ± sphalerite, baryte, witherite and little chalcopyrite
4 C zone: essential copper sulphides ± galena, baryte, witherite.

The presence of pyrite and marcasite, especially in the F Zone, is noted, but the published version does not deal with carbonates other than witherite in the paragenesis. As already noted (pp. 89, 91) Small also determined silver and antimony in galenas, and yttrium in fluorites from his collecting areas. His determinations of sodium: potassium weight ratios in rapid leaching experiments on fluorite, quartz, baryte and sphalerite are referred to later on p. 103.

Meanwhile to complete the Geological Survey investigation the opportunity has been taken to examine all the localities not considered in 1939–42, and to revisit certain other key sites, in order to collect tailings samples from which representative bulk compositions of the oreshoots (less the galena removed) could be calculated.

Zonal pattern

The zonal pattern for the Stainmore to Craven region is presented in Figure 21, which may be compared with pl.ii in Dunham (1948). There are some important differences between the two halves of the orefield. In the present region, so few veins contain fluorite without barium minerals that a zone comparable with the widespread inner fluorite zone of the Alston Block cannot be established in the present region. The fluorite zone here almost everywhere carries witherite or baryte interbanded or intergrown with the fluorite. Witherite is, or has been before oxidation, surprisingly common both in the F and B zones, particularly in areas 11 and 13, considering its rarity elsewhere in the world. Most of the largest lead producers lay in the fluorite zone, within a mile or so of its margin, a situation that closely reflects that established in the northern region. Thus the most significant line in the zonal pattern is the outer margin of the F zone at high level in the three principal areas of Lownathwaite – Arkengarthdale, Grassington – Buckden, and Skyreholme – Greenhow. In each of these areas, and only in this zone, the vertical range of mineralisation is substantial. In the North Swaledale Belt, fluorite has been proved from the Simonstone Limestone up to the Crow Chert, in oreshoots covering a vertical range of over 750 ft (230 m) (p. 135). At Greenhow, 500 ft (152 m) is exceeded. The height of the Merryfield oreshoot (p. 207) in Ashfold Side Beck is 380 ft (116 m) while the maximum vertical ranges at Grassington, 330 ft (101 m) at Yarnbury and 280 ft (85 m) at Coalgrove Beck are still large for a field where oreshoots rarely exceed 70 ft (21 m) in vertical dimension. The zone boundary as mapped on Figure 21 lies topographically high for most of its course (solid black line). The only substantial change from the picture as published in 1952 for this boundary is the detachment of the area of feeble mineralisation in the Cracoe reef knolls, an area to which the same important status cannot be accorded. It is, however, now recognised that it is not unreasonable to use the feeble occurrences of fluorite, for the most part in limestones low in the Brigantian, to recognise that the fluorite zone also comes to surface in deep valleys such as Wensleydale and Swaledale, as Small has advocated. Thus in Wensleydale the minor West Burton, Seata, Roger Wood, Worton, Raygill Farm and Mossdale Beck occurrences are joined with previously identified inliers of the zone at Keld Heads, Wet Grooves and Stags Fell Groove (broken black line on Figure 21). All the sites are in Asbian or early Brigantian strata save Stags Fell which is in Main Limestone on the downthrow side of a substantial fault. Where the veins can be traced to higher levels up the valley sides, as at Keld Heads, Wet Grooves and Warton, they pass into baryte gangue. Similarly in Swaledale, Lover Gill Vein, which shows a little fluorite from the Hardrow up to the Five Yard Limestone, is tentatively linked with the fluorite-bearing flat at East Arn Gill (p. 124), and the showings in Gayle Limestone at Marble Scar near Gunnerside with Friar Intake Mine. We hesitate to join the latter with Copperthwaite near Reeth, the western workings of which contained fluorite; this appears to us from its position in Main Limestone and cherts to be a detached high-level occurrence, but not one of much significance.

The second zone is dominated by baryte, or where replacement deposits occur, by dolomitic carbonates. Hurst and Prys, both useful lead producers typify the former, Whitaside – Apedale the latter. Witherite, or baryte after witherite (p. 173) is common in small amounts, and at least one deposit in this zone (Barrass) has been worked in a small way for this mineral. Towards the outer margin, lead values and the size of oreshoots fall off, and the succeeding calcite-dominated zone is one mainly of small, unprofitable trials. Appletreewick is one of the few exceptions; it was a lead producer and later one vein was found to contain such pure calcite that it was mined for this mineral. It must be emphasised that calcite is present in veins of all the zones, but predominates only in the C zone. As already noted, Dickinson, writing about Skyreholme – Greenhow, has postulated a separate calcite zone, but implied that it is in fact the lower or inner zone. This is understandable: Appletreewick is topographically, though not stratigraphically, lower than the fluorite-zone workings farther north, and in the Skyreholme deposits, a number of fluorite oreshoots passed down into calcite, a mineralogical change noted in many parts of the field when an oreshoot is about to pinch downwards. The examination of the order of crystallisation shows that in most deposits, calcite was the first mineral to crystallise, lining the walls, but the galena, fluorite and barium minerals were introduced during episodes of reopening of the fracture system. Calcite was also normally the last primary mineral to form in the veins, but galena is truly interbanded with, and enclosed in, this mineral, particularly in the C zone. Some of the best lead deposits in the C zone were the flots of Harkerside and Grinton How; other feebler examples include the Kining Vein above Gunnerside, representing the outermost zone from the Lownathwaite – Old Gang centre; and many feebler deposits in the valleys cutting the Richmond plateau, and the higher reaches of Walden and Coverdale. The few small veins on the gently-inclined top of the Askrigg Block (p. 182) also belong here.

Veins marked 'R' on Figure 21, belong to the Richmond copper deposits. Chalcopyrite is very widely present throughout the whole region, or was before oxidation to carbonates, as witness the malachite or azurite stains found at many mines or prospects; but the quantity is always low, probably less than 0.1 per cent of the veinstuff. The five R deposits are different; they include four from which copper ore has been produced: Feldom, Middleton Tyas, Merrybent and Billy Banks; and though the total quantity was no more than a few thousand tons, the copper values were high in some of the deposits (p. 162), as a result of the presence of coarse primary chalcopyrite and bornite, and of some secondary enrichment.

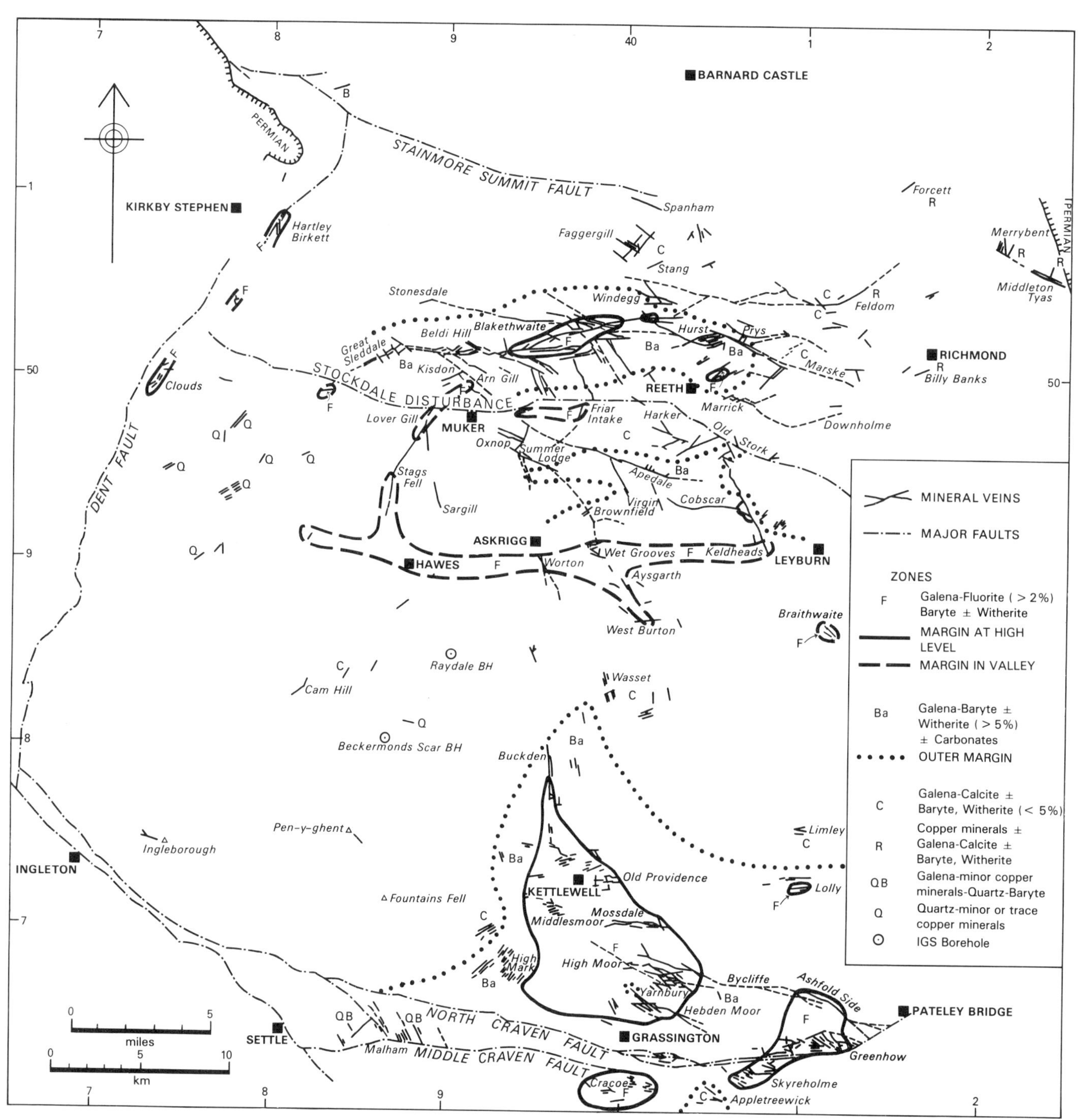

Figure 21 Map to illustrate the distribution of primary minerals between Stainmore and Craven.

However the gangue minerals, calcite, dolomite, some baryte and witherite at Feldom and Forcett, link these deposits with the outer, not the inner zone of the North Swaledale Belt. Before erosion there can be little doubt that they were within a few hundred feet of the sub-Rotliegendes (Permian) unconformity, and in highly permeable strata, but if it is supposed that they were affected by this, it must also be borne in mind that the intergrowth of chalcopyrite with witherite, at one time excellently seen at Forcett, positively links these deposits with the Northern Pennine Orefield. Pending further discussion they must be accepted as outer zone deposits.

The feebleness of all the small veins marked 'Q' in and adjacent to the Mallerstang fells (to which a reference by Goodchild and Clough was quoted above) has previously led to a similar suggestion as to their significance (Dunham, 1974) but this has now been reconsidered in the light of Small's (1978) contention that the zonal pattern focusses on the largely-unworked Zone (1) or Q-zone of his list. Specimens of quartz from the only substantial veins in Small's suggested focal area, those on Cumpston Hill (Hanging Lund Scar) have been subjected to thermometric analysis (p. 103) revealing formation temperatures below the minimum temperature for the fluorite zone. Thus it is impossible to accept these fractures, and the few other minor veins marked 'Q' on Figure 21 as indicating that the Mallerstang fells were the source area for the strong Swaledale mineral belt. The former view is maintained, that these occurrences are outer marginal deposits showing a small recrudescence of copper mineralisation, and that the principal feeders lie beneath Lownathwaite – Old Gang, Upper Wensleydale and Greenhow.

Quartz is far more restricted in occurrence in the Askrigg half of the northern Pennines than in the Alston half, where it is widespread. Here it is restricted to small occurrences in a north – south belt lying west of the major areas of mineralisation. In the north, it occurs at Augill, North Stainmore (Dunham, 1948, p. 140) with baryte, galena and sphalerite. Then come the 'Q' occurrences of the Mallerstang and adjacent fells, with the small, quartz-free vein-complexes along the Dent Zone to the west of them, and the major North Swaledale and Swale/Ure belts to the east, again devoid of free quartz. There is a linking deposit at Great Sleddale (p. 119), almost the only place where fluorite and quartz occur together, with associated chalcopyrite. This may be regarded as the western end of the North Swaledale belt, but certainly not as its source. From West Baugh Fell southward to the Malham – Settle district, no veins are known, but in the latter district, far to the west of the Greenhow and Grassington centres, the galena – chalcopyrite – baryte veins carry quartz and are accompanied by intense silicification of the adjacent limestone. Sphalerite may have been abundant in some of these, having regard to the widespread smithsonite now found.

The extent to which movement of mineralising fluids took place in the major boundary faults is an interesting question, with important possible implications (p. 113). There appear to have been minor centres at Hartley Birkett and Clouds on the Dent line; perhaps beneath Malham on the Mid-Craven, and a major centre at Greenhow – Skyreholme on the North Craven Fault.

In the tables that follow, proximate mineral compositions calculated from analyses of samples representative of tailings heaps, in most cases of material coarser than 1/8th in (3 mm), are presented. High accuracy was not a consideration in most of the analyses, having regard to the obvious sampling error but where insufficient components were determined to give a unique calculation, check by XRD was provided by Mr Young. It will be noted that zinc was found in almost every analysis showing that sphalerite was ubiquitous in small quantities, save in the 'Q' veins. This is often not obvious on the ground owing to the pale colour of the low-iron sphalerite (p. 90) and the easy oxidation of the sulphide. Zinc ore has been produced only from Whitaside in sulphide form, and as carbonate, only from Malham and Cobscar Rake. It is doubtful whether a low tenor of zinc was exceeded at more than one or two places (Black Jack Vein near Blackside in Arkengarthdale might be one). There is little evidence in this field of a downward change from galena to sphalerite, but there are some clear instances of increase of fluorite with depth.

The tailings assays give numerical forms to trends of variation that are not always easy to appreciate visually, owing to the generally white or colourless nature of the gangue minerals. Not only is zinc universally present, but there are always small amounts of fluorine even when no fluorite can be seen, and barium is almost never absent. Necessarily, therefore, the zone-boundaries on Figure 21 are arbitrary lines; the outer limit of the fluorite zone is drawn to correspond with a content, difficult or impossible to see in the tailings with the unaided eye, of 2.0 per cent CaF_2. The outer limit of the barium zone is drawn where barium minerals fall below 5 per cent of the vein-filling. In Area 11 west of Gunnerside Gill, the upper workings at Lownathwaite, Swinnergill, Beldi Hill and the Keldside mines up to Lane End all visually belong to the barium zone. The samples begin with the deep root below the Lownathwaite lode, where Friarfold, Gorton, Old Rake and Watersykes veins converge (Assay 1). Fluorite and witherite markedly diminish upward (2–4), the witherite probably as a result of secondary alteration (p. 108) but the fluorite as a primary effect; parts of the Old Rake and Friarfold dumps indicate that the barium zone was approached and in places reached, though the weighted average indicates the outer part of the fluorite zone. The fluorite-content increases eastward (5–10) to reach a maximum at Damrigg (11) before diminishing again both in the direction of Turf Moor (17, 18) which links through Fell End with the barium zone at Hurst (31); and also through Blackside (14), Justice and Folly veins (15, 16) to Tanner's Rake and Sun Gutter, where the barium zone is also entered (22, 25, 26, 30). In Area 13 there is a similar upward and lateral change from Keld Heads (43) north-north-west through Cobscar Rake to Bolton Park, Virgin and Whitaside (42–38) and, less closely documented, from the fluorite zone at Wet Grooves (44) through Brownfield Vein to Summer Lodge in the barium zone (37, 26) and Beezy where the calcite zone is entered (35, 34).

In Wharfedale, fluorite is at its maximum on Grassington High Moor at Ringleton, Turf Pits, West Peru and Sarah shafts (54–57) diminishing in all directions in favour of baryte. The zonal change is well seen by comparing Yarnbury (63–67) with Bolton Gill Mine on the same main vein (68). The Bycliffe – Merryfield – Prosperous vein system passes from the fluorite zone at Grassington (48–52) into the barium zone beneath Hebden Moor, but east of Merryfield Hole (69, 70) it re-enters fluorite mineralisation (71, 72) which increases south-eastwards towards Greenhow.

The yttrium-content of fluorite (Table 11) does not fit in well with this general zonal pattern. Values over 100 ppm are confined to Great Sleddale and to six 'valley inlier' localities in Upper Swaledale and Upper Wensleydale, all in deposits of minor or negligible productivity. They do not correspond with main centres as postulated here. The studies of Smith (1974) on good underground exposures of large fluorspar oreshoots showed wide variation, but indicated that high values correspond in detail with feeders. Virtually all the results in Tables 15–17 are on dump material, where critical material could easily be missed. Nevertheless, the possibility of a concealed feeder in the Buttertubs vicinity [SD 875 962] is worth bearing in mind. The antimony figures (Table 10) bring some but not all the same localities into question if a downward limit of 1000 ppm is adopted, but they add Friarfold, Wet Grooves, Keld Heads and the 'inliers' of Area 10 to the picture. No comparative data is available for the southern part of the region. The great range and very high silver figures at Lover Gill and Raygill Farm must raise the question of possible secondary enrichment of this element; there is otherwise a possible north-western area of enhanced silver values, like the ill-defined area between Clargill and Hudgill Burn near Alston.

Table 15 Proximate mineral compositions and metal values of mine tailings, Area 11

Locality	NGR	Analysts	CaF_2	$BaCO_3$	$BaSO_4$	$CaCO_3.MgCO_3$	$FeCO_3$	$CaCO_3$	Pb	Zn	Ni*	Cu*	FeO.OH	Rest†
Fluorite Zone Lowanthwaite to Arkengarthdale														
1 Sir Francis Level, Gunnerside Gill, Lownathwaite Vein	NY 9399 0001	DSM	19	32	15				1.5	5				25
2 Old Gang Mines, Old Rake East	NY 950 014	COH	8	1	45			17	4	0.3	46	116		24
3 Old Gang Mines, Old Rake West	NY 944 013	COH	4	1	51			17	4	0.7	49	138		20
4 Old Gang Mines, Friarfold Vein, Smithy Area	NY 946 016	COH	2	1	63			10	3.5	1.5	46	140		17
5 Old Gang Mines, Friarfold Vein West	NY 957 022	COH	4	1	72			4	3.5	0.7	49	177		12
6 Old Gang Mines, Friarfold New Whim	NY 9640 0240	COH	2	1	72			14	1	0.3	32	70		10
7 Old Gang Mines, Friarfold Forefield Shaft	NY 9680 0249	COH	8	4	40			29	2.5	0.7	53	144		14
8 Surrender Mine, High and Low shafts	NY 972 025	COH	6	2	44			29	3	1	52	161		19
9 Moulds Top Mine, Jacobs North Vein, Fourth Whim	NY 980 023	EW	22.6	11.5	37.5			10	1.5	0.7				14
10 Moulds Top Mine, Jacobs North Vein, Jacobs Whim	NY 985 022	EW	24.6	0.4	57.8			—	3	0.7	59	156		12
11 Arkengarthdale Northside mines, Dam Rigg, North Vein	NY 979 028	EW	61.5	—	23			5.1	2	0.5	42	93		4
12 Arkengarthdale Northside mines, Martin Vein	NY 980 028	EW	30.7	2	46.7			2	3	0.5				12
13 Arkengarthdale Sunside mines, Stemple and Dodgson veins	NY 990 024	JE	38.4	tr	34.5			1.5	3.3	0.3				23
14 Arkengarthdale mines, Underedge Level	NY 991 030	EW	51.3	—	3.2	0.9		1	0.5	0.7	32	115	1	39
15 Arkengarthdale Sunside mines, Justice Level	NY 991 027	EW	36.9	—	31.2			0.9	1.7	0.2				29
16 Arkengarthdale Sunside Mines, Folly Vein	NY 991 025	JE	41.7	tr	5.3			3.2	1.1	0.9				47
17 Turf Moor Mine, head of hush	NY 993 022	JE	16.1	tr	49.1			1.8	2.4	0.8				28
18 Turf Moor Mine, Butts Shaft	NY 9898 0226	JE	17.5	tr	56.4			2.2	2.5	1.3				19
19 Moulds Old Level	NY 997 026	EW	22.6	1.2	45.4	1.2		9.8	0.6	1.0				20
20 Dam Rigg Dressing Floor, Arkengarthdale, east dump	NY 988 036	EW	30.8	2.9	27.6	2.6	1.5	1.6	0.6	0.6	44	127		32
21 Dam Rigg Dressing Floor, Arkengarthdale, west dump	NY 985 037	EW	26.7	3.0	20.2	2.5	2.0	1.4	0.6	0.5				43
22 Tanners Rake Mine, junction with Blackside Vein	NZ 013 032	EW	4.7	—	58.5			3.6	4.3	2.3				25
Barium Zone Blakethwaite, Barras, Arkengarthdale, Hurst														
23 Blakethwaite Low Level, Blakethwaite Vein	NY 9392 0234	JE	tr‡	8.4	29.8			10.5	5.1	2.4				43
24 Barras High Level, Barras East Vein	NY 9876 0123	EW	1.3	6.3	64.1	0.9		0.6	5	0.8				20
25 Tanners Rake Mine, Upper Level	NZ 017 030	EW	1.9	—	39.2			12.0	0.4	1.4				45
26 Tanners Rake Mine, Doctor's Level	NZ 018 028	EW	1.4	—	67.3			3.0	2.2	3.4				21
27 Windegg Mine, Cocker Rake	NZ 009 037	EW	tr	1.5	81.5	0.3		—	0.1	0.03			0.3	16
28 Windegg Mine, Middle Level, Windegg Vein and Flots	NZ 010 041	EW	tr	—	42.8	23.3		14.1	3.5	2.5			7.5	3
29 Windegg Mine, Washy Green Level, Windegg Flots	NZ 021 037	EW	1.2	—	53.5			16.7	4.5	0.2				25

Table 15 *continued*

Locality	NGR	Analysts	CaF_2	$BaCO_3$	$BaSO_4$	$CaCO_3.MgCO_3$	$FeCO_3$	$CaCO_3$	Pb	Zn	Ni*	Cu*	FeO.OH	Rest†
30 Sun Gutter Level, Scatter Scar Vein	NZ 0193 0291	EW	tr‡	—	73.4			8.5	0.8	1.8				15
31 Prys Mine, Low Level, Prys, Shaw and Wallnook veins	NZ 069 026	EW	0.6	—	74.2			1.2	2.2	1.3				21
Calcite Zone Faggergill														
32 Faggergill Mine	NY 989 070	EW	0.2		3.6	19.9	2.3	15.6	4	2.3			14.3	40

Analysts DSM Dunford, Smith and Moore; COH C. O. Harvey, 1942; JE J. Esson, 1977; EW E. Waine, 1978 [D. Peachey, B. P. Allen (Ca, Pb, Zn, Fe, Mg); J. L. Chapman (CO_2); M. E. Stuart (F); T. K. Smith (total Ba by XRF); D. Hutchison, C. H. Branch (Cu, Ni and acid-soluble Ba by AAS to check for the presence of witherite)].

All the samples used in the determinations were taken, using hand tools, from exposed tailings of both gravels and less coarse types. More recently the Arkengarthdale dumps have been sampled by Minex Ltd, making deep trenches into the heaps by mechanical means. The order of results is generally similar, save that our results give higher absolute values for F, Ba, Pb and Zn since the coarse rock content was not included in our samples.

In the 1978 analyses, F, Ba, Ca, Pb, and Zn were determined in all samples, CO_2 in all but nine, and Mg and Fe in all 26 samples that contained significant amounts of carbonate.

— signifies that the constituents of the mineral concerned were determined, but were not present in sufficient amounts to make the mineral.

* Values for Ni and Cu are shown in parts per million.

† Mainly SiO_2 from chert, but may include small amounts of S and Al_2O_3; also MgO and Fe oxides where not otherwise determined.

‡ tr in the CaF_2 column indicates fluorine present in the range 0.017–0.1 per cent.

Table 16 Proximate mineral compositions and metal values of mine tailings, Area 13

Locality	NGR	Analysts	CaF_2	$BaCO_3$	$BaSO_4$	$CaCO_3.MgCO_3$	$FeCO_3$	$CaCO_3$	Pb	Zn	Ni*	Cu*	FeO.OH	Rest†
Calcite Zone South Swaledale, Wensleydale														
33 Sargill Mine, Sargill–Providence Vein	SD 895 928	EW	weak tr‡	—	3.2	13.7	3.9	28.1	1.2	1.8			9	37
34 Beezy mines, Beezy Level, East Vein	SD 947 942	EW	tr	—	0.5	6.2	7.4	50.8	1.6	6.9			3.2	15
Barium Zone Swale–Ure watershed														
35 Beezy mines, West Vein	SD 945 943	EW	tr	—	40.1	2.3	1.8	9.2	1.6	7.4			7.4	29
36 Summer Lodge Mine, Shafts to Summer Lodge veins	SD 955 950	EW	weak tr	2.9	3.8	74.2	3.3	3.2	1.2	0.4			—	3
37 Summer Lodge Mine, Summer Lodge Level	SD 962 951	EW	tr			52.5		7.2	3.4	1.3	50	45		13
38 Whitaside Mine, shafts to flots, Apedale Vein	SD 998 954	EW	tr	—	43.4	15.9		1.8	10.5	8.0				8
39 Virgin Mine, Virgin Level	SE 003 934	EW	weak tr	4.7	3.1	3.1	1.3	35.7	0.7	1.4			7.2	47
40 Bolton Park Mine, Upper Level	SE 030 929	EW	0.2	—	29.9	43.6	0.0	6.1	1.1	2.5			6.4	10
41 Cobscar Mine, Redmire Shaft, Cobscar Rake	SE 050 930	EW	1.5	—	76.7	0.2	1.5	0.3	1.6	2.5				14
Fluorite Zone Wensleydale														
42 Cobscar Mine, shafts E of Smelt Mill	SE 060 930	EW	6.6	—	43.6	0.9		3.3	3.7	10.2			2.3	21
43 Keld Heads Mine, Chaytor Rake	SE 080 908	EW	22.6	—	25.6			2.0	0.8	2.9	61	86		45
44 Wet Grooves Mine, residue from worked-over dumps	SD 987 902	EW	22.6		4.5			21.5	1.8	3.9				44
45 Worton Hall Level	SD 956 901	EW	13.3	—	3.9	62.1	0.0	2.8	0.1	1.2				17

Analysts EW E. Waine, 1978 [D. Peachey, B. P. Allen (Ca, Pb, Zn, Fe, Mg); J. L. Chapman (CO_2); M. E. Stuart (F); T. K. Smith (total Ba by XRF); D. Hutchison, C. H. Branch (Cu, Ni and acid-soluble Ba by AAS to check for the presence of witherite)].

— signifies that the constituents of the mineral concerned were determined, but were not present in sufficient amounts to make the mineral.

* Values for Ni and Cu are shown in parts per million.

† Mainly $SiO_2$from chert, but may include small amounts of S and Al_2O_3; also MgO and Fe oxides where not otherwise determined.

‡ tr in the CaF_2 column indicates fluorine present in the range 0.017–0.1 per cent.

Table 17 Proximate mineral compositions and metal values of mine tailings, areas 14 and 15

Locality	NGR	Analysts	CaF_2	$BaCO_3$	$BaSO_4$	$CaCO_3.MgCO_3$	$FeCO_3$	$CaCO_3$	Pb	Zn	Ni*	Cu*	FeO.OH	Rest†
Fluorite Zone Wharfedale, Grassington														
46 Cam Pasture mines, white heaps	SE 962 744	EW	41.1		12.4			43.8	3.4	1.2				
47 Old Providence Mine, Kettlewell, Middle Level	SE 992 728	EW	24.6	—	54.1	2.7	0.8	3.3	4.9	2.3				8
48 Green Bycliffe Mine, Bycliffe Vein	SE 0185 6844	GCL	19.1	—	65.3				2.9	0.7				21
49 How Gill Shaft, Bycliffe Vein	SE 0229 6817	GCL	14.7	—	54.9				4.3	0.3				25
50 How Gill Dressing Area, Bycliffe Vein	SE 022 682	EW	16.2	—	52.9			8.3	6.0	0.3				15
51 Richards Shaft and Levels, Bycliffe Vein	SE 0287 6787	GCL	16.8	—	54.9				2.9	1.2				25
52 Bycliffe Mine, Bycliffe Vein	SE 028 679	EW	24.6	—	36.4				1.9	1.9	30	127		35
53 Palfrey Vein opencut	SE 018 680	GCL	19.7	—	48.5				4.4	0.6				30
54 Ringleton Low and High shafts	SE 028 674	GCL	65.0	—	8.3				0.9	0.2				25
55 Turf Pits Shaft	SE 0278 6755	GCL	43.5	—	24.8				2.5	1.0				28
56 West Peru Shaft	SE 0322 6756	GCL	39.6	—	18.4				0.8	1.1				39
57 Sarah Shaft	SE 0357 6735	GCL	47.7	—	28.1				1.9	0.8				22
58 Chatsworth, Taylor area	SE 028 670	GCL	26.8	—	55.6				1.8	1.3				15
59 Coalgrove Head Shaft, Coalgrove Head Vein	SE 0318 6703	GCL	18.7	—	45.7				2.6	1.1				34
60 Old Moss Shaft, Middle Vein, slimes	SE 0346 6702	EW	22.6	—	28.2			2.4	1.7	4.9				40
61 Coalgrove Beck Dressing Area	SE 029 668	GCL	16.6	—	67.1				5.0	1.2				9
62 Coalgrove Beck Flotation Mill tailings	SE 029 668	EW	20.5	—	15.2	1.2		0.7	1.5	2.2			2.6	44
63 Good Hope Shafts, Yarnbury	SE 0090 6646	GCL	13.9	—	58.9				2.1	2.0				21
64 Yarnbury Dressing Area near incline	SE 016 660	GCL	9.6	—	64.5				3.6	1.1				20
65 Beevers and Union shafts, Yarnbury	SE 0203 6556	GCL	21.9	—	15.7				2.2	1.2				60
66 Experimental jig product, Yarnbury, coarse	SE 020 656	EW	18.4	—	64.4			14.7	2.9	0.1				8
67 Experimental jig product, Yarnbury, sand	SE 020 656	EW	20.5	—	57.1			8.5	5.0	0.4				8
Transition to Barium Zone Hebden Moor, Ashfold Side Beck														
68 Bolton Gill Mine	SE 0314 6538	GCL	3.2	—	56.1				1.5	0.8				37
69 Stony Grooves Mine, Stony Grooves Vein	SE 095 666	GCL	2.9	—	68.5				5.6	1.0				21
70 Merryfield Hole and Stony Grooves Shaft	SE 1013 6656	GCL	4.3	—	78.3				1.4	1.1				16
Fluorite Zone Ashfold Side, Greenhow, Skyreholme														
71 Merryfield Mine, Merryfield—Prosperous Vein	SE 108 664	GCL	19.8	—	59.6				1.0	1.4				17
72 Providence Mine, Merryfield—Prosperous Vein	SE 1197 6592	GCL	20.4	—	49.1				2.0	1.7				26
73 Galloway Pasture, excavation in gulf	SE 110 636	GCL	67.8						1.5				SiO_2	6.4+
74 Greenhow Rake, channel sample	SE 1092 6420	GCL	50.7		4.6			36.8	0.6					2.8+
75 Cockhill Mine, Waterhole North Vein	SE 1120 6435	GCL	87.7		1.9			1.2	0.3					3.2+
76 Blackhill Old Engine Shaft, Blackhill Vein	SE 0844 6322	GCL	63.3						5.4					5.1
77 Gill Heads Mine, untreated product 1977	SE 0660 6205	EW	89.2						0.3	0.1				6

Analysts EW E. Waine, 1978; GCL Laboratory of the Government Chemist, 1942

— signifies that the constituents of the mineral concerned were determined, but were not present in sufficient amounts to make the mineral.

* Values for Ni and Cu are shown in parts per million.

† Mainly SiO_2 from chert, but may include small amounts of S and Al_2O_3 also MgO and Fe oxides where not otherwise determined.

Summarising: (1) there is a clear relationship between the outer part of the fluorite zone, the innermost part of the barium zone and the most productive lead oreshoots; (2) only a few inner deposits approach fluorite-dominance such as to yield high-grade fluorspar oreshoots; (3) lead oreshoots diminish in size and importance outward in the barium zone and throughout the calcite zone; (4) the Richmond copper district appears to lie in the outermost zone; (5) an agnostic attitude is adopted towards the economically negligible quartz veins of Mallerstang, but trace-element variation in fluorite and galena may indicate a concealed mineralisation centre between Upper Swaledale and Upper Wensleydale; (6) the clearly identifiable centres nevertheless lie beneath Lownathwaite and Dam Rigg under the eastern part of Grassington High Moor and beneath Greenhow.

Combining conclusion (6) with the evidence of stratigraphical control of oreshoots (pp. 85, 86) the hypothesis is proposed that mineralising fluids rose from depth at the centres enumerated until reaching tubular openings formed by the intersection of minor faults with certain brittle hard beds in the Asbian, Brigantian and early Pendleian sequences, and then spread laterally, emplacing ribbon and replacive oreshoots. South-east of the line where the intra-E1 unconformity cuts out the Main Limestone, the oreshoots became confined to successively lower beds until the fluids were able to break through to Grassington Grit and (at Ashfold Side Beck) higher strata.

WALLROCK ALTERATION

Adjacent to the veins, varying degrees of chemical reconstitution of limestone have occurred, but alteration seldom extends more than 1 to 2 m from the vein except where there are closely-spaced strong veins, or where metasomatic flots have spread out into the wallrock.

Dolomite is the commonest new mineral, usually pale buff or brown in comparison with the white or grey of the unaltered limestone. Dolomite accompanies most veins of any substance, and occasionally it accompanies fractures that are barren of any other fillings; examples can be seen among the limestone scars of Upper Wharfedale. The dolomite as seen under the microscope forms an interlocking aggregate of rhombs which may reach several millimetres across. The identity of the carbonate has been checked by Mr R. K. Harrison, who finds no significant variation from refractive index $\omega = 1.685$ in material from the following localities: Fourth Whim, Jacob's North Vein; Dam Rigg East; Windegg Middle Level; Faggergill Mine; Prys Mine; Summer Lodge shafts; Summer Lodge Level; Bolton Park; Sargill Mine; Warton Hall Level; Old Providence Mine. Map references and partial analyses will be found in Tables 14 to 17, pp. 94–100, analyses nos. 9, 20, 28, 32, 31, 36, 37, 40, 33, 45, 47. The refractive index measurements point to very low $FeCO_3$ contents in the dolomite. If a little had been present at an earlier stage, it might account for the brown colour of the mineral, now probably due to finely-dispersed goethite; but in the case of fresh material from Warton Hall Level, no $FeCO_3$ is present in the dolomite rhombs. In addition to relict calcite from the limestones, recrystallised late **calcite** is also generally present; at Fourth Whim and Faggergill it predominates over dolomite, but elsewhere it is subordinate in amount. A common mode of occurrence is as fillings of the pore-spaces created during dolomitisation. Refractive index measurements showed no variation from $\omega = 1.658$. Carbonates of the **ankerite** series (Mg, Fe)$CO_3.CaCO_3$ are present, but as already noted, they are relatively uncommon here. The most conspicuous occurrence forms a total replacement of a slice of Main Limestone exposed between the Middle and South veins in Lownathwaite Hush (p. 125). Microprobe analyses of the macrocrystalline rock are given on p. 93. At Prys Mine, Mr Harrison reports ankerite with $\omega = 1.740$, suggesting a composition closer to the ferro-dolomite end-member than any of the common ankerites from the Alston Block (Dunham, 1948, pp. 91, 92), but only in subordinate amount. The least common of the metasomatic carbonates is **siderite**.

Though its presence is suggested by several of the proximate analyses, it has been identified replacing limestone only at Old Providence Mine, where the index $\omega = 1.780$ suggests a magnesian variety; and at Hartley Birkett.

Fluorite is much less common than the carbonates in metasomatic deposits here. Fluoritised limestone occurred at East Arn Gill (p. 124) replacing Underset Limestone, and is also present there as large crystals in cavities. At Seata (p. 180), Raygill (p. 166), Worton Hall Level (p. 167), at Great Sleddale and at Hartley Birkett limestone more or less completely replaced by the mineral can be found. The fluorite, even though it has presumably grown in the solid state, exerts its cubic faces producing a microscopic texture which could be described as porphyroblastic (Dunham 1952, pl.i, fig. 1 illustrates such a texture). No flot workable for fluorspar has, however, been found in the present region. **Galena**, **sphalerite** and **pyrite** also tend to form euhedral crystals in altered limestone, but there is an early generation of pyrite in the form of minute framboids (Vaughan and Ixer, 1980), particularly in carbonaceous limestones, which probably represents sedimentary or diagenetic constituents.

In spite of the abundance of chert of primary or diagenetic origin in the Carboniferous sequence, especially in the late Brigantian and early Pendleian strata, implying **silicification** on a regional scale during or soon after sedimentation, this type of alteration as an accompaniment of the mineralising process is rare, and is almost confined to the western belt of QBa and Q veins (Figure 21). Particularly in the Malham subarea (pp. 227, 228) a considerable development of silicification is combined with dolomitisation in the early Asbian limestones, apparently directly related to the mineral veins. The wide 'quartz dikes' of Cumpston Hill and Grisedale are in part composed of silicified Main Limestone, clearly connected with the vein-fractures. Quartz replaces Main and Underset limestones in Great Sleddale along the western faults of the North Swaledale Belt, extending beyond the immediate vicinity of the small copper prospects. Generally, however, the mineralising fluids of the Askrigg Block, unlike those of Alston, were not silica-carriers and it must be assumed that their pH was too low to make this possible. Even where the veins traverse the thick chert sequences, any recrystallisation of the chalcedony is on a microscopical scale only. Mr Harrison reports that silicified limestone from Great Bell, Hartley (Figure 24) and Raygill Level (Figure 30, loc. 5) has been impregnated with later fluorite; at Raygill coarser (0.1 mm) quartz crystal aggregates lie adjacent to the fluorite, suggesting remobilisation of silica. At Ladthwaite ankerite veins cut through silicified limestone and carry sphalerite, galena and fluorite.

Silicification, where it is present, is probably usually an early effect of the mineralisation. This may also be true of the widespread dolomitisation, but dolomite – calcite veinlets are also common as the latest stage of the paragenesis. The attempt, however, to produce a 'standard' paragenetic sequence for the metasomatic processes, or for the vein-fillings is probably not a profitable one, since there is plain evidence of many surges of mineralisation, causing repetitions in the order of deposition of the minerals. An impression is nevertheless gained that where fluorite and baryte occur in association, the fluorite crystallised first since it presents euhedral faces towards the baryte, whereas the faces of the latter, if developed, are directed away from the fluorite. In metasomatic deposits such as the Summer Lodge deposits, baryte appears to have mainly formed later than the ubiquitous dolomite; it is only rarely found as a direct replacement of limestone. Calcite and aragonite are everywhere the latest minerals to form, other than secondary products such as iron hydroxides.

Oreshoots are rarely found between sandstone wallrocks in the region, except at Grassington and along Ashfold Side Beck, where the Grassington Grit becomes productive. The 1.2 per cent of dolomite in the Coalgrove Beck flotation tailings may have come from dolomitisation of sandstones having a calcite matrix, but generally introduction of limonite in the oxidation zone has obscured primary alteration in these rocks. No evidence of wallrock alteration of shale has been obtained.

ORIGIN AND AGE OF THE DEPOSITS

Historical

The early literature of the southern part of the Northern Pennine Orefield contains remarkably little speculation about the possible origin of the mineral veins, but evidently the phenomenon of 'bearing beds', which must have been recognised from very early times, had led some to the belief that the minerals in the veins were leached out of the adjacent strata. This view was emphatically rejected, together with any suggestion that the deposits originated contemporaneously with the sediments, by John Phillips (1836, p. 89) who maintained that 'the real dependence, in this region, is on the consolidation of the walls, and the openness of the fissure'. The authors of the Mallerstang memoir (Dakyns, and others, 1891) in their chapter on metalliferous mining, concur with this view, but offer no suggestion as to the source of the metals. Carruthers and Strahan (1923, p. 26) recognised that in Swaledale and Arkengarthdale, erosion had cut right through the mineralised zone but Varvill's (1920) contribution showed that this is not the case at Greenhow Hill. In a review of the genetic problem (Dunham, 1959) it was maintained that the most likely explanation of the zonal disposition of the minerals is that heated waters (hydrothermal solutions) were involved, and the following possibilities were examined using such quantitative data as was available at that time: (1) juvenile water from crustal igneous magma; (2) sub-crustal mantle sources; (3) resurgent metamorphic water; (4) connate and sub-vadose groundwater. With regard to (1) it was suggested that the 1×10^6 tons of lead in the Askrigg Block could be provided by the separation of the metal from a granite 3.3 cubic miles (13.5 km^3) carrying a normal 30 ppm Pb. In connection with (4) it was noted that if Finlayson's (1910) figure of 15 ppm Pb for the limestones of the Pennines is correct, it is not at all likely that these could have supplied the required quantity; but that Hudson (1938, pp. 349–351) had concluded that the Harrogate mineral waters, the most saline of which contains 8928 mg/l Cl; 601 mg/l CO_2; 69 mg/l K; 67.7 mg/l Ba; 89.6 mg/l S, with traces of F and Cu, were derived from connate waters diluted with meteoric water. In a second review (Dunham, 1970), written after the proving of the pre-Carboniferous Weardale Granite, but before Bott's (1967) results had been tested in Wensleydale, considerably more emphasis was placed upon the importance of deep formation waters having the character of hypersaline brines. In the two most recent contributions to the subject, Small (1978) has invoked connate brines to explain the zoning in the northern part of the Block, whereas Rogers (1978) in his thermometric study considers that heated brines were expelled from the Craven Basin to the south and from the Stainmore Trough to the north into the fissured block. In the meanwhile Mitchell and Krouse (1971) had favoured an igneous-juvenile origin for the fluids at Greenhow – Skyreholme on the basis of stable isotope studies.

Temperature of formation and heat-flow

Rogers' investigation consisted of the measurement of homogenisation temperatures in about 3000 primary two-phase fluid inclusions in fluorites from all parts of the present area. Using a pressure correction equivalent to 15°C for Wharfedale and Wensleydale, and 8°C for Swaledale, the range of formation temperatures indicated was 108 to 164°C for Greenhow – Skyreholme, 92 to 160°C for Grassington and Upper Wharfedale, 101 to 154°C for Wensleydale, 115 to 155°C in the Dent zone, and 97 to 153°C in Swaledale – Arkengarthdale. Virtually all the specimens came from dumps and sampling across veins was usually impossible; nevertheless the number of crystals examined was sufficient to give a good indication of the range. No systematic variation within the exposed F zone (Figure 21), such as Sawkins (1966) found in the Alston Block, could be discerned but it should be recalled here that this zone in the Askrigg Block is really only the equivalent of the intermediate zone III of Alston. Only one determination on baryte was made, from Haw Bank (Carperby) giving a mean temperature of 132°C, but in the mixed fluorite-baryte paragenesis, it is to be expected that the two minerals crystallised at much the same temperatures. Thermometric measurements of fluorites from six localities by Smith (1974b with additional unpublished data) and from eleven by Small (1977) all in Wensleydale, Dent and Swaledale, give a range from 98° to 150°C if the pressure corrections used by Rogers are applied, thus broadly in agreement with Rogers' more extensive survey.

Dr T. J. Shepherd undertook measurements on material needed to clarify certain critical localities. From a depth of 49.85 m in the Beckermonds Scar Borehole (p. 186), colourless fluorite from a presumed flat in Danny Bridge Limestone showed, on the basis of 16 primary or pseudosecondary inclusions, an uncorrected range of minimum formation temperatures from 172° to 207°C, with a mean value of 190°C. Secondary inclusions gave a mean value of 130°C. For comparison with Rogers' data, the corrected minimum temperature would be 210°C, the highest recorded from the Askrigg Block. Material from several 'Q' localities of Figure 21 was tested, and one yielded quartz with measurable two-phase inclusions. This was Hanging Lund Scar, from the Cumpston Hill South Vein, Small's locality for tetrahedrite. Dr Shepherd's (1979) results on primary or pseudosecondary inclusions gave a mean temperature of 88°C, a little lower than the minimum temperature of fluorite deposition, and not consistent with Small's identification of the 'Q' veins as the focus of the zonal pattern (p. 97).

To investigate post-Carboniferous heat-flow through the region, Creaney, Allchurch and Jones (1979) have measured reflectances of vitrinite from coals and carbonaceous sediments over much of the Askrigg Block, as Creaney (1980) did over the Alston Block, where he was able to show that temperatures exceeding 180°C were attained prior to the intrusion of the Whin sills. The reflectance profiles in the Raydale and Beckermonds Scar boreholes (Creaney, 1982) are anomalously high at 0.48 and 0.54 per cent per 100 m respectively, indicating abnormal heat-flow since the deposition of the Danny Bridge Limestone. However, the similarity of the profiles in these two borings shows that heat was not being channeled through the Wensleydale granite cupola, as it was through the cupolas at Rookhope and Tynehead. Nevertheless, the isoreflectance contours (two of which are reproduced on Figure 22) parallel the outline of the granite as interpreted by Bott (1967), the area of greatest reheating being displaced west from the Raydale 'high'. Critical evidence of the exact timing of the reheating here is lacking, but it is reasonable to argue from the analogy of the Weardale granite and Whin sills that it took place shortly before 296 Ma and was waning when the mineral deposits were emplaced, many of them at temperatures below the maxima reached during the reheating. It may further be suggested that the reheating of the Wensleydale granite provided the heat engine which set in motion the circulation of the hydrothermal fluids which eventually caused widespread mineralisation. These were, nevertheless, not the means by which heat was imparted to the Carboniferous sediments. The wide spacing of the veins, and the total absence of known deposits from the most heated area precludes this, and conduction from below must be postulated as the means by which heating took place.

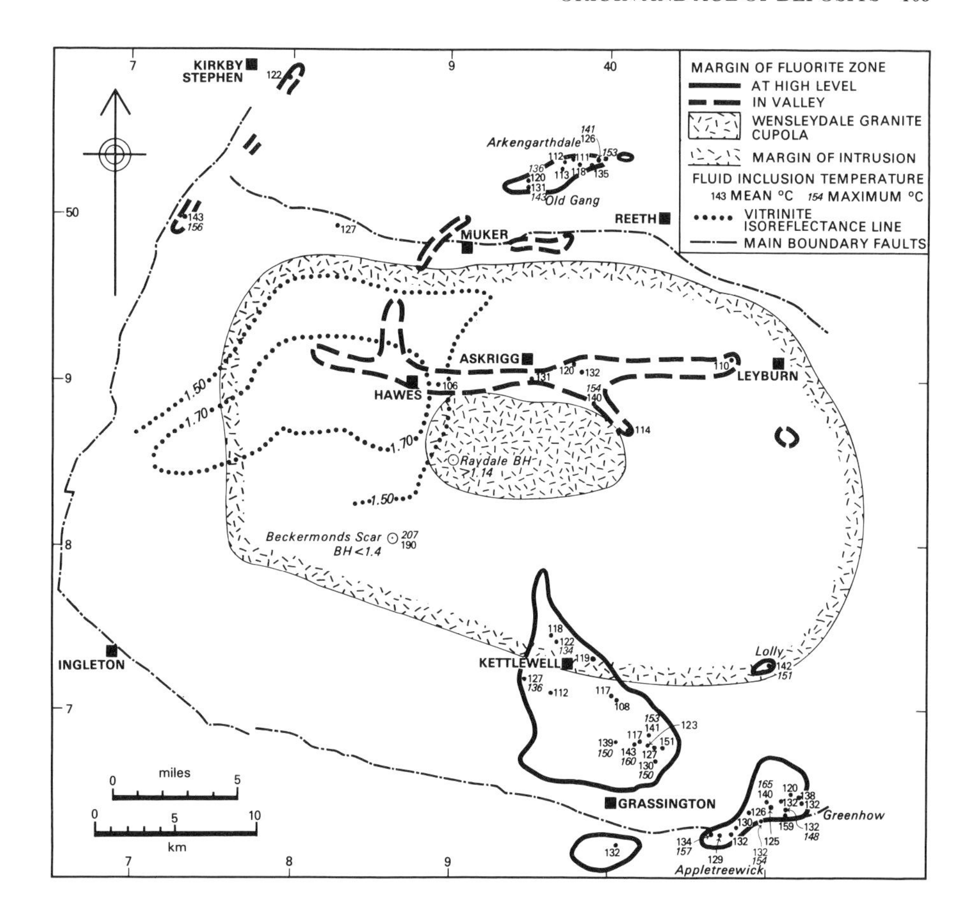

Figure 22 Formation temperatures of fluorite and quartz in relation to the Wensleydale Granite and to vitrinite reflectance contours. Data from Bott (1967); temperature data based on fluid inclusion measurements by Rogers (1978) and Small (1978); reflectance data from Creaney (1980) and Creaney, Allchurch and Jones (1979).

Nature of the mineralising fluid

The data already presented make it clear that hydrothermal solutions were involved as mineralising agents, but more information about them is forthcoming from the fluid inclusions, which in effect preserve samples of the fluids. The highest temperature inclusions in the Beckermonds flat fluorites carry, according to Dr Shepherd's measurements, a solution containing the equivalent of 23.8 wt per cent NaCl, and the secondary cavities contain brine only slightly less saline at 23.5. This fluid is, then, more than six times as saline as sea water. Rogers' (1978) measurements on primary inclusions from 31 localities in all parts of the region showed a range from 16.3 to 24.3 wt per cent and averaged 21.1. The brines thus appear to have maintained a fairly uniform salinity over the large areas involved. At Cumpston Hill, however, Dr Shepherd found first melting temperatures after freezing of $-40°C$ as opposed to normal NaCl solution temperatures about $-21°C$, and suggested that the brine probably contains $CaCl_2$ in appreciable amount.

For the other outer-zone copper deposits at Middleton Tyas, Dr Shepherd reports that the fluids in inclusions in calcite have low salinities, comparable with those of the secondary inclusions at Cumpston Hill, which freezing experiments showed contained only 0.2 to 1.3 equivalent wt per cent NaCl.

A further approach to the chemistry of the mineralising fluids has been the determination of alkalis in leachates obtained from the crushing of minerals under carefully controlled conditions. Sawkins (1966) made the pioneer experiments in the northern half of the orefield, showing a progressive fall in K/Na ratio from 0.9 in the centre of the zonal system to 0.3 at the margins. For the present region, determinations by Small (1978, and unpublished data 1977) and Rogers (1978) are available. Fluorites from the North Swaledale centre yielded a reasonably consistent range from 0.035 to 0.1. Those for western deposits included Hartley Birkett 0.048 to 0.99, Clouds 0.24, Great Sleddale 0.18. For 12 deposits in Area 13, Small found a range from 0.067 to 0.96, but the high figure here, for Wet Grooves fluorite, differs markedly from that obtained by Rogers of 0.069; there is also a difference between the two observers for Warton between 0.75 and 0.034. The technique is necessarily an inaccurate one, and large differences such as those mentioned may be caused by differences in the abundance of secondary inclusions in the minerals examined. Gill Heads fluorite, for example, yielded to Rogers ratios of 0.018 and 0.13 in two samples; his range for Nidderdale and Wharfedale runs from 0.018 to 0.18 and averages 0.08. It is difficult to know what conclusion can be drawn from these figures except that the fluids in the Askrigg fluorite centres carried substantially less potassium relative to sodium than in the Alston Block. For baryte, witherite and calcite, only determinations by Small are available. Baryte gave ratios varying from 0.018 at Feldom and 0.025 at Hurst, up to 0.84 in Arkengarthdale and 1.55 at Braithwaite. Five samples of witherite gave a range from 0.9 to 2.45, while six calcites ranged from 0.12 to 2.2, the two extreme ratios being from the adjacent localities of Stang and Faggergill. The K/Na ratio for sea water being 0.034, it appears that the ore fluids were generally mildly enriched in potassium, and in some deposits strongly enriched. The high ratios for some samples of baryte and calcite are not easy to understand, and require further investigation.

Sulphur isotope-ratios on 16 samples of galena from Greenhow, Black Hill, Burhill and Gill Heads, measured by Mitchell and Krouse (1971) gave a range of delta ^{34}S (-0.6 to -3.8) so small that these authors concluded that for that centre the brines were isotopically homogeneous, and probably of igneous juvenile origin. They recognised, however, that only a small part of the Askrigg orefield is covered by these results. The isotopic composition of the sulphate sulphur here also remains unknown. In the Alston and Derbyshire fields, the wide range of sulphide sulphur ratios and the narrow range in sulphate sulphur point to major contributions from formation waters, and probably to mixing of diverse fluids (Solomon, Rafter and Dunham, 1971).

Some comparisons

In recent years the lead – zinc – fluorite – baryte deposits of the English Pennines have from time to time been classified with those of the numerous lead – zinc districts of the Middle West of the United States as belonging to a 'Mississippi Valley' type. Common characteristics of the eight principal districts in the Mississippi Valley have been summarised by Heyl, Landis and Zartman (1974) as follows:

1 Most deposits are largely structurally controlled, open-space fillings; replacements are of less importance . . . All are epigenetic in their present form; all have low-temperature wallrock haloes.
2 Although the deposits may favour certain carbonate beds for major commercial deposits, they are not strictly stratabound or stratiform. They occur in all Palaeozoic strata present in each district.
3 The mineralogy of the orebodies and associated trace element patterns can no longer be called unusually simple but is subtly complex, especially locally. The most abundant minerals are galena, sphalerite, baryte, calcite, dolomite (with fluorite east of the Mississippi River), and cryptocrystalline quartz.
4 Fluid inclusion data . . . indicate that the hydrothermal solutions were mostly heated basin brines (evolved connate waters) with salinities in excess of 20 wt per cent. Temperatures of ore deposition were about 200°C or less.
5 All districts contain . . . J-type radiogenic lead with $^{206}Pb/^{204}Pb$ ratios that are 20 or greater. A shallow crustal source for nearly all the lead is indicated.
6 Delta ^{34}S indicates a crustal source for most of the sulphur in the sulphides and sulphates in the deposits.
7 Each district has distinctive patterns in lead and sulphur isotopes that reflect directions of solution flow, buried heat sources and areas of localisation of ore fluids. The ores were deposited from a hydrothermal fluid that resulted from the mixing of at least two different sources of water.
8 Oxygen isotope data from wallrocks outside the alteration haloes show that they were never exposed to abnormally high fossil geothermal gradients.
9 The mineral zonation in each district is distinctive and district-wide; vertical zonation is present in places but is less widespread than lateral change.
10 Post Pre-Cambrian igneous rocks are sparse except in two districts.
11 All the major districts show evidence that part of the solutions moved through . . . basement in deep-cutting faults and fracture systems.
12 Silver is present in most districts; the ores in two districts contain economic quantities.

All the above could equally well have been written about the three orefields of Alston, Askrigg and north Derbyshire, save that in the Pennines the $^{206}Pb/^{204}Pb$ ratios, though according to Mitchell and Krouse (*op.cit.*) of J-type, do not quite reach 20; and no oxygen isotope data on the country rocks is yet available. Vitrinite reflectance has been used in the northern Pennines to demonstrate former higher-than-normal geothermal gradients. The Pennine fields clearly belong to the world-wide type, but like individual Mississippi Valley districts, each has certain distinctive characters of its own. Some of these are compared in Table 18. This comparison underlines the point made by Heyl and others that although there is a general community of type, each district has its individual features. A remarkable fact, stressed by these authors, but equally striking in the Pennines, is the evidence of a district-wide flow of fluids. Perhaps this is more obvious in the Alston and Derbyshire orefields than in the Askrigg where, owing to the large almost unmineralised area between Wensleydale, Western Wharfedale and Settle – Malham, the continuity is perhaps less clear, but the links are nevertheless present. In the American Middle West, one explanation that has been given for the district-wide flow is that the hydrothermal solutions moved through widely extensive sandstone aquifers. An excellent case can be made for the Cambrian Lamotte Sandstone as the channel of flow into the great lead – zinc fields in the overlying carbonate formations in south-east Missouri, and equally for the St Peter Sandstone in the northern Illinois – Wisconsin field. To our knowledge a similar case has not been made for the three orefields that have most in common with those of the Pennines, the major fluorspar-producing southern Illinois – Kentucky district, the smaller central Kentucky field and east Tennessee. Veins are more important in these three districts than elsewhere in the Middle West, and lateral zoning with many features in common with the Pennine patterns is displayed; in the first-mentioned district, the large metasomatic mantos zone outwards from fluorite – galena – sphalerite near the feeders to baryte at the margins, (Heyl, 1969). The possible role of sandstone aquifers which permitted hydrothermal flow without precipitating the solutes needs to be considered for these orefields and for their English counterparts. A case could, for example, be made for the Tuft (the 'Water Sill' of Forster, 1809) in the Alston Block, or for the sandstone between the Main and Underset limestones in the North Swaledale Belt, but it must be admitted that we have no positive evidence to offer for this, other than the manifest fact (p. 85 and Dunham, 1948, p. 81) that the largest number of oreshoots bottom at or near these horizons.

The common factor that increasingly binds together the Mississippi Valley and Pennine ore deposits is the evidence, from the chemistry of fluid inclusions, that the hydrothermal fluids were hypersaline brines with about six times the salinity of sea water. Actualistic evidence of the existence of brines of this sort carrying substantial amounts of zinc at the present day comes from the northern Alberta basin in Canada, to which the major Mississippi Valley-type ores of Pine Point may well be related (Billings, Kessler and Jackson, 1969; Jackson and Beales, 1967). These contain 19 mg/l Zn, but much more highly mineralised brines exist today in Upper Mesozoic sediments at depths of 8000 to 13 000 ft (2.34 to 3.96 km) over an area of 2000 sq. miles (5120 km^2) beneath central Mississippi (Carpenter, Trout and Pickett, 1974), containing 110 mg/l Pb and 360 mg/l Zn. Such basin brines are to be regarded as in the general class of evolved connate waters in the sense of White (1968); for the Western Canada brines, Hitchon, Billings and Klovan (1971) have traced the chemical processes. Adjacent to the northern Pennines, the Durham coalfield basin contains brines at depth that are enriched relative to sea water not only in chloride, but in barium, lead and zinc (Edmunds, 1975; Dunham, 1948, p. 329) but the deeper basins of Northumberland, the North Sea, the Stainmore Trough and Craven were almost certainly also being dewatered at the time of mineralisation. Rising into the Pennine 'highs', the ejected brines became involved, at least in part, in deep convective circulation brought about by the reheating of the Weardale and Wensleydale granites; they also mixed with sulphate brines of more local origin (Solomon, Rafter and Dunham, 1971). That they may have received some juvenile contribution, for example of fluorine, is not ruled out by White and Heyl for the Mississippi Valley and is positively advocated for the Alston Block by Russell and Smith (1979) on the basis of high yttrium in fluorite (as well as on more speculative grounds). However this may be, the Pennine orefields, as seen in 1980, owed their mineralisation mainly to brines of basin origin, and are not basically igneous-juvenile phenomena.

Table 18 Comparison of orefields in the Pennines

	Northern		Central	Southern
	Tyne-Stainmore	Stainmore-Craven	Craven Basin	North Derbyshire
Country rocks	Westphalian B to base Asbian, cyclothemic	Kinderscoutian to Asbian, cyclothems on massive limestone	Asbian and ?earlier limestones	Early Brigantian and Asbian limestones
Post-Devonian igneous rocks	Dykes 58 Ma Whin sills and dykes 296 Ma	None	None known	Asbian to Westphalian lavas, necks and sills
Basement	Ordovician slates Weardale Granite 410 Ma	Ordovician-Silurian slates Wensleydale Granite 400 Ma	?	Ordovician slates and ?lavas
Principal vein direction in full tension	ENE E-W	WNW E-W	ENE	ENE E-W
Mineralisation temperatures (°C)	210 to below 70	164 to below 88 (one location 210)	178 to below 122	170 to below 70
Salinity of primary fluids (equivalent wt % NaCl)	20–27	15–25		18–25
Sulphide (^{34}S %)	− 5.9 to + 15	− 3.8 to − 0.6*	n.d.	− 23 to + 7
Sulphate (^{34}S %)	+ 17 to + 22.2	n.d.	n.d.	+ 4 to + 23
^{206}Pb/^{204}Pb	18.27 to 18.33 (4)	18.46 to 18.67 (16)		18.41 to 18.50 (4)
Average Ag in Pb (ppm)	150	40		below 25
Pyrrhotite	in zone GSV 1	absent	absent	absent
Sulpharsenides	present	very rare		
Mercury	absent	very rare		very rare
Fe in sphalerite (ppm)	500 to 3200	500		
Cd in sphalerite (ppm)	250 to 400			Up to 1500
Y in fluorite (ppm)	100 to 1200	15 to 280		
Fluorite colour	Purple, green, amber	Pale mauve or yellow; colourless	Purple	Colourless, or dark purple banded
Fluorite fluorescence	Strong	Weak or absent		None
Baryte	Mainly platy	Fibrous or platy	Fibrous, platy	Mainly cawk
Witherite	Zones IV, V	Areas 11–13	Type locality at Anglezarke, quantity small	Absent
Production (million tonnes)				
PbS, recorded	3.01	0.56	Small	0.78
Estimated total PbS	4	1	Small	2.5
ZnS	0.28	Very small	None	0.06
Other materials produced				
Copper ore	√	√		√
Fluorspar	√	√		√
Barytes	√	√	√	√
Others	Witherite Bilbao-type iron ore Quartz	Calamine Calcite		Calcite

Data from Derbyshire from: P. J. Rogers (1977), temperatures and salinity; P. G. Coomer and T. D. Ford (1975), B. W. Robinson and P. R. Ineson (1979), sulphur isotopes; S. Moorbath 1962), R. H. Mitchell and H. R. Krouse (1971), lead isotopes.

* Greenhow–Skyreholme only.

Age of the deposits

John Phillips (1836) clearly recognised that the veins of Wensleydale and district post-dated his Yoredale Rocks. No further significant contribution was made until Wager (1931) wrote:

'A pre-Permian age for most of the mineralisation (of the Great Scar Limestone in Craven) is suggested by the fact that lead veins are scarcely ever found in the Permian strata. H. C. Versey (1925) has confirmed this by showing that detrital fluorspar and barytes, which seem to have been derived from the lead-veins, occur in the Saxonian Yellow Sands at the base of the Magnesian Limestone in Yorkshire. Although the lead veins cut the Whin Sill, there are now good reasons for regarding that sill to have been intruded early in the interval between the Coal Measures and the Permian (Holmes and Harwood, 1928); so that the lead veins can well have been pre-Saxonian though later than the Whin Sill.'

If the reasonable assumption is made that the reheating of the Wensleydale Granite took place a little prior to the intrusion of the Whin Sill in the Alston Block, at the same time as the reheating of the Weardale Granite (Creaney, *op. cit.*) the Stephanian – Thuringian (Upper Permian) interval is the most likely date for the main surge of mineralisation in the present area. Acceptance of basin brines as the hydrothermal fluids responsible may imply that resurgences could take place later than this.

Mitchell and Krouse (1971) determined ^{204}Pb, ^{206}Pb, ^{207}Pb in 16 samples of galena from Greenhow – Skyreholme. The ratios determined using Kanasewich's decay constants (Kanesewich and Slawson, 1964), yield model ages from the Holmes – Houtermans equation ranging from 164 to – 82 Ma. They attempt no case for a late Mesozoic or Tertiary age, but accept that J-type radiogenic lead is involved here, as elsewhere in the Pennines.

SECONDARY PROCESSES

The Oxidation Zone

Many of the primary minerals so far discussed require reducing conditions for their formation and become unstable in the presence of waters carrying oxygen, dissolved carbon dioxide or sulphuric acid. Above the water table, downward moving meteoric water normally contains O_2 and HCO_3, derived from atmospheric sources, and oxidation of sulphides may yield SO_4 in solution. In addition, phosphate, probably from organic sources in the soil, is occasionally present. Table 19 lists the primary minerals and their alteration products.

In valleys the base of the vadose zone of the present morphological cycle comes to surface at stream level while under the valley sides, the water table rises more gently than the topography. In the southern region, figures showing its exact position are more difficult to derive than in the northern region, where it can be shown to lie 150 to 350 ft (46–106 m) below surface. In area 11 at Old Gang mines, the Old Rake and Friarfold workings from surface shafts and from the Watersykes level system probably never reached below the oxidation zone at their maximum depth, approaching 250 ft (106 m) below surface; the Bunting – Hard level systems may have been in unoxidised ground, Sir George and Sir Francis levels undoubtedly were. Here a depth similar to or slightly greater than in the Alston area is suggested. Mines such as Faggergill (maximum depth of workings, 355 ft, 108 m) and Windegg (perhaps 250 ft, 76 m) seem not to have reached unoxidised ground. In Area 13, the extensive flots worked between Whitaside and Apedale mines under the local watershed were highly oxidised under a cover of at least 300 ft (91 m). At Greenhow, the Hargate End and Woodhouse veins, examined in 1942 at 405 ft (123 m) below surface, were completely free from oxidation. The vein section from Grassington shown in Plate 1 consists of galena entirely unoxidised in Grassington Grit country rock; the estimated depth is 250 ft, (76 m). It by no means follows that these figures, suggesting depths of 350 to 400 ft (107–122 m) beneath the interfluves, necessarily correspond with the present water level. Evidence has been presented elsewhere (Dunham, 1952) for levels far deeper than those obtaining at present, perhaps corresponding to drier periods prior to Pleistocene times. This subject is referred to again below, when cavernisation is considered.

Secondary minerals

Cerussite, $PbCo_3$; Anglesite, $PbSO_4$; Minium, Pb Pb_2O_4 Galena crystals exposed to oxygenated water become coated with a thin but dense skin, usually of cerussite in the predominantly limestone environment, and this protects them from further decomposition. The oxidation zone has been very little depleted in lead values, if at all, for this reason. As separate crystals, cerussite is rare; the Grinton How flots in Area 13 yielded specimens (MI 1539)† and they have been found at Malham Tarn, and at Grassington, but to our knowledge, no deposit mainly of this mineral has been exploited. The only record of minium is of pseudomorphs after cerussite, found by Mr A. W. G. Kingsbury at Turf Pits, Grassington, and of another occurrence found by him at Old Moss Mine (1968)‡. The lead recorded in appreciable amounts in the mine tailings throughout the region (Tables 15 to 17) is in part in oxidised form, and it may well be that the rather high losses during dressing arose from the fact that cerussite and anglesite are difficult to see against a background of white spar, while galena is very easy. The national collections contain no crystals of anglesite from this part of the Pennines.

Table 19 Primary and secondary minerals

Primary minerals	Secondary minerals	Active radical in vadose water
Galena	Cerussite	HCO_3
	Anglesite	O_2
	Minium (very rare)	O_2
	Pyromorphite	PO_4
	Plumbojarosite	OH, SO_4
	Argentojarosite	OH, SO_4
Sphalerite	Smithsonite	HCO_3
	Hemimorphite	SiO_2
	Hydrozincite	OH
	Parahopeite (rare)	PO_4
	Spencerite (rare)	PO_4
Sphalerite, Chalcopyrite	Aurichalcite	OH, CO_2
	Rosasite	OH, CO_2
Chalcopyrite (Bornite), Tennantite-Tetrahedrite	Malachite	OH, CO_2
	Azurite	OH, CO_2
	Chrysocolla	SiO_2
	Cuprite	O_2
	Native copper	
	Covelline	$CuSO_2$ in solution reacting with sulphides near water table
	Chalcocite	
	Digenite	
Pyrite, Marcasite, Bravoite	Limonite, Goethite	O_2
	Jarosite	SO_4, K
Cinnabar	Calomel (very rare)	HCl
Siderite, Ankerite	Limonite, Goethite	O_2
Witherite, Barytocalcite	Baryte*	SO_4
Calcite	Aragonite	HCO_3
	Gypsum	SO_4
	Crandallite	PO_4
Fluorite	(etched)	?acids
Quartz	unaltered	

* The IMA-approved spelling is baryte

† MI numbers refer to the Mineral Inventory of the Geological Museum
‡ Specimens of the numerous rare minerals found by Mr Kingsbury at Grassington and Greenhow are now in the main collection of the British Museum (Natural History).

Pyromorphite, $Pb_5Cl(PO_4)_3$ This mineral, with its characteristic yellowish green colour, is also uncommon. Good material came from the Merryfield Vein in Ashfold Side Beck (BMNH) and A. T. Small analysed pyromorphite in collaboration with R. G. Hardy and A. Carr (Small, 1982) from Old Rake Vein, Gunnerside Gill with the following results: Pb, 76.38; Cl, 2.61; P, 6.85; As, Ca, Nil; O, 13.63; Total, 99.47. This has almost precisely the standard formula.

Plumbojarosite, $PbFe_6(OH)_{12}(SO_4)_4$; Argentojarosite Both minerals are known only from Mr Kingsbury's collecting at Turf Pits and West Turf Pits dumps, Grassington respectively. The jarosites tend to occur as incoherent yellow to brown powders and may well be commoner than is generally recognised, for material of this character is often seen associated with limonite in oxidised ores from the veins. Having regard to the low primary silver content of the galena, the chances of finding argentojarosite must be very small. Ordinary jarosite occurs at the Ringleton shafts.

Smithsonite, $ZnCO_3$; Hemimorphite, $Zn_4(OH)_2Si_2O_7.H_2O$; Hydrozincite, $Zn_5(OH)_6(CO_3)_2$ Unlike the other metals with which it is associated in the ores, zinc is taken into solution in carbonated groundwater and may be transported some distance before being deposited. Thus concentrations of oxidised zinc minerals may form away from the primary sphalerite. This process, which leads to a secondary concentration of zinc, has been well known since G. F. Loughlin's work at Tintic, Utah (1914). It is well illustrated here by the calamine** filled caverns worked near Malham Tarn (Raistrick, 1947; 1954) and by concentrations of zinc in smithsonite and hemimorphite far in excess of the normal sphalerite as indicated by the tailings assays at a number of places in the region. The flots at Windegg (p. 144) contained a coarsely crystalline impure smithsonite that is probably pseudomorphous after calcite (Table 20). The hemimorphite from Hag Gill Level may also have replaced a coarse carbonate.

Certain of the high zinc assays in the tailings (Tables 17 to 19) are also known to be due to concentrations of oxidised zinc minerals having been encountered, though it is likely, in some cases at least, that they were not recognised as such. These include Faggergill (smithsonite 4.4 per cent, cerussite 5.2) in the run-of-mine tailings; Beezy West (mainly hemimorphite 13.3); Beezy Level (smithsonite 14.2); Whitaside – Apedale flots (smithsonite 15.4); Cobscar Rake (smithsonite 15.1). Tests carried out by the Laboratory of the Imperial Institute on large samples collected by Dr Earp in 1942 from the Moulds and Damrigg dumps in Area 11, and from Devis Hole, Grinton How and Grovebeck in Area 13 showed that all but a very small proportion of the zinc is oxidised. The sample from Devis Hole was found to contain 73.3 per cent hemimorphite and the zinc-contents of 13.1 and 9.6 per cent respectively for Grinton How and Grovebeck were considered to be due mainly to this mineral.

Specimens of hemimorphite had already begun to appear in the collection of Sir Arthur Russell (now in the British Museum (Natural History)) in the 1920's. The first specific identification by the Geological Survey was on material collected from a shaft on Copperthwaite Mine [NZ 0558 0013] and examined by Dr Phemister, who found that the white, yellow and brown crystals of hemimorphite were optically positive with 2E approximately 60°, refractive indices $\beta = 1.616$, $\gamma = 1.635$, gelatinising with acid and giving strong zinc reaction. The hemimorphite forms chains of crystals and mamillary growths resting on siderite ($\omega = 1.830$), sphalerite, witherite, baryte and galena in cavities, and with a generation of calcite later than the hemimorphite. Localities in the Russell Collection additional to those already mentioned include Hartley Birkett, Wet Grooves and Virgin; it is present in addition to smithsonite in Cobscar Rake. Among smithsonite specimens, the 'dry bone' type is most common and the only crystals showing the green colour characteristic of the best specimens came from Chaytor Rake and are in the Russell Collection.

Hydrozincite has not been identified specifically from this region, but its copper-bearing analogue aurichalcite has, as described below. It is however extremely probable that the post-mine white coatings commonly seen include this mineral. It is not known here in massive form.

** The reader may care to be reminded of the position regarding the nomenclature of zinc carbonate and hydroxysilicate minerals. The ancient name *calamine* was widely used for the 'dry bone' and similar ore mined for use in making a variety of brass; both minerals were present in at least some of this ore. Greg and Lettsom (1858) in the standard work on British mineralogy followed some earlier authors in applying calamine to the carbonate, and smithsonite to the silicate, but added (p. 428) 'The name Smithsonite is given to this species in honour of the chemist Smithson. Several continental mineralogists, however, give this name to the carbonate of zinc . . . : but as Smithson was one of the first whose labours were directed to explain . . . the silicates, it is with a silicate that his name should in propriety be associated.' International mineralogical usage has not accepted this view, and the terminology applied here is the standard one. Calamine may apply to either mineral.

Table 20 Analyses of impure 'calamine'

Analyst E. Waine, 1978

All the results are percentages

— signifies that the constituents of the mineral concerned were determined, but were not present in sufficient amounts to make the mineral

Locality	NGR	CO_2	F	Ba	Ca	Pb	Zn	Fe	Mg	Rest*
78 Windegg Flots, Arkengarthdale	NZ 014 044	33.9	0.02	0.99	15.0	0.09	25.8	1.8	0.54	9
79 Malham Tarn Calamine Mine†	SD 874 640	28.1			0.2		40.0			19.6‡
80 Walls Vein, Satron Side	SD 945 968	3.4	0.04	1.1	1.5	6.0	33.0	2.5	0.23	23
81 Hag Gill Level	SE 065 963	0.7	0.5	1.68	0.73	11	40.0	0.95	0.06	

Proximate mineral compositions

	$ZnCO_3$	$Zn_4(OH)_2Si_2O_7.H_2O$	$CaCO_3.MgCO_3$	$CaCO_3$	CaF_2	$BaSO_4$	PbS	SiO_2	FeO.OH
78 Windegg Flots	47.0	—	3:7	35.5	tr	1.6		—	2.9
79 Malham Tarn	76.7	—	0.8		—	—		19.6	—
80 Walls Vein	3.8	55.6	2.0	2.6	tr	1.9	7		4.0
81 Hag Gill Level		71.8	0.4	0.3	1.1	2.7			1.6

* Remainder from the mineral calculation. † Analysts: D. Hutchison and C. Branch, 1979. ‡ SiO_2; also present 0.24% insolubles and 0.3% H_2O.

Parahopeite, $Zn_3(PO_4)_2.4H_2O$; Spencerite, $Zn_4(PO_4)_2(OH)_2$ $3H_2O$ The discoveries of zinc phosphate minerals are due to Mr Kingsbury, the first-named at Turf Pits Mine, Grassington and at Cockhill Mine, Greenhow, the second (only the second record of the species), also at Turf Pits, in aggregates of iron-stained pearly crystals (Embrey, 1977).

Aurichalcite, $(Zn,Cu)_5(OH)_6(CO_3)_2$; Rosasite, (Cu_3Zn_2) $(OH)_5.2.5CO_3$ The first of these minerals has been identified by Mr Young using X-ray diffraction as a green coating on witherite, baryte, fluorite, pyrite and goethite at Brandy Bottle Incline on Friarfold Vein. A post-mine origin is suspected. Rosasite, in effect a zincian malachite, was found by Mr Kingsbury on the dumps of Coalgrove Head and Turf Pits shafts at Grassington.

Malachite, $Cu_2(OH)_2CO_3$; Azurite, $Cu_3(OH)_2(CO_3)_2$ The two copper carbonates are widespread in every zone, but except in Area 12 and at Great Sleddale Head, Long Rigg and Cumpston Hill, in minor amounts only. These testify to the widespread presence of copper, probably mainly as chalcopyrite, in trace or very minor amount in the primary ores; the colours, green and azure blue, of the copper carbonates being so strong as to make very small quantities conspicuous. There are perhaps more than the normal amounts in the Malham district, where several trials for copper took place. We have no certain evidence that carbonate ores were shipped from Malham, or from the Westmorland Pennines, but carbonates certainly contributed to production at Middleton Tyas (Hornshaw, 1975).

Covelline, CuS; Chalcocite, Cu_2S; Digenite, $Cu_{1.97}S$; Cuprite, Cu_2O; Native Copper The sulphides in this list are typical of the secondary enrichment situation and it may be supposed that their occurrence in Area 12 is the result of this process involving downward moving copper sulphate solutions. Wells (1958) recorded chalcanthite ($CuSO_4$) crystals from East Layton but such crystals are ephemeral. The first records of bornite, covelline and chalcocite from Middleton Tyas are due to Deans (1951); Small (1977) identified digenite in small amount. The appearance of cuprite and native copper near the outcrops of the deposits may be regarded as illustrating the opposite process to downward enrichment, that of residual concentration. The ores worked at Middleton Tyas during the 18th century were remarkably high grade (p. 156) presumably as a result of the enrichment processes.

Goethite, Limonite, Fe_2O_3,xH_2O Hydrous ferric oxide is widely present in the oxidised ores, but has only been found in sufficient quantities to be workable, albeit on a small scale, at or near Hartley Birkett, where Dakyns and others (1891, p. 107) note that copper was obtained in working to the bottom of a mass of hematite following a fault in the disturbed mass of Great Scar Limestone east of Kirkby Stephen. If the locality is the Eden Valley Mine at Long Rigg, the iron mineral was goethite, not hematite, from its yellow streak, and here it appears to be an oxidation product of siderite. However, near the boundary of the Permian red sandstones farther west, irregular pockets, joint-fillings and staining with hematite has taken place; according to Trotter (1939) this effect, which is a widespread one around the Vale of Eden, is related to Permian weathering. The principal sources of goethite (and limonite, the name given to the supposedly amorphous variety) are the carbonates of the ankerite series, siderite and the sulphides pyrite and marcasite. Boxworks can be used to determine the primary source in some cases; marcasite for example is now represented by platy goethite boxworks at Faggergill and other mines, while the rhombic form of ferriferous carbonates in some cases persists in the hydroxides.

A picked sample of goethite-rich veinstuff from the Sloat Hole workings at Faggergill, collected by Dr Earp, was found to contain 67.9 per cent Fe_2O_3 (analysis by C. O. Harvey, 1941) but a representative sample of the tailings contained only 18.7. The absence of massive deposits of limonitic iron ore from the southern region is a striking difference from the northern, and emphasises that while ankerite is the dominant carbonate there, here dolomite and calcite are the important carbonates, and ankerite and siderite occur in minor amount only. Nor is there any evidence of concentrated deposits of any iron sulphide, or of goethite derived from them.

Baryte, $BaSO_4$ This mineral is abundant in secondary as well as primary form, the source of the secondary material being the alteration of witherite (and, no doubt, barytocalcite) by sulphate waters derived from the oxidation of sulphides. The form of the secondary generation of baryte is easy to recognise; it is very fine-grained, usually heavily stained by limonite, and forms stalagmitic masses full of reticulated tubes. Occasionally this material contains cores of witherite, but usually the carbonate has been totally destroyed, the form not being pseudomorphously preserved. The variety of baryte contrasts markedly with the crystallographically distinctive one already described (p. 91) as characteristic of a late stage of sulphatisation of witherite during the primary mineralisation. The very widespread occurrence of both varieties, especially in Areas 11 and 13, points to the fact that before the deposits were oxidised, witherite was widespread and in places—like Old Rake and Friarfold Rake—abundant, probably to at least as great an extent as in the northern region, excepting only the Settlingstones and South Moor concentrations.

Aragonite, $CaCO_3$; Gypsum, $CaSO_4.2H_2O$ Sulphate waters acting on calcite or on limestone also give rise to gypsum, which can occasionally be found at the old mines as fibrous clusters. Aragonite, besides occurring in the primary paragenesis, is also found as a late, post-cavernisation mineral. From Arkengarthdale there are spectacular specimens from Sloat Hole, Faggergill, and Windegg mines in the Natural History Museum collections, consisting of white reticulating tubes, the variety known as 'flos ferri'. Other localities include Danby Level, Friar's Intake, Old Rake and Victoria Level.

Post-mineralisation cavern development

Interest in the sport of caving during the present century has made the Yorkshire Dales very well known for the numerous cavern systems in the Great Scar Limestone, in the Hawes Limestone and in most of the limestones of Brigantian age, as well as in the Main Limestone. The five-volume guide by Brook and others (1977) documents the many hundreds of systems explored, and includes surveys of some of them. Geomorphological aspects have been studied by Sweeting (1950) who has recognised caverns related to at least two erosion surfaces (around Ingleborough at 1300 ft and 950 to 1000 ft (396 and 290 – 305 m) OD) above the stream valleys of the present cycle. Extensive caves below some of the present streams provide underground drainage. The joints in the limestones, especially the strong north-north-west set, have exerted a powerful control on the courses followed by the caverns, but they nevertheless frequently jump from one set to another, producing highly reticulated patterns, in marked contrast with the laterally-continuous mineral veins. Caves have been found in many of the mines, and where these are adequately documented it is certain that the caverns post-date the primary mineralisation. The relationship illustrated from the plans and sections of the Grassington High Moor mines in Figure 23 is typical of the situation where the caverns developed across the courses of the lead veins, which outcrop on the roof, sides and floor of the caves. In this case, galena was collected from the residual clay on the floor of the caverns; the

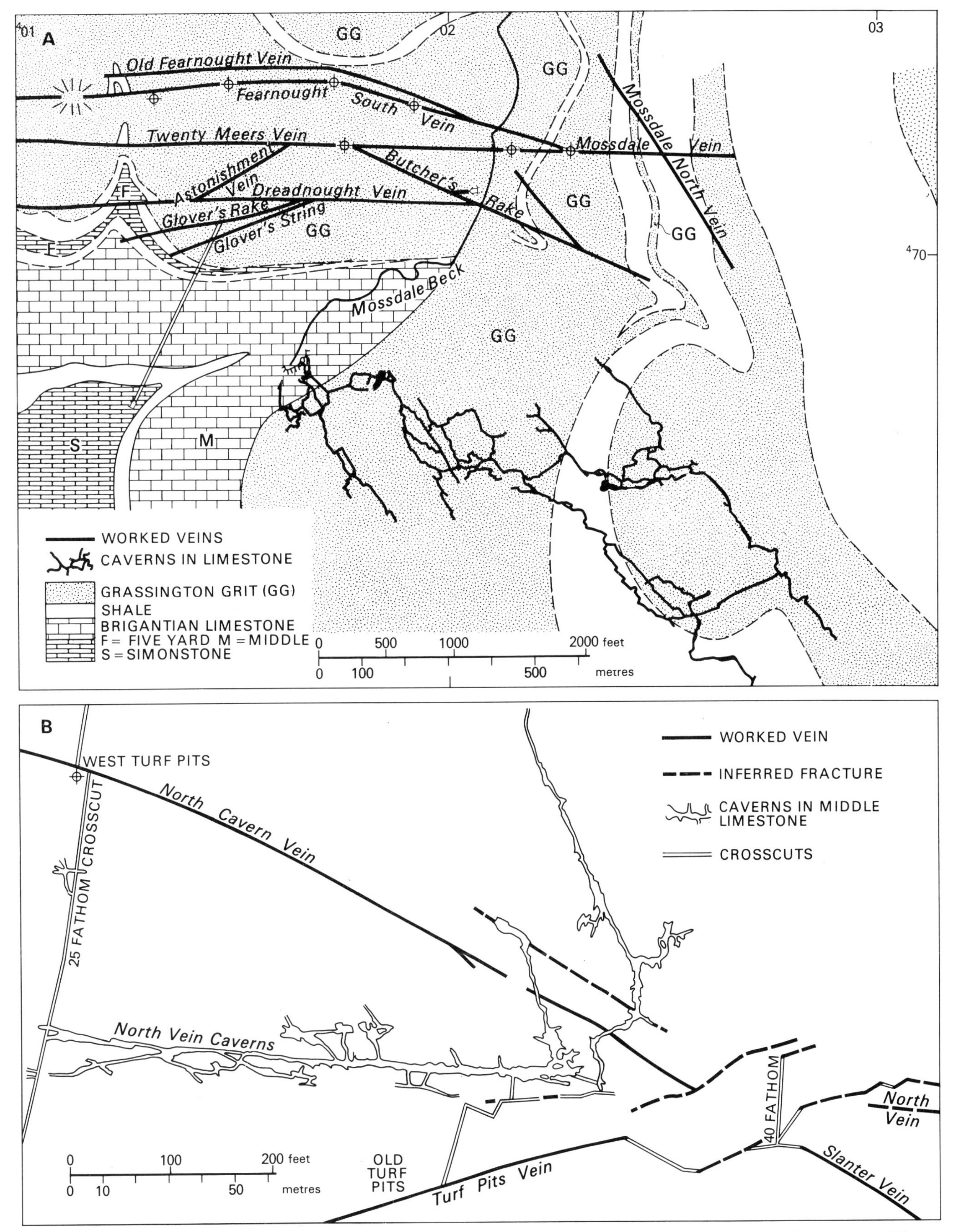

Figure 23 Patterns of veins and cavern systems: the Mossdale area (A) and the Turf Pits area, Grassington (B). A section through the Grassington veins and caverns appears in Carruthers and Strahan (1923, pl. 1). Cavern Vein area from Grassington mine-plans, Mossdale Cavern survey supplied by the British Speleological Association, 1942.

very numerous shallow shafts west of Old Turf Pits and West Turf Pits shafts were mostly sunk off the veins, with the object of reaching the caverns. Another situation is illustrated by the Windegg Mine Caverns (Brook and others 1977, p. 72) where a level incorrectly marked 'Windegg Lead Mine Level' on the six-inch map, actually Alcock's Level (p. 144), gave access to a very extensive cavern system lying between the Clay or Harker's Vein and Alcock Sun Vein. Here it seems probable that flots already existing in the Main Limestone were extensively eroded by the cavern system. This was certainly true at Sloat Hole, Faggergill (p. 145), where the thin galena streaks along the bedding of the limestone have been cut through by the later caverns along the joint system (Figure 23, p. 109). Finally, in a number of cases, limestone-solution related to the present karst cycle has opened up the courses of mineral veins and flat leaders. The extensive Faggergill workings were almost certainly facilitated by this process, and Devis Hole Mine in Area 13 (p. 179), a plan of which is given by Brook and others (1977, p. 58), provides another remarkable instance. The mine level, driven south-south-west near the base of the Main Limestone, passed through a maze of reticulated caverns following north-north-west, west-north-west and east-north-east joints and beyond this reached Wellington Vein, followed by a linear cavern for over 2000 ft (610 m). The productive workings on this vein are understood to have been at the top of the limestone and possibly in the overlying Red Beds Limestone, in flots rather than in the vein. Another series of caverns marked on the primary geological survey map is shown linking these workings with those of Crina Bottom Mine, at the Red Beds horizon (p. 179).

In spite of widespread access by cave explorers to the subsurface in most parts of the present region, very few exposures of mineral veins not known to the lead miners have come to light. Although many caves have been explored in Bishopdale, Coverdale, Dentdale and Garsdale, as well as in the better-known country around Ingleborough, the strong impression that these are very poorly-mineralised parts of the district has been confirmed rather than the reverse. Caverns have, however, provided sites for the deposition of secondary minerals, especially 'calamine', for example at Malham Tarn (drawings of which are included in Raistrick, 1954), Devis Hole and Windegg *sensu stricto*.

Perhaps the most remarkable manifestations of post-mineralisation limestone solution are the 'gulfs' of the Greenhow Hill and Skyreholme districts (Dunham and Stubblefield, 1945; Dickinson, 1970; Wilson, Dunham and Pattison, in preparation). These are, in effect, swallow-holes and caverns that have developed across or along the productive veins. Their fillings include clay, sand from the overlying Grassington Grit, partly-rotted limestone fragments and an insoluble residue mainly of fluorite with cerussite-coated lumps of galena. It may be suspected that at the bottoms of the gulfs, the natural concentration process produced residual deposits with very high values in ore and spar. The sites of such cavernisation were often, though not always, found where a cross fracture cut through the vein. The walls were plainly water-worn, either smooth or fluted, showing the white powdering appearance produced by partial solution of the carbonate. The vertical dimensions of these features may exceed 350 ft (107 m), as illustrated by Dunham and Stubblefield (1945, fig. 6).

CONCLUSION

The mineral deposits of the southern, Stainmore to Craven, half of the Northern Pennine Orefield have many features in common with those of the northern half, but the Askrigg field is not a mirror image of the Alston field. The most strongly mineralised area, the North Swaledale Belt, is more obviously related to the deep Stainmore Trough than to the granitic core of the Askrigg Block; for the Alston field, Closehouse may occupy a comparable position, and the veins of the detached Haydon Bridge district (Area 8) seem more likely to be related to the Northumberland Trough than to the Weardale Granite convective system. The feeders of the Alston field are all underlain by granite; the three principal feeders here identified in the Askrigg field are at best peripheral to underlying granite. In Derbyshire, no hotspot has yet been identified. All that can be concluded in the present state of knowledge is that various different routes were followed by brines probably being expelled from basins adjacent to the rigid block. The main period of flow coincided generally with the Hercynian earth-movements, but it continued into the Permian, perhaps contributing metals to the Zechstein lagoon (Dunham, 1964) particularly while the Kupferschiefer or Marl Slate was being deposited (Deans, 1950; Hirst and Dunham, 1963).

The source or sources of the elements carried by the hypersaline brines can only be matters for conjecture. Black shales could well be the most favourable (for example, Dunham, 1961) and here, metal source-rocks and petroleum source-rocks might under some circumstances be the same. Leaching of basement rocks—slates and greywackes as well as granite—must also be considered. The present host-rocks of the oreshoots—largely limestones and sandstones, many of Brigantian or early Pendleian age—are the least likely sources for the introduced elements.

Study of the Whin sills and dykes leads to the conclusion that before these were intruded, the country rocks were already highly lithified. Thus the mineralising fluids came into an environment of low initial permeability, except perhaps in sandstones, and free movement of fluids was inhibited until widespread fracturing associated with north–south tension imparted a considerable secondary permeability to the rocks. They were not, therefore, comparable in this respect with those of Pine Point (Beales and Jackson, 1966) where reef lithology provided high primary permeability; reefs and crinoid banks in the present region failed to provide such environments.

It appears that the plumbing system, established by Hercynian earth-movements may have continued to be effective during Permian times, but it did not outlast that epoch, or there would be evidence of continued mineral emplacement of the Pennine type during the Mesozoic era, and this is lacking. Only in Pleistocene or Flandrian time did a new wave of fluids—this time, oxygenated meteoric water of surface origin—arrive to produce the cavern systems, and to set in motion the secondary processes, largely resulting in alteration of the existing mineral suite, deposited at near-neutral pH and under reducing conditions.

REFERENCES

BEALES, F. W. and JACKSON, S. A. 1966. Precipitation of lead-zinc ores in carbonate reservoirs as illustrated by the Pine Point ore field, Canada. *Trans. Inst. Min. Metall.*, Vol. 75, pp. B278–285; discussion, *ibid.*, Vol. 76, 1967, pp. B175–177.

BILLINGS, G. A., KESSLER, S. E. and JACKSON, S. A. 1969. Relation of zinc-rich formation waters, northern Alberta, to the Pine Point ore deposit. *Econ. Geol.*, Vol. 64, pp. 385–391.

BOTT, M. H. P. 1967. Geophysical investigations in the northern Pennine basement rocks. *Proc. Yorkshire Geol. Soc.*, Vol. 36, pp. 139–168.

BRADLEY, L. 1862. *An inquiry into the deposition of lead ore in the mineral veins of Swaledale, Yorkshire.* (London: E. Stanford.) 40 pp.

BRAITHWAITE, R. S. W. and GREENLAND, T. B. 1963. A mercury mineral from the British Isles. *Nature*, Vol. 200, p. 1004.

BROOK, D., DAVIES, G. M., LONG, M. H. and RYDER, P. F. 1977. *Northern Caves.* Vol. 1 – The Northern Dales. (Clapham: Dalesman.) 211 pp.

CARPENTER, A. B., TROUT, M. L. and PICKETT, E. E. 1974. Preliminary report on the origin and chemical evolution of the lead- and zinc-rich oilfield brines in Central Mississippi. *Econ. Geol.*., Vol. 69, pp. 1191–1206.

CARRUTHERS, R. G. and STRAHAN, A. 1923. Lead and zinc ores of Durham, Yorkshire and Derbyshire, with notes on the Isle of Man. *Spec. Rep. Miner. Resour. Mem. Geol. Surv. G. B.*, Vol. 26.

COOMER, P. G. and FORD, T. D. 1975. Lead and sulphur isotope ratios of some galena specimens from the Pennines and north Midlands. *Mercian Geol.*, Vol. 5, pp. 291–304.

CREANEY, S. 1980. Petrographic texture and vitrinite reflectance variation on the Alston Block. *Proc. Yorkshire Geol. Soc.*, Vol. 42, pp. 553–580.

— 1982. Vitrinite reflectance determinations from the Beckermonds Scar and Raydale boreholes, Yorkshire. *Proc. Yorkshire Geol. Soc.*, Vol. 44, pp. 99–102.

— ALLCHURCH, D. M. and JONES, J. M. 1979 Vitrinite reflectance variation in northern England. *C. R. 9e Congr. Int. Stratigr. Geol. Carbonif. Abs.*, pp. 46–47.

DAKYNS, J. R., TIDDEMAN, R. H., RUSSELL, R., CLOUGH, C. T. and STRAHAN, A. 1891. The geology of the country around Mallerstang with parts of Wensleydale, Swaledale and Arkendale. *Mem. Geol. Surv. England and Wales.* (London) 211 pp.

DEANS, T. 1950. The Kupferschiefer and associated lead-zinc mineralisation in the Permian of Silesia, Germany and England. *Rep. 18th. Int. Geol. Congr. (London.)*, Pt. 7, pp. 340–352.

— 1951. Notes on the copper deposits of Middleton Tyas and Richmond. Abstr. in *Mineral. Soc. Not.* No. 74.

DICKINSON, J. M. 1970. The Greenhow lead mining field (A historical survey). *North. Cavern Mine Res. Soc.*, Individ. Surv. Ser. No. 4.

DUNHAM, K. C. 1934. Genesis of the North Pennine ore deposits. *Q. J. Geol. Soc. London*, Vol. 90, pp. 689–720.

— 1937. The paragenesis and colour of fluorite in the English Pennines. *Am. Mineral.*, Vol. 22, pp. 468–479.

— 1944. The production of galena and associated minerals in the northern Pennines; with comparative statistics for Great Britain. *Trans. Inst. Min. Metall.*, Vol. 53, pp. 181–253.

— 1948. The geology of the Northern Pennine orefield: Vol. 1, Tyne to Stainmore. *Mem. Geol. Surv. G. B.* (First Edition).

— 1952. Fluorspar. *Spec. Rep. Miner. Resour. Mem. Geol. Surv. G. B.*, Vol. 4, 4th Ed. London, 141 pp.

— 1959. Epigenetic mineralisation in Yorkshire. *Proc. Yorkshire Geol. Soc.*, Vol. 32, pp. 1–29.

— 1961. Black shale, oil and sulphide ore. *Adv. Sci.*, Vol. 18, pp. 1–16.

— 1964. Neptunist concepts in ore genesis. *Econ. Geol.*, Vol. 59, pp. 1–29.

— 1970. Mineralisation by deep formation waters: a review. *Trans. Inst. Min. Metall.*, Vol. 79, pp. B127–136.

— 1974. Epigenetic minerals. Pp. 294–308 in *The geology and mineral resources of Yorkshire* RAYNER, D. H. and HEMINGWAY, J. E. (Editors). (Leeds: Yorkshire Geological Society.) 405 pp.

— and DINES, H. G. 1945. Barium minerals in England and Wales. *Geol. Surv. Wartime Pam.*, No. 46.

— and STUBBLEFIELD, C. J. 1945. The stratigraphy, structure and mineralisation of the Greenhow mining area, Yorkshire. *Q. J. Geol. Soc. London*, Vol. 100, pp 209–268.

EDMUNDS, W. M. 1975. Geochemistry of brines of northeast England. *Trans. Inst. Min. Metall.*, Vol. 84, pp. B 39–52.

EMBREY, P. G. 1977. Fourth supplementary list of minerals *in* Facsimile reprint of Greg and Lettsom, 1858 (q.v.).

FINLAYSON, A. M. 1910. Problems of ore-deposition in the lead veins of Great Britain. *Q. J. Geol. Soc. London*, Vol. 66, pp. 299–328.

FORSTER, W. 1809. *A treatise on a section of the strata, commencing near Newcastle upon Tyne, and concluding on the west side of the mountain of Cross-Fell. With remarks on mineral veins in general.* (Newcastle: Forster.) 156 pp.

GREENWOOD, D. and SMITH, F. W. 1977. Fluorspar mining in the northern Pennines. *Trans. Inst. Min. Metall.*, Vol. 86, pp. B181–190.

GREG, R. P. and LETTSOM, W. G. 1858. *Manual of the mineralogy of Great Britain and Ireland.* (London: Van Voorst.) 437 pp. Facsimile reproduction 1977. EMBREY, P. G. (Editor) with additions. (Broadstairs: Lapidary Publications.) 437 + 67 pp.

HAY, J. 1939. *Ore deposits of Grassington and Greenhow Hill.* Unpublished thesis, University of Leeds.

HEYL, A. V. 1969. Some aspects of genesis of zinc-lead-barite-fluorite deposits in the Mississippi Valley, U.S.A. *Trans. Inst. Min. Metall.*, Vol. 78, pp. B148–160; discussion, *ibid.*, Vol. 79, 1970, pp. B91–96.

— LANDIS, G. P. and ZARTMAN, R. E. 1974. Isotopic evidence for the origin of Mississippi Valley – type mineral deposits. *Econ. Geol.*, Vol. 69, pp. 992–1006.

HILLEBRAND, W. F. 1907. The vanadium sulphide patronite and its mineral associates from Mina Ragra, Peru. *Am. J. Sci.*, Vol. 24, p. 151.

HIRST, D. M. and DUNHAM, K. C. 1963. Chemistry and petrography of the Marl Slate of S. E. Durham, England. *Econ. Geol.*, Vol. 58, pp. 912–940.

HITCHON, B., BILLINGS, G. K. and KLOVAN, J. E. 1971. Geochemistry and origin of formation waters in western Canada sedimentary basin—III. Factors controlling chemical composition. *Geochim. Cosmochim. Acta*, Vol. 35, pp. 507–598.

HOLMES, A. and HARWOOD, H. F. 1928. The age and composition of the Whin Sill and related dykes of the north of England. *Mineral. Mag.*, Vol. 21, pp. 493–542.

HORNSHAW, T. 1975. Copper mining in Middleton Tyas. *North Yorkshire County Record Office Publ.*, No. 6.

HUDSON, R. G. S. 1938. The Harrogate mineral waters. *Proc. Geol. Assoc.*, Vol. 49, pp. 349–352.

IXER, R. A. 1978. The distribution of bravoite and nickeliferous marcasite in central Britain. *Mineral. Mag.*, Vol. 42, pp. 149–150.

— STANLEY, C. J. and VAUGHAN, D. J. 1979. Cobalt-, nickel- and iron-bearing sulpharsenides from the north of England. *Mineral. Mag.*, Vol. 43, pp. 389–395.

INESON, P. R. 1976. Ores of the northern Pennines, the Lake District and North Wales. Pp. 197–230 in *Handbook of stratabound and stratiform ore deposits,* Vol. 5, *Regional Studies and specific deposits 2.* WOLF, K. H. (Editor). (Amsterdam: Elsevier.) 319 pp.

JACKSON, S. A. and BEALES, F. W. 1967. An aspect of sedimentary basin evolution: the concentration of Mississippi Valley-type ores during late stages of diagenesis. *Bull. Can.*

Petrol. Geol., Vol. 15, pp. 383–433.
KANESEWICH, E. R. and SLAWSON, W. R., 1964. Precision inter-comparison of lead isotope ratios. *Geochim. Cosmochim. Acta*, Vol. 28, pp. 541–549.
KINGSBURY, A. W. G. 1968. Demonstration of minerals from Greenhow and Grassington, *Mineral. Soc.*, March 14 (Unpublished).
LOUGHLIN, G. F. 1914. The oxidised zinc ores of the Tintic district, Utah. *Econ. Geol.*, Vol. 9, pp. 1–19.
MITCHELL, R. H. and KROUSE, H. R. 1971. Isotopic composition of sulfur and lead in galena from the Greenhow – Skyreholme area, Yorkshire, England. *Econ. Geol.*, Vol. 66, pp. 243–251.
MOORBATH, S. 1962. Lead isotope abundance studies on mineral occurrences in the British Isles and their geological significance. *Philos. Trans. R. Soc. London*, Vol. A254, pp. 295–360.
PHILLIPS, J. 1836. *Illustrations of the geology of Yorkshire. Part II. The Mountain Limestone District.* (London: John Murray.) 273 pp.
RAISTRICK, A. 1936. The copper deposits of Middleton Tyas, N. Yorks. *Naturalist,* pp. 111–115.
— 1938. The mineral deposits *in* HUDSON, R. G. S. (Editor). Geology of the country around Harrogate. *Proc. Geol. Assoc.*, Vol. 49, pp. 343–341.
— 1947–49. The Malham Moor mines, Yorkshire. *Trans. Newcomen Soc.,* Vol. 26, pp. 69–77.
— 1954. The calamine mines, Malham, Yorks. *Proc. Univ. Durham Philos. Soc.*, Vol. II, pp. 125–130.
— 1973. *Lead mining in the Mid-Pennines. The mines of Nidderdale, Wharfedale, Airedale, Ribblesdale and Bowland.* (Truro: Bradford Barton.) 172 pp.
— 1975. *The lead industry of Wensleydale and Swaledale.* Vol. 1 *The Mines.* (Buxton: Moorland.) 120 pp.
ROBINSON, B. W. and INESON, P. R. 1979. Sulphur, oxygen and carbon isotope investigations of lead-zinc-baryte-fluorite mineralisation, Derbyshire, England. *Trans. Inst. Min. Metall.*, Vol. 88, pp. B107–117.
ROGERS, P. J. 1974. Bravoite inclusions in fluorite from the Derbyshire and Askrigg ore-fields. *Bull. Peak Dist. Mines Hist. Soc.*, Vol. 5, p. 333.
— 1977. Fluid inclusion studies in fluorite from the Derbyshire orefield. *Trans. Inst. Min. Metall.*, Vol. 86, pp. B128–132.
— 1978. Fluid inclusion studies on fluorite from the Askrigg Block. *Trans. Inst. Min. Metall.*, Vol. 87, pp. B125–131.
RUSSELL, M. J. and SMITH, F. W. 1979. Plate separation, alkali magmatism and fluorite mineralisation in northern and central England. *Trans. Inst. Min. Metall.*, Vol. 88, p. B188 (abstract).
SAWKINS, F. J. 1966. Ore genesis in the north Pennine orefield in the light of fluid inclusion studies. *Econ. Geol.*, Vol. 61, pp. 385–399.
SHEPHERD, T. J. 1979. Microthermometric analysis of quartz-sulphide veins, Cumpston Hill, Yorkshire. *Rep. IGS Isotope Geology Unit*, No. 79/3.
SIEGEL, F. R. 1958. Effect of strontium on the aragonite-calcite ratios of Pleistocene corals. *Bull. Geol. Soc. Am.*, Vol. 69, p. 1643 (abstract).
SMALL, A. T. 1977. Mineralisation of the Stainmore Depression and northern part of the Askrigg Block. Unpublished Ph.D. thesis, University of Durham.
— 1978. Zonation of Pb-Zn-Cu-F-Ba mineralisation in part of the North Yorkshire Pennines. *Trans. Inst. Min. Metall.*, Vol. 87, pp. B9–14.
— 1982. New data on tetrahedrite, tennantite, chalcopyrite and pyromorphite from the Cumbria and North Yorkshire Pennines. *Proc. Yorkshire Geol. Soc.*, Vol. 44, pp. 153–158.
SMITH, F. W. 1974a. Factors governing the development of fluorite orebodies in the north Pennine Orefield. Unpublished Ph.D. thesis, University of Durham.
— 1974b. Yttrium content of fluorite as a guide to vein-intersections in partially developed fluorspar orebodies. *Trans. Soc. Mining Eng.*, AIME. Vol. 255, pp. 95–96.
SMYTHE, J. A. and DUNHAM, K. C. 1947. Ankerites and chalybites from the northern Pennine orefield and the North-East Coalfield. *Mineral. Mag.*, Vol. 28, pp. 53–74.
SOLOMON, M., RAFTER, T. A. and DUNHAM, K. C. 1971. Sulphur and oxygen isotope studies in the northern Pennines in relation to ore-genesis. *Trans. Inst. Min. Metall.*, Vol. 80, pp. B259–276; Discussion, *ibid.*, Vol. 81, pp. B172–177; Vol. 82, p. 46.
SPRINGER, G., SCHACHNER-KORN, D. and LONG, J. V. 1964. Metastable solid solution relations in the system FeS_2-CoS_2-NiS_2. *Econ. Geol.*, Vol. 59, pp. 475–491.
SWEETING, M. M. 1950. Erosion cycles and limestone caverns in the Ingleborough district. *Geogr. J.*, Vol. 115, pp. 63–78.
TROTTER, F. M. 1939. Reddened Carboniferous beds in the Carlisle basin and Edenside. *Geol. Mag.*, Vol. 76, pp. 408–416.
VARVILL, W. W. 1920. Greenhow Hill lead mines. *Mining Mag.*, Vol. 22, pp. 275–282.
— 1937. A study of the shapes and distribution of the lead deposits in the Pennine limestones in relation to economic mining. *Trans. Inst. Min. Metall.*, Vol. 46, pp. 463–559.
VAUGHAN, D. J. 1969. Zonal variation in bravoite. *Am. Mineral.*, Vol. 54, pp. 1075–1083.
— and IXER, R. A. 1980. Studies in sulphide mineralogy of north Pennine ores and its contribution to genetic models. *Trans. Inst. Min. Metall.*, Vol. 89, pp. B99–110.
VERSEY, H. C. 1925. The beds underlying the Magnesian Limestone in Yorkshire. *Proc. Yorkshire Geol. Soc.*, Vol. 20, pp. 200–214.
WAGER, L. R. 1931. Jointing in the Great Scar Limestone of Craven and its relation to the tectonics of the area. *Q. J. Geol. Soc. London*, Vol.87, pp. 392–424.
WELLS, A. J. 1958. The stratigraphy and structure of the Middleton Tyas—Sleightholme Anticline, North Yorkshire. *Proc. Geol. Assoc.*, Vol. 68, pp. 231–254.
WHITE, D. E. 1968. Environment of generation of some base-metal ore deposits. *Econ. Geol.*, Vol. 63, pp. 301–335.
WILSON, A. A., DUNHAM, K. C. and PATTISON, J. (in preparation). The stratigraphy of recent boreholes in the Greenhow and Grimwith areas, Yorkshire.

CHAPTER 8

Mineral deposits

Details, Area 10 MALLERSTANG and STAINMORE

The first area for detailed description covers deposits in the former County of Westmorland (now Cumbria) lying adjacent to the great fault-system that forms the western margin of the rigid block of the northern Pennines. The pass of Stainmore forms the northern limit of the present account, part of the ground here now being in County Durham. The depression is believed to overlie an E – W belt of Lower Carboniferous rocks substantially thicker than those of the Alston Block to the north, or the Askrigg Block to the south, which continues into Ravenstonedale to the west of the Pennines. Stainmore, at the stratigraphical levels exposed (Reading, 1957), contains very few mineral deposits. From the west end of the pass a short distance west of Augill springs the curviform belt of steeply-dipping strata known as the Dent Line or Stainmore Monocline (Turner, 1927; Miller and Turner, 1931), representing a probable pre-Permian uplift of what is now the Howgill Fells and Ravenstonedale relative to the Askrigg Block and Stainmore (p. 73). In post-Triassic times this situation was reversed by powerful faulting along the course of the monocline, operative at least as far to the south-south-west as Low Stennerskeugh [743 017], 12 miles (19.3 km) from Augill. Small ore deposits, generally carrying galena, baryte and fluorite, and with chalcopyrite a little more conspicuous than in the North Swaledale or Swale/Ure mineral belts, occur in places within or to the west of the monocline but in spite of the highly tectonised nature of the belt, no large ore deposit has yet been found (Dunham, 1952, p. 66). At Low Stennerskeugh the monocline converges upon the Dent Fault proper, which increases in displacement towards the south-south-west but is nowhere known to be mineralised. The western two-thirds of the area of the Hawes (50) Sheet is in fact almost devoid of mineralisation, in spite of the presence within it of all the familiar bearing horizons of the Asbian, Brigantian and Pendleian stages. However, in Garsdale and in Grisedale, Kirkby Stephen (40) Sheet, and the upper reaches of the River Eden a number of small veins carrying quartz with uneconomic amounts of copper minerals occur. Dakyns and others (1891, p. 167) noted the occurrence of these in Mallerstang and adjacent parts of Yorkshire, for the most part outside the region of productive lead veins. For the sake of completeness, all the instances known to us are summarised in this chapter, the sites being shown on Figure 24 which also indicates the positions of the lead and copper deposits in relation to the Dent Line Monocline. Finally a few small lead deposits lying on the south side of the Stainmore pass drainage are included here, but they should be regarded as fringe deposits of the North Swaledale Belt, to which Chapter 9 is devoted.

Fell End Clouds Scrins Lead ore

SD 79 NW; Westmorland 36 NW
Direction N25-40°E, probably no throw

Johnson's Vein Lead ore

SD 79 NW; Westmorland 36 NW
Direction N20°E

Clouds Vein Lead ore

NY 70 SW; Westmorland 36 NW
Direction near E-W, small downthrow S

Clouds North Vein Copper ore?

NY 70 SW; Westmorland 36 NW
Direction N75°E, small N downthrow

The hill known from south to north as Fell End Clouds, Clouds and Stennerskeugh Clouds lies on the west side of the Dent Fault, here close to the road to Sedbergh from Kirkby Stephen, 7 miles (11.2 km) to the north. The southern part of the hill is an anticline plunging S20°W, splendidly exposed in Great Scar Limestone. To the east, the Robinson and higher limestones of Yoredale facies come on with the rising ground, dipping in that direction off the structure. To the north the structure may close into a dome, but here the exposure is inadequate to make this certain. The scrins occur on the southern flank of Fell End Clouds and are, in effect, mineralised joints in the pale grey limestone carrying a little galena, baryte and calcite. They form a branching system which has been followed by shallow trenching (Plate 4), but which appears not to have been regarded as sufficiently encouraging for a level to be driven to give access at depth. Probably working occurred here before the 17th century, for Dr G. A. L. Johnson found a bole hill at [7357 9931] with remains of metallic lead as well as galena. Farther north, Johnson's Vein runs from [7358 9965] for about 800 ft (243 m) towards the NNE, almost along the principal axis of the structure. Widths up to about 3 ft (1m) seem to have been worked, and cuttings from dressing of the ore contain baryte, colourless fluorite, oxidised copper minerals, calcite and galena. Extraction again seems to have been confined to opencut working. Clouds Vein is marked by a line of old shafts 1500 ft (456 m) long; the productive ground seems to have been chiefly above the Robinson Shale, yielding galena, baryte, colourless and white fluorite and copper sulphides. This is perhaps the deposit which Goodchild and Clough mention as the principal copper producer of the area (Dakyns and others, 1891, p. 168) but the output cannot have been more than a few tons. The primary map by de Rance shows a northern branch, downthrowing north, but no mineralisation appears to have been found in this. A shaft into the Robinson Limestone about 350 ft (106 m) S of Clouds Vein does however show some baryte and galena, though the direction of the deposit is not evident. Finally a weak fault or joint, here called North Vein, can be followed through the grikes of Great Scar Limestone on the west flank of the hill, but seems to have been mineralised only around [746 007] where there are bell-pits close to a cross-fault.

Fothergill Sike Vein

NY 70 SE; Westmorland 30 NW, NE
Direction probably N10°E

A fault trending NNE crosses the cutting a short distance north of the north entrance to the Birkett Tunnel on British Railways

Kirkby Stephen to Settle line at [7715 0356]. This fault or a more northerly-trending branch apparently carried some galena north of the cutting, where it was tested by shafts and by a level driven SE from Fothergill Sike. The country rock is Great Scar Limestone, here more or less vertical in the Dent Line Monocline. Dolomite appears to have been the only associated mineral.

Dalefoot Vein Lead ore

NY 70 SE; Westmorland 30 NE
Direction N23°E, probable W downthrow

Great Bell Vein

NY 70 SE; Westmorland 30 NE
Direction N35°W,

Old workings situated on the south slope of Great Bell, south-east of Nateby on the north-east side of the River Eden have explored deposits in the upper part of the Great Scar Limestone. The fracture here called Dalefoot Vein, tried by means of a hush 500 ft (152 m) long along it may represent the most southerly appearance of the Argill Fault (Burgess and Holliday, 1979, p. 77). Dalefoot Level [7858 0422], driven NE across this structure found pale purple and amber fluorite with galena, dolomite and calcite, but only in modest quantity. The development cut black shale, probably on the footwall of the fault, possibly the shale beneath the Robinson Limestone, or between the Robinson and Peghorn limestones. As the fracture is followed north, the top of the Great Scar Limestone comes in on the hangingwall side. At 160 ft (49 m) NE of the level portal, a cross fracture perpendicular to the Dalefoot Vein was tested in an opencut. Immediately north of the summit of Great Bell, a shaft appears to be on the vein mapped by R. H. Tiddeman and tested by another shaft 510 ft (155 m) to the SE, calcite, dolomite and limonite being found. A level started 850 ft (258 m) SW of the mid point between these shafts can hardly have reached vein judging from the dump, which contains only limestone and calcite. The workings listed in this section are those mentioned (Dunham, 1952, p. 66) as at 'Bells', where the view is expressed that they offer little hope of production.

Ore Hill Vein

NY 70 NE; Westmorland 30 NE
Probable direction N5°E.

Ore Hill is a knoll-like mass of limestone belonging to the Great Scar, lying west of the Dent Line Monocline [788 061]. Shafts on the east side are marked Stephen Grove Mine on the 1860 edition of the Ordnance Survey six-inch map, but this name is no longer given on the current edition. The alignment of the pits suggests a fracture sympathetic with the boundary of the monocline, but dips in the limestone vary; between NE and NW. The operation can only have been a minor prospect.

Plate 4 Scrins at Fell End Clouds.

Small subparallel veins in Danny Bridge Limestone have been worked by open-cut. In the distance, beyond the Dent Fault, stand the Howgill Fells, chiefly in Silurian greywackes.

Ladthwaite Vein Lead ore (fluorspar) (baryte)

NY 70 NE, 80 NW: Westmorland 23 SE, 30 NE
Direction N28°E, probable downthrow W

Hartley Birkett Veins Lead ore (baryte)

NY 80 NW; Westmorland 23 SE
Direction near N – S

From Ladthwaite Level [7968 0721] the Dent Line Monocline is mineralised for a distance of nearly 1 mile (1.6 km) to the north-north-east. This is one of only two areas where appreciable deposits occur within the monocline, the other being Augill in Area 1 (Dunham, 1948, p. 140). In both cases the mineralisation occurs in many fractures through the rocks, and is of the disseminated, low-grade type. The present area was worked by the (now abandoned) *Hartley Birkett Mine* which was still smelting lead ore in a small way up to 1877. The area is structurally complex, this was shown during the primary geological survey, a century ago, when three independent six-inch versions were produced (respectively by W. J. Aveline, J. G. Goodchild and W. H. Dalton) and Dalton also made a 1:2500 map. In our view a completely satisfactory interpretation could not be made without drilling, but a following general view follows.

To the north-west, the fine feature of Longrigg Scar exposes the Robinson Limestone[1] for 1½ miles (2.4 km), the scar facing in the direction of the dip of the limestone, which swings round from 25° to ENE to 50° to ESE opposite Hartley Birkett. This outcrop of the Robinson Limestone forms the east flank of the Hartley half-dome (p. 73). Between Longrigg Scar and Hartley Birkett hill there is lower, drift-covered ground through which must run the Argill Fault, continued southward from the Brough Sheet, where the structure has been described by Burgess and Holliday (1979). On the east side of the low ground, a west-facing scar [around 804 084] exposes limestone in which Mr J. S. Turner (personal communication) first found *Girvanella* and which we take to be the Hawes Limestone. This has an inverted dip of about 70° to the west, well attested by the fact that the shale member of the cyclothem overlies the sandstone where this is exposed in the beck to the south. The ground between the Robinson and Hawes limestones thus contains beds dipping steeply towards the supposed line of the Augill Fault from either side. There may be a concealed crop of the Hawes Limestone west of the fault, and a bed that we regard as the Birkdale Limestone of Burgess and Mitchell (*op. cit.*) appears in a hush (described below) east of the fault, this limestone lying between the Hawes and Robinson. Hartley Birkett hill itself is a mass of tectonised limestone which from its steep westerly dip must be some hundreds of feet thick. It is not, however, likely that the main mass of the Great Scar Limestone is involved, and the hill is interpreted as Robinson Limestone, thickened by faulting, perhaps with Birkdale and Hawes limestones also involved, and the clastic partings faulted out. The eastern contact of this limestone mass with shale and sandstone is exposed on the west side of the great artificial valley that may be called Birkett Hush; it is highly irregular due to fault-intersections, and there can be little doubt from the substantial tonnage excavated that some of the lead ore came from this contact and from the contacts of an 'island' of unidentified limestone tectonically engulfed in the clastics. On the east side of Birkett Hush there is a smaller hush where Hartley Birkett Vein is considered to run, imparting a sideways shift to the Hawes Limestone, W side S.

[1]At Hartley Birkett the high Asbian sequence characteristic of the Alston Block has begun to appear in that the Robinson Limestone can be recognised as a split off the upper part of the Danny Bridge Limestone. However, the Hawes Limestone has not yet divided into the Peghorn and Smiddy limestones, though it does so to the north around Augill (*vide* Burgess and Mitchell, 1976, fig. 1).

This vein or a branch swings to the NW on the north side of the hushes, terminating the outcrop of tectonised limestone of Hartley Birkett hill close to the site of the former smelt mill. The clastics stratigraphically below but structurally above the Hawes Limestones, already mentioned, appear north of this fault. The whole monoclinal belt is roughly 2000 ft (606 m) wide, with normal steep easterly dips on either side of a central zone about 600 ft (182 m) wide where dips are overturned to the west. The central zone lies immediately east of the postulated post-Triassic Argill Fault. It can safely be assumed that as a result of the intense movements, every contact has been jostled, and parts of the limestone converted into microbreccia, now mainly cemented with calcite.

Mineralisation occurs in two main linear zones, each one containing vein fractures of no great continuity, and opened-up contacts within the limestones and between limestones and clastics. The

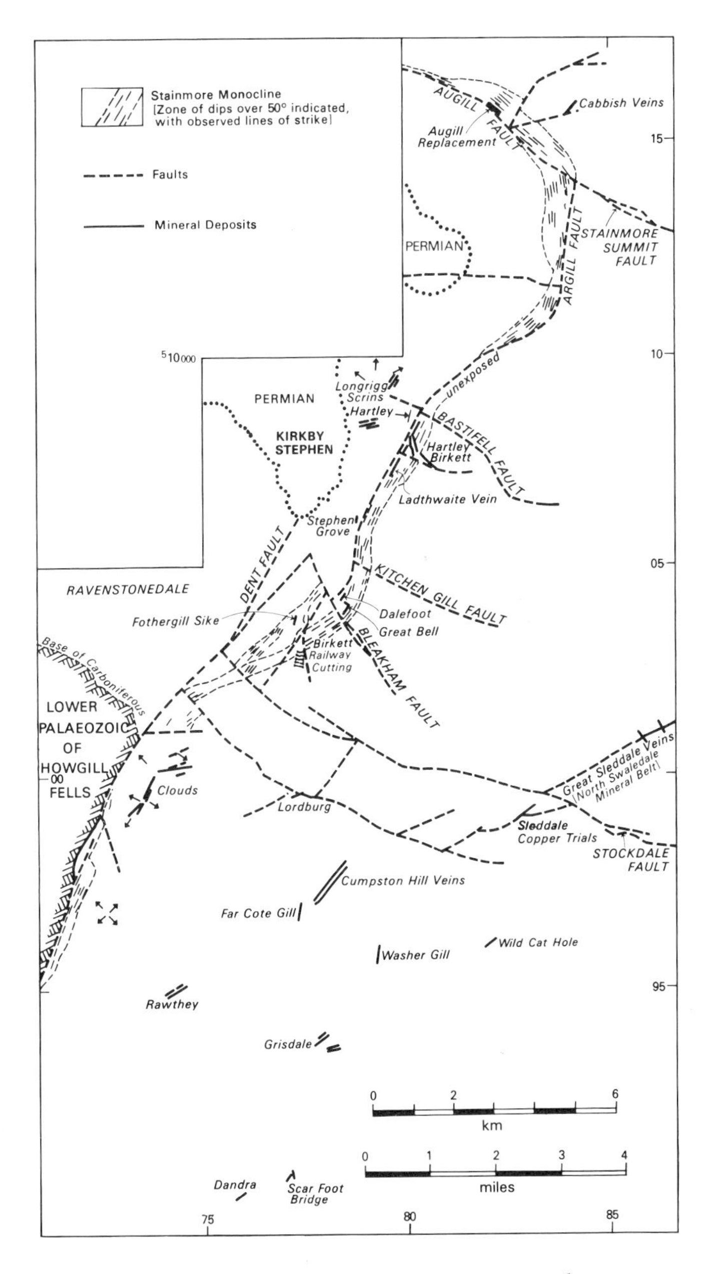

Figure 24 Key map, Area 10, Mallerstang and Stainmore

Ladthwaite Zone, the existence of which we deduce from the alignment of shafts from near Ladthwaite Level to the hill slope close to the smelt mill, carries amber, purple and colourless fluorite, yellowish baryte, calcite, dolomite, a little sphalerite and chalcopyrite accompanying galena. Smelt Mill Level [8023 0815] probably driven across the end of Hartley Birkett hill, must have passed through the Ladthwaite Zone. The zone apparently shifted N side W by the fault from Hartley Birkett Vein, and what we regard as its continuation is seen in the hush west of the Hawes Limestone scar, NE of the Smelt mill. Here up to 5 ft (1.6 m) of baryte with cavities containing rounded crystals and small purple fluorites is exposed, succeeded to the west by 3 ft (1 m) of yellowish brown fluoritised limestone, probably the Birkdale Limestone. A crushed mixture of the same minerals on the slopes outside the hush shows where galena was dressed out. The vein continued north to the cross hush leading to the Hawes Limestone outcrop, but beyond this it seems to have been lost, for four shafts in the next 750 ft (280 m) found nothing, but may not have got through the drift.

The second mineralised zone commences south of the head of Birkett Hush with three short N – S veins, worked from shafts, yielding mainly baryte and galena. The hush continues this trend to the north but the ore probably came mainly from sheared contacts between clastics and limestone. This zone converges on the Ladthwaite Zone and the Hartley Birkett Vein at the tail of the hush. The eastern zone yielded rather discoloured baryte, fluorite, dolomite and limonitised limestone in addition to galena; there is evidence of replacement of limestone or dolomite by all these minerals. Taken as a whole, however, the deposit gives the impression of being very patchy and low grade. Having regard to the steep dips, it could continue to considerable depths, beyond the range of the several shallow shafts from which it was explored west of the shale-limestone contact.

Though not so strongly mineralised as some sites described in later chapters, Hartley Birkett is the most interesting one, on present evidence in Area 10. There could be said to be a belt of fracturing and replacement at least ½-mile (0.8 km) long and 200 ft (66 m) wide with the prospect of extension in depth. It is disappointing therefore to have to record that an airborne electromagnetic (AEM), magnetic and radiometric survey over an area of 12 km² including Little Longrigg Scar, Hartley Birkett, Birkett Common and Kitchen Gill failed to reveal any anomaly which might indicate the presence of strong sulphide mineralisation in depth at Hartley Birkett (Cornwell, Patrick and Hudson, 1978). Slightly more encouraging results were obtained in the neighbourhood of Kitchen Gill, near Great Bell (p. 114), where AEM anomalies were followed up with Turam investigation on the ground. It was concluded that the anomalies do not indicate extensive mineralisation, either near the surface or at depth. However, though much of the EM response appeared to be due to conductive shale horizons, some was attributed to fractures, beneath drift, that may carry some mineralisation of a sporadic nature. A scatter of small AEM anomalies was also discovered over Birkett Common, south of Kirkby Stephen, but ground investigation by VLF and Slingram methods led to the conclusion that the EM anomalies were caused either by stratigraphical conductors or fault crush-zones without significant metallic conductors.

High Longrigg Scrins

Lead ore, copper ore

NY 70 NE; Westmorland 23 SE
Direction N25°E

Within the Hartley half-dome, about 1 mile (1.6 km) ENE of Hartley village, shallow opencuts and numerous pits mark the site of High Longrigg Mine, that worked at least four small veins in the Great Scar Limestone, here dipping 10°N. The maximum length of the mineralised area is about 600 ft (182 m) and the width 400 ft (122 m). Malachite and azurite are more conspicuous here than anywhere else in Area 10, accompanied by galena, pale fluorite and baryte, with calcite and very little chalcopyrite. To the south-east, a shaft at [7999 0908] may have been sunk to search for the Bastifell Fault, which makes a conspicuous shift, down north, of the outcrop of the Robinson Limestone in Long Rigg to the south-east; the tip contains only pale limestone with dolomite. Goodchild and Clough (Dakyns and others, 1891, p. 167) report of High Longrigg that the quantity of copper 'was insufficient to warrant more than a few shallow workings'. Copper was also met with in working to the bottom of a deposit of hematite along a fault traversing the same mass (of disturbed limestone), but is said to have given place to galena a short distance down. The exact site is uncertain, but this was probably the ground worked by the Eden Vale Mining Company, credited in the official statistics with the production of 135 tons lead concentrates in 1877, containing 5 oz recoverable silver per ton of lead metal.

Much of the inner part of the Hartley half-dome is concealed by drift, only the Robinson Limestone giving continuous outcrops; there could be other deposits in the Great Scar Limestone. However, the excellent exposures at Hartley limestone quarry, [centred at 789 083] reveal four small faults trending N80°E, with throws, from south to north respectively of 40, 15, 10 and 30 ft (12, 4.5, 3 and 9 m) N, tending to neutralise the gentle southerly dip of the beds. Lenticular calcite bodies, in one case as much as 6 ft (2 m) wide occur along these, but no other mineralisation.

Dukerdale Head Vein

NY 80 NW; Westmorland 23 SE, 30 NE
Direction N45°W, small N downthrow

The Main Limestone and associated late Brigantian and early Pendleian beds, both in the high west-facing fells east of the Dent Line (described by Scanlon, 1955; Rowell and Scanlon, 1957), and in the low ground are remarkably lacking in evidence of mineralisation, in spite of reasonably good outcrops[1]. One of the few instances lies below Coldbergh Side at Dukerdale Head, where a small vein, tested by a cut through the Main Limestone outcrop and a level at its base yielded a little lead ore with dolomite and calcite. Some 1¼ miles (2 km) to the north, trials were also made on the Greenfell Moss Fault where it shifts the Main Limestone, but without success.

Scarfoot Bridge Veins

SD 79 SE; Yorkshire 64 NE
Directions N35°E, N10°W

Dandra Scar Vein

SD 78 NE; Yorkshire 64 NE
Direction N60°E [756 897]

Attention is now turned to the small quartz veins, with minor or trace amounts of chalcopyrite which occur sporadically in the western dales, including Garsdale, Cotterdale, Uldale and upper Edendale. In Garsdale 900 ft (274 m) downstream from Scarfoot Bridge [7720 9057] a minor NE fault crossing the sandstone under the Hardrow Scar Limestone contains a quartz-cemented breccia, while 150 ft (46 m) upstream of the bridge one trending NNW at the waterfalls over the Hardrow Scar Limestone contains 1 ft (0.3 m) of quartz with traces of chalcopyrite. Another NE fault from which A. Strahan recorded a little chalcopyrite occurs in the Simonstone Limestone in Assey Gill [7711 9022]. The veins have not been regarded as worth prospecting.

[1] The successful barytes mine at Peniston Green, North Stainmore, on NY 81 NW is described under Area 1 in Dunham, 1948, also the Augill deposit in the Great Scar Limestone (Figure 24)

Rawthey Faults

SD 79 NW, SW; Yorkshire 49 SE
Directions N65°E

In the Rawthey valley [743 949] the Main Limestone and the overlying Main Chert are affected by four faults, the northern pair having throws of a few feet only, a third 90 ft (27 m) upstream throwing 20 ft (6 m) SE and a fourth 120 ft (36 m) farther on throwing 40 ft (12 m) NW. Russell (Dakyns and others, 1891, p. 69) remarks: 'In all these faults quartz in finely crystallised hexagonal pyramids with a little copper-ore throughout forms the vein-stuff. . . . One of the smaller veins was tested in a trial hole, but did not yield nearly enough ore to be worth working.'

Wild Cat Hole Vein

SD 89 NW; Yorkshire 50 NE
Direction N45°E

J. E. Scanlon (1955) reported similar mineralisation in a small vein cutting Main Limestone in upper Cotterdale [821 961].

Washer Gill Fault

SD 79 NE; Yorkshire 50 NW
Direction N7°E, throw 30 ft (9 m) E

Near the fell wall above Low West End Farm, at [7929 9577] there is quartz mineralisation where the sandstone under the Main Limestone is faulted against the lower part of the limestone to the east. Up to 2½ ft (0.76 m) of quartz is seen replacing the limestone, and an adit, now collapsed, was driven a short distance north from the bed of the gill. Large boulders of fairly coarse quartz, somewhat limonite-stained, and with rare flecks of malachite, can be seen in the wall surrounding the farm garden, but any debris from the adit has been washed away.

Cumpston Hill North Vein, South Vein

SD 79 NE; Westmorland 36 NE, SE
Direction N47°E, small SE and NW throws respectively

The two veins cut through the sandstone underlying the Underset Limestone and are exposed in the bed of the River Eden, 2000 ft (608 m) S of Aisgill Bridge in the case of the South Vein. This can be followed up the hillside to the north-east; it makes a rib of silicified Underset Limestone up to 9 ft (3 m) wide behind Intake Farm. Where this vein enters the Main Limestone a shaft shows quartz, highly silicified limestone, azurite, malachite and a little yellow sulphide. Across the higher ground the vein forms a distinct low feature, tested in several places with similar results. The North Vein also forms a wide belt of silicification which stands up as it crosses the limestone; in places this is over 20 ft (6 m) wide. At some remote period in the past shallow opencuts were made to test this, but they are overgrown; however, a shaft at [7832 9768] shows vein-stuff identical with that on the South Vein. A. T. Small (1977) found tetrahedrite among the minerals here and our sampling of the small dump showed 0.74 per cent Cu. The two veins thus show evidence of mineralisation over a length of at least ¾ mile (1.2 km), of at least 300 ft (91 m) in vertical range and with substantial widths; but the grade gives no encouragement for further prospecting.

Far Cote Gill Vein

SD 79 NE; Westmorland 36 SE
Direction N-S

On the opposite side of the valley, in the gill above the waterfall over the Underset Limestone, a hush 400 ft (122 m) long has been made to explore a vein which, as indicated by material from two shafts, was associated with black silicified limestone with traces of yellow sulphide. No oxidised copper minerals are, however, to be seen here.

Grisdale East House Veins

SD 79 SE; Yorkshire 49 SE
Direction N45°E, South Vein throws 15 ft (4.6 m) NW

High Ings Veins

SD 79 SE; Yorkshire 49 SE
Direction near E – W

The four veins cross the bare plateau made by the Main Limestone on the north-east side of Grisdale. Though they do not line up with the Cumpton Hill veins—the courses of the East House veins lie considerably south of these—the resemblance is very close. Each vein makes a small ridge feature as it crosses the limestone outcrop, and in the case of East House South Vein, this can be followed down the hillside as far as the outcrop of the Underset Limestone. Almost the only introduced mineral is quartz. Dakyns and others, (1891, p. 167) state that amethystine quartz has been found, and mention a small proportion of copper pyrites; our observations suggest that this is a very minor component indeed. The only trial shaft was on High Ings North Vein at [7801 9352] but no ore was found.

Lordburg Veins

SD 79 NE; Westmorland 36 NE
Direction and throw uncertain

The Hall Hill Fault, which is indirectly linked to both the Great Sleddale veins at the west end of the North Swaledale Mineral Belt (Chapter 9) and with the Stockdale Disturbance (p. 76), crosses the upper Eden valley close to Hall Hill Farm, and west of this it terminates the Main Limestone outcrop of Angerholme Wold and High Bank, and drops this formation down about 300 ft (92 m) to the north, almost to the level of the railway. There is a noticeable rise in the base of the limestone as the fault is approached from the south, so that at Lordburg Rigg beds below the limestone occur on the footwall of the fault, which is here following a nearly E – W course. At Lordburg Rigg a hush, with its reservoir at [7735 9959] has been excavated for 400 ft (122 m) down the hillside across the course of the fault. It is now much overgrown and shows only quartzitic sandstone with small baryte and calcite veinlets, but there was evidently much more to see when de Rance made the primary survey. He shows the fault here enclosing a horse of rock including, east of the hush, some limestone; the northern element of this he marks as mineralised, and shows another vein coming in through the hush from the WSW to meet it. This prospecting, probably done long ago, was evidently regarded as unsuccessful.

North Spanham Vein Lead ore

NY 91 SE, NZ 00 NW, 01 SW; Yorkshire 23 NW
Direction N70-82°W, throw as much as 150 ft S in the portion worked

Hazel Bush Veins

NY 91 SE, NZ 01 SW; Yorkshire 23 NW
Directions N50°W and N50°E

White Fell Veins

Direction N20°W, N45–50°W

As already noted, the Main or Great Limestone outcropping east of the Dent Line Monocline shows very little evidence of mineralisation. Across the southern side of the Stainmore – Cotherstone Syncline which marks the boundary between the Alston Block area covered by Dunham (1948) and the present area, there is no continuous outcrop of this limestone, since it is cut off by the Stainmore Summit Fault (Figure 15). However it reappears in the general area of Gilmonby Moor, in the south-west corner of the Barnard Castle district (Sheet 32). From here a broad outcrop across Bowes Moor on the north flank of the Sleightholme – Middleton Tyas Anticline is also devoid of mineralisation until the small copper deposits of Area 12 are reached at East Layton. However, the outcrop of the south flank of the anticline, through Stang Forest, contains some lead-bearing veins that are best regarded as fringe deposits of the North Swaledale Belt (Chapter 9). They are included here because they lie within the Stainmore – Cotherstone pass drainage, flowing to the Tees and not to the Swale.

The North Spanham Fault is discussed by Mills and Hull, (1976) where it is shown that its throw reverses in the neighbourhood of Eller Beck [998 103], west of which it becomes 150 ft (45 m) N. The principal workings in this fault-vein were at Eller Beck, where it was exposed in a hush in the Main Limestone and a level was driven westward at the base of this limestone at 1250 ft (380 m) OD. To the south-west of the hush, shallow shafts explored the small Hazel Bush Hills veins in chert above the limestone, finding calcite and a little galena. Between Eller Beck and Spanham West Hill the Main Limestone has been eroded off on the north side of the fault, but it reappears to form West and East Spanham hills, both prominent features. In the south side of the East Hill, almost exactly 1 mile (1.6 km) E of Eller Beck, there is a second group of workings, including Spanham Hush [009 099] where Little Limestone and associated beds (Mills and Hull, 1976, p. 47) are faulted against Main Limestone. Some lead ore was evidently raised here since there are scattered heaps of tailings, always dominated by calcite but with a little aragonite, galena and limonite. At 600 ft (182 m) N of the hush a level at 1200 ft OD was driven at the top of the Underset Limestone; this may have penetrated Main Limestone on the hangingwall of the vein, which yielded some production at this level. Calcite is again the principal gangue. Probably this level communicated with the shafts in the bottom of the hush. At 1450 ft (441 m) E of these, a short level from the south side tried the vein close to its junction with an ENE-striking branch from the South Spanham Fault, but little seems to have been found. The easternmost test was by means of an adit driven NW for 760 ft (231 m) from Scargill Beck, in shale beneath the sandstone under the Main Limestone; the dump contains chert, coarse sandstone, shale, ironstone nodules and limestone, but only a little calcite veinstuff. The official statistics contain no record of production from the Spanham mines.

South of these workings, the South Spanham fault-system, throwing N about 150 ft (45 m) creates a trough about 1500 ft (0.5 km) wide. The southern faults, crossing poorly-exposed ground, seem not to have attracted the attention of prospectors, but a long level was driven S40°W, starting south of the faults at [0095 0920]. This drive, the Elsey Crag Level of Scargill Mine, intersected at 2060 ft (526 m) from the portal a southward-running system of caverns in the Main Limestone that after 1000 ft (304 m) direct distance SSW gave access to what appears to be the continuation of the North Vein from Faggergill Mine (p. 145). About 600 ft (180 m) of workings on this under Round Hill evidently yielded the tailings, about 100 tons of calcite with a little galena and limonite, seen near the mouth of the level.

The Spanham fault-trough dies out eastward, and the outcrop of the Main Limestone swings away at the summit of Stang (Arkengarthdale) Forest to the east-south-east. At White Fell several slightly mineralised faults have been investigated, particularly by means of a hush in Main Limestone at [040 075], where at least three veins were cut. A little galena was found, with calcite and minor baryte. The main vein was followed south into the cherts, but no mine was developed. About ¾-mile (1.2 km) SW of White Fell Hush a fault trending WNW across the fell north of Arndale Beck was tested where it brings Main Limestone against shale beneath the Ten Fathom Grit to the south. A little galena in calcite was found. As indicated on the Richmond Sheet (41) a NE fault may link this locality with the White Fell veins, but although these are marked as mineral veins, none of them has so far proved to contain anything worth working.

REFERENCES

Burgess, I. C. and Holliday, D. W. 1979. Geology of the country around Brough-under-Stainmore. *Mem. Geol. Surv. G. B.* 131 pp.

— and Mitchell, M. 1976. Viséan Lower Yoredale limestones on the Alston and Askrigg blocks, and the base of the D_2 zone in northern England. *Proc. Yorkshire Geol. Soc.*, Vol. 40, pp. 613–630.

Cornwell, J. D., Patrick, D. J. and Hudson, J. N. 1978. Geophysical investigations along parts of the Dent and Augill faults. *Rep. Miner. Reconn. Program. Inst. Geol. Sci.*, No. 24.

Dakyns, J. R., Tiddeman, R. H., Russell, R., Clough, C. T. and Strahan, A. 1891. The geology of the country around Mallerstang, with parts of Wensleydale, Swaledale and Arkendale. *Mem. Geol. Surv. G. B.* 213 pp.

Dunham, K. C. 1948. Geology of the Northern Pennine Orefield. Volume 1 – Tyne to Stainmore. 1st edition. *Mem. Geol. Surv. G.B.* 357 pp.

— 1952. Fluorspar. 4th edition. *Spec. Rep. Miner. Resour. Mem. Geol. Surv. G.B.*, Vol. IV

Miller, A. A. and Turner, J. S. 1931. The Lower Carboniferous succession along the Dent Fault and the Yoredale beds of the Shap district. *Proc. Geol. Assoc.*, Vol. 42, pp. 1–28.

Mills, D. A. C. and Hull, J. H. 1976. Geology of the country around Barnard Castle. *Mem. Geol. Surv. G. B.* 385 pp.

Reading, H. G. 1957. The stratigraphy and structure of the Cotherstone syncline. *Q. J. Geol. Soc. London*, Vol. 113, pp. 27–56.

Rowell, A. J. and Scanlon, J. E. 1957. The Namurian of the north-west quarter of the Askrigg Block. *Proc. Yorkshire Geol. Soc.*, Vol. 31, pp. 1–38.

Scanlon, J. E. 1955. The Upper Limestone Group and Millstone Grit of the Askrigg Block from Swarth Fell to Rogan's Seat and Summer Lodge Moor. (Unpublished Ph.D thesis, University of Leeds.)

Small, A. T. 1977. Mineralisation of the Stainmore Depression and the northern part of the Askrigg Block. (Unpublished Ph.D thesis, University of Durham.)

— 1978. Zonation of Pb-Zn-Cu-F-Ba mineralisation in part of the North Yorkshire Pennines. *Trans. Inst. Min. Metall.*, Vol. 87, pp. B9–14.

— 1982. New data on tetrahedrite, tennantite, chalcopyrite and pyromorphite from the Cumbria and North Yorkshire Pennines. *Proc. Yorkshire Geol. Soc.*, Vol. 44, pp. 153–158.

Turner, J. S. 1927. The Lower Carboniferous succession in the Westmorland Pennines and the relations of the Pennine and Dent faults. *Proc. Geol. Assoc.*, Vol. 38, pp. 339–374.

CHAPTER 9

Mineral deposits

Details, Area 11 THE NORTH SWALEDALE MINERAL BELT

Crossing through the northern tributaries of the River Swale is the most highly mineralised zone of the whole area covered by this Volume (Figure 25). The oreshoots, chiefly of vein type, were emplaced in a linked system of faults in which the dominant trends are E – W and ENE, though there are also numerous NNW fractures, and some trending NW. The zone is regarded as beginning in the west with the copper prospects in the upper reaches of Great Sleddale[1]. Thence it runs east-north-east for 3 miles (4.8 km) through the Keldside mines to the junction with Whitsundale. Here it turns eastward and runs for 4 miles (6.4 km) to Gunnerside Gill, crossing East Gill and Swinner Gill en route. Although a strong vein-system continues east from Gunnerside Gill, the main strength is carried east-north-east through Friarfold, Surrender and Blackside veins into the Arkengarthdale drainage, where many branches are given off. The main linkage, however, is through Tanner's Rake and Fell End Vein to the Hurst complex, 7 miles (11.3 km) beyond Gunnerside Gill. East of this the fractures become more widely spaced but the zone can be clearly identified as far east as the Feldoms, making a total length from west to east of 18 miles (29 km). To this zone the term North Swaledale Mineral Belt is here applied.

Structurally and for the continuity of mineralisation along it, the belt is unique in the northern Pennines. The Ettersgill – Coldberry – Wiregill – Sharnberry belt of Teesdale is perhaps the nearest analogue, but that is simple by comparison.

Cutting through the belt near the Sleddale Head prospects, the Stockdale Disturbance ('Vein') can be followed on an east-south-east course giving place to a more easterly direction for at least 9.5 miles (15.3 km) to near Healaugh. At Gunnerside Gill it lies 2 miles (3.2 km) S of the main belt. Beyond Healaugh it may link up with the Great Stork Vein of Area 12, but although it has been tested at various points west of Healaugh, it has nowhere been proved to carry workable mineralisation. North of the main belt, a number of branches are given off. Of these the Blakethwaite Vein is the most important. This links with an elaborate fault-system in the Namurian cover rocks lying north and west of Rogan's Seat [918 030]. Most of the fractures have not been tested beneath the intra-Pendleian unconformity, though one was found to be mineralised when tried from the remote Stonesdale Moor Mine in West Stonesdale (p. 140). In the other direction, Blackside Vein, after being shifted by Tanner's Rake [013 032] diverges north-eastwards to link with the strong Moresdale Vein which, however, is only feebly mineralised in spite of (or perhaps, because of) the persistent minor folding that accompanies it. In the north-east part of the belt the oreshoots have become few and scattered before Feldom is reached; there were few productive workings in the chert table-land country north of Marske.

[1]Grid references are given with the detailed descriptions below

Throughout the belt, the stratigraphical range of the oreshoots, with only one important exception, is limited to the beds from the Underset Limestone up to the Crow Chert, a total range of about 350 ft (107 m), less the vertical displacement of the vein-fracture. It is, however, exceptional for the full range to be ore-bearing, the most favoured range being from the base of the Main Limestone to near the top of the Richmond Cherts, averaging about 150 ft (46 m) along the western part of the belt, but east of Arkengarthdale increasing to 175 ft (54 m); Bradley (1862); Dakyns and others (1891); Wells (1955).

At present, no mining activity is taking place within Area 11 other than the treatment of dump material from Friarfold and Old Gang mines at the plant in Hard Level Gill for the recovery of barytes. Fluorspar was recovered in recent years from a tailings heap in Arkengarthdale (p.153). The North Swaledale Belt as far east as Fell End, including the whole of Arkengarthdale, lies within the Yorkshire Dales National Park, while the far eastern part of the area, east of Rake Beck and Throstle Gill, in the headwaters of Marske Beck, comes within the Feldom Field Firing Ranges and is a danger area.

In the detailed account which follows, the principal veins are described from west to east; branches and cross-courses on the south side are discussed before those on the north.

Keldside Veins Copper ore, Lead ore

SD 89 NW, NY 80 SW,SE; Yorkshire 35 SW, SE
Subparallel fractures trending N55–70°E, total displacement about 100 ft (30 m) NW.

In the headwaters of Great Sleddale one of the fractures is followed by the stream where it brings Main Limestone to the south-east against shale and sandstone (Ten Fathom Grit) to the north-west. A shaft on the south-east side of the stream [8265 9888] sunk in 1912–14 tested the vein for copper ore. The base of the limestone on the footwall must be about 30 ft (10 m) below surface, but the exact depth of the shaft is not known. The dump contains vuggy veinstuff with pale mauve and amber fluorite, brown oxidised carbonate (?former ankerite), malachite, azurite and small amounts of chalcopyrite. The prospect is the type locality of the inner F zone of Small (1978). The fault forms the cliff behind the shaft, striking N54°E and dipping 72°NW. Heavy silicification of the limestone accompanies the mineralisation. To the north the stream turns east, following a branch vein, tested by a short adit [8299 9904] which found silicified limestone with specks of sulphide. The main fracture was tested by an adit driven from the Hush Gutter [8325 9935], passing from Main Limestone into chert, probably the Crow Chert. Silicified limestone with tiny sparse specks of sulphide and a little fluorite was found. At or near the place where the Stockdale 'Vein' crosses the Keldside fracture [834 995] several shafts were sunk, but it is doubtful whether the limestone was reached. The first trials for copper at Great Sleddale Head predated the first edition of the six-inch ordnance survey, since 'Old Copper Mine' is shown. There is, however, no record of any production of the metal. Copper is certainly present, but there is no indication that it exists at workable grade.

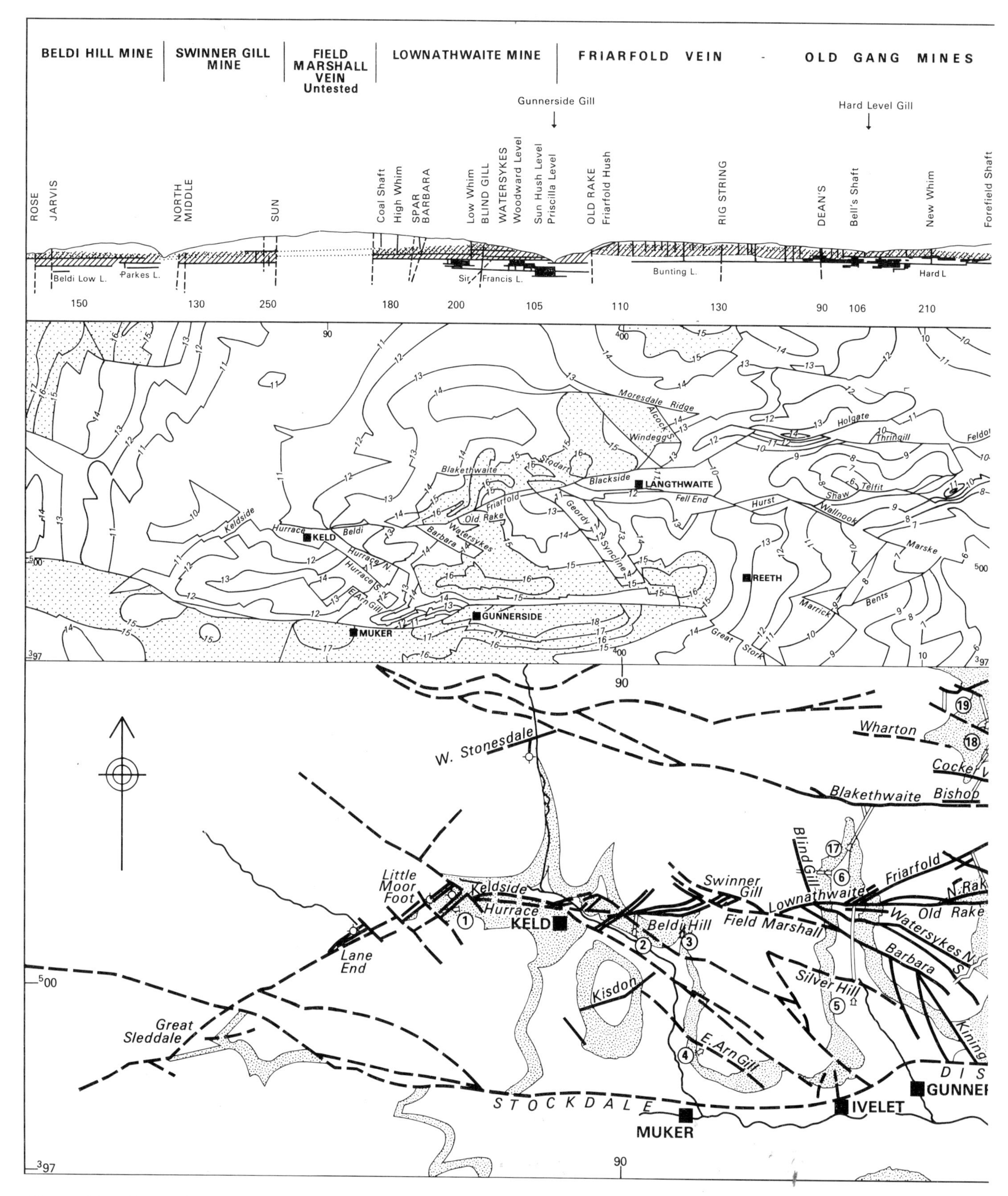

Figure 25 Key and structure map, Area 11, North Swaledale Mineral Belt. Contours on the base of the Main Limestone in feet above Ordnance Datum. Longitudinal section from Beldi Hill along Friarfold and Blackside veins to North Rake, Arkengarthdale

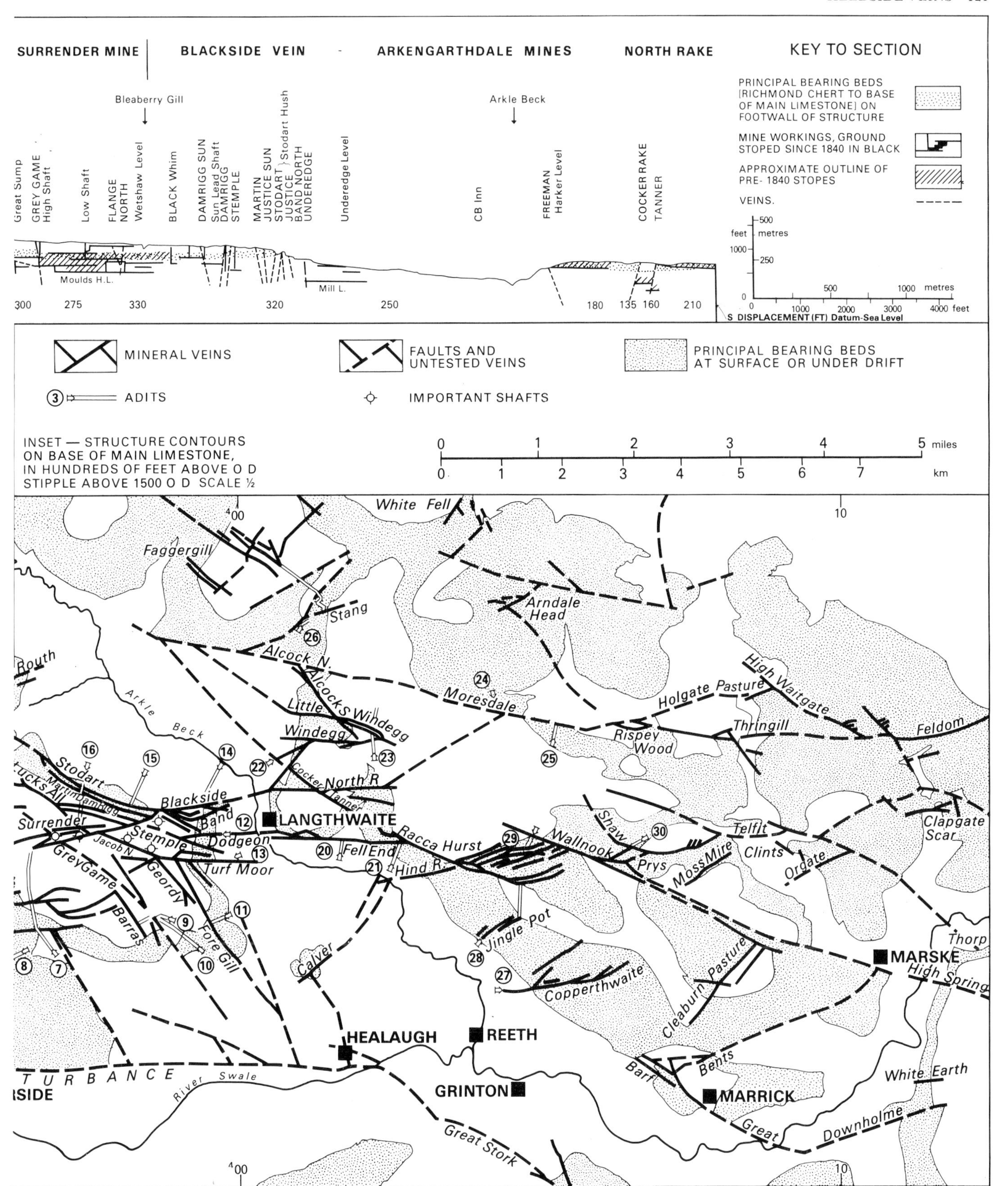

Key to numbered levels **1** Rumble Pool Level; **2** Beldi Hill Low Level; **3** Parkes Level; **4** Adelaide Level; **5** Sir Francis Level; **6** Blind Gill Level; **7** Hard Level, Old Gang; **8** Victoria Level; **9** Barras End High Level; **10** Barras End Low Level; **11** Fore Gill Level; **12** Moulds Horse Level; **13** Moulds New Level; **14** Smelt Mill Level; **15** Dam Rigg Level; **16** Danby Level; **17** Blakethwaite Lead Level; **18** Punchard Level; **19** Routh Level; **20** Storthwaite Hall Level; **21** Haggs Level; **22** Windegg Lead Level; **23** Washy Green Level; **24** Porter Level; **25** Moresdale Ridge Level; **26** Stang Level; **27** Fremington Edge Level; **28** Jingle Pot Level; **29** Queen's Level; **30** Prys Level

To the east-north-east, the main fracture has not been tested along the drift-covered northern slope of the valley, but it re-enters the stream near Hags Gutter. It is now in three parts. The southern section, according to J. R. Dakyns' field notes, throws at least 50 ft SE. The middle section has been tested, apparently unsuccessfully, by a shaft on the south bank of Sleddale, opposite the Birkdale Beck confluence. The northern section was worked from *Lane End Mine*, the shaft of which [8553 0082], collared at 1220 ft (372 m) OD, starts at the hangingwall position of the base of the Ten Fathom Grit. The dump shows that it reached limestone, and assuming that this is the Main, it must be at least 175 ft (53 m) deep; Raistrick and Jennings (1965) say it was '45 or 50 fathoms deep by 1801'. Unfortunately we have been unable to trace any plan or section of these workings. The dump shows galena, associated with coarse baryte, witherite, calcite and small amounts of sphalerite, pyrite and chalcopyrite. Raistrick (1975) mentions that the first recorded working hereabouts was in 1753 when there were mines at High and Low Birkdale; the latter may have been Lane End. In 1825 Mr Jaques, who was also active at Old Gang, leased the Swaledale ground west of Keld, a condition of the lease being that he installed pumping equipment, since the mines in Sleddale were drowned out. This is not surprising in view of the proximity of all of them to the river and the permeable nature of the exposed strata. In 1828 his Little Moor Foot and Keldside Mining Company brought an 80 h.p. steam engine, second-hand from Ashton Green Colliery, to Lane End, where it worked for ten years.

About 1000 ft (305 m) ENE of the shaft, the Lane End block terminates against a cross fault trending N40°W, apparently throwing north-east. Beyond this, one of the Keldside fractures runs near the south bank of the river, bringing Ten Fathom Grit, very strikingly jointed, parallel and perpendicular to the fault, against the top 12 ft (3.7 m) of the Richmond Cherts, overlain by the dark mudstone beneath the Ten Fathom Grit. In this block the only mining trial was a shaft south of Jenny Hill, the remains of which have completely disappeared; Dakyns believed that this tested a fracture situated 460 ft (137 m) SE of the one in the river. Having regard to the overall effect of the Keldside faulting, this one should throw north, but it is no longer exposed. This block terminates eastward at a second cross fault, mapped by Dakyns as parallel to the first, but shown on the A.D. Company plan (Northallerton ZLB 41/1–25) running N–S. This crosses the stream near Firs Farm, throwing south-west. Around 300 ft (91 m) ENE of the farm, two diverging fractures can be seen in the face of a low cliff in Richmond Cherts, throwing north since the overlying mudstone comes into the stream. A little fluorite is present, but the mineralisation is very slight. However, about 200 ft (61 m) to the south-east lies the vein worked from *Little Moor Foot Mine* [8676 0135]. The shaft, near Hoggarth House has now collapsed and though good plans of the workings are extant, there are no sections. The shaft was sunk near the junction of the third of the cross faults with the productive vein, collared at about 1200 ft (366 m) OD, where the fault brings Ten Fathom Grit to the north-east against the underlying mudstone to the south-west. The shaft was drained to about 1125 ft (343 m) by means of a level starting at Rumble Pool [8732 0131] (Figure 25, loc.1) which followed the vein worked at Keldside Mine for 480 ft (146 m), then continued west for 750 ft (228 m) to cut the third cross fault mentioned above, continuing along the rather sinuous course of this to reach the shaft after a further 1100 ft (335 m) of drivage. This level communicated with the top level on the vein, driven only 150 ft (46 m) WSW. The Rumble Pool Level probably starts in a shale within the Richmond Cherts but by the time the vein was reached near Little Moor Foot Shaft, it was probably in shale above the Main Chert. The relative positions of the levels show that the vein dips steeply north-west and at surface it brings Richmond Cherts against the overlying mudstone. The plans show two main levels below the Top Level. The bottom level was driven 600 ft (183 m) ESE and it may be inferred that it terminated where the base of the Main Limestone, rising gently westward, came in on the footwall side. If this inference is correct, the bottom of the shaft must have been at about 1050 ft (320 m) OD. The main working level lay between the Top and Bottom levels, and was much the most extensive, running 1750 ft (533 m) WSW and 300 ft (92 m) ENE. It may be supposed that this was entirely in Main Limestone except perhaps after it passed through the second cross fault (if it did so). From the workings ENE of the shaft, a short crosscut NW discovered a parallel vein or string, hading towards the main vein and therefore presumably throwing in the opposite sense. An additional level was developed above the main level from rises in this part of the mine. A water wheel may have been used to drain these extensive workings, but no evidence of it remains on surface, nor is it known whether it was placed underground so as to discharge into Rumble Pool Level. No production data are available, but with Lane End and Keldside mines, this property was able to keep a small smelter going. Probably production ceased with the Little Moor Foot and Keldside Company and its immediate successors; an elaborate scheme to drain the mines when the A.D. Company held the lease in 1865 never came to fruition.

The fourth and final block on the Keldside veins, lying between the third cross fault and the West Stonesdale Fault, was entered by the easternmost workings from Little Moor Foot, but not for more than 300 ft (91 m). The most important fracture in this block is the vein worked at *Keldside Mine* [8750 0145], to which reference has already been made. The productive workings lay east-north-east of Rumble Pool Level. Old Engine Shaft, 750 ft (229 m) from the level portal, was 65 ft (20 m) deep[1] and there was a second shaft known as Little Engine, 40 ft (12.2 m) deep, the total length of the worked ground being about 600 ft (183 m). The presence of coarse sandstone on the dumps suggests that the former shaft was sunk through the Main Limestone, at least on the footwall side of the vein. Minerals on the dumps include witherite, sphalerite and calcite as well as galena. From Rumble Pool Level a winze was sunk 42 ft (13 m) in the same vein, probably into the Main Limestone, 1780 ft (542 m) WSW of Engine Shaft; there was also a branch from the level driven on the same fracture to reach a shaft some 600 ft (183 m) farther SW, still open on the west bank of Ash Gill. In the gill the fracture is exposed in mudstone underlying the Ten Fathom Grit, dipping 50–60°S, and exposing the top of the Richmond Cherts on the north side. A further test of this fracture was made from a shaft giving access to a south crosscut in Little Ash Gill. None of these prospects appear to have been successful. To the north-east, all the Keldside veins and fractures are considered to terminate against the Whitsundale Fault, which could be regarded as the fourth cross fault of the series. This throws about 100 ft (30 m) SW. One of its branches was tested from shafts linked by a crosscut south-east of Keldside Mine, but no veinstuff is present on the heaps.

Hurrace 'Veins'

NY 80 SE; Yorkshire 36 SW
Direction E–W, to N65°W

Springing from the footwall of the Whitsundale Fault near Keldside Mine, the northern fracture of this group pursues a sinuous eastward course along the south side of the Swale, crossing the river at Morrill Hole; at Berry Hole Scar [8968 0141] where it crosses East Gill, it changes direction abruptly to east-south-east. Two short levels near Keldside Mine tested the fracture without success; there were probably prospects in it near Keldside Smelt Mill, and a level driven beneath the Underset Limestone opposite Morrill Hole

[1] ZLB 41/1–25. ZLB numbers refer to the Radcliff – Schuckburgh – Hesketh papers and plans in the North Yorkshire County Record Office, Northallerton.

also found nothing workable. From Berry Hole it runs across the face of Beldi Hill, cutting through the strongly mineralised E – W and ENE veins there, and repeating the outcrop of the Main Limestone. The fault is exposed near the Beldi Hill smelt mill in Swinner Gill [9095 0063] where, as Clough noted (Dakyns and others, 1891, p. 176) it is a vertical-sided breccia of limestone fragments about 3 ft (1 m) wide, cemented with calcite. The displacement on the North Hurrace fault increases eastward, until on Beldi Hill it may reach as much as 150 ft (46 m) to the south-west. South-east of Swinner Gill the fault can be followed for a further mile, repeating the outcrop of the Lower Howgate Grit, but the throw diminishes towards Gunnerside. West of Keld, a second persistent fracture lies about 150 ft (46 m) S of the first. Near the village of Keld, this also turns away to the east-south-east, diverging from the northern fault until on Ivelet Moor they are ½ mile (0.8 km) apart. The southern fracture downthrows north, and the strata on the south side dip north more rapidly than the regional dip. From Berry Hole a fracture with a few feet of south-west displacement runs across to Cat Rake, linking the two faults. As Raistrick (1975, p. 46) reports, the A.D. Company started in 1864 an ambitious scheme to dewater the Keldside mines by means of a tunnel, the *Sir George Level* starting from the south side of Catrake Force [8925 0126] at approximately 1000 ft (305 m) OD. The level followed the linking fracture for 672 ft (204 m) WSW until the main southern fault was reached but it was then abandoned, presumably because no payable ground was found. Had it proved possible to continue the drive to Little Moor Foot Shaft, at least 1½ miles (2.4 km) distant by the most direct route, it could only have been about 100 ft (30 m) below the Rumble Pool adit. Although tested at many points, the Hurrace fractures must be taken to represent a barren stretch in the North Swaledale Belt.

Rose North Vein Lead ore (Zinc ore)
Rose Vein
Jarvis Vein

NY 90 SW; Yorkshire 36 NW
Approximately parallel veins trending N60°E, small SE displacements

Beldi Hill – Swinner Gill Lead ore

North, South and Middle veins

NY 90 SW; Yorkshire 36 SW
General trend, E – W, becomes ENE east of Swinner Gill. Overall displacement at least 200 ft (60 m) S; Bradley (1862) states 240 ft

The six veins were worked at the extensive *Beldi Hill Mines*, from the Main Limestone scar adjacent to the crossing of the North veins by the northern Hurrace Fault [9025 0110] eastward across the hill to Swinner Gill. Lead ore probably came to outcrop at the scar, where it would be discovered in ancient times. The earliest workings were opencuts and hushes on the three ENE veins and the Beldi Hill North Vein. Below the scar, the Oldfield Hush Gutter follows more or less the course of Hurrace Vein towards the now-ruined Crackpot Hall Farm. Raistrick (1975) states that it was already being worked in 1738, and that some reworking took place as recently as 1846. The ground lends itself, however, to access by levels, and most of the ore was won underground, all the levels entering from the hangingwall side of the structure. Two such levels west of the Hurrace Fault intersection were unsuccessful: White Wallace's Level at about 1060 ft (323 m) OD which tested the Rose – Jarvis fracture and the main E – W structure in the Main Limestone of the down-dropped block south-west of Hurrace Vein, and New Level, driven on top of the Main Limestone, driven 725 ft (220 m) NNE through the structure to cut the Hurrace Vein, probably against a footwall of sandstone under the Underset Limestone. These two trials, coupled with the extensive exposures on the west side of Beldi Hill make it clear that mineralisation does not persist west of the Hurrace northern fault, nor is the fault mineralised except with calcite. Moving eastward, the first productive levels are the Rose Vein Level, driven from the hush, beneath the Ten Fathom Grit, and the Crow Beds or Watty Will Level, above this sandstone. Both gave access to the ENE vein workings, but probably of greater importance was Tommy Milner's Level [9035 0110] which, starting in shale under the Ten Fathom Grit would give access to the Main Limestone on the footwall of Beldi Hill North Vein and in the Rose and Jarvis veins, at a little above 1400 ft (427 m) OD. Crackpot Hall Level [9065 0097] also at 1400 ft (427 m) OD was, however, the most important of the working adits, giving access to extensive ground in the Richmond Cherts on the Middle Vein, and in the Main Limestone on North Vein. A crosscut led to Jarvis Vein at a point 1400 ft (427 m) ENE of Rose Vein level mouth, and from Jarvis Vein the Plate Crosscut was driven 1050 ft (320 m) N in shale below the Main Limestone, probably in an attempt to reach the Field Marshall veins (p. 124) proved at Swinner Gill Mine. The interpretation of the main E – W belt in terms of three fractures is probably an over-simplification; the old plans (ZLB 41/1–3, DM 2040[1] and Raistrick (1975, map 1) show a complex pattern of workings. However, those on North Vein are plainly identifiable not only in the western workings, but for 750 ft (229 m) adjacent to Swinner Gill, where a level at 1350 ft (411 m) OD gave access to them. The Middle Vein may be conceived as a branch starting from North Vein near Tommy Milner's Level, and lying about 200 ft (60 m) S of it in Swinner Gill; while the Beldi Hill Sun Vein forms two great loops on the south side of the other two. On Sun Vein the bearing beds, at Beldi Hill very clearly ranging from the base of the Main Limestone up to the Crow Chert, are at their lowest level, and Landy Level [9012 0096] at 1136 ft (346 m) OD was the principal means of access, entering from the hangingwall side on the top of the Main Limestone. Two long levels were driven to drain the mines below the Main Limestone, and to test lower horizons: Parkes Level [9107 0089] (Figure 25, loc. 3), from Swinner Gill at 1127 ft (344 m) OD started in the shaly beds above the Underset Chert, 60 ft (20 m) below the Main Limestone, cutting North Vein at 1425 ft (434 m) with the Twentyseven Fathom Grit (Sandstone below Underset Limestone) on the footwall; and Low Level, 2500 ft (762 m) (Figure 25, loc. 2) to the west, at a similar stratigraphical level, reaching the structure at 1300 ft (396 m). It appears as if little if any ore extended down to these levels since very little driving was done (Carruthers and Strahan, 1923). Near Low Level some crushing was done, perhaps of ore sent down from higher workings, and witherite and sphalerite can be seen in the cuttings. According to Bradley (1862), the Rose and Jarvis veins (he lists them as Raws and Garvis) carried no rider, that is, there were no gangue minerals; and this was also true of Middle and Sun veins. However, North Vein was ridered, and may be pictured as a wide vein with calcite, pink baryte, witherite, galena, pyrite and sphalerite, the last-mentioned being a pale brown variety. Hemimorphite and hydrozincite were both present, and generally zinc minerals are more noticeable here than at many of the North Swaledale mines. Raistrick and Jennings (1965) and Raistrick (1975) indicate that the Beldi Hill mines were leased by the Parkes brothers and Leonard Hartley in 1742, and that Parkes Level was driven in 1746–49. It must have been one of the earliest deep trials in the district, and it must be assumed that at least some of the productive adits at higher levels already existed. Crackpot Hall Level was not, however, started until 1773, though the Crow Beds Level to the north of it probably already existed. Raistrick has given a graphic account of

[1]DM numbers refer to the collection of plans of abandoned mines at Baynards House, W2, administered by the Health and Safety Executive.

the dispute, during the years 1767 and 1772 between the Parkes Company and Lord Pomfret over the tenure of the mining area, and the violent actions between the workmen of the two interested parties. The settlement, in favour of Parkes, may have created the conditions for an ambitious development such as the Crackpot Hall drivage, and possibly Low Level also. Hunt's statistics record only 6 tons of lead from Beldi Hill in 1882–83, but some production may have been included with that of the A.D. Company, who held the lease after 1873, and who reopened Parkes Level. What is clear is that the main production from these old mines was before and during the 18th century, when they first provided ore for the Spout Gill smelt mill, and after 1771 kept their own mill going (see also Jennings, 1959; Raistrick, 1926, 1955).

East of Swinner Gill the North, Middle and Sun veins are still recognisable, but they begin to turn NE or ENE after entering the area of the *Swinner Gill Mines*. The Main Level [9199 0120] at 1271 ft (387 m) OD driven on top of the Main Chert followed the Sun Vein for some 2000 ft (603 m) to the ENE before encountering the southern element of a powerful fracture belt known as Field Marshall Vein. The workings were carried ESE along this until, after almost 1000 ft (305 m), what was identified as Sun Vein turned off to the NE and was worked for 500 ft (152 m) further. The Middle and North veins were reached from crosscuts from Main Level, and from two higher levels, Smiddy and North. These veins likewise cut off against the WNW continuation of the Field Marshall Vein, and west of the shifted position of the Sun Vein, two more NE veins were found, though not in the right spacing to suggest that they must be the offset continuations of the two veins. The threefold arrangement of fractures is nevertheless a curious feature of the western part of the North Swaledale Belt. The fractures were cut off in turn by what may be regarded as the northern element of the Field Marshall zone, not here mineralised. This zone is well exposed in the headwaters of Swinner Gill, where the southern and Middle (or Dore Vein) sections downthrow NE but the northern fracture throws SW, the beds here affected being the Ten Fathom Grit and Crow Limestone. The net effect on the zone is a small SW downthrow, but the interesting question is the extent to which transcurrent movement has taken place. Swinner Gill mines were held and worked by Lord Pomfret and his heirs from the 18th to the later 19th century; separate production figures are available only for the closing stages from 1873 to 1888, when the output was 4253 tons of lead concentrates.

Kisdon Vein — Lead ore

SD 89 NE, 99 NW, NY 90 SW; Yorkshire 36 SW
Mean direction N60°E, small NW downthrow

South Kisdon Vein

SD 99 NW; Yorkshire 36 SW
Direction N86°E

East Arn Gill Vein — Lead ore

SD 99 NW; Yorkshire 36 SW
Direction N45°W, small NE downthrow

Crossing the summit of Kisdon, the striking hill isolated by the late-Glacial diversion of the River Swale from Skeb Skeugh to its present course below Beldi Hill, there is a weak vein which has been worked from a level on the east side of the hill [9049 0013] driven WSW for 1080 ft (329 m). There was a surface shaft at 350 ft (107 m). The vein must pass through or near the large Harker Pot Hole, which lets down the Ten Fathom Grit capping of the hill. On the west side of the hill Alderson's Shaft [8958 9967] tested the vein in the cherts and Main Limestone, and there was a level 450 ft long (ZLB 41/32) but with little success, and no veinstuff can be found where the fracture crosses the limestone outcrop farther west. Almost at right angles to Kisdon Vein, the East Arn Gill vein impinged on it 270 ft (82 m) WSW of the portal of the eastern adit. A row of shafts and an adit tested it as it crosses the Main Limestone, but only coarse calcite remains on the heaps. On the other side of the Swale valley, however, this vein was well mineralised at the *East Arn Gill Mine* [915 994]. Here a level driven SE and several shafts tested the vein in the Underset Limestone, where it was evidently productive. In 1865 the A.D. Company began to drive Adelaide Level (Figure 25, loc. 4) at 1118 ft (340 m) OD in sandstone beneath the Underset Limestone. After crossing the vein 290 ft (88 m) from the portal, it entered the limestone and encountered a belt of replacement deposits related to ramifying NW, E–W and ENE fractures, and followed them for 500 ft (152 m) to the NE. South of these workings, High Level at 1216 ft (370 m) OD was driven SE along the vein opening up 525 ft (160 m) of productive ground (ZLB 41/36). The dumps from Adelaide Level contain amber fluorite, galena and pyrite; those from High Level only calcite and limonitised limestone. No separate returns of concentrates produced during the mid 19th century are available, but Raistrick (1975) records that one flat yielded ore worth £12 000. The metal price in 1865–70 averaged £18 per ton (Dunham, 1944) so this represents about 600 tons of concentrates if that was the period of working. An attempt was made to work a small flat from 1918 to 1920, when 30 tons were obtained, but drainage problems prevented further work; there may thus be ore left, down dip, below Adelaide Level. In the next side valley north of East Arn Gill, called West Arn Gill, trials were made on the southern fracture of the Hurrace group in 1811 and again in 1864, but no ore was found. The small South Kisdon Vein, which cuts the Main Limestone 1400 ft (427 m) S of Kisdon Level was also tested in 1867 and yielded some galena; fluorite is present in the tailings.

Lownathwaite North Vein — Lead ore

NY 90 SW; Yorkshire 36 SE

Old Gang Old Rake = Alderson's Vein — Lead ore Baryte

NY 90 SW, SE; Yorkshire 36 SE

Moor House Vein

NY 90 SW, SE; Yorkshire 36 SE
Interpreted as parts of a single fracture system with general trend E–W, but in short stretches swinging to ENE or NE. Displacement S, very small at east end; averaging 36 ft (11 m) according to Bradley (1862); increasing to 70 ft (21 m) on east side Gunnerside Gill, and reaching over 175 ft (53 m) W of junction of Barbara and Watersykes veins.

Lownathwaite Middle & Sun Veins; Bunton Vein, Moss Vein, Merryfield Vein — Lead ore

NY 90 SW; Yorkshire 36 SE
Branches and loops of the North Vein – Old Rake system

Watersykes Vein — Lead ore

NY 90 SW, SE, SD 99 NE; Yorkshire 36 SE
Direction N60°–70°W; throw 125 ft (38 m) SW at Gunnerside Gill, decreasing SE

Blind Gill Vein — Lead ore

NY 90 SW; Yorkshire 36 SE
Direction N20°W, throw 12 ft (3.6 m) W

Barbara Vein Lead ore

NY 90 SW, SE, SD 99 NW; Yorkshire 36 SE
Direction N65-75°W, throw 60 ft (20 m) NE decreasing to zero west of Gunnerside Gill; continuing in three branches to west-north-west: **Spar Vein,** N70°W, **Great Break**, N75°W, in addition to the linear continuation of Barbara Vein. Estimated aggregate downthrow, 80 ft (29 m) SW. **Kining Vein**, direction N30°W, throw not known.

North Rake Lead ore

NY 90 SE; Yorkshire 36 SE
Direction N57°E, throw 18 ft (5.5 m) at Hard Level, dying out to east.

Reformer Vein

NY 90 SE; Yorkshire 36 SE
Direction E – W, throw minimal

With these veins we return to the main axis of the North Swaledale Belt. Between the eastern foreheads of Swinner Gill Mine, and the westernmost workings of the *Lownathwaite* (or Lownithwaite, ZLB 41/17, or Lownethwaite, ZLB 41/22) *Mine*, a stretch 2200 ft (670 m) long remains untested. This seems to be the ground referred to by Raistrick (1975, p. 41) as having been leased by Lord Pomfret to the London Lead Company for 31 years in 1742, terminating eastward at the forefield of Lownathwaite Level, but a connection was never established between Swinner Gill and Lownathwaite mines. In the westernmost workings from the Priscilla Level of Lownathwaite mine, a fault called by the miners the Great Break came in from the south-east side, cutting off the ore. Clough (*in* Dakyns and others, 1891, p. 176) states that it was of considerable width, and full of black 'dowk', very unpromising for ore, and that its effect was to bring sandstone below the Underset Limestone on the footwall into contact with the Main Limestone on the hangingwall. Clough notes that the fault may be the same as the one which brings in the pebbly (i.e. Lower Howgate Edge) grit at the surface against the Crow Chert but adds that 'the base of the Millstone Grit is readily traceable between Lownathwaite hushes and the east and west gills to the south of them, and no great fault could cross it without observation'. After a re-examination of the evidence, we suggest that the impression that the large displacement was brought in by a fault from the south-east was illusory, and that the figure of 176 ft (55 m) SW for the displacement is in fact the total displacement of the Lownathwaite structure. This is borne out by a coal shaft (ZLB 41/71) sunk south of the structure to test the coal beneath the Lower Howgate Edge Grit, from which this total displacement can be calculated. No doubt a fault did impinge upon the Lownathwaite main vein as stated, but the displacement need be no more than 20 ft (6 m) SW in a vertical sense. This fault we believe to be the continuation of the Field Marshall zone from Swinner Gill, and the possibility of transcurrent movement on this has already been referred to (p. 124). The direct line can be followed back to join up with, or form one branch of Barbara Vein which, however, must change from a NE displacement east of Gunnerside Gill to a SW displacement on the watershed west of the gill. Another WNW fault was cut underground from the Dolly Level at Lownathwaite. Called the Spar Vein by the miners, this also is best regarded as a branch of Barbara Vein, the direct continuation of which can also be traced from the first appearance of Lower Howgate Edge Grit south of the Lownathwaite veins proper. Altogether, the three branches of Barbara Vein add 80 ft (24.3 m) of throw to the Lownathwaite structure. The evidence is nevertheless not consistent with the faulting of Ten Fathom Grit against Lower Howgate Edge Grit south of the Lownathwaite veins and the appearance to which Clough alludes is best explained by a sideways shift to the Lownathwaite structure by the powerful Watersykes fault-vein, which itself adds over 100 ft (30.5 m) of displacement to the whole structure. These relationships are illustrated in Figure 26.

From the western extremity where the Field Marshall – Great Break movement plane terminated the oreshoots, there is evidence of more or less continuous workable mineralisation for over 3 miles (4.8 km) eastward, to beyond Hard Level Gill, including some of the most productive ground in the region. The Lownathwaite property extended as far east as Gunnerside Gill and beyond that the veins passed into the Old Gang royalty. The threefold hush at Lownathwaite (Plate 5) is probably at least as old as the 16th century, for when Philip Swale and Robert Barker leased the mines about 1676 'a good deal of the ground which was entered had been worked by the old man' (Raistrick and Jennings, 1965, p. 155), and it may reasonably be assumed that as much ore as possible would have been got by means of hydraulic opencasting before going underground. Unfortunately no survey was ever made of the ancient workings, and there is no section of the ground stoped for several centuries before the deeper developments in the late 18th and 19th centuries, nor is any record of the wallrocks in these workings available. The same is true for the whole length of Old Rake in the Old Gang property, and for part of Friarfold Rake. The working was by small partnerships of miners, from shafts rather than from levels. The spoil from the shafts, rich in baryte, carrying some witherite, usually with a little galena left in it indicates the nature of the ground, and fragments of the wallrocks, chiefly limestone and chert rather than sandstone, leave little doubt as to the horizons that were most favourable. When the driving of long levels began in the 18th century, a very complex pattern developed at Lownathwaite. The account which follows is, quite frankly, an interpretation of this pattern, allied with the surface exposures in the hushes, gills and beck, and with the 19th century records of underground geology (ZLB 41/17–24, 37, 39, 48, 64, 68, 71, 72, 79). None of the workings is safely accessible.

On the north wall of North Hush, the base of the Main Limestone is seen at 1600 ft (488 m) OD and the base of the Ten Fathom Grit at about 1750 ft (572 m) OD on the footwall of Lownathwaite North Vein (Figure 26). Woodward Level [9363 0132] is the highest recorded underground working, reaching the vein after crosscutting 760 ft (232 m) at N84°E and following it 510 ft (155 m) WSW to, or perhaps beyond, its intersection with Watersykes Vein. Priscilla Level [9382 0141] at 1343 ft (408 m) OD driven from the gill through the footwall, was on this vein for 950 ft (290 m) before encountering Watersykes Vein; it then turned WNW along the latter vein for 375 ft (114 m) before reaching the westward continuation of the Lownathwaite vein-system. The relative positions of these levels in North Vein indicate a dip of the vein of 50°S, but comparison with its position in Sir Francis Level, here at 1120 ft (341 m) OD shows that the dip increases with depth to about 60°. Watersykes Vein, from a comparison of its position in Priscilla and Sir Francis levels, must also be dipping at 55–60° SW, and the evidence suggests that it effects a shift of the North Vein, west side north. West of this intersection, the North Vein may continue on its old course to the WSW, but the main interest is in a great loop on the hangingwall side, reminiscent of the Sun Vein loops at Beldi Hill. The fracture or fractures concerned are evidently steep, for the Dolly Level workings 80 ft (24.4 m) above Priscilla Level are displaced only a short distance to the north, and farther west cross and re-cross those at Priscilla Level, possibly after the Great Break mentioned above has impinged on the structure. Dolly Level [9378 0087] was driven from the south side of the structure, starting beneath the Main Limestone at 1421 ft (433 m) OD, running 1550 ft (472 m) NW to Sun Lead Shaft and thence NNW to reach the North Vein west of its shift by Watersykes Vein. In the later stages of the mine it was the main access to the upper workings west of Watersykes Vein. At Dolly Level these continued to within

225 ft (69 m) of the western forehead of Priscilla Level, mentioned above and it is believed that the Great Break had been penetrated by them. At 840 ft (256 m) from the portal of Dolly Level, the link between Barbara Vein and Great Break crossed the level; its position is given on the existing plan, but no details of the beds affected. From here two branches are considered to be given off towards the NW: the Great Spar Vein found in a crosscut 180 ft (54 m) WSW of Sun Lead Shaft; and a more northerly branch which serves to bring in the Lower Howgate Edge Grit at the surface, as described by Clough (*in* Dakyns and others, 1891, p. 176). All three western branches from Barbara Vein downthrow SW. The sections on Figure 18 give a reconstruction of the footwall and hangingwall geology of the Lownathwaite structure. It is regrettable that no record remains of the stoped ground prior to the middle of the 19th century; the workings shown on the section belong to the closing stages of the mine, under the A.D. Company. This Company's development of Sir Francis Level [9399 0001] (Figure 25, loc. 5) starting at 1083 ft (330 m) OD, 4350 ft (1326 m) S of the place where North Vein crosses Gunnerside Gill enabled working to be continued below the gill, a shaft (Woodward's) sunk from the west bank of the stream north of Priscilla adit having failed to achieve this object. The driving of the level, initiated by Sir George Denys has been described by Crabtree and Foster (1963) and Raistrick (1975). It starts in the sandstone beneath the Five Yard Limestone, cuts Barbara Vein at 2650 ft (808 m), passes through Watersykes Vein at 3400 ft (1036 m) dipping 50°S, and reaches Lownathwaite – Old Rake at a point 200 ft (61 m) E of the bottom of the gill. From the level head a drift was driven 1725 ft (525 m) due W and 800 ft (244 m) SW to the intersection with Watersykes Vein, then WNW along this vein, giving access to stoped ground on the Middle or Sun veins, and also to Blind Gill Vein, described below. When it reached the Lownathwaite veins, Sir Francis Level was nearly 500 ft (152 m) below the footwall position of the Main Limestone, and 100 ft (30.5 m) below the base of the Fourth Set (= Five Yard) Limestone. On the south side of the vein, however, thanks to the 125 ft (38 m) upthrow on Watersykes Vein, which cancels out the effect of the gentle northward dip of the beds, it was only 28 ft (8.5 m) below this limestone, indicating a S displacement by Lownathwaite Old Gang Vein here of 72 ft (21 m). The vein proved to be well mineralised with galena, witherite, baryte and sphalerite, the ore passing into the level sole over a distance of 450 ft (137 m) W of the gill, and beyond that rising into higher strata as shown on Figure 18. An internal shaft was sunk 120 ft (37 m) below Sir Francis Level and equipped with a hydraulic winding and pumping engine which is said still to be in the mine (Lodge, 1966); the water main leading to it can still be seen at surface, emerging from a small shaft. A level at 80 ft (24 m) below Sir Francis continued to yield ore from sandstones above the Middle Limestone from 1880 until 1882, when these workings were flooded. Ore appears to have been left in the bottom over a length of at least 400 ft (122 m), but no record as to its quality exists. The low metal price obtaining in the 1880's discouraged any further investment to recover this working. Lownathwaite is unique among the mines of the North Swaledale Belt, and indeed among all the mines of Region II for the great vertical extent of the mineralisation. From the footwall position of the Ten Fathom Grit at 1792 ft (546 m) down to the Sir Francis sublevel at 1038 ft (316 m) OD lead ore was extracted. This range of over 750 ft (229 m) is more than twice that found anywhere else in the belt, save at Surrender. It must be explained in terms of the complex structure, which may well have made Lownathwaite one of the foci from which mineralising fluids were distributed (Dunham, 1952)

On the footwall side of North Vein a steep vein trending NNW was discovered late in the history of the mine. At Sir Francis Level it leaves North Vein close to the junction with Watersykes, and for a time was regarded as a continuation of that vein, until its very different trend became apparent. It was worked from both Priscilla and Sir Francis levels, and also from a level at Blind Gill [9352 0183] (Figure 25, loc. 6) which gives its name to the vein. The extent of the worked ground and its stratigraphical setting are shown in Figure 18. It had been hoped that this vein would prove sufficiently productive at Sir Francis Level to enable that level to be driven up to the western forehead of Blakethwaite Mine (pp. 139–140) where it might have provided drainage in depth for a further 150 ft (46 m), but it proved impracticable to extend it more than 2100 ft (540 m) beyond Lownathwaite.

The possibility of further ore beneath Lownathwaite was tested by a single borehole (Gunnerside G.1.) drilled by COMINCO from the south side of the structure at [9385 0126] in 1968 (Plate 5). Veinstone carrying fluorite with a little sphalerite and traces of chalcopyrite was passed through from 225.22 to 229.46 m. If, as seems probable, the limestone below the vein-intersection is the Simonstone, the vein cannot be North Vein since the displacement would be insufficient, and it must be assumed to have been one of the branches in the hangingwall that pass through Sun Hush. If, on the other hand, the limestone is the lower part of the Hardrow Scar, then it could be North Vein. The test by means of a single borehole is, in any case, inconclusive. Sun Hush Level [9384 0131] at 1397 ft (425 m) OD, driven about the middle of last century, gave access to branches and strings on the hangingwall of North Vein beneath the hush, but no details of stopes are available. One such branch must cross Gunnerside Gill to link up with the weak Bunton and Moss veins there. A branch may also link with Watersykes Vein before reaching the gill. No separate figures of production of lead ore from Lownathwaite and Blind Gill are available; the extensive surface and underground workings suggest, however, that it may have been considerable. The three branches considered here to represent Barbara Vein on the west side of Gunnerside Gill contributed nothing to this production. Clough (*in* Dakyns and others, 1891), from eyewitness evidence, states that the Great Break 'nearly 500 ft below the surface at a point 340 yds WSW of Lownathwaite House . . . was of considerable width and full of black 'dowk', looking very unpromising for ore' (p. 172). The same may well have been true of the other branches, though the Spar Vein presumably had some baryte, witherite or calcite in it. The small Silver Hill Vein, parallelling Barbara about ½-mile (0.8 km) to the south probably yielded some production to the Lownathwaite total. It was reached by Silver Hill Level [9366 0116] driven directly into the vein against the upthrow position of the base of the Main Limestone, and also from Harriet Level, a crosscut at the top of the Richmond Cherts starting 1300 ft (396 m) to the north-west. It may be assumed that the worked ground lay between these two levels, which cut the vein 1040 ft (317 m) apart. Some baryte is present on the dumps.

Attention may now be turned to the east side of Gunnerside Gill, to the *Old Gang* property, a brief history of which has already been given. The ground occupied was from Gunnerside Gill through Hard Level Gill to a boundary with the Surrender Royalty through Great and Little Pinseat, and across Surrender Moss. The principal oreshoots worked lay on the powerful Friarfold Vein, which branches ENE off the Lownathwaite – Old Rake Vein a short distance east of Gunnerside Gill, and which is described, with the veins associated with it, in a later section (pp. 129–131); and also on Old Rake itself. A great belt of hushes, opencast workings and extensive heaps of debris from shaft mining marks the course of the outcrop of Old Rake, its branches Merryfield Vein, Moor House Vein and the ENE North Rake across the felltop between the two gills, a distance of 1½ miles (2.4 km). The surface strata range from Three Yard Limestone in Gunnerside Gill to the lower part of the Ten Fathom Grit on the high ground, while on the south side, Crow Beds are also present. Much chert is present in the large cuttings heaps, indicating that the Main and Richmond cherts were extensively mineralised, and it is probable that much of the ore had been got by independent miners wo-king from small shafts before 1670, and by the operation of deeper shafts equipped with horse whims for much of the 18th century. Drainage problems now made it essential to drive levels from the gills. From the Gunnerside Gill working, Bunting Level (Bunton Level on the 1:10 560 map), [9403 0121] at 1372 ft (418 m) followed the weak Bunton strings ENE for 700 ft (213 m) then turned N and cut through Old Rake in the next 200 ft (61 m) continuing N to give access to Friarfold Rake (p. 129). A branch was driven east in Old Rake for 3850 ft (1173 m) until in the year 1828 it overlapped with, and was linked by hoppers to Hard Level, 30 ft (9 m) below, which had been driven westward from Hard Level Gill. It may be estimated approximately that the base of the Main Limestone on the footwall dropped in this stretch from 170 ft (52 m) above Bunting Level to 135 ft (41 m) above at the east end, with the southward displacement by Old Rake slowly decreasing from its Gunnerside Gill figure of 42 ft (12.5 m) obtained by direct measurement in Old Rake Hush, east of the intersection of Friarfold Vein. Bunting Level thus had command of the beds from beneath the Underset Limestone on both sides of the vein upwards. It is most unfortunate that the poor records kept by the lessees of Old Gang at this period contain no section of the stoping; a few rises above the level are shown on the plan, but it is impossible to gain any satisfactory picture of how much ground had been left intact beneath the 'old man' workings of earlier centuries. The volume of cuttings, however, from Bunting Level is clearly much less than those left across the fell top from the old shaft workings. Nevertheless, ore must have been continuing in the sole of this level, for in the early 1850s another level, Sir George Level [9399 0113], was started from the lowest possible site in the valley near the vein, at 1291 ft (393 m). A cross-measure drive of 700 ft (213 m) was required to reach the vein, and the drive continued east in Old Rake for 3350 ft (1021 m). With this development the whole of the arenaceous series between Underset and Three Yard limestones became accessible for exploitation, and it is impossible to believe that Sir George Level would have been driven so far to the east had these beds been barren, or that the Old Gang Company would have been willing to put up four-fifths of the cost of Sir Francis Level (see pp. 4–5) when Sir George Denys proposed his scheme. Sir Francis Level not only gave access to the deep workings of the A.D. Company, but was driven 2100 ft (640 m) eastward in Old Rake from the level head. As with the more ancient workings, the Old Gang Company preserved no record of the stoping adjacent to the beds between the Three Yard and Five Yard limestones, but it is difficult to believe that the drivage would have been continued so far in barren ground. The plans show no evidence of further sinking though the workings from the A.D. shaft were left in ore at the Old Gang boundary, beneath Sir Francis Level. The A.D. records persistently refer to the vein as Friarfold Vein, and Raistrick also mentions this vein as the objective of the deep level, but the Old Gang plans make it quite clear that the drive to the east was on Old Rake; no drift on the diverging Friarfold Vein is shown.

Hard Level [9712 0068] (Figure 25, loc. 7) at 1310 ft (399 m) OD, at first known as Foss Level, was started in the year 1785 under Lord Pomfret, in the sandstones and shales beneath the Underset Limestone. It runs NW beneath the gill for 2150 ft (655 m) then turns N and reaches the Old Rake near Dolphin's Shaft [9660 0146], 1200 ft (366 m) ahead. A geological section of this level in the Old Gang records (ZLB 41) probably, since it is dated 1814–18, by Frederick Hall, shows that the northward dip of the beds carries the base of the Underset Limestone into the level sole at 1110 ft (338.3 m) from the portal, while at 2022 ft (616.3 m) the base of the Main Limestone, and at 2802 ft (854 m), the top of the Main Limestone go into the sole. There is a shallow syncline before the Old Rake hangingwall is reached, so that at that point the top of the Main Limestone is 40 ft (12 m) above the level sole. No record survives of the strata on the footwall at this point, but it is clear that the throw of Old Rake can only be small. Hard Level was driven north from Dolphin's Shaft as Pedley's or North Crosscut later, reaching Friarfold Sun Vein in 1861, but the only record of the geology is of the northern 900 ft (274 m) where it was above the Main Limestone. From Dolphin's Shaft the main Hard Level drift turned W along Old Rake. At 893 ft (272 m), at Hill Top Whim near Level House, a crosscut was driven NNW to North Rake and Friarfold Vein, starting 22 ft (7 m) above the level. At this point the base of the Underset Limestone was 40 ft (12.1 m) above Hard Level sole on the footwall, dipping steeply north, so that the North Crosscut was designed to be driven in shale above the Underset Chert. The top of the Underset Limestone here is shown 10 ft (3 m) below Hard Level on the hangingwall, dipping S, indicating a substantial southward throw on the Rake. No more geological detail is available for the remaining 3250 ft (991 m) of the level to its connection with Sir George Level, but calculating from the surface geology, it must have had command of the Main Limestone and underlying measures down to the Underset Limestone on both sides of the vein; the question that cannot be answered is how much ore remained in this ground after the extensive shaft workings from Hill Top and Alton's Whims. Sublevels above Hard Level were, however, still being mined in 1858–62. South of Old Rake, the weak Moor House Vein runs parallel in the stretch, and was reached from a crosscut at 1480 ft (451 m) W of Hill Top Whim. Another crosscut started nearby from Moor House Vein and was carried 1660 ft (506 m) SSE, to prospect the ground; its horizon is not known, but it appears to have stopped short of the Healaugh Side Vein (below). Old Rake, like other main structural elements in the North Swaledale Belt, is remarkable for the great loops developed on it. One starts from the head of Old Rake Hush, above Gunnerside Gill and extends to beneath Merryfield House ruins; another starts near here and continues for 1700 ft (815 m) to the east; the ground inside the loop was tested by crosscuts, but only weak veins (Merryfield and Freeman's veins in the case of the eastern loop) were found, in a chordal position. The extensive dumps from Old Rake were sampled in 1941 (Table 15); at least 100 000 tons are estimated to have been scattered along the course of the vein. Baryte is the dominant mineral, with subordinate fluorite, witherite, baryto-calcite and calcite. Crude dressing led to considerable losses of lead, partly as galena, but mainly in oxidised form. Chert forms roughly 25 per cent of the tailings, indicating the importance of the chert formations above the Main Limestone as host rocks. Recovery of barytes from these heaps at a small plant in Hard Level Gill (Messrs Shevels & Woodward) has been in progress since the 1950s; an account of the work has appeared (Mitchell, 1977). Although no reliable figures for the width of the oreshoots on Old Rake are obtainable, it is likely that the vein was a wide one, like Friarfold (p. 129), and that the material brought to surface by the 16th to 18th century miners, before the introduction of mechanised dressing, was obtained from narrow slits along the galena-bearing portions of the vein. It was, in fact, picked material in which galena might make from one-third to over one-half the total. Some sorting of veinstuff underground was usually practised, the waste being used as back-fill. Hence it is possible that substantial tonnages of unworked vein and of backfill may remain underground between Gunnerside and Hard Level gills, though this requires substantiation by newly-made entries into the workings. When last examined by mining engineers in the early 1950s, Hard Level was accessible to a short distance beyond Hill Top Whim and it was possible to reach Brandy Bottle Vein (p. 129) but not Friarfold by this route; at that time, however, the prospects did not appear sufficiently encouraging to justify the cost of reaching the old stopes.

Old Rake dies out eastward by splitting into a number of weak branches. North Rake, given off the footwall ½-mile (0.8 km) W of Hard Level Gill trends ENE and downthrows NW, so that there is an uplifted block exposing the top of the Main Limestone (dipping east from 1520 to 1460 ft (463–445 m) OD) between its footwall and that of Old Rake. This vein was no longer productive where Pedley's Crosscut cuts it. East of Dolphin's Shaft, another great loop, called Alderson's Vein is given off on the north side, but except near its origin, where a branch called Lucky String provided some ore between 1902 and 1905, it appears to have yielded little. It was tried in the cherts from Raw's Level, which started 2300 ft (485 m) NNW of Hard Level portal, without success; this prospecting level was driven ½-mile (0.8 km) by stages to the NNW, but found only some weak NNE strings. The direct continuation of Old Rake (called Reformer Vein) was also tested unsuccessfully by this crosscut. Finally Raw's Vein (next section, below) may also be regarded as part of the fingering out of Old Rake. Looking eastward from Old Gang Moor, where the displacement on Old Rake is still 40 to 50 ft (12–15 m), the line of the Rake and its several branches can be followed with the eye towards Crag Willas on the opposite skyline, exposing the Lower Howgate Edge Grit; in this there is no break or shift; Old Rake as a structural feature has completely died out.

As the Old Rake workings became exhausted in the middle of the last century, the Old Gang Company turned its attention to the WNW veins crossing Winterings Moss and Standard Moor, south of the Rake. Watersykes Vein was tried, without success, from Watersykes Level [9430 0107], high on the east side of Gunnerside Gill at 1590 ft (485 m) OD; the level was driven on to investigate Old Rake in the cherts, and to connect with West Whim. From Bunting Level, starting 2475 ft (754 m) E of the gill, Jock's Crosscut was driven cutting the vein 1740 ft (530 m) S of Old Rake; here it was more promising, and development was undertaken towards the SE, a connecting shaft [9498 0087] being sunk from surface. A test of the Main Limestone was now made by driving a new main drift, Victoria Level [9658 0064] (Figure 25, loc. 8) SW at 1470 ft (448 m) OD from the Hard Level valley in the year 1860. This reached Watersykes Vein at 1860 ft (567 m) from the portal and a good oreshoot 2000 ft (610 m) long was worked in the limestone, the minerals being galena, baryte and calcite. The workings were linked up with Jock's Crosscut, which was driven 590 ft (180 m) farther to the SSW, discovering a parallel vein which became known as Watersykes Sun Vein. A crosscut at the Victoria level horizon enabled an oreshoot about 600 ft (183 m) long to be developed under Ash Pot Holes, while the main Victoria Level was driven on, and tests were made down to the Underset Limestone, probably only with limited success. Lastly Barbara Level [9419 0055] at 1438 ft (438 m) OD from Gunnerside Gill on the base of the Main Limestone, gave access to Barbara Vein and its several branches (Figure 26). The dumps give the impression that neither this vein, nor its SSE branch known as Kining Vein, entered from Kining Level above Gunnerside Village, were very well mineralised; calcite being the principal mineral associated with galena. It is interesting to note that towards their eastern ends, Watersykes, Barbara and Kining veins turn SSE into a direction that becomes prominent both to the east, at the Barras mines, and in Area 13. The displacement of Watersykes Vein decreases in this direction, from 125 ft (38 m) at Gunnerside Gill, to only about 20 ft (6 m) in the Victoria workings. Barbara Vein, as noted above, is likewise believed to change the direction of its downthrow from south to north.

The last working at Old Gang Mine was in a NNW branch from Alderson's Vein, directly beneath Crag Willas. A section by Edward Cherry (date probably 1903) shows the base of the Main Limestone 16 ft (5 m) above the sole of Hard Level, and an oreshoot extending 570 ft (173 m) in the limestone. The thickness of beds between the base of Main Limestone and Lower Howgate Edge Grit is thus 415 ft (126.5 m) approximately.

Stockdale 'Vein'

SD 99 NW, NE; Yorkshire 51 NE
General direction E – W, displacement NE of Gunnerside, 550 ft (168 m) N

Healaugh Side Veins Lead ore

NY 90 SE; Yorkshire 36 SE
Direction E – W, to WSW, little or no throw

Raw's Vein Lead ore

NY 90 SE; Yorkshire 36 SE, 37 SW
Direction N80°E–N60°E

Raw's Sun Vein

NY 90 SE; Yorkshire 37 SW
Direction N70°E

Knott's Veins Lead ore

NY 90 SE, SD 99 NE; Yorkshire 36 SE, 37 SW
Three parallel fractures trending N23°W, small W throw

These veins were of no economic importance and would probably not have been prospected had there not been a substantial mine nearby. They are situated in the south and south-east part of the former Old Gang mining area, in the manor of Healaugh. The structural evidence on the Stockdale 'Vein' has already been discussed (p. 76); it remains here to record that it was tested, from the upthrow side, by a level driven north-east from Gunnerside Gill beneath the Hardrow Scar Limestone; and by shafts at White Hill and Brownsey, north-east and east-north-east of Gunnerside village, all without success. The only thing that could be said in its favour is that in the Carboniferous of midland Eire, major zinc-lead and copper orebodies are associated with faulting of this magnitude in reef and massive limestones and no attempt has anywhere been made to test this fault in the Great Scar Limestone, or to look for geochemical indications along its lengthy outcrop. The Healaugh Side veins are a pair of weak fissures that cut through the Main Limestone; the Herring Rake Level, on the base of the limestone, passed through them in its northward course from the bottom of Hard Level Gill and proved other weak fissures beyond them. A test was also made of the south fissure from Roger Level, near the top of the limestone, west of the Gill. The Raw's Vein was found in the course of very extensive exploration at the east end of the Old Rake, but it is difficult to be certain from the plans exactly at what horizon these were conducted. The most likely interpretation is that Hard Level was driven E, NE and E from Dolphin's Shaft (p. 127) along Alderson's Vein and associated strings for a total distance of 2675 ft (815 m); Alderson's Vein having died out, crosscuts were then driven N for 680 ft (207 m) without finding anything, and S for 985 ft (300 m), encountering Raw's Vein directly beneath Healaugh Crag, the continuation of Crag Willas. This gradually changed direction as it was followed to the NE for 2000 ft (604 m). Presumably it must have afforded some production or promise. The plans show what may be assumed to be a belt of natural caverns that run 1000 ft (304 m) NNW from Raw's Vein before linking across to Barras West Vein (p. 132) on a highly irregular eastward course. These may therefore be taken as an indication that the workings on Raw's Vein were in limestone and almost certainly in the Main Limestone. A further crosscut to the SSE for 490 ft (149 m) found the Sun Vein, close to the chimney of the Old Gang Smelt Mill and nearly 1000 ft (305 m) of driving was done in this vein, as well as a crosscut to and considerably beyond Healaughside Vein to the south. The Raw's veins were also tested from Barras High Level. New Raw's or Craw or Crow Level [9726 0110], starting 1550 ft (472 m) NNE of Hard Level Portal, was driven NNE to Raw's Vein probably in the Crow Limestone; baryte and calcite occur on the dump. Almost beneath this drift, another was driven at the Hard Level horizon back to Healaugh Side Vein, and connected to a shaft, no doubt to ventilate these extensive drives. The Knott's Veins, named for one of the partners in the 19th century Old Gang

Company, form a belt of weak fissure investigated from the NNW Spence Level driven along one of them under the Underset Limestone, starting near Hard Level portal at 1306 ft (398 m) OD which was continued up to and beyond the Healaugh Side veins. The fissures continue southward on the west side of Hard Level Gill, where they were cut in Knotts Level (under 'Knotts Hush' on the 1:10 560 map), again in the Underset Limestone. The eastern fissure, found 300 ft (91 m) SW of the portal, was developed for 650 ft (198 m) to the SSE and shafts were sunk at several points in the ½-mile (0.8 km) of ground ahead of the forehead. Some of the heaps show a little veinstuff with baryte, but no mine was found. The middle fissure was evidently unpromising, and the western one perhaps turned away W of S since the level was continued 1600 ft (488 m) in that direction.

Friarfold Vein Lead ore Baryte (Fluorspar)

NY 90 SW, SE; Yorkshire 36 SE
Direction N60°–65°E, throw 60 to 230 ft (18–70 m) SE

Gorton Vein Lead ore

NY 90 SW; Yorkshire 36 SE
Direction N60°E, throw 24 ft (7 m) SE

Gorton North and South Strings

NY 90 SW; Yorkshire 36 SE
Direction N65°E, forming a small fault-trough dropping beds about 25 ft (7.6 m)

Rig Vein Lead ore

NY 90 SW, SE; Yorkshire 36 SE
Direction N40°W

Rig String Lead ore

NY 90 SE; Yorkshire 36 SE
Direction N55°E

Deans Vein, Cherry's Vein Lead ore

NY 90 SE; Yorkshire 36 SE, NE
Direction N50–60°W, throw 12 to 18 ft (3.7–5.5 m)

Friarfold Sun Vein Lead ore

NY 90 SE; Yorkshire 36 NE, NW
Direction N60–75°E, throw 160 to 200 ft (49–61 m) SE

Friarfold is probably the most continuously mineralised vein in Area 11 and indeed in the whole of the region herein described. At the surface, it can be seen to branch off Old Rake at the junction of Friarfold and Old Rake hushes (Figure 26). The exact inclination of the junction is uncertain, but it is noticeable that the crosscut from Sir George adit portal strikes the vein 170 ft (52 m), vertically below the surface position. Friarfold Vein was not pursued at Sir George Level, but it was tested again by means of a N crosscut, 835 to 930 ft (277–285 m) farther NE, but again without success. It must be concluded that Friarfold Vein has no separate downward extension below the Underset Limestone such as Old Rake displays. As already noted, Sir Francis Level did not find it, or it was completely barren. Bunting Level at about 1400 ft (427 m) OD was the principal 19th century development from the west, being driven 4650 ft (1417 m) ENE in the vein, to overlap with a drift at the Hard Level horizon. Even so, it is doubtful whether the vein yielded much ore from this drive, for there were very extensive ancient workings throughout the south-western 1 mile (1.6 km) of the vein, probably the work of partnerships of independent miners before the year 1670. The estate plan of 1848 identified no less than 41 shafts in this stretch, and among the great quantities of debris left on the surface (Dunham and Dines, 1945, fig. 11) clear evidence that the Main Limestone and overlying cherts had been extensively stoped is to be found. The vein stuff carries galena, baryte, pale amber and mauve fluorite, a little sphalerite and traces of chalcopyrite, with the oxidation products of the sulphides and carbonates of barium and calcium. Some of the shafts must have reached or perhaps exceeded 300 ft (92 m) deep if they reached the limestone on the downthrow side; most were sunk through Ten Fathom Grit, but there is little evidence that this sandstone was a favourable host. No record has been found either of the shapes of these considerable workings, nor of stoping related to Bunting Level save, in the case of the latter, on a cross-cutting vein known as Rig Vein, 4000 ft (1219 m) ENE of Gunnerside Gill. Friarfold Shaft, sunk or reopened at the junction, marks the position. A level was driven NNW at about 1630 ft (437 m) OD for 1360 ft (415 m) to find an oreshoot averaging 350 × 100 ft (107 × 30 m) in Main Limestone and overlying chert. To the south-south-west, a drive at Bunting Level encountered at 500 ft (152 m) from Friarfold Vein the subparallel Rig String, which was stoped for 950 × 40 ft (290 × 12 m) in the Richmond Chert east of the String, and for 300 × 60 ft (91 × 18 m) to the west in Main Limestone. The Underset Limestone, though accessible above Bunting Level, was not developed. It is tempting to conclude that these stopes were the only substantial outcome of the long drive of Bunting Level on Friarfold Vein, but this is not certain. Nineteenth century stoping, as recorded on the Old Gang Company's mine-section, begins 6400 ft (1950 m) ENE of Gunnerside Gill and continues beneath and beyond Hard Level Gill as far as Forefield Shaft (Figure 25). This ground was reached, as part of *Old Gang Mine*, by three principal workings: (i) the Black Crosscut of the Hard Level system (p. 127) 2350 ft (716 m) long, driven in shale above the Underset Chert and reaching Friarfold Vein at 1370 ft (418 m) OD beneath the Gill; this was driven as a small water level into which Bell's Shaft on the south side of the vein pumped water from workings extending nearly 100 ft (30 m) below; (ii) Brandy Bottle Incline, 735 ft (224 m) long at an average inclination of 20° from the right bank of the Gill, reaching the vein near Bell's Shaft; this was the main haulage exit for the ore; (iii) Pedley's Crosscut (p. 127), driven north at Hard Level horizon from Dolphin's Shaft to the west end of the Friarfold Sun Vein workings, being connected from here by Alderson's Crosscut to the bottom of New Whim Shaft on Friarfold Vein, 1400 ft (427 m) ENE of Hard Level Gill. The first two entrances belong to Frederick Hall's time in the 1820's, the Pedley's Crosscut to the 1850s. An unfinished section of the last-mentioned (ZLB 41/80) shows the northern 900 ft (274 m) of the crosscut, with the top of the Main Limestone dropping from the level of the roof of the crosscut until it is 58 ft (17.7 m) below the sole at the north end. The drawing also shows, correctly, the horizon of Mould's Horse Level (p. 132) as 180 ft (55 m) below Pedley's Crosscut. The geological information has important structural implications, as will be discussed below.

Geological information on the wallrocks along Friarfold Vein is confined to surface exposures (Plate 5), the accurate mapping of key boundaries at surface by W. Gunn a hundred years ago, F. Hall's section of Black Level, Bell's Shaft and Brandy Bottle Incline, Pedley's Crosscut and the geological stope sections of Rig Vein, Rig String, Deans Vein and Cherry's Vein. Dean's Vein crosses Friarfold Vein 1020 ft (311 m) WSW of Bell's Shaft; drives were carried 1670 ft (509 m) NW at the base of the Main Limestone and a stope 700 × 80 ft (213 × 24 m) was worked in the limestone and chert. The wallrock geology for the rest of the very extensive Friarfold workings has to be inferred from the sources listed, but a fair picture is obtainable, showing gentle disharmonic folding on opposite sides of the veins. In Friarfold Hush at the

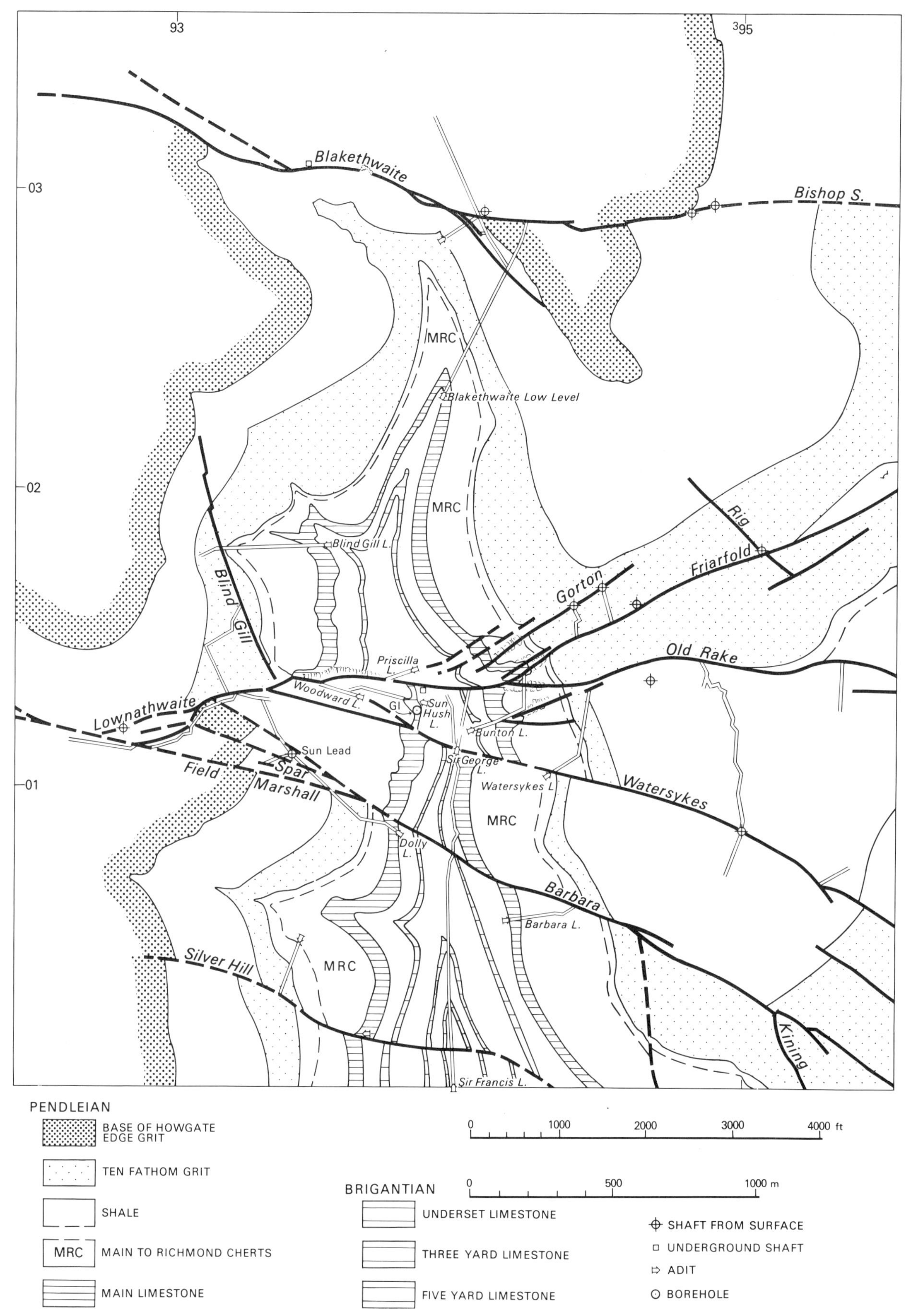

Figure 26 Geology and mine workings in Gunnerside Gill. Geological lines are taken mainly from the primary survey by C. T. Clough; mine workings are taken from the plans of the A.D. Company

western extremity, the measured throw is 37 ft (11.2 m) SE, but this is in effect cancelled out by a branch in the footwall which throws in the opposite sense, but which joins the vein farther east. At the Rig Vein crossing, the throw of Friarfold Vein has increased to 130 ft (40 m), but this diminishes to 110 ft (33 m) in the vicinity of Bell's Shaft according to Hall's section. Farther east-north-east, the throw must increase markedly, so that after the Sun Vein has appeared, the total displacement across the structure is 400 ft (122 m). The throw of the Sun Vein, from the evidence of Pedley's Crosscut, comes in so abruptly that a sharp monoclinal fold, or less probably, a cross-fault in the beds on the hangingwall side must be invoked. The surface geology enables some constraint to be placed on the distribution of displacement between the Sun and main veins; Clough (*in* Dakyns and others, 1891, p. 159) remarked 'The fairly large outlier of Millstone Grit on Surrender Moss has a roughly triangular shape . . . the western part of the north boundary is the well-known Friarfold Vein . . .'; however, he mapped this boundary on a line diverging in a more easterly direction east of New Whim shaft and not along the line of the main vein. His line, when compared with the underground workings is clearly that of the Sun Vein, indicating that its average inclination is no more than 50°S. Since the Lower Howgate Edge Grit does not appear north of the surface position of Sun Vein, it follows that the throw of this vein must increase eastward, while that of the main vein decreases; that of the latter may well have become zero before Grey Game Fault (p. 132) is reached, explaining the termination of the workings. This interpretation of the structure accords well with the fact that continuity with the Blackside Vein (p. 133) of Arkengarthdale is established through the Great Sun Vein of Surrender, but Clough's suggestion of a transfer of throw from Friarfold to Blackside by way of the Grey Game Fault is not supported. Sun Vein does not appear to have been very productive in the Old Gang property, or on Surrender ground west of Grey Game Fault. The termination of the main Friarfold workings leaves a gap in the ore-bearing ground, but this appeared again in strength east of the fault.

In the early 1950s it was possible to examine Friarfold Vein underground from the Brandy Bottle Incline. It is a wide vein of 6 to 8 ft (1.8–2.4 m) with clean walls, filled with white spar (baryte, witherite, baryto-calcite, calcite) which had carried bands with galena, these having been removed along narrow slits. Tests of the white spar by Messrs Shevels and Woodward showed that it was difficult to make an acceptably heavy barytes concentrate from it, perhaps because of the carbonate content, and mining development was not proceeded with. The dumps, or some of them, are, however, being treated successfully, and it is a reasonable conjecture that if entrance were gained elsewhere to this vein, useful raw material for barium minerals or even fluorspar production might be found.

From 1817 to 1867, the total output of the Old Gang mines was 55 689 tons of lead (Fieldhouse and Jennings, 1978) requiring the production of 79 556 tons of 70 per cent lead concentrates. From 1868 to 1911 Old Gang produced 28 162 tons but the main period as a serious producer ended in 1878 when production fell for the first time below 1000 tons per year. The total output can hardly have been less than 150 000 tons of concentrates, and may have reached 200 000.

Plate 5 Hushes in Gunnerside Gill viewed from the head of Friarfold Hush.

Differing levels of the cliffs in Main Limestone (right and left foreground) show the displacement by the Friarfold Vein. Beyond Gunnerside Gill a fan-shaped pattern of hushes radiates towards the observer. Gunnerside G1 Borehole, sunk within this plexus, was sited in the exact centre of the photograph, above the cliff in Underset Limestone.

Grey Game 'Vein'

NY 90 SE; Yorkshire 36 NE, 37 NW, SW
Direction N40–50°W, maximum displacement, 140 ft (43 m) NE

Surrender Far North Vein Lead ore

NY 90 SE; Yorkshire 36 NE
Direction irregular, E – W, displacement 36 ft (11 m) N

Surrender North Vein, Wetshaw Head Vein Lead ore

NY 90 SE; Yorkshire 36 NE, 37 NW
Direction N50–60°W, probably SW downthrow

Surrender North Middle or Labour Vein Lead ore

NY 90 SE; Yorkshire 36 NE
Direction N75°E

Flange Vein Lead ore

NY 90 SE; Yorkshire 36 NE, 37 NW
Swings round from ENE to NW, throw probably NE

Surrender Great Sun Vein Lead ore, Baryte

NY 90 SE; Yorkshire 36 NE, 37 NW
Direction N75°E, throw 180 ft (55 m) S

Minor Veins and Strings: Royal Exchange, South Middle, Eyle Tail, Jackson's, Commodore, Metcalf's

Barras West Vein Lead ore, Witherite

NY 90 SE; Yorkshire 37 SW
Direction N25–30°W

Barras East Vein

NY 90 SE; Yorkshire 37 SW
Direction N25°W

Surrender Mine worked a stretch of only about ½-mile (0.8 km) long, on the North Swaledale Belt, lying between the Old Gang and Arkengarthdale properties, but it was very highly mineralised, with numerous veins. Access was gained initially through two shafts, the Old or High Shaft [9706 0248], collared at 1875 ft (572 m) OD, and the New or Low Shaft [9739 0256] at about 1830 ft (558 m) OD, respectively 420 ft (128 m) and 400 ft (122 m) deep. Eventually the operations were carried on to the level of Mould's Horse Level (p. 136) at about 1150 ft (350 m) OD giving this mine the greatest vertical range of working next to Lownathwaite in the region. Moulds Level came from the Arkengarthdale side, starting no less than 8960 ft (2.73 km) away at [9968 0257]. The details of the geological structure would be of great interest, but unfortunately they can only partially be inferred from the records that remain. These include a strata section for High Shaft (ZQX 1.9.2) which suggests by comparison with the section at Wetshaw Fourth Whim (Figure 20 and Dakyns and others, 1891, p. 116) that about 90 ft (27 m) of beds have been cut out by faulting between the Ten Fathom and Lower Howgate Edge grits. There are two detailed plans of the workings, one for those at or about the horizons of the Crow Limestone and cherts (ZLB 41/44) and another covering lower horizons down to the Underset Limestone (ZLB 41/43) on which the strata identified at 41 different points in the mine are indicated, together with the altitudes of 36 of these above the Horse Level. This information, though valuable, is not enough to permit a unique solution to the structure, even when a geological section of the veins around High Shaft (which has been used in constructing Figure 18) is taken into account. It must be borne in mind therefore that the complete accuracy of the interpretation which follows cannot be established.

The Grey Game fracture, lying a short distance west of High Shaft but probably not passing through it, seems to have formed the original western boundary of the Surrender workings. This was cut north of Surrender in the Danby Mine (p. 139) so that it does not terminate by virgation with Friarfold Vein as Clough mapped it. Southwards it was cut several times in the Barras levels, and on the surface here it can be seen to effect a major shift in the outcrop positions of the stratigraphical boundaries from base of Lower Howgate Edge Grit down to and including Underset Limestone. Bradley (1862) records it as mineralised in the Main Limestone and Chert at Surrender, but apart from the fact that a main sump was sunk from a level close to the bottom of High Shaft down to Surrender Level at 1280 ft (390 m) OD, there is little evidence of this from the plans. About the year 1850 an adjustment was made by agreement between the Surrender and Old Gang syndicates, so that some 500 ft (152 m) of additional ground west of Grey Game Vein became available for working from Surrender. In this ground it is possible to establish the exact relationship between the Old Gang and Surrender working levels, thanks to a fragmentary document preserved in the series ZLB 41/124–173. Surrender Level, mentioned above, was driven forward in the Sun Vein which is offset about 30 ft (9 m) W side N by Grey Game Vein. The drive was in Main Limestone on the hangingwall side of Sun Vein, probably close to the top of the limestone, for a sump was sunk close to the new boundary to explore it beneath the level. The working level from the Old Gang side in Sun Vein was 125 ft (38 m) above Surrender Level, and plans were made to link the two by an incline 390 ft (119 m) long. Whether this was ever driven is not certain, but in view of Raistrick's statement (1975, p. 64) that it was possible to walk through underground from Gunnerside Gill to Moulds Horse Level perhaps the connexion was actually made. At 740 ft (226 m) W of the new boundary, the level in Sun Vein was linked to a rise which came up 50 ft (15 m) from Hard Level. The cross-section at Surrender High Shaft shows the relation between Surrender Level and Moulds Horse Level, which lies 140 ft (43 m) below, but which was never driven through to connect directly with Hard Level. These relations having been established, it is possible to state that the best interpretation of the dip slip on Grey Game Vein is that it amounts to between 60 and 80 ft (18–24 km) E at Surrender, somewhat less than the figure 2 miles (3 km) to the south at Barras.

The Great Sun Vein between Grey Game Vein and the eastern boundary of the Surrender ground must be regarded as the major structural element, worked at various levels between the Crow Beds, 405 ft (123 m) above Moulds Level, and the latter level, which followed it more or less closely across the property. From the strings, particularly Jackson's and Commodore worked on the hangingwall side of Sun Vein, it can be established that the base of the Main Limestone on that side dropped to over 50 ft (15 m) below the Horse Level, though there is no evidence of sump workings on Sun Vein. In the absence of stope sections, no definite conclusions about the distribution of ore can be reached. Sun Vein is considered to be continuous eastward with the Blackside Vein (p. 133), though no workings actually connect through. The Moulds Level link with the Arkengarthdale ground comes by way of Jacob's North Vein, which appears to have pursued a very irregular course after entering the Surrender ground, until it joined up with the Sun Vein on the hangingwall side. On the footwall side of Great Sun Vein, the beds rose steeply in the vicinity of High Shaft, and the next strong vein, Flange Vein, downthrows and dips north. It follows a curviform course, in the western part of the mine more or less paralleling North Middle Vein, but eastward curving to parallel the North Vein. Its dip can be established by comparison of the positions in the Crow and Main horizons as N, becoming NE, to the east, and decreasing to around 50°. A series of workings related to Flange Vein, but from the abrupt rectangular turns probably following caverns in the limestone, which apparently rise from 132 ft (40 m) above Horse Level to 282 ft (86 m) above before linking with the vein, 350 ft (107 m) WSW of Low Shaft. East of these workings a series of main crosscuts traversed the structure from Sun Vein northwards to North Middle Vein: (i) The Shale Crosscut, 102 ft (31 m) above Moulds Level; (ii) the Crow Beds Crosscut, 454 ft (138 m) above Moulds Level (iii) Hutchinson's Crosscut, 408 ft (124 m) above. Beyond these a fourth crosscut apparently reached the Underset Limestone on the footwall of Flange Vein, 144 ft (44 m) above Moulds Level. The principal developments, however, in Flange and North Middle veins seem to have been from a level 282 ft (86 m) up, commanding the Main Limestone. Under the name Labour Vein, North Middle Vein was also worked at the Crow Beds horizon, either at 405 or 454 ft (123 or 138 m) up, while, as already noted there was also a short working on Flange Vein. On the east side of the worked area, the North Vein, striking NW forms the eastern boundary of the Lower Howgate Edge Grit outlier at surface, and must thus throw SW. The main level on this is shown at 180 ft (55 m) above Moulds Level, in limestone. Part of the implied displacement of 80 to 100 ft (24–30 m) is possibly transferred to North Middle Vein, but the direct line of North Vein may be correlated with the Wetshaw Head Vein of Danby Mine in Arkengarthdale; and this seems to be replaced *en échelon* by the Lucks All Vein, throwing 9 ft (3 m) NE (p. 139). The Far North Vein was originally found from shafts sunk in or through the Lower Howgate Edge Grit, within the Surrender property. However, its northward dip of about 45° was sufficient to carry it into Arkengarthdale ground at the Main Limestone and Chert horizons, and a long crosscut driven N to the property boundary from High Shaft either in the Richmond Cherts at 1510 ft (460 m) OD or in Ten Fathom Grit and Crow Beds at 1660 ft (506 m) OD—the evidence is conflicting— failed to find it. The Danby workings establish the throw of this vein as 36 ft (11 m) N, and its effect is to bring in a small outlier of the Lower Howgate Edge Grit on the downthrow side. Between this vein and Labour or North Middle Vein, the grit is not present and Crow Beds must come to within 60 or 70 ft (18 or 21 m) of the surface. It was this fact that enabled the Surrender complex to be discovered by the shaft miners of times long ago; there is no evidence that the Howgate Edge Grit was itself mineralised.

The dumps from the two Surrender shafts are among the most extensive in the belt; the mineral suite includes, besides galena, baryte, witherite, sphalerite, calcite and traces of chalcopyrite. The large vertical extent of the workings (for this area) is a function of the complex structure, but it seems that here, unlike Gunnerside Gill, there was nothing to lead the miners downward below Moulds Level, where on Sun Vein, sandstones below the Underset Limestone must have been in contact with Main Limestone. At the same time it is difficult to feel satisfied that this promising structure has received the full mining investigation that it deserves. Analyses of tailings from Surrender are given in Table 15. Separate figures of production of lead concentrates are available for 1858 to 1882, the aggregate production being 1882 tons; since the beginning of this period nearly coincides with the adjustment of the western boundary, it is likely that this was the yield from the new area west of Grey Game Vein. Fieldhouse and Jennings (1978) give the lead output from 1800–11, 1817–85 as 27 587 tons. This must have represented at least 40 000 tons of galena concentrates.

The *Barras End Mine* explored ground south of the Surrender royalty, though both its principal levels entered this property. Though the greater part of the workings lay within Old Gang ground, they are included here as relevant to Surrender. Two subparallel NNW veins, with the Grey Game Fault cutting across between them, were extensively examined here. The High Level [9876 0123] (Figure 25, loc. 9), incorrectly named Barras End Low Level on the current six-inch topographical map, starting in shale above the Crow Chert at 1413.6 ft (430.9 m) OD was driven in a general WSW direction from Bleaberry Gill 475 ft (145 m) W of the portal. The upthrow gave the level command of the Main Limestone; the East Vein was cut after another 300 ft (91 m) of driving, and the West Vein 480 ft (146 m) beyond this. A drive was carried NNW in the West Vein for 3340 ft (1018 m) cutting several ENE veins on the west side of the workings, including the Raw's South strings and Raw's Vein (p. 128), the latter being developed to the point previously reached by Hard Level from the west. This level must have been at about 1360 ft (415 m) OD, near the base of the Main Limestone, and Barras End High Level would here be at 1425 ft (434 m) OD, not far below the top of the limestone. The caverns mentioned on p. 128 were cut by High Level 400 ft (122 m) N of Raw's Vein. The forehead of the level, according to the document relating Old Gang and Surrender levels was in Red Beds Limestone, so that there must be a fairly rapid northward dip, north of the caverns. The possibility of linking the Surrender workings west of Grey Game Fault with this forehead was under consideration in the 1850s; a drive of about 3000 ft (914 m) would have been required, and as Barras End High Level forehead must be 145 ft (44 m) above Surrender Level (not 95 ft (29 m) as roughly estimated in the document) nothing would have been gained and it is not surprising that the work was not undertaken. The absence of indications of rise workings on the plans and the small size of the tailings heaps at the mine combine to suggest that productivity was low on the West Vein, but it may be noted that the witherite found here during a reopening about 1925 (Dunham and Dines, 1945, p. 28) was probably in the West Vein.

The East Vein converges northward upon Grey Game Fault, and an ambitious development was undertaken during last century to test this vein on the downthrow side of the fault. The Barras End Low Level [9932 0069] (Figure 25, loc. 10) at 1237 ft (377 m) OD was driven 2425 ft (739 m) as a NW crosscut beneath the Main Limestone until it was nearly underneath the High Level. Here it turned west, shortly cutting the Grey Game Fault, and was continued 600 ft (183 m) to test the West Vein, but this was not proceeded with. The drive was carried NNW in East Vein, again passing through the fault which may shift the vein N side NW about 100 ft (30 m), but the vein was found and followed 1850 ft (563 m) to the NNW, until the Waterblast Vein from Turf Moor Mine (p. 136) was encountered. According to the longitudinal section of Waterblast Vein, the Barras End Low Level, at or above 1250 ft (381 m) OD, must by this stage have been considerably above the Main Limestone. About 1300 ft (396 m) NNW of the point where the level passed through the fault, a crosscut was driven 425 ft (130 m) WSW to test it again, but apparently it was still barren. An adit from surface starting 1550 ft (472 m) SE of the portal of High Level tested the Grey Game Fault between the upthrow and downthrow positions of the Main Limestone, but again nothing workable was found. Altogether nearly 2¼ miles (3.6 km) of development were carried out at Barras, for very little result; the NNW direction of fracturing, though structurally continuous over considerable distance, is neverthless the least promising of directions within the North Swaledale Belt. The Forefield veins, about ½-mile (0.8 km) to the east seem to have proved little better. KCD

Great Blackside Vein Lead ore (Fluorspar)

NY 90 SE, NZ 00 SW; Yorkshire 37 NW
Direction N80°E, E – W, N60°E, throw 270 ft (82 m) S at Black Whim, 320 ft (98 m) opposite Stodart Hush, 420 ft (128 m) S north of Justice Vein rising to 550 ft (167 m) near Crag End Quarries. After Freeman's Vein leaves, 280 ft (85 m) S, falling to 140 ft (43 m) at Tanner's Rake, increasing to 270 ft (82 m) S, 1000 ft (305 m) to east.

The Great Blackside Vein may be regarded as the main axis of the Arkengarthdale Royalty, worked by the C.B. Mines Company (p.

4) from 1821 to their closure in 1912. It is the eastward continuation of the Great Sun Vein of Surrender Mine, and by comparison with all the other veins of the Belt, has the largest and most persistent vertical displacement. This varies in response to the gentle differential folding and fracturing of the strata on either side of the fault. Only two fractures are known to cross and displace Blackside Vein: Stodart – Justice Vein, which shifts Blackside Vein N side E about 150 ft (46 m); and Cocker Rake – Tanner's Rake, which shifts it 200 ft (61 m) E side SE. Near Scar House, a major branch from Blackside Vein known as Freeman's Vein runs NNE, carrying away nearly half the displacement; but Cocker Rake, which converges northward upon Freeman's Vein does not shift it sideways. A number of workings opened up the vein in the western 1 mile (1.6 km) of its course through Arkengarthdale ground. Crow Beds Level of Surrender Mine, starting at [9795 0243] entered through beds belonging to the Lower Howgate Edge Grit or higher, but presumably encountered Crow beds on the footwall of the vein, at about 1660 ft (506 m) OD. A level 181 ft (55 m) above Moulds Level also penetrated into Arkengarthdale ground along Blackside Vein, possibly at an intermediate position of the Main Limestone in a downfaulted slice within the fault. As at Surrender, the fault is not a single fracture, but a wide belt enclosing dipping masses of wallrock. One section, formerly in the C.B. records (MLD 9214)[1], representing workings near the footwall intersections of Dam Rigg and Black Jack veins (p. 137), shows the disrupted blocks of strata dipping south, as in Figure 18, but cut by minor faults associated with the Blackside main fault which all dip north, while the main fault itself is shown as vertical to 85°S dip. Notwithstanding the abundance of fractures, the general position of substantial faults in the northern Pennine orefield, that they rarely offer the best conditions for the formation of oreshoots, seems to have applied to Blackside Vein; there is no reason to believe that it was a major producer. It was the numerous veins of much smaller displacement that diverge from both footwall and hangingwall that carried the most productive oreshoots. Black Whim [9795 0257], 342 ft (104 m) deep from 1728 ft (527 m) OD, and Sun Lead Shaft [9833 0270], depth unknown, are among a number of shafts into Blackside Vein, and there were also connections from Danby Level at 1490 ft (454 m) OD, Dam Rigg Level at 1325 ft (404 m) OD and Turfmoor Level at 1197 ft (365 m) OD. No plan shows any lengthy development of the vein, and it must be concluded that oreshoots such as there were, must have been small and patchy. North-east of the intersection with Stodart Vein, two deep trials of Blackside were made during the 19th century. Smelt Mill Level, starting at [9964 0365] (Figure 25, loc. 14) at 930.5 ft (283.6 m) OD was driven 2570 ft (783 m) S36°W, and then 500 ft (152 m) due S to cut the footwall approximately 70 ft (21 m) below the base of the Middle Limestone; the drive was then continued 1200 ft (366 m) along the vein to the west. The paucity of veinstuff on the large heap below the Smelt Mill, and the absence of subsequent development indicate that this trial, stratigraphically the deepest one in Arkengarthdale, was not successful, neither as far as Blackside Vein was concerned, nor in finding ore in the several veins that here lie adjacent to the hangingwall (p. 137). Above this, a drift 320 ft (98 m) long tested the vein at Moulds Old Level horizon, about 1140 ft (347 m) OD.

Dumps from shallower workings in the vein and in adjacent veins in this stretch contain more fluorite, always pale amber, pale mauve or colourless, than anywhere else in the district. Proceeding eastward from the mouth of Underedge Level (p. 137) down the hillside past the ruined smelter flues, the workings on Blackside Vein are seen to die out, and it appears that no effort was made to test it near surface from the position of the Three Yard Limestone downwards. The vein must cross the Reeth – Tan Hill road almost beneath the C.B. Inn, but nothing can be seen until Scar House Wood on the east side of the valley is reached. From the hangingwall position of the Main Limestone upwards into the area marked 'Great Black Lead Vein' on the six-inch map, old shallow workings, shafts and levels can be seen but these do not appear to have yielded major tonnages of ore. In this stretch, Freeman's Vein diverges to the NNE but the workings on this vein only become substantial above 1500 ft (457 m) OD, opposite the footwall position of the Main Limestone. At 1300 ft (396 m) OD Joseph Harker's Level [0062 0320] was driven obliquely ENE through the vein-zone, reaching beds a little below the Main Limestone on the footwall side. A total of about 600 ft (183 m) of driving was done along the vein, rises were put up into the limestone, and stringers in ENE and NNW directions were examined from crosscuts to the north and south, but little ore seems to have been found. Completed in 1875, this must have been one of the last tests of the vein. The structure however continues strongly to the east, through the hushes leading to Tanner's Rake, where it brings Crow Chert to the south in contact with Richmond Cherts to the north; some production seems to have been obtained here. The shift by Cocker = Tanner's Rake has already been mentioned; it can be inferred from the relative positions of the large opencast excavations. East of Tanner's Rake, the equivalent of Blackside Vein is considered to be North Rake, worked in the hush of that name, against a footwall of Main Limestone that dips appreciably towards the Slei Gill valley. In numerous small heaps left from the dressing of lead ore, baryte is the principal mineral. A NE crosscut from Tanner's Rake Level appears not to have been driven right through North Rake, but the dip is uncertain. The position of the fault as it crosses Slei Gill can easily be traced from the appearance of the Main Limestone on the north side; it then continues up North Gutter, where a level has been driven in it, gradually changing direction to N65°E. Near the head of the Gutter, a crosscut from Sun Gutter Level (p.143) must have cut the vein, but it was not developed. About 1½ miles (2.4 km) ahead, the Blackside – North Rake fracture is regarded as terminating against the Moresdale Ridge Fault on Fair Seat.

By tradition West Arkengarthdale has been regarded as divided by the Blackside Vein into a Sunside district, south of it, and a Northside district beyond it. The Sunside group of veins will be described first.

Fore Gill = Geordy Vein Lead ore

NY 90 SE; Yorkshire 37 SW
Direction N44–50°W, throw 40 to 150 ft (12–46 m) SW

Toppy's String Lead ore

NY 90 SE; Yorkshire 37 SW
Direction N20°W, throw 10 ft (3 m) NE

Cobbler's Vein Lead ore

NY 90 SE; Yorkshire 37 NW, SW
Direction N40–50°W, throw small to 40 ft (12 m) NE

Moulds Sun, Moulds, Moulds North Veins Lead ore (Baryte)

NY 90 SE; Yorkshire 37 SW
Direction close to E–W, overall displacement 130 ft (40 m) N

Waterblast Sun, Waterblast, Waterblast North Veins Lead ore

NY 90 SE; Yorkshire 37 SW, NW
Direction N8–60°E, 18 to 60 ft (5–18 m) N overall displacement

Jacob North Vein Lead ore

NY 90 SE; Yorkshire 37 NW
Direction sinuous, N15–30°W, throw small to 30 ft (9 m) N

[1]MLD numbers refer to photographic copies in BGS records

Stemple Vein Lead ore

NY 90 SE; Yorkshire 37 NW
Direction N60–70°W, throw small to N

Dodgeon Vein Lead ore

NY 90 SE; Yorkshire 37 NW
Direction N80°E

Justice Sun Vein = New Rake = Chip Vein Lead ore

NY 90 SE; Yorkshire 37 NW
Direction N55°W, throw 0 to 20 ft (6 m)

Justice = Folly Vein Lead ore (Fluorspar)

NY 90 SE; Yorkshire 37 NW
Direction N55–65°W, throw 30 to 70 ft (9–21 m)

Fore Gill or Geordy Vein may be regarded as the counterpart, in West Arkengarthdale, of the Barras veins in Surrender – Old Gang. Between Barras and Fore Gill lies a synclinal trough, here named the Geordy Syncline, having a maximum amplitude of about 300 ft (91 m), and a width of about ½-mile (0.8 km). The main development on Geordy Vein was the Fore Gill Level [9971 0117] (Figure 25, loc. 11) driven SW beneath the Main Limestone, reaching the vein at about 1160 ft (354 m) OD, 650 ft (138 m) from the portal. Masses of close-grained ankerite lie about on surface in the gill where the vein crosses. The workings from the levelhead in Geordy Vein were carried 3250 ft (990 m) NNW until they overlapped with workings proceeding SSE from the Turf Moor Mine. Contemporary agents' reports refer to floats in the limestone associated with this vein, as at Barras, but it is unlikely that the vein itself carried much apart from minerals of the dolomite – ankerite series. Toppy's String, which diverged into the footwall a short distance north of the end of the adit was also vigorously pursued both from Foregill adit and from Turf Moor. Cobbler Vein, tried by a small hush and level in the Main Limestone, seems to have yielded little and is classed as unproductive by Bradley (1862). The vein was cut by the SW crosscut from Turf Moor Level (below) which gave access to the northern workings on Toppy's and associated strings in the

Plate 6 North Rake Hush, near Booze, Arkengarthdale.
The Main Limestone is exposed in the north wall of the hush.

Main Limestone, but it was not developed. This crosscut was eventually extended to Geordy Vein, but there was also access to this from the western end of Turf Moor Mine at 1280 ft (330 m) OD, thus over 100 ft (30 m) above Fore Gill Level. Even on the footwall side of Geordy Vein, most if not all the Main Limestone must have been below Turf Moor Level, while if Bradley's figure (1862, p. 28) of 156 ft (48 m) for the throw of this vein applies to the ground between Turf Moor and Fore Gill part of the limestone must have been below the latter level on the hangingwall side and may not have been explored. Bradley states that Geordy Vein was productive in the Main Limestone and Chert and in the Black Beds, but it carried no 'rider' (veinstuff other than galena; but Bradley seems not to have recognised dolomite and ankerite as veinstuff).

The Moulds group of veins form a strong E – W zone with northward downthrow of similar order to the northward downthrow of the Geordy zone; when the two zones meet in the western part of Turf Moor Mine, virgation occurs and though the fractures continue to the west and north-north-west, the displacements are minor only. The spectacular Turf Moor hush, near the Low Row – Langthwaite road, must have worked mainly on the Sun Vein (Plate 2). The footwall exposes beds from the Underset Limestone up to the Main Chert. Evidently the main fault also runs through the hush, for on the hangingwall side, the Main Limestone top is opposite the top of the Underset on the footwall. It may be assumed that the Moulds structure was first discovered here; certainly the hush predates all the underground mining. At the head of the hush, Moulds Hush Level gave access to a stretch 750 ft (229 m) long on Sun Vein, at 1330 ft (405 m) OD, containing patchy oreshoots. About 50 ft (15 m) N of Sun Vein lay Mould's Vein, the main producer; in Figure 20, Dr Earp's reduction[1] of the mine section is reproduced. North Vein was about the same distance north again. The beds dip west into the Geordy Syncline and the oreshoots in the Main Limestone in places extending up to the Red Beds, plunge gently with the dip. The distribution of throw between the three veins can be estimated from surface levelling and from the geological details given for the north-east wall of each vein; Sun Vein 10 ft (3 m) or less; Moulds Vein, 120 ft (37 m), North Vein, 20 ft (6 m) maximum. The ore-bearing ground against the footwall side of Moulds Vein was reached by means of Turf Moor Level [9940 0216] at 1284 ft (331 m), a straight level in the footwall. The 1800 ft (549 m) long oreshoot on Moulds Vein seems to have terminated against Geordy Vein; a NNE-striking fissure farther east seems to have terminated the smaller run of ore on North Vein. The three-fold zone was succeeded to the west of Geordy fissure by a vein not exactly corresponding in position to any of the Moulds veins, to which the name Jacob North Vein was given. This had apparently been discovered from surface at some remote period in the past, for very extensive working from shafts had taken place before the driving of the deep levels, from the *Moulds Top Mine*. Four deep shafts can be identified here, known (from east to west) as Jacobs Whim, Second Whim, Third Whim and Fourth Whim (also called Wetshaw Fourth Whim). Sections of Jacobs Whim [9869 0228], probably 408 ft (124 m) deep, give partial details, and Fourth Whim [9802 0238], which is fully recorded, is the standard section for the district (Dakyns and others, 1891, p. 116). It should be noted, however, that the full depth of this shaft is probably 427.5 ft (130.3 m) and that the section cited down to the Three Yard Limestone is derived from rises and other workings in the mine. As the shaft is not shown on the 19th century plans as connected to the levels then being driven, it must be taken to have been sunk at least as early as the 18th century. There is no record of the depths or stratigraphy of Second and Third whims, but the large spoil heaps from all four shafts at Moulds Top indicate that there must have been very extensive ancient workings in the Richmond Cherts, and also probably in the Crow Beds on Jacob North Vein. At Jacobs Whim, which was connected to the late workings, the vein was associated with three other subparallel faults (Figure 25).

Returning now to the Moulds veins, the continuation of ore into the sole of Turf Moor Level led in 1801 to the initiation of a new and lower main level, *Moulds Horse Level* [9964 0256] (Figure 25, loc. 12), strategically placed at 1130.4 ft (344.5 m) OD to give access to the Main Limestone over almost the whole of the Sunside district. This became the principal centre of operations during last century, with a total of at least 2 miles (3.2 km) of drivage at this horizon in Arkengarthdale property, in addition to a lengthy extension into Surrender ground (p. 132). On Moulds Vein extensive stopes were worked in the downfaulted Main Limestone, and a small oreshoot was found opposite the sandstone below the Underset Limestone on the footwall side. There is, however, no indication of any attempt to sink below Moulds Horse Level here. However, the ground below was partly tested by the driving, in the 1870s, of Moulds New Level [9999 0226] (Figure 25, loc. 13) westward along the course of the Moulds zone, 102 ft (31 m) below Horse Level, in sandstone and shale between Three Yard and Five Yard limestones. The total length was 2375 ft (423 m), so that this level had just come below known ore-bearing ground in the Main Limestone when it was abandoned. It was evidently very disappointing, but another 1000 ft (305 m) of driving to the west would have been needed, with some crosscutting, to make it a decisive test. Nevertheless, on the existing evidence, the Moulds veins and Jacob North seem to have carried ore principally from the footwall position of the Underset Limestone up to the top of the Richmond Cherts, and possibly in the Crow Beds. Baryte, fine to coarsely crystalline, was the predominant mineral; coarse calcite and witherite were both present, but little if any fluorite is to be seen on the heaps. A minor amount of sphalerite was associated with the galena, which was the only mineral sought.

The Waterblast veins diverge WSW and SW from Jacob North Vein near the point where Geordy Vein crosses through, close to Jacobs Whim. These carried small oreshoots, that on Middle Vein continuing in a patchy fashion for 700 ft (213 m) against the Main Limestone as it dips into the Geordy Syncline at 7°. The total displacement across the Waterblast zone is of the order of 60 ft (18 m) N at the east end, but this probably decreases westward; the Sun Vein is stated by Clough to have a throw of 18 ft (5 m) N as it approaches Barras Vein, this being sufficient to bring in the Tan Hill Coal near surface.

North of the Moulds veins, the next important vein is Stemple, marked by a line of surface workings with fluorite, baryte, calcite and fragments of galena. From Moulds Horse Level an oreshoot in Main Limestone and Chert 900 ft (274 m) long was worked but this may well have run into very old workings to the west-north-west. A further attempt was made to investigate this vein beyond Stemple Whim by means of crosscuts at the Turf Moor and Moulds Level horizons, perhaps following the split-up remnants of the Geordy Vein. An orebody 450 ft (137 m) long in Main Limestone and Chert was found, terminating about 250 ft (76 m) from Blackside Vein. Stemple High Whim [9867 0263], about 200 ft (61 m) NE of the probable surface position of Stemple Vein, 370 ft (113 m) deep, reached the top of the Main Chert, but its purpose is in doubt since no plan survives showing mine workings connected with it. It may well have given access to workings in the Richmond Cherts and above, on Stemple Vein. Considered in relation to the wallrock geology of Stemple Vein, it indicates that the beds rise towards Blackside Vein. Stemple Whim [9888 0244], nearer the Moulds Level workings is also not connected. A SW hade for the vein is suggested, but a section, formerly at Scar House, no doubt somewhat diagrammatic, shows that it consists of three fissures, one downthrowing SW about 20 ft (6 m), the other two with small N displacements.

[1]Made in 1941 at Scar House, the former headquarters of the C. B. Company. The original plans have since been lost or destroyed, apart from a very poor copy.

Moulds Horse Level was driven on a strong but unmineralised fault trending W–E, which is mapped as linking up with the important Fell End structure towards the east (p. 141). In Moulds Level the downthrow is about 25 ft (8 m) N, but at about 900 ft (274 m) from the portal, this splits into two, a WNW fissure which becomes the important Justice Vein, and a WSW branch, Dodgeon which links with Stemple Vein near Stemple Whim. Surface workings mark the courses of both branches, but neither appear to have yielded any ore to Moulds Level; they may have been worked out before the level was started. However, Justice or Folly Vein carried an oreshoot farther north-west, extending over 700 ft (213 m) of strike, in Richmond Chert worked from Justice Level [9912 0274] at 1400 ft (427 m) OD, much of the ground being worked below the level. A sump was sunk to 1200 ft (366 m) OD, passing through the full thickness of the Main Limestone but the plan of this level gives the impression that many old man workings were in existence before it was driven. A crosscut south-west however discovered a parallel vein, New Rake, Justice Sun Vein or Chip Vein, that was probably largely virgin. The effect of this vein in faulting up the Crow Chert and Limestone on the hillside can be well seen in the area of surface devastation known as Hungry Hushes. Another fracture was cut 125 ft (38 m) beyond New Rake but was not developed. As no stope section of New Rake has survived, it is impossible to say whether this vein was ever tested in the Main Limestone; the discovery crosscut must have been in Richmond Cherts. Returning to Justice or Folly Vein, this fracture appears to pass through and shift Blackside Vein, and to become Stodart Vein north-west of it. Whereas, however, Justice Vein throws NE, Stodart downthrows at least 40 ft (12 m) SW. The crossing, at the mouth of the great Stodart Hush, is undoubtedly complex and it may be that Peatstack Vein (below) is the dying-out continuation of Stodart Vein south of Blackside.

Peatstack Vein Lead ore

NY 90 SE; Yorkshire 37 NW
Direction N75°W, throw 12 ft (3.7 m) N

Band Sun or Northwest Vein

NY 90 SE; Yorkshire 37 NW
Direction N65°W, throw 10 ft (3 m) NE

Band Vein Lead ore

NY 90 SE; Yorkshire 37 NW
Direction N75°E, throw 18 ft (5.5 m) N

Band North Vein Lead ore (Fluorspar)

NY 90 SE; Yorkshire 37 NW
Direction N85–65°E, throw 18 ft (5.5 m) N

Underedge Vein

NY 90 SE; Yorkshire 37 NW
Direction E–W, throw 38 ft (12 m) N

This group of veins, still forming part of the Sunside district, occupies the angle between Justice and Blackside veins. The surface measures range from the Ten Fathom Grit down to the Main Limestone, and these horizons were probably worked out before the days of driving of levels. One was started from the bottom of Band Hush, but was abandoned after only 350 ft (107 m) of drivage. The section showing the expected veins ahead, with their wallrocks, prepared at the time is, however, of interest (MLD 9211). Underedge Level [9913 0298], at 1244 ft (379 m) OD, was driven SW through Underedge and Band North veins, and seems to have produced some ore, in which fluorite is common, to dress. Moulds Level, north branch passed beneath all these veins save Band Sun at about 1140 ft (347 m) OD in Main Limestone, but found nothing workable. Beneath this, Smelt Mill Level (p. 134) at about 935 ft (285 m) OD should have had 27-Fathom Grit on the hangingwall as it proceeded west along Blackside Vein, but apparently no interesting intersection from the present group of veins was noticed.

Luke Vein and Strings, Far North Vein Lead ore

NY 90 SE; Yorkshire 37 NW
Direction E–W, throw 36 ft (11 m) N on Far North section

Dam Rigg Sun Vein Lead ore

NY 90 SE; Yorkshire 36 NE, 37 NW
Direction N70°W, throw 10 ft (3 m) N

Black Jack Vein ?Zinc ore

NY 90 SE; Yorkshire 37 NW
Direction ?E–W, throw not known but small

Dam Rigg Vein Lead ore

NY 90 SE; Yorkshire 36 NE, 37 NW
Direction N65–70°W, throw 72 ft (22 m) SSW

Martin Vein Lead ore

NY 90 SE; Yorkshire 36 NE, 37 NW
Direction N65°W, throw 50 ft (15 m) NE

Stodart Vein Lead ore

NY 90 SE; Yorkshire 36 NE, 37 NW
Direction N75°W, turning westward to N65°W, throw 36 to 45 ft (11–14 m) S

The Arkengarthdale Northside district is now considered. Luke's Vein and other short, ramifying strings in the neighbourhood of Black Whim [9798 0257], north of Blackside Vein, were probably worked from this shaft and other shallower shafts nearby including Smith's Shaft. There was also an adit starting about 100 ft (30 m) S of Black Whim. No detail of the workings is available, but Luke's Vein was tested from Dam Rigg Level (below) which here controls the Main Limestone. Luke's Vein probably continues westward across Surrender property, through North Vein to become the Surrender Far North Vein. As already noted, this was worked in the Crow beds in Surrender ground, but owing to the low northward dip of the fracture (Figure 18), the Main Limestone workings lay in Arkengarthdale ground. An oreshoot 1240 ft (378 m) long × 90 ft (27 m) high was extracted here from Main Limestone and Chert, while immediately west of the intersection with Grey Game Vein another, 320 ft (98 m) long was got in Great Limestone.

The principal entrances to the Northside mines were Dam Rigg Level [9835 0342] (Figure 25, loc. 15) at 1320.2 ft (402.4 m) OD, and Danby Level [9754 0352] (Figure 25, loc. 16) at 1472 ft (449 m) OD. Both levels enter from the north-facing slope of the upper part of the dale. Dam Rigg Level seems to have been driven about the beginning of the 19th century; it is not shown on the detailed plan of Dam Rigg Vein and adjacent veins dated May 30, 1794 (ZLB 41/84) and although this plan gives some stratigraphical information, no geological record of the beds cut by Dam Rigg Level is known to have survived. It must, however, start in the sandstones and shales between the Three Yard and Five Yard limestones, and remain in this group throughout the 2330 ft (710 m) of straight drivage at S29°W to the underground position of Dam Rigg Vein. In this stretch, Stodart Vein was passed through at about 1430 ft (436 m) from the portal, but no development appears to have been done. Martin Vein or perhaps a subsidiary fracture associated with it was found 200 ft (64 m) ahead, and driven on for 850 ft (259 m) to the ESE, but the course of this drive by no means parallels the surface position of the vein. Dam Rigg Vein, the next vein to the south, may be assumed to have been worked out before the level was driven. Very extensive workings are recorded on the 1794 plan referred to above. These were reached from Sun Lead Shaft on Blackside Vein where Dam Rigg Vein impinges on it; Drawing Whim, 460 ft (140 m) farther WNW, where the base of the Main Limestone in the sump on the footwall was opposite the base of the Main Chert on the hangingwall (depth of the shaft 252 ft, 77 m); and 320 ft (98 m) beyond this, South Whim sunk 252 ft (77 m) to the top of the Main Chert on the hangingwall. The beds involved thus ranged from the base of the Main Limestone up to the top of the Richmond Cherts, and from the numerous rise workings shown on this plan, the impression is gained of a fairly continuous major oreshoot, some 2000 ft (610 m) long and at least 90 ft (27 m) high, extracted before the end of the 18th century. A north string yielded additional ore from the footwall side, while south of the vein, crosscuts communicated with workings in Chip North Vein, presumably part of Dam Rigg Sun Vein, as well as with this vein itself. At surface, along this vein, the Ten Fathom Grit is at outcrop, starting at about 1000 ft (305 m) WNW of Sun Shaft. At 1775 ft (541 m) WNW of this shaft, a shaft 144 ft (44 m) deep is shown as in Red Beds, while to the south of the vein another shaft 138 ft (42 m) deep reached the top of the Ten Fathom Grit in the down-dropped block. In 1794 no connection is shown between the Sun Lead Shaft and Black Whim (p. 137) workings on Luke and Blackside veins; but such a connection seems to have been made by way of Black Jack Vein when Dam Rigg Level was driven.

Returning now to the surface evidence, it may be noted that lines of very old, shallow shafts mark the positions of Dam Rigg Sun and Dam Rigg veins; in the tips, the quantity of fluorite appears to diminish with increasing distance from Blackside Vein, but baryte remains common. Martins Vein was extensively worked in an opencut 1300 ft (396 m) long, mainly WNW of Dam Rigg Level intersection, with chert, probably including Main Chert on the footwall (S) and Richmond Cherts on the hangingwall. Baryte and chert fragments compose the cuttings heaps remaining from dressing the lead ore. Stodart Vein (we here use the Ordnance Survey spelling, though many of the plans spell the name Stoddart) has been extracted in one of the most impressive hushes in the present district, a great gash 100 or more feet wide excavated through Underset, and Main limestones and overlying cherts, about 1750 ft (533 m) long (Plate 7). At the western end, this great excavation communicates with a fine example of a hush made for prospecting,

Plate 7 Stodart Hush, Arkengarthdale.

Densely jointed Main Limestone capped by Main Chert forms the prominent cliff; it is faulted by the Stodart Vein against Black Beds of the Richmond Cherts (cliff, top left, and screes on left.

running down the hillside from Martin Vein in a NNE direction, known as Dam Rigg Cross Hush. Perhaps this was the earliest working of all on this hillside. In the east end of Stodart Vein, a number of short levels were driven southward to explore this vein beneath the Main Limestone. There is a recorded section of one of these, Black Hills Level, which is interesting in that it shows, in a stretch 230 ft (70 m) long beyond the hangingwall of the vein, seven sympathetic fractures faulting the base of the limestone variously up or down and again emphasising the composite structural nature of the veins here, few of which are single clean fractures. Little ore seems to have been found beneath the Main Limestone on the vein.

Danby Level was driven to command the Main Limestone as it dips gently westward, during the period 1830–41, by which time it was not considered necessary to have command of the Underset Limestone. This adit is therefore 150 ft (46 m) above Dam Rigg Level. It runs S14°W for 2300 ft (701 m) to the main forehead, cutting the continuation of Stodart Vein at 700 ft (213 m) from the portal (not developed) and Martin Vein at 1160 ft (354 m), a drift being driven ESE along this vein almost to its intersection with Blackside Vein. Relative to the surface position of the vein, this drift is partly north, partly south, indicating a varying hade. Dam Rigg Vein, cut by Danby Level at 1800 ft (549 m) from the portal, proved to have changed direction to nearly E – W. The southward downthrow of 72 ft (22 m) still persisted, and the vein was mineralised for 900 ft (274 m) to the west of the level, at which point it became unproductive. At 700 ft (213 m) W of the level, a SSW crosscut was started from the vein, approximately parallel to the level, cutting at 350 ft (107 m) a NW vein to which the name Lucks All was given; 100 ft (30 m) beyond this, Wetshaw Head (the probable continuation of Surrender North Vein, discussed above) was found, and at the 1835 forehead, an irregular NW string, Marsh String. Evidently Lucks All Vein appeared most attractive, and this was followed north-west for 3250 ft (990 m). Although this fracture has a NE throw of 9 ft, it may be regarded as the continuation, *en échelon* of the Surrender North and Wetshaw Head fracture system. The forehead had been reached by 1841 and strings leaving it towards the west at 1750 and 2240 ft (533, 683 m) respectively from the start of the drive on the vein had been tested; but no record remains of the productivity of this ground or whether floats were found here. The mapped position of the Ten Fathom Grit shows that along the course of Dam Rigg Vein west of Danby Level, beds at higher horizons are faulted against the grit to the south, the base of the Main Limestone here being at about 1550 ft (473 m) on the hangingwall. If the NW drift on Lucks All Vein was in Main Limestone, as might be expected, it must have started from a rise, or perhaps from Dobson's Shaft, substantially above the Danby adit horizon. Even so it is strange that there is no record of the Dam Rigg fissure passing through it, yet the surface evidence suggests that this must raise the base of the limestone to above 1600 ft (488 m) OD. The westward continuation of the line of Dam Rigg Vein is discussed in the next section, below. A drive was carried SSE in Wetshaw Head Vein, and another NW in March String, but work then seems to have been suspended until 1880, when the crosscut was extended from March String to reach Far North Vein and Grey Game Vein, with the results already described.

The principal workings of the Northside mines have now been described. In the later part of their history, the raw ore was carted to two dressing floors on the flatter part of the valley at [985 027] and [988 036]. These have now been largely worked over for fluorspar and barytes; the analyses quoted in Table 15 perhaps give a representative average for the gangue minerals in the ore mined from Northside veins during last century. No separate production figures for the two main Arkengarthdale districts are available, but they were the main contributors to the grand total of 34 342 tons of lead concentrates from 1868 to 1890, the best single year being 1878 with 2459 tons. Between 1848 and 1867 the C. B. Mines contributed 41 731 tons of ore to the 98 177 tons recorded for Swaledale in the official statistics, but it will be realised that even by 1848, the maximum production from these mines had almost certainly been passed. After 1890, what effort remained was in the more remote northern part of Arkengarthdale, at Faggergill and Stang, described in the next-but-one section below. KCD, JRE

Bishops Sun Vein Lead ore

NY 90 SE; Yorkshire 36 NE
Direction N86°W, downthrow N? small

Bishop North Vein

NY 90 SE; Yorkshire 36 NE
Parallel to Sun Vein, 175 ft (53 m) to N, throw 9 ft (2.7 m) N

Blakethwaite Vein Lead ore

NY 90 SW; Yorkshire 36 NE
Direction E – W, stretches alternating with N75°W stretches; throw 24 to 50 ft (7–15 m) N

Blakethwaite Sun Vein Lead ore

NY 90 SW; Yorkshire 36 NE
Direction N45°W

Stonesdale Moor Vein Lead ore

NY 80 SE; Yorkshire 36 NW
Direction N63°E, throw 3 to 4 ft (0.9–1.2 m) N

Surface mapping by C. T. Clough and W. Gunn, coupled with the mining information now available, requires the presence of an approximately E – W fracture east of Lucks All Vein, throwing south as already described, and, west of Lucks All Vein, on what appears to be exactly the same line, a fault throwing north, with progressive increase of displacement westward. An attempt was made by means of a shaft to find Dam Rigg Vein west of Lucks All at about the point where March String should intersect it, but this does not appear to have been successful. Beyond this there is an untested gap, with Crow Beds and Ten Fathom Grit at surface, to the old Bishop workings which start 2860 ft (872 m) from the shaft. Here Bishop Shaft gave access directly to North Vein and the Sun Vein was reached by crosscuts. Also, a south crosscut from Cocker Vein (next section, below) described by Clough (*in* Dakyns and others, 1891, p. 175) as a boat level, intersected North Vein 1325 ft (404 m), and Sun Vein 1525 ft (465 m) from Cocker Vein beneath the Main Limestone. North Vein is recorded by Bradley (1862) as productive, though without 'rider', in Main Chert and Limestone, but he states that Sun Vein was unproductive. The former vein was worked for a length of 1400 ft (426 m), the latter tested for a somewhat less distance. On the east slope of Little Punchard Head, two shafts searched for the continuation but, though there is clear evidence of a shift of the base of the Crow Limestone, now indicating a northward downthrow, no ore was obtained. However on the west side of this valley the shafts at Heigh Punchal had found an orebody at least 750 ft (229 m) long by the year 1811 (ZLB 41/1–17) in A. D. mining ground, and a search at the remote head of Gunnerside Gill had revealed strong mineralisation in Crow Limestone on the same line. This was the discovery of the Blakethwaite Mine. A short level was driven to the vein from the steep side of Gunnerside Gill on top of the Ten Fathom Grit, and shafts were sunk from surface on the vein. Nathan Newbold, well known for his work at Greenhow Hill (Chapter 13) drew the 1811 plan and marked on it the positions of the Ten Fathoms Plate, Red Beds, Black Beds, Main Chert, Great Lime, Under Grit and Under Plate. By 1831, the Low Level (Blakethwaite Lead Level of the six-inch map) had been driven beneath the Main Limestone, south of Eweleap [9392 0234] (Figure 25, loc. 17) at 1510 ft (460 m) OD for

2350 ft (716 m) at N28°E, and 300 ft (91 m) turning to N to cut the vein. The beds dip north in the upper part of Gunnerside Gill and it is estimated that most of the Main Limestone was already below the Low Level horizon when the vein was reached. Nevertheless, the vein was stoped to 70 ft (21 m) above the adit at intervals along a stretch 1025 ft (312 m) long eastward to where the vein is recorded as having 'turned black', that is, filled with shale gouge. From the adit level a shaft was sunk 204 ft (52 m) deep, with levels at 60 ft (18 m) and near the bottom, respectively controlling the Main and Underset limestones on the footwall, at a point 700 ft (213 m) W of the original level head, and a branch was driven from the adit starting 1660 ft (506 m) from the mouth in a NNW direction to connect with the shaft. On the main vein the Main Limestone oreshoot was stoped continuously for 1800 ft (549 m) E of the shaft, and the adit level was continued 650 ft (198 m) farther east until it ran into the old man workings at Heigh Punchal. Some ore also appears to have been raised from the Underset horizon. Adjacent to the NNW part of the adit runs the Blakethwaite Sun Vein, from which some ore was raised in the Main Limestone and overlying cherts. At the east end of the mine, what was believed to be the continuation of Dean's Vein from Old Gang Mine (p. 129) was explored for about 600 ft (183 m). Meanwhile development was proceeding westward. After the Sun Vein united with the main vein about 200 ft (61 m) W of the Pump Sump Shaft to which reference has already been made, the dip of the measures to the west increased to a little over 4° and at 1760 ft (536 m) W of the shaft, the adit had reached shale above the Richmond Cherts on the footwall. Several sumps had been sunk into the Main Limestone, but what success these had is not known. A second underground shaft, Blakethwaite Engine Shaft, was now sunk, the section being as follows: (ZLB 41/13)

Collar at about 1530 ft (466 m) OD Site: [9342 0309]

	ft	m
Plate (shale)	15	4.5
Richmond Cherts: Red Beds	15	4.5
Black Beds	34	10.4
Main Chert	28	8.5
Main Limestone	72	21.9
Sandstone	30	9
Shale	29	8.8
Underset Chert	16	4.9
Underset Limestone (top)		

The engine at this shaft is believed to have been a hydraulic machine, such as was later used at Sir Francis Level. The 240-foot level was driven westward, opening up highly productive ground, until at 2674 ft (815 m) the top of the Main Limestone on the footwall went into the sole, and the throw was presumably such as to bring shale against the overlying cherts of the footwall. This was the position reached by 1850 (ZLB 41/1–9). A sublevel or 'rulleygate' 36 ft (11 m) above the 240-level was driven for approximately 900 ft (274 m) W from the shaft, but no doubt this became unnecessary as the beds dipped down. Both sides of a large loop made by the vein 300 to 1500 ft (91–457 m) W of the shaft were investigated, but the south side was most productive. From the shaft workings, ore worth £100 000 is stated in the records to have been raised; at the prevailing price of lead (Dunham, 1944, p. 198) this perhaps represented about 9700 tons of concentrates with 70 per cent Pb. The possible total production of the mine might have reached 20 000 tons lead concentrates, lead output 1817–74 amounting to 13 205 tons (Fieldhouse and Jennings, *op. cit.*). The cuttings from the dressing of the ore brought out at Low Level contain a large amount of chert, indicating that the vein was productive where at least one wall was in this rock. Witherite is a noticeable constituent (Table 15) with baryte; Clough records that he found some fluorite, but we were unable to do so. A little sphalerite is present.

The position of the west forehead of the 240-foot level is roughly on line with the prolongation of Blind Gill Vein from Lownathwaite, and as already noted, there was a plan to drive Sir Francis Level through to Blakethwaite on Blind Gill Vein that had to be abandoned with just over 5000 ft (1524 m) left to drive (p. 126). This would have given about 175 ft (53 m) of additional depth; at constant dip, this would have enabled 3000 ft (914 m) of additional ground in the Main Limestone to have been exploited. In fact, however, the dip slackens westward, for a bold project to explore for the Blakethwaite Vein which had been started in the 1850s, had proved its position. A shaft was sunk at Starting Gill, West Stonesdale [8856 0361] collared at 1330 ft (405 m) OD, with the following section: (ZLB 41/13)

	ft	m
Ten Fathom Grit sandstone	10	3
Shale	48	14.6
Richmond Cherts: Red Beds	18	5.5
Black Beds	36	10.9
Main Chert	27	8.2
Main Limestone	72	21.9
Sandstone	27	8.2
Shale	27	8.2
Underset Chert, into	5	1.5

The total depth, 270 ft, given by this record differs from that cited by Clough (*in* Dakyns and others, 1891, p. 177), who gives a figure of 299 ft (91 m) and states that the section started below the Ten Fathom Grit. The adjacent gill is named Great Bridge Gill on the six-inch map. From the shaft a level was driven N8°E for 1190 ft (363 m) where it encountered the NE vein here called Stonesdale Moor Vein. A rise was put up to the Main Limestone, and a drift was carried 1360 ft (415 m) NE in ground which is said to have yielded about £12 000 worth of lead ore, perhaps 1000 tons of concentrates. Fieldhouse and Jennings (1978) record 716 tons of lead produced from 1855–60. At the end, a fault throwing 42 to 54 ft (12.8–16.4 m) N and striking WNW was found; this was considered to be the continuation of Blakethwaite Vein, 12 400 ft (3.78 km) beyond the western forehead at the mine. This view was confirmed by C. T. Clough when he carried out the primary six-inch survey in 1870, but he also showed that a branch is given off the fault on the north side about 1 mile (1.6 km) from the Blakethwaite forehead, which should be about 1000 ft (305 m) N of the fault in West Stonesdale. From Stonesdale Moor Mine, 600 ft (183 m) of driving was done in Blakethwaite Vein, between wallrocks of Main Limestone and Chert, but it proved to be barren and the trial was abandoned. The western workings at Blakethwaite lie more deeply under the Namurian cover than any others in the present district. Directly above them, shales under the Fossil Sandstone (Figure 11) are faulted against shales under the Pickersett Edge Grit. Moreover, another fault, downthrowing north, leaves Blakethwaite Vein in the vicinity of the loop mentioned above, and crosses the moorland country in a west-north-west direction, eventually turning west to join up with the other branch already mentioned. The short drive from Stonesdale Moor Mine cannot be regarded as an adequate test, for the Stonesdale Moor Vein, carrying sphalerite, witherite and baryte as well as galena, is far removed from the main centre of mineralisation in North Swaledale indicating that this continues under the cover rocks.

In Dakyns and others (1891, pp. 177–178) Clough discusses the stratigraphical evidence for the faults shown crossing this moorland country, and Rowell and Scanlon (1957) in their resurvey fully confirmed the essential picture of the faulting. KCD

Cocker Vein, Cocker String Lead ore

NY 90 SE; Yorkshire 36 NE

Direction N75–85°W, vein throws 12 ft (3.7 m) N

Shakes Vein (Lead ore)

NY 90 SE; Yorkshire 36 NE
Direction N0–15°W

Wharton's String

NY 90 SE; Yorkshire 36 NE
Direction N80°W, S downthrow

Routh Veins

NY 90 SE; Yorkshire 36 NE
Direction WNW, branching

Punchard Gill Vein

NY 90 SE; Yorkshire 36 NE
Direction N60°W, throw 40 ft (12 m) NE, dying westward

Fox Holes Veins

NY 90 SE; Yorkshire 36 NE
Direction N65–70°E, two fractures 400 ft (122 m) apart, downthrowing S

Cocker Vein and, as already noted, the Bishops veins, were worked from Punchard Level [9605 0373] (Figure 25, loc. 18) starting at about 1470 ft (448 m) OD from Little Punchard Gill at the base of the Main Limestone. This level was probably begun earlier than 1800, and it was first driven SSW for 1050 ft (320 m) to where it cut Cocker Vein. This was worked eastward for 400 ft (122 m) until it diminished into a string running towards Lucks All Vein (p. 139). No connexion was made with the Danby Level workings. A strong string 90 ft (27 m) N of Cocker Vein was then followed westward from the main level; at 1450 ft (442 m) it cut Shakes Vein and this was followed for 200 ft (61 m); the main forehead on Cocker String lies 375 ft (114 m) W of this intersection. Cocker Vein and perhaps the string were worked from Cocker Low and High shafts on the western slope of the valley; at these the considerable quantity of spoil including baryte and chert suggests that the vein was more productive in the Richmond Cherts than in the Main Limestone. A shaft into the vein from the valley bottom, Gill Shaft, seems to have had less success. As elsewhere in the North Swaledale Belt, it is almost certain than these whim shafts were sunk prior to the driving of the level. The crosscut to the Bishops veins takes off south from Cocker Vein 275 ft (84 m) E of the main level head. This crosscut cut strings at 450 ft (137 m) and 1030 ft (314 m), and reached Bishops North Vein at 1450 ft (442 m). It was continued 700 ft (213 m) farther, the 525 ft (160 m) beyond Bishops Sun Vein finding nothing.

Wharton's String is a fault discovered in hushes on the north slope of the main gill, north of the Cocker workings. A level running south on the west side of the hush doubtfully reached it; a little baryte, sphalerite and traces of chalcopyrite occur on the dump. Although it strikes towards the Main Limestone scars in Little Punchard Gill, apparently it does not extend to them. Punchard Vein appears from the mapping to be the joined-up continuation of Martin's and Stodart veins from the Northside mines (p. 137). These veins throw respectively north and south before uniting, but the combined downthrow in Punchard Gill is to the north, repeating the Main Limestone outcrop; the displacement dies away to the west. Several small levels east of Punchard Gill tested both veins, without success; it seems clear that they are effectively barren west of Danby Level. The Routh Level [9611 0442] (Figure 25, loc. 19), started in 1855 from Punchard Gill at about 1340 ft (408 m) OD beneath the Main Limestone, follows an irregular course NW for 1150 ft (350 m) to cut the branching Routh veins, and to follow them to a point 1900 ft (579 m) W of the level head, where a crosscut was driven south for nearly 800 ft (244 m) to test Punchard Gill Vein, here apparently broken up into a belt of weak strings. The lack of any tailings outside Routh Level suggests that little ore was obtained from this extensive exploration. Kipling's Level, higher on the hillside in Black Beds, seems to have failed even to reach the veins. South-west of Routh Quarry two faults, here named the Foxholes veins, displace the outcrop of the Main Limestone. Shallow shafts and a short level found traces of baryte and a little calcite, but no substantial mineralisation. The last trials hereabouts were Fox's Level, driven to test Punchard Gill Vein under Low Whites Hill, east of the Gill, started about 1860, and Agnes Level, already mentioned in connexion with Wharton's String, driven 1907. Both were unsuccessful.

The description of these outlying workings completes the survey of the veins situated on the western slopes of Arkengarthdale. Attention will now be devoted to the eastern side of the dale, starting at Booze and Langthwaite and working north.

Fell End Vein Lead ore

NY 90 SE, NZ 00 SW; Yorkshire 37 NW, NE
Direction near E–W but varying a few degrees; throw 60 to 250 ft (18–76 m) N

Wellington Vein Lead ore

NZ 00 SW; Yorkshire 37 NE
Direction variable about N85°W, throw N

Tanner's Rake = Primrose Vein = Slack Vein Lead ore

NZ 00 SW; Yorkshire 37 NW
Direction generally N60°W, but with loops and branches

Blucher Vein Lead ore

NZ 00 SW; Yorkshire 37 NE
Direction N60°W

Scatter Scar Vein Lead ore

NZ 00 SW; Yorkshire 37 NW, NE
Direction nearly E–W, eastward turning to SE

This group of veins was worked at *Fell End Mine*, and from several adits in Slei Gill. Fell End Vein is considered to be the eastward continuation of the Justice and Dodgson's veins after the union near the mouth of Moulds Horse Level (p. 136). The fracture, now running E–W is mapped as crossing the drift-covered low ground north of Langthwaite, but on Scotty Hill, near Booze, a looping system of faults can be seen to bring Underset Limestone on the south side against beds above the Main Limestone on the north side; clearly the displacement has increased as the fracture crossed the valley. Several old shafts tested the faults on Scotty Hill, but with little success. Booze Wood Lead Level, starting 92 ft (280 m) west of the confluence of Slei Gill with Arkle Beck, may have been started as a working for tilestones in the sandstone beneath the Five Yard Limestone, but about 1870 it is probable that it was extended north to explore for lead. Baryte and galena, with some calcite and a little sphalerite, are present on the extensive dump, and it is possible that this level cut Fell End Vein beneath the hamlet of Booze; this cannot, however be confirmed as no plan has been found. From Spout, in Booze, the single main fracture can be followed down a shallow hush where a little mineralisation occurred, to cross Slei Gill and enter the Fell End Hush, one of the most spectacular mining excavations in the whole district (Plate 8). This has been cut through beds approximately from the Three Yard Limestone up to the Main Limestone on the footwall (south) side, and from the Underset up to the Crow Limestone on the hangingwall side, part

Plate 8 Fell End Hush, Arkengarthdale.

The observer stands on the downthrow side of the Fell End Vein upon an outcrop of Underset Chert. In the distance cliffs of Underset and Main limestones can be seen, displaced by the vein. The plateau on the left side of the vein is in Crow Limestone. that on the right in Richmond Cherts. The dumps of Fell End Mine are to be seen in the middle distance, right.

of this latter section being very well exposed (Figure 11). Breccia forming part of the Fell End Vein remains in place opposite the hangingwall positions of the Main and Crow limestones, showing that the vein was a broad fracture-zone cemented, mainly, with baryte, intermixed with galena. The principal underground entrance was Fell End Level [0215 0231] close to 1070 ft (326 m) OD. This starts as a NE crosscut through the Twentyseven Fathom Grit of the footwall, reaching the vein at 240 ft (73 m). From here, workings followed an irregular but generally eastward course along Fell End Vein for 2200 ft (671 m) to where it joins the WNW-trending Wellington Vein. About 125 ft (38 m) above this level, Compass Level, starting directly on the vein in the hush, opposite beds just below the footwall position of the Main Limestone, was also driven through to Wellington Vein, a distance of 1675 ft (511 m); it is noteworthy that this level diverges in its course farther and farther to the south of the position of the lower level, until it is more than 150 ft (46 m) S of Fell End Level at the junction with Wellington Vein. Assuming the underground survey to be accurate, the implication is that the inclination of Fell End Vein decreases from near 80° beneath the hush to 45° before reaching Wellington Vein. This was probably a highly mineralised stretch, but no stope sections are extant. At 1240 ft (377 m) from the head of the Fell End Level crosscut, a north crosscut was driven to investigate veins crossing from the higher reaches of Slei Gill. The horizon was probably the Main Limestone on the hangingwall side of Fell End Vein, and thus a favourable level for exploration. At 240 ft (73 m) Slack's Vein, a strong vein dipping north was found; working to the east, this was cut off by Blucher Vein, perhaps equivalent to Wellington Vein but not on exactly the same line. To the west, the Slack's Vein workings linked up with Primrose String, described below. Fell End Vein was evidently sufficiently encouraging to be worth testing in depth for in addition to the problematical Booze Wood Level, two other deep adits were driven. East of the main hush, Storthwaite Hall Level [0178 0217] (Figure 25, loc. 20), east of Slei Gill at 830 ft (253 m) OD (about the same topographic level as Booze Wood Adit) was driven N and NNE for 900 ft (274 m) to cut the vein beneath the footwall position of the Five Yard Limestone; the vein was followed east for 510 ft and a north crosscut was then tried, but although some baryte and calcite occur on the heap, it is clear that no productive mine was found. The same was true of Haggs Level [0254 0204] (Figure 25, loc. 21), driven NE on top of the Five Yard Limestone at 1020 ft (311 m) OD for 1600 ft (488 m) to cut Fell End Vein 3100 ft (945 m) E of the Storthwaite Hall Level intersection, and to cut Wellington Vein beyond that. The wallrock on the hangingwall side must have been the Underset Limestone or Chert, but evidently the juxtaposition of this with Five Yard Limestone did not provide favourable conditions. It must be concluded that at Fell End Mine, the range of favourable ground was not greater than from the Underset Limestone on the footwall up to the Crow Limestone on the hangingwall and thus not much exceeding 150 ft (96 m).

In Slei Gill the next vein-system of importance is Tanner's Rake, already mentioned as one of the two faults which shift Blackside Vein (p. 133). Its continuation north of Blackside Vein will be discussed later; but south-east of the intersection, it has been exposed in a major hush on the west side of Slei Gill, and followed by Doctor's Level [0180 0283], beneath the Main Limestone, and Tanner's Rake Level [0172 0294] in the cherts. The courses of these levels suggest that they were following a wide belt of fractures, and this is borne out by a surviving section of a crosscut driven north from the Rake Level, showing the beds successively stepped upward in that direction. This mine was still producing in the second half of the 19th century, but no separate figures are available for it. Baryte was the chief mineral associated with the galena (Table 15). As Tanner's Rake approaches Slei Gill it splits into two main fractures, the northern one being known as Primrose Vein east of the gill; the two reunite before linking with Slack Vein, already mentioned above. On the east side of the gill, the principal mine working was Sun Gutter Lead Level [0193 0291], running NE for 780 ft (238 m) beneath the Main Limestone to cut the E – W Scatter Scar Vein. This vein had also been tested by the crosscut from Tanner's Rake Level, in the Main Chert, where the downthrow was 22 ft (6.7 m) S; however, east of Slei Gill, the direction of the fracture changed to SE, and the inclination—and presumably downthrow—to NE. It was followed in this direction at the Sun Hush horizon for nearly 2000 ft (610 m) until it died out; it makes no shift in the scarp formed by the outcrop of the Lower Howgate Edge Grit, whereas this scarp terminates abruptly on the line of Blucher and Wellington veins to the south. These veins are in fact the main continuation of the North Swaledale Belt to the east, connecting up with the extensive series of veins at Hurst (p. 146). Before these are discussed, however, a digression will be made to cover the veins of north-eastern Arkengarthdale, lying north of the Blackside Vein. As already noted, Blackside Vein—now called North Rake, swings away NE on the east side of Slei Gill, and a test, probably in the Main Limestone on the downthrow side, made by means of a crosscut from Sun Gutter Level, gave no encouragement to follow it in that direction.

Cocker Rake Lead ore

NZ 00 SW; Yorkshire 37 NW
Direction N45°W, throw about 30 ft (9 m) NE

Freeman's Vein Lead ore

NZ 00 SW; Yorkshire 37 NW
Direction N25–50°E, throw 70 to 90 ft (21–27 m) SE

Windegg Vein Lead ore

NZ 00 SW; Yorkshire 37 NW, NE
Direction within a few degrees of E – W, throw 60 to 160 ft (18–49 m) S

Little Windegg Vein

NZ 00 SW; Yorkshire 37 NW, NE
Direction N70°W, changing westward to near E – W, throw 70 ft (21 m) N

Alcock Sun Vein

NZ 00 NW, SW; Yorkshire 37 NW
Direction N30–40°W, throw 70 ft (21 m) NE

North of Scar House, Langthwaite, Arkengarthdale has a steep eastern side with the tributary Shaw Beck or Shaw Gutter running at its foot. The scarp, formed by Brigantian and Early Pendleian strata, is cut through by a number of strong faults in which limited amounts of mineralisation have been found. Cocker Rake (not the same as Cocker Vein, pp. 140, 141) is, as already noted, the north-westward continuation of Tanner's Rake between the head of Tanner's Hush and Freeman's Vein, where it crosses the Richmond Cherts. A series of old shafts with associated hand dressing places mark its line; it was evidently exhausted in the cherts and probably in the Main Limestone at an early period. Freeman's Vein, which terminates Cocker Rake to the north-west, became productive north-north-east of the intersection, and substantial shafts were sunk into it. The gangue mineral along Cocker Rake was chiefly baryte, but north of this vein, dolomite and calcite predominate. Windegg Vein is a strong fracture, splitting up near the escarpment into two E – W and two SW fractures, the former pair well exposed in a prominent opencut through the Main Limestone. Windegg Lead Level [0083 0388] (Figure 25, loc. 22) seems to have followed one of the SW branches, in shale above the Underset Chert, but no plan is available; probably it reached the main vein about 1500 ft (467 m) NE of the portal. Chert, dolomite and a little baryte occur on the dump. Between Windegg Vein and Little Windegg Vein, about 1200 ft (366 m) to the north, Main Limestone with an outlier of Main Chert forms the summit of the fell on which stands Horrocks' Cross [0100 0418]. This is, in effect, a horst, and though the bounding faults were poorly mineralised, the limestone in the horst contained the most extensive area of floats in the district. At least five adits were driven in from the scarp face, and on the fell top the sites of at least 42 shafts, sunk into the limestone, can be identified on the ground and from the air photograph. These seem to arrange themselves into ENE lines, suggesting the direction of the leaders from which the floats were mineralised. The spoil shows that dolomite was the main mineral associated with the galena, together with calcite. To the east the limestone dips beneath younger beds, or is perhaps faulted down by a continuation of Alcock's Sun Vein; but the area beyond is marked 'Brass Pump Floats' on the six-inch map, so it appears probable that the replacement ore continued under cover; there are, however, no remains of shafts here. Probably this area was reached from Washy Green Level [0215 0368] (Figure 25, loc. 23) driven beneath Slack Wife Gill, the headwater of Slei Gill at about 1285 ft (392 m) OD. This was driven N15°W starting in shale above the Richmond Cherts, and reaching Windegg Vein at 1020 ft (311 m), here upthrowing 160 ft (49 m) and giving command of the Main Limestone in the succeeding 800 ft (244 m) to where Little Windegg Vein was encountered. On Windegg Vein drifts were carried 600 ft (183 m) W, and 360 ft (110 m) E; mapping on surface suggests that about 900 ft (274 m) ahead of the latter forehead, the two Windegg faults virgate and die out. The area of virgation was probably tested by a level in the Ten Fathom Grit still visible when W. Gunn carried out the primary survey, but no mineral was found. Little Windegg Vein evidently offered no encouragement for development from Washy Green Level. West of the level, Alcock Sun Vein may be regarded as a NW branch off Little Windegg Vein, but though tried south of Alcock Hush, it nowhere proved to be worth working. Summarising it may be said that in north-east Arkengarthdale, the mineralisation becomes markedly poorer north of Blackside Vein, the only oreshoots of interest at Windegg being the floats; and it may be inferred that most of the ore from these was extracted by small partnerships in ancient times.

Alcock North Vein = Moresdale Ridge Vein Lead ore

NZ 00 NW, SW, SE; Yorkshire 37 NW, NE
Direction N75°W, changing near Rispey Wood to N70–80°E; throw 180 to 240 ft (55–73 m) N changing to S

Alcock Sun Vein forms a link between the Windegg veins to the south and a powerful nearly E – W belt of faulting to the north, the junction being near Alcock Hush [012 052]. The main fault, here called Alcock North Vein, carries the Main Limestone down from the top of the scarp into the valley to the north; in the hush the beds can be seen to turn down in this direction at 30 to 40°. Alcock Hush Level (incorrectly marked Windegg Lead Level on the six-inch map) was the last underground working in the Windegg area. It was begun in 1870 and driven on the base of the Main Limestone at about 1490 ft (454 m) OD on the south side of the hush, and consists of a straight crosscut for 600 ft (183 m), cutting a small offshoot of Alcock North Vein here named Harker's Vein. This was followed east for about 900 ft (274 m) until 1877, when a crosscut was driven NE to test the main vein. Another crosscut was driven south to test the Sun Vein, but both veins were too poor to work. Small floats were found on the south side of Harker's Vein and were worked until 1893, when the mine closed. Extensive caverns were found, a plan of which appears in Brook and others (1977, p. 72).

Alcock North Vein appears to line up with the strong fault which follows Moresdale Ridge, south of Moresdale Gill, over the watershed to the east, but if the fractures are the same, the throw has reversed, for the downthrow here is to the south. The fracture is situated in an axial position along an anticline. The presence of the fold is not evident in the highest reaches of Moresdale Beck, where the solid geology is heavily obscured by peat; however, near Fair Seat Hill, where Porter Level [0438 0478] (Figure 25, loc. 24), driven from Moresdale Beck, cut Moresdale Ridge Vein (Plan ZWX 13)[1] close to its junction with North Rake (Blackside Vein), the rise of the beds to the south is clear. The vein was cut 900 ft (274 m) S of the portal; in this stretch the base of the Main Limestone rises from 1200 to 1360 ft (366–415 m) OD. A little calcite and white baryte was found, but nothing workable. Another test of the vein was made by means of a line of shafts across its course, 1 mile (1.6 km) farther west. These appear to show that chert and limestone to the north is faulted against Ten Fathom Grit to the south, and thus that the reversal of throw takes place nearer to Alcock Hush; but mineralisation was not proved. East of Porter Level, Moresdale Ridge Level, driven 950 ft N10°E from the south side of the ridge at [0523 0400] (Figure 25, loc. 25) started near the top of the Richmond Cherts at 1220 ft (372 m) OD, but (Plan ZWX 11) reached the vein below the Main Limestone without finding ore. The structure is well exposed in Rispey Wood and on Hollin Brow, ¾ mile (1.2 km) farther east, where the south limb of the anticline is now the most important. A breccia zone, dipping southward at 56°, a lower angle than the main fault on the downthrow side, is exposed on the hillsides on both sides of the river, but it contains only limonite and calcite. The various fractures gave some hope to early prospectors, but none of their shallow shafts found a mine. The Moresdale Ridge structure, apparently so favourable, and with the best wallrock horizons readily accessible, is disappointingly barren.

Faggergill Old Vein Lead ore

NY 90 NE; Yorkshire 23 SW
Direction N55°W, low SW dip

First To Fifth Faggergill Strings Lead ore

NY 90 NE, NZ 00 NW; Yorkshire 23 SW
Directions mainly N50–55°W, but sinuous to N25°W, little if any throw

Faggergill North Vein Lead ore

NZ 00 NW; Yorkshire 23 SW
Direction N55°W, throw 3 ft (1 m) SW

Main Crossing

NY 90 NE, NZ 00 NW; Yorkshire 23 SW
Direction N40°E, small NW throw

Pounder's Vein Lead ore

NZ 00 NW; Yorkshire 23 SW
Direction N30–50°E, throw 14 to 20 ft (4–6 m) NW

Black Vein

NZ 00 NW; Yorkshire 23 SW
Direction N70°E, turning to N50°E, throw where proved, 100 ft (30 m) NE

Stang Vein Lead ore

NZ 00 NW; Yorkshire 23 SW
Direction N45°E, turning to E – W, throw N, variable

[1] North Yorkshire County Record Office

Faggergill and *Stang Mines*, the extensive workings of which lie beneath and adjacent to Hoove Hill in the north-easternmost part of Arkengarthdale, represent the workings farthest removed from the central axis of the North Swaledale Belt. They were also the last substantial workings in the Belt, finally closing in 1913. Although Faggergill Old Vein may have been discovered in the cherts above the Main Limestone before the 19th century, the main period of development, during which at least 5 miles (8 km) of tunnelling were accomplished, lay between 1840 and 1881. Faggergill No. 1 Level [9895 0710] at about 1355 ft (413 m) OD was driven SE into Old Vein beneath the base of the Main Limestone, and followed this vein to a point 3180 ft (969 m) from the portal; here it was turned NE to become an exploratory crosscut running N46°E for 2100 ft (640 m), then N13°W for 340 ft (104 m) and finally approximately N40°W for 2150 ft (655 m), with a step of 80 ft (24 m) to the SE at 1150 ft (350 m) from the beginning of this last stretch. The level had command of the Main Limestone throughout its length, and this bold exploration led to the discovery of at least six strings, weak veins that nevertheless had considerable lateral continuity. The last stretch of the exploratory crosscut was driven along a fracture known as the Main Crossing, but it is also possible that the earlier parts of the development also followed small faults. Probably the crossing was not ore-bearing, but the strings, on which altogether not less than 5015 ft (1.53 km) of drivage north-west of the crossing, and 4580 ft (1.4 km) SE of it was undertaken must have yielded sufficient ore to justify the work. Cut from No. 1 String, at the outset of the N40°E stretch of the crosscut, the SE drive encountered a strong vein trending NE, 1600 ft (488 m) from the crossing, which, as Pounder's Vein, became for a long period the mainstay of the mine. It proved to be continuously mineralised over nearly the full height of the Main Limestone for a distance of 4700 ft (1.43 km); the Main Chert was also productive in part, where faulted against the limestone. However, the vein must have cut off clean above this level, for no attempt to follow it into the Richmond Cherts or higher beds was made. The limited surface evidence on the east side of Hoove suggests that the Richmond Cherts contain much more shale than in the ground to the south and east. One test of Pounder's Vein by means of a winze to the Underset Limestone was made at a point 360 ft (104 m) NE of No. 1 String intersection, where the base of this limestone was found 79 ft (24 m) below the base of the Main, but the trial was unsuccessful. The Pounder's Vein oreshoot is thus an ideal illustration of the ribbon type of shoot so characteristic of the Northern Pennine Orefield; the detailed dimensions were: 4800 × 60 (average) × 4+ ft (1463 × 18 × 1.2 m) estimated to contain, at 14 cubic feet per tonne, 82 300 tonnes. No means exists of estimating the average grade. At the northern end of Pounder's Vein workings, another NW vein, North Vein was cut and this proved productive for 320 ft (98 m) SE of Pounder's Vein, and 2040 ft (622 m) NW, the ore again being confined to the Main Limestone, the oreshoot averaging 50 ft (15 m) in vertical dimension. The ventilation of these extensive workings was achieved without shafts to the surface, suggesting that there must have been open, cavernised ground on the strings. To improve the situation, an air level was driven, starting 2225 ft (678 m) ENE of the portal of No. 1 Level, at the base of the Coal Sill Sandstone, connected to a rise from the north-westernmost workings on 1st String. In the presence of a well-developed sandstone between the Main and Richmond cherts, the Faggergill valley differs from all the rest of the North Swaledale Mineral Belt, resembling more closely the Alston Block at this level. In Carkin Crag, on the west side of the valley, the sandstone must reach at least 50 ft (15 m) thick. It is not clear whether the Coal Sill Sandstone was tested from the Air Level; a short vein called Harker's Vein nearby was more probably tested in the Main Limestone.

On the west side of Faggergill, No. 2 Level, starting from the low side of the large spoil heaps, 1250 ft (381 m) SW of No. 1 Level, was a prospecting drive in sandstone and shale beneath the Main Limestone running 1960 ft (598 m) N16°E, presumably driven in search of Old Vein on this side of the valley; nothing was found. No. 3 Level, starting 450 ft (137 m) NNE of No. 1 was begun at about 1390 ft (424 m) OD from the gill bottom in 1888, and driven irregularly northward, cutting at 1200 ft (363 m) N of the portal a vein running WNW (perhaps the continuation of one of the strings worked on the east side of the valley) cut by a strong NW string. The vein was worked for 1850 ft (564 m) W of the level head, the forehead being suspended here in 1909. Farther south on the west side of the valley, Sloat Hole Level starting 1980 ft (604 m) WSW of No. 1 Level portal at about 1395 ft (425 m) OD was begun in 1908. The orebodies exploited here may have been discovered in two small levels driven from swallow holes in the top part of the Main Limestone, Sloat Hole and Nut Hole. There seems to have been a small fault (Scott's Vein) running through these swallows, parallel to the NE veins of Faggergill and this formed the south-eastern limit of the float-like orebodies mined here. The workings, a network of drives and crossings oriented NW and NE perhaps illustrate the nature of float workings in Swaledale; no comparable plan remains available. The dump shows a mixture of limonite and dolomite against limestone wallrock (analysis, Table 15, No. 32). Limonitic dolomite was, in fact, the principal associate for the galena at Faggergill; hardly any baryte occurs and no fluorite was found. At one time, fine specimens of aragonite could be obtained from cavities (p. 94). Dr F. W. Smith, who has examined these deposits underground informs us that galena which can be seen replacing limestone parallel to the bedding is no more than 1–2 cm thick.

East of Hoove, the Stang workings were in operation during the middle period of Faggergill operations. The Black Vein is a main fault that flies off the Alcock North fault-vein, west of Shaw Beck, running at first ENE. Shaw Level [0512 0499], driven almost duc N from the west bank of Shaw Beck for 2030 ft (619 m), starting in arenaceous beds between the Five Yard and Middle limestones but perhaps entering beds above the former limestone after passing through the north-westerly continuation of Little Windegg Vein, was designed to reach the junction of Black Vein with Alcock North. From this junction, a drift followed Black Vein for 1225 ft (373 m) to the ENE, then ran N as a crosscut for 600 ft (183 m) and finally turned ENE for a further 1650 ft (503 m). The geology of this extensive exploration can only be surmised, since no extant records have been found; but the final stretch, which cuts across the line of Faggergill Old Vein without finding it, may have been about 100 ft (30 m) below the base of the Main Limestone as this formation dips westward off Black Vein. There is no veinstuff on the dump from Shaw Level, the deepest test stratigraphically in the Faggergill – Stang ground. Black Vein farther north turns NNE, and its effect can be seen west of the road to the Stang Forest summit, where the Ten Fathom Grit dips steeply westward on its hangingwall. Stang Level [0101 0591] (Figure 25, loc. 26) started at almost the same topographic level as Faggergill No. 1 Level, beneath the Main Limestone. Pursuing a winding course NE for 1100 ft (335 m) it entered Stang Vein, a branch from Black Vein and followed this for only 375 ft (114 m) ENE before turning to the north as a crosscut, again following a very irregular course. Black Vein was cut 950 ft (284 m) due N of the start of the crosscut, and after passing through it (presumably it was quite barren) a NW fissure named '1st String N. of Black Vein' was found and followed for 2300 ft (701 m). Further NE faults and blocks of tilted strata parallel to Black Vein were found but these yielded no orebodies. The string, however, north-west of the second of these, was productive, for it was developed both at the Stang Level horizon and, by way of a crosscut from Faggergill No. 4 string, and Faggergill No. 1 horizon, here 32 ft (10 m) above Stang Level as a result of its greater distance from the point of origin. Such ore as was won from Stang Mine must have come from this working, but the quantity of tailings, chiefly dolomite and limestone, is small. The Arkengarthdale production of lead concentrates from 1891 to 1902, amounting to 3469 tons, came largely or wholly from Faggergill and might perhaps represent a quarter of the total output of this outlying area. Stang contributed 240 tons in 1907–9, while Sloat Hole (Nut Hole) produced 265 tons from 1910 to 1912, possibly the total yield of the flots. At Arndale Head, 2½ miles (4 km) E of

Stang Mine, ENE and NE faults in Richmond Cherts and Ten Fathom Grit have been investigated, disclosing calcite, some of it pink.

Copperthwaite North Vein Lead ore

NZ 00 SW, SE; Yorkshire 37 SE
Direction N59°E, throw 15 ft (4.6 m) NW

Copperthwaite Vein Lead ore

NZ 00 SW, SE; Yorkshire 37 SE
Direction N85°E, swinging eastward to N55°E, throw 24 ft (7 m) S

Copperthwaite Sun Vein Lead ore

NZ 00 SE; Yorkshire 37 SE
Direction N 56°E

From the outlying northern part of the North Swaledale Mineral Belt at Faggergill, a return is now made to the south side. The Copperthwaite veins are the southernmost group worked by the former *Hurst mines* which extracted numerous deposits on the plateau extending north-eastward from Fremington Edge, opposite the town of Reeth. Reference has already been made to the fact that there is limited evidence of Roman working in this area. The main Copperthwaite Vein cuts through the outcrops of the Underset and Main limestones on Fremington Edge and has been worked continuously for 1¼ miles (1.9 m) across the outcrops of the Main and Richmond cherts, almost reaching the outcrop of the Main Limestone on the west side of the valley leading to Hurst hamlet. At the western end it is divided into three main parts, forming a mineralised belt 150 ft (46 m) wide, but as the direction changes, the branches unite to form a single fissure. The workings belong largely to the period of shallow shafts; the debris from the hand dressing of the ores contains baryte, calcite and, at the western end, a little fluorite. Samples collected [0551 0008] by Dr Earp in 1941 and petrographically examined by Dr J. Phemister proved that a brown to yellow mineral in fibrous and radiating clusters was hemimorphite, and this mineral is not uncommon along the vein. A little sphalerite, from which it was no doubt derived can also be found. Traces of malachite also occur, but in spite of the name of the vein, there is no evidence that it was ever a source of copper ore. Levels were driven into the vein from both sides of the plateau, but the evidence of the small dumps suggests that by the time this was done, most of the ore had already been mined from shafts. From the East Level [0635 0048] a sample rich in hemimorphite collected by Dr Earp assayed 19.2 per cent Zn (Geological Survey Lab. No. 1201). At the western end of the range of workings, a string is given off NE and, according to notes on the primary six-inch map by W. Gunn, flots were mined in connexion with this string. The principal level shown on the 19th century plan of the mine (N. Yorks. County Record Office No. ZWX 17) started from Fremington Edge approximately 50 ft (15 m) above the top of the Main Limestone. The eastward rise of the measures here brought this horizon into the level 970 ft (295 m) from the portal and the base of the limestone entered at 2630 ft (802 m). Patchy stopes are shown in the last 1020 ft (311 m) in the limestone and overlying beds, and there was a level 60 ft (18 m) below the main adit. Bradley (1862) states that the vein was productive in the Underset Limestone and Chert. The final phase of work at Copperthwaite was the driving, about 1870, of a level from Fremington Edge [0434 9998] (Figure 25, loc. 27) on top of the Five Yard Limestone. The drive passed through considerable slipped ground, but appears to have entered solid about 1000 ft (305 m) from the portal; the vein was then found by means of short south crosscuts, and some 300 ft (91 m) of drivage on vein was done, reaching beneath ground productive in the Main Limestone. This trial in depth evidently failed to produce a mine. Copperthwaite Sun Vein is a weak string worked only on the east side of the ground, while Copperthwaite North Vein, pursuing an unvarying straight course north of the curviform main vein, failed to reach the Main Limestone outcrops either on the east or west sides of the plateau. However, it yielded some lead ore from shafts over a stretch about ½ mile (0.8 km) long, associated with baryte, calcite and a little pale fluorite. A cross fracture running NNW may have terminated the oreshoot to the east.

Jingle Pot North Vein Lead ore

NZ 00 SW; Yorkshire 37 SE
Direction N74°E, small S downthrow

Jingle Pot Vein and String Lead ore

NZ 00 SW; Yorkshire 37 SE
Direction N55°E, branching N80°E; throw 12 ft (3.6 m) N

These veins cut through Fremington Edge about ½ mile (0.8 km) NW of the Copperthwaite workings. Workings, mainly shafts, extend over a length of 2400 ft (730 m) and yielded galena, baryte, calcite and some hemimorphite from the Main Limestone and overlying cherts. Jingle Pot Level [0399 0081] (Figure 25, loc. 28) at 1280 ft (390 m) OD started half-way up the Main Limestone (Plan ZWX 15) but owing to the eastward rise of the beds, most of it was in the sandstone and shale beneath that limestone. Underset Shaft was reached at 1050 ft (320 m), the name suggesting that a trial was made below the level. Thereafter the nearly E–W branch of the vein was followed until the String joined it from the SW. The mineralisation probably became impoverished 1300 ft (396 m) beyond Underset Shaft, and the level was then turned to nearly due N and driven for 2050 ft (625 m) beneath the shallow valley separating Fremington Edge from Hurst (Plan DCRM 14, dated 1852), where its course can be followed by a line of 12 shafts. This bold exploration, known as Langstaff Trial, unfortunately revealed no new veins, but the level must have given access to the southernmost of the Hurst veins proper, the Grinton Dam or Petticoat Vein, in the Underset Limestone (Figure 27). A level was also driven along the String in the Main Limestone from the eastward-facing scarp at the back of Fremington Edge, but probably not before much of the ore had been won from shafts. No attempt was apparently made to test the veins in depth from the main Fremington scarp, but the shaft mentioned above probably indicates that a trial was made to the Underset Chert and Limestone.

Hind Rake = Golden Vein Lead ore

NZ 00 SW; Yorkshire 37 SE
Direction N80–65°E, throw 54 to 30 ft (16–9 m) N

Racca = Grinton Dam Vein Lead ore

NZ 00 SW; Yorkshire 37 NE, SE
Direction N60–80°W, throw 200 + ft (61 + m) (Racca); small N throw (Grinton)

Gladding Garth Vein Lead ore

NZ 00 SW; Yorkshire 37 SE
Direction N85°W

Petticoat Vein Lead ore

NZ 00 SW, SE; Yorkshire 37 SE
Direction N80°W swinging to N45°E

Barf Vein Lead ore

NZ 00 SW, SE; Yorkshire 37 SE
Direction N85°E swinging to N40°E

Stodart or Standard Vein Lead ore

NZ 00 SW, SE; Yorkshire 37 SE
Direction N75°E, modest N throw

Mole Vein Lead ore

NZ 00 SW; Yorkshire 37 SE
Direction N75°E, throw 6 ft (2 m) N

Wagget or Scott Vein Lead ore

NZ 00 SW, SE;Yorkshire 37 SE
Direction N75°E, turning to E – W, throw 30 to 36 ft (9–10.4 m) N

Blindham = Redshaft Vein Lead ore

NZ 00 SW, SE; Yorkshire 37 NE, SE
Direction N70°E, turning to E – W, throw 90 ft (27 m) N

Cleminson Vein Lead ore

NZ 00 SW, SE; Yorkshire 37 NE, SE
Direction N70°E, throw 54 ft (16 m) N

Woodgarth Vein Lead ore

NZ 00 SW; Yorkshire 37 NE, SE
Direction near E – W, throw 36 to 42 ft (10–12 m) N

Trench Vein

NZ 00 SW; Yorkshire 37 NE, SE
Direction N70°E, throw 18 ft (5.5 m) N

Wallnook Vein

NZ 00 SW, SE; Yorkshire 37 NE, SE
Direction N70°W, throw 20 ft (6 m), reversing to 245 ft (75 m) NE south-eastward

The veins listed above carried the major deposits worked at the *Hurst Mine* proper, but numerous minor veins, strings and flots were also exploited. The structure (Figure 27) is one of unusual interest, epitomising a tendency for WNW fracturing to intersect and give place to ENE fissures at angles of about 40° and 140°, seen elsewhere along the North Swaledale Belt. Here the Racca Vein forms a significant link with the Fell End veins (p. 141) and may be regarded as the continuation of the line of Cocker Rake – Tanner's Rake – Primrose, Blucher and Wellington veins, though it is evident that a major NE throw is transferred to it from Fell End Vein, reaching over 200 ft (60 m). Indeed Gunn regarded Racca as the continuation of Fell End Vein rather than of the others, though the mining evidence does not justify the gradual swing round of Fell End into the Racca direction. As Racca or its continuation, Grinton Dam Vein reaches the higher ground south of Hurst, its throw dies away, and it splits into a series of curviform weak veins, including Barf, Petticoat and many associated strings. Meanwhile, parallel to Racca Vein and ½-mile (0.8 km) to the north-east, Wallnook Vein changes from a minor to a major fault as it continues towards the south-east. The gap between the two is bridged by the complex north-facing monocline of Hurst, riddled with closely spaced fissures, which formed the locus for the lead ore deposits. As might be expected, the total displacement effected by the Hurst structure is comparable with the maximum throws of the Racca and Wallnook faults; the base of the Main Limestone falls from 1230 ft (374 m) OD on the south side of Stodart Vein to between 900 and 960 ft (274–292 m) OD north of trench Vein. Surface geological mapping did not, and does not enable the displacements of the numerous veins to be determined in most cases; but fortunately the questionnaire circulated by L. Bradley to mine operators prior to the publication of his book in 1862 elicited the amounts and working horizons, but not the directions of displacement. Since, however, the sum of the throws of the principal veins (276 ft (84 m) maximum) compares so closely with the total displacement in the structure, the direction must be consistently to the north. It seems certain that lead ore came to crop out on the Hurst hillside, both in Main Limestone and in the overlying cherts. No doubt the earliest workings were shallow opencuts, but the gentle topography did not lend itself to the excavation of large hushes: North Hush (on the strung-out continuation of Mole and Scott veins) and Sun Hush (following Sun Strings which branch from Stodart Vein) are modest affairs compared with Fell End Hush, and the excavation on Golden Vein is only a shallow trench. Much of the early working and, it may be suspected, a high proportion of the ore won here, came from shafts operated by small partnerships. The first edition of the six-inch Ordnance map shows a distribution of bell-type shafts as follows: Hind Rake 9, Golden Vein 13, Blindham 6, Redshaft 18, Wagget 16, Mole 12, Stodart 18, Barf and Petticoat 12 each; and not every shaft visible on the ground is on the map. The Main Limestone disappears below ground north of Stodart Vein, and north of Redshaft Vein was probably beyond reach of shallow shaft mining. The tips from the shafts in this part of the area make it clear that much ore was won from the Main and Richmond cherts, probably largely from the latter. The earliest adits were probably the Nungate [0530 0235] and Opencut levels, designed to reach the ground beneath North Hush and Hurst Level [0439 0220], but the lowest of these left the Main Limestone undrained in the northern half of the mining area. Probably late in the 18th century, a water level was driven up from Washfold, where there is a small outcrop of Main Limestone with its base at 990 ft (302 m) OD, the terminal point apparently being Wallnook Shaft [0488 0248], but the portal of this had been lost by the mid-19th century (its position is still unknown), for a part of the level was revealed in a swallow hole or subsidence in the hamlet near Padley Beck. The Water Level is 138 ft (42 m) below surface at Wallnook Shaft and thus at about 1000 ft (305 m) OD. Queen's Level [0501 0248] (Figure 25, loc. 29) at 1097 ft (334 m) OD, presumably started about 1837, became the principal entrance to the mine in the 19th century (Plans ZWX 8–10); its course is shown on Figure 27. The eastern branch, driven southward through the whole structure, must have reached beds below the Underset Limestone towards the south end; Bradley records that Wagget Vein and the Sun strings were productive in this limestone and the overlying chert. The middle branch followed Cleminson Vein, probably in Richmond Cherts, to Redshaft Vein where lower beds including Main Limestone would be accessible on the footwall. Rises linked the level with Nungate Level south of Redshaft Vein. The more important branch, however, was the south-western one, driven along Woodgarth Vein to Cat Shaft [0455 0226] near Hurst Hall, the principal operating shaft during last century. Queen's Level was continued to Middle Whim [0404 0220] and ended 595 ft (181 m) WSW of this shaft according to the mine-section dated 1881 (N. Yorks. Records, ZWX 19). This document shows that Cat Shaft was 240 ft (73 m) deep, not 300 ft (91 m) as stated in a report made during the first World War. As a result of the structural situation, the workings on Woodgarth Vein above Queen's Level were in the upper part of the Richmond Cherts; in addition, ten rises between Cat and Middle Whim shafts gave access to workings at the Crow Limestone and Chert horizons. With the most promising of the bearing-beds below Queen's Level, another level, known as Brown's Low Rulleyway was developed 72 ft (22 m) below, at about 1025 ft (312 m) OD, connecting Cat Shaft with Smith's Sump near the portal of Queen's Level. From Cat Shaft the Rulleyway continued 1650 ft (198 m) to connect with Middle Whim. Drainage was provided by the old Water Level already mentioned, which was extended from Wallnook Shaft via Smith's Sump (ZWX 10) to form the bottom level from Cat Shaft at about 1000 ft (305 m) OD. The mine-section (ZWX 19) nevertheless makes it clear that the workings from these levels were wholly above the Main Limestone, except on Golden Vein, where a sublevel 450 ft (137 m) long, driven eastward from a sump below Middle Whim at about the horizon of the Water Level, gave access to the top one-third of the limestone. The sublevel was probably not connected to the Water Level but was drained by a water swallow marked on the section, but no doubt the water found its way into the level. It seems therefore that on the north side of the Hurst structure, the Main Limestone is still unexplored on some of the principal veins such as Woodgarth, Trench and much of Golden. The base of the limestone is shown as reaching its lowest point at 893 ft (272 m) OD, more than 100 ft (30 m) below the Water Level, between Cat Shaft and Middle Whim. Potentially productive ground may, therefore, remain in this part of the mine.

Towards the west, the Hurst veins link up with Fell End (Figure 27). The principal development in this ground was from Hurst Horse Level [0439 0220] (ZWX 3–6) 180 ft (55 m) above Queen's Level at 1277 ft (389 m) OD. This was driven via Blindham Vein and adjacent strings for some 1800 ft (549 m) WSW to meet the Racca – Grinton Dam trend, where its direction changed through E – W to WNW. At March Shaft [0372 0202], The Golden Vein cuts this trend and continues WSW as the weak Hind Rake, eventually reaching Fremington Edge where there were workings in the Main Limestone and Chert. The Horse Level now follows Racca (= Wellington) Vein 2650 ft (808 m) to Wellington Shaft, close to the western margin of the Hurst property. The productive ground above the Horse Level was mainly against a hangingwall of Crow Limestone and Chert, made accessible by numerous rises. There was a small number of sumps below the Horse Level, some of which reached the Red Beds Limestone of the Richmond Cherts, but none penetrated the Main Limestone or even the lower part of the Richmond Cherts. At the western side of the property, the ground was developed in depth by Wellington Level at 1208 ft (368 m) OD running 1020 ft (310 m) ESE from Wellington Shaft [0302 0233]. This level is marked 'from Mill Level' on Plan ZWX 19, probably indicating a connexion with the top or Compass Level of Fell End Mine. There was a sublevel at 1247 ft (380 m) OD between the Horse Level and Wellington Level. At greater depth, the main Fell End Level was continued along Wellington Vein into Hurst ground for 1270 ft (387 m) at about 1080 ft (323 m) OD. The 1881 mine section shows clearly that this level commanded most of the Main Limestone on the hangingwall of Wellington Vein; as a result of the considerable displacement transferred to this vein from Fell End Vein (about 200 ft—61 m) the beds in contact with the Main Limestone, Main Chert and Little Limestone on the footwall side must have been the Underset Chert and Limestone, and the underlying sandstone. No stope section has been preserved, but the numerous rises up to the hangingwall position of the Little Limestone from this part of Fell End Level suggest a good oreshoot here. The position of the base of the Main Limestone on the footwall must have corresponded with the 1247 ft (380 m) OD sublevel mentioned above and what most probably was an upper oreshoot lay between beds from the base of the limestone up to the lowest part of the Richmond Cherts on the footwall, and the Ten Fathom Grit, Crow Limestone and Crow Chert on the hangingwall. A short sublevel near the top of the hangingwall position of the Main Limestone carried the lower oreshoot 150 ft (46 m) ESE of the forehead of Fell End Level, but there is then an unexplored gap nearly 2000 ft (610 m) long back to the sump and sublevel on Golden Vein already mentioned. The upper oreshoot, accessible from Hurst Horse Level, continues through this ground as already noted. These data again suggest that the Hurst veins have not yet been fully exploited. It is in this ground that the ENE veins one after another take over the displacement from the Wellington – Racca – Grinton Dam trend, a structurally interesting prospective situation. Of the closing stages of mining here, Raistrick (1975) records that Cookson & Co., who took over in 1885, 'had great good fortune and soon brought the output up to 2000 bings (800 tons) a year', but the falling price of the metal and a law suit brought abandonment in 1890. The Mineral Resources Development report of 1917 also refers to frequent breakages of inadequate machinery as a cause of the shut-down and adds 'Old miners state that the ore increases in quantity with depth and also that 200 tons of lead ore were left broken in the bottom level'. From 1852 to 1890 (with no production 1881–85) the total yield of lead concentrates was 13 375 tons. Raistrick (*op. cit.*) gives the annual production as 400 tons a year in 1814 when Thomas Stapleton of Richmond took over and this was still the yield in 1851. The total output since the 18th century can hardly have been less than 40 000 tons and if the earlier period of shaft working is taken into account, the grand total must be substantially greater. The widely-scattered small tailings heaps contain some baryte of good quality, and white witherite is occasionally seen. Unrecovered but partly oxidised galena is widely present, but fluorite is only a minor or trace mineral. The general impression gained is that gangue minerals were only locally present in substantial quantity, but it has to be borne in mind that the miners of old times were accustomed to slitting the veins so as to leave the gangue as far as possible, and to sort the ore underground to avoid bringing it out.

Wallnook North Vein

NZ 00 SE; Yorkshire 37 SE, 38 SW
Direction N70–75°W, throw 160 ft (49 m) NE

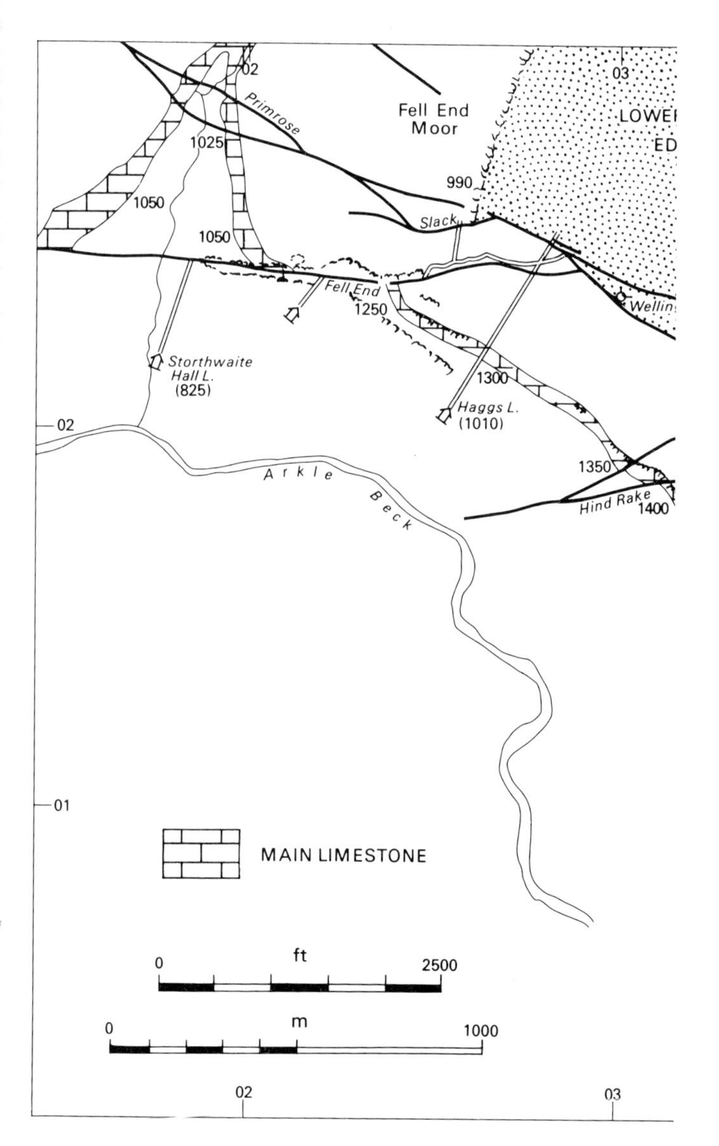

Figure 27 Geological map of the Fell End-Hurst mines

Prys North Vein Lead ore

NZ 00 SE; Yorkshire 37 SE
Direction N50°E, throw 30 ft (9 m) NW

Prys Sun Vein Lead ore

NZ 00 SE; Yorkshire 37 SE
Direction N30°E

Shaw Vein

NZ 00 SE; Yorkshire 37 NE, SE, 38 SW
Direction N30°W changing eastward to N84°E, throw 0 to 130 ft (34 m) NE

The main Wallnook 'Vein', forming the north-east margin of the Hurst vein complex has been tested at a number of places; Bradley (1862) lists it as unproductive and there is no reason to think that it is other than a barren fault. In Roan or Langstaff Hush, which runs south-west from the head of Padley Beck, the fault zone was identified in shale above the Crow Chert; the limited surface evidence suggests a small SW throw here. Farther ahead, the fault was apparently not identified in the workings of the coal seam lying a short distance below the Lower Howgate Edge Grit. Some driving in the fault zone was done from Wallnook Shaft [0488 0248] but the branching nature of the drives suggest that the ground was not encouraging. Hodgson's Shaft, 425 ft (130 m) to the ESE probably reached the Crow beds, but found nothing. Where the fault passes under the Hurst road, a north branch is given off, and Underset Shaft was sunk 500 ft (152 m) ESE of this point to test it. A drainage level (Shaw Level) was driven up 1400 ft (427 m) from Washpool at 990 ft (302 m) OD passing through a shallow syncline in Main Limestone 580 to 1200 ft (177–366 m) inbye. The Underset Limestone would be accessible on the footwall of the fault, but there is no evidence to suggest that ore was found. The entrance to Shaw Level (ZWX 14) can be identified only with difficulty, and the stream seems to have washed away most of the spoil. However, strong mineralisation appears in a block on the north-east side of the Wallnook North Vein at the *Prys* (or Pryse) *Mine*, in the two Prys veins, and in a series of flots related to Shaw Vein. Here the Main Limestone crops out on the downthrow side of the Wallnook structure, and no doubt the original discoveries were made from surface, but development in depth was presumably impeded by water. In 1859 a long adit was driven from Shaw Beck below White Scar. Two alternative courses are marked on document DCRM 14, but examination of the site shows that Prys Level was driven from [0664 0252] (Figure 25, loc. 30) and not from the alternative site 800 ft (244 m) to the NE of this point. Valuable information on the

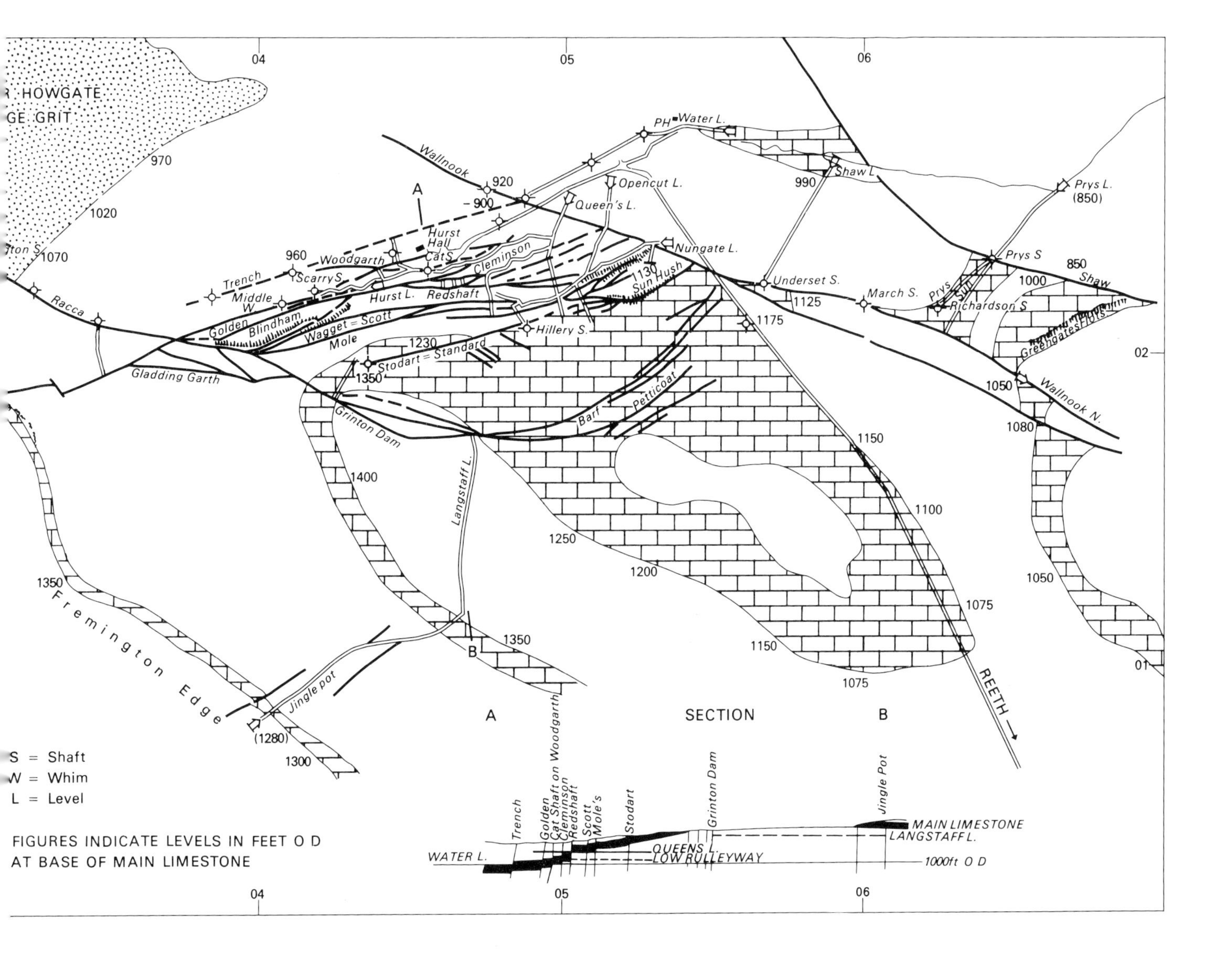

workings was supplied to Dr Earp by Mr W. S. Rider in 1941, and North Yorkshire County Records contain a plan and detailed geological section of the mine (ZWX 16). Prys Level starts at 850 ft (259 m) OD in shale of the Black Beds and Little Limestone in a gentle structural sag, reaching the fault known as Shaw vein 830 ft (252 m) from the portal. A slice of Main Limestone was passed through in the fault-zone, but beyond it the level entered grit and shale beneath the limestone. The drive had been aimed at East Close or Prys Shaft (not the Prys Whim of the primary geological survey, that being the same as March Shaft mentioned below). Close to the shaft the level cut Prys North Vein, which had no doubt been discovered long before at surface in the Main Limestone. It was continued along this vein for some 1100 ft (335 m) until it swung round and united with Wallnook North Vein, and the level connected to March Shaft [0600 0215] on that vein. The oreshoot on Prys North Vein, mainly developed after 1859, proved to have a greater vertical extent than most in this district. An underground shaft was sunk 600 ft (183 m) SW of East Close Shaft, and levels were driven at 810 ft OD (Water Level), 738 (18 Fathom), 700 (28 Fathom), 628 (38 Fathom) and 587 (45 Fathom) (respectively 246, 224, 213, 191 and 179 m OD). At each level the mineralised ground appears to have covered nearly the full distance between Shaw Vein footwall and Wallnook North Vein. The dip of Prys North Vein at 57° NW is lower than normal, but it steepens to 67° towards the SW; perhaps the curviform fracture favoured the creation of open space at the time of mineralisation. Stratigraphically the oreshoot was adjacent to beds from the Main Limestone down to the Five Yard Limestone which was reached in a single sump below the 45 Fathom Level. No detailed record of the stoping remains, but the numerous rises from each of the levels leaves little doubt of the continuity of mineralisation over a section-area approximately 1100 × 300 ft (335 × 91 m). Structurally the vein cuts across a gentle syncline of amplitude about 70 ft (21 m) lying between the two NW veins. Sun Vein was presumably worked; it was not developed below Prys Level. The underground shaft was equipped with a hydraulic engine, the water draining from Prys Level, not the so-called Water Level which may have been intended for the alternative adit mentioned above. The tailings, of which only about 2000 tonnes remain, are mainly a mixture of chert and baryte (Table 15). Along Shaw Vein south-east of the mine, extensive runs of flots in the outcropping Main Limestone were worked at Greengates and Cuddy's Grove, no doubt long before Prys Level was driven. No separate official figures for the production from Prys are available, but the mine is said to have yielded about 140 tons of concentrates per year for many years. The most recent work was in 1938 when an attempt was made to continue Prys Level north-westward along Wallnook Vein. After driving beyond March Shaft, a sump was sunk on the downthrow side of the vein, cutting a string of ore in the Underset Limestone. Owing to difficulty with water this project was soon abandoned.

Hurst and Prys mines may be regarded as the last substantial mining operations encountered in proceeding eastward along the North Swaledale Belt. Thereafter, the known structures are mainly faults of large displacement which, though flots and belts of scrins have been found in association with them in a few places, have for the most part failed to carry workable oreshoots; Shaw Vein is a good case in point. Nothing was found in it where is crosses Padley Beck at Washfold, and to the north-west its throw dies out. A trial made by W. Garthwaite in 1795 by driving north in shale above the Main Chert from Washfold may not have reached the vein, but two shafts sunk into it from the base of the Ten Fathom Grit probably did reach it, but found only ferruginous limestone. In Prys Mine there was ample opportunity to examine the fault over a considerable vertical range, but nothing in the records suggests that anything was found other than flots in the Main Limestone, already mentioned. Shaw Vein can be traced for over 1 mile (1.6 km) E of the intersection at Prys Adit, its direction having meanwhile changed through 90° from the course north of Padley Beck. North and north-east of Munn End [0777 0210], scrins (closely-spaced mineralised fractures or joints) running NE and ENE were found in association with it. From here, W. Gunn found that it was linked by a curviform fault downthrowing nearly 300 ft (91 m) N with the Telfit workings, described below. To the south of this middle range of faults, the Wallnook veins continue, the North Vein becoming the major fault known as the Marske Fault, southward linkages from which connect with the complex but poorly mineralised ground around Marrick Barf (q.v.). North of Shaw Vein, the major fault from Moresdale Ridge and Rispey Wood (p. 144) splits into two, and through a complicated series of linkages eventually connects with the Feldom Vein, the first of the Richmond copper veins (Chapter 10). A fair amount of prospecting has been done through the years in the section of the North Swaledale Belt lying east of Prys Mine, perhaps enough to show that, though there is a well-developed fracture system, this was penetrated by only a meagre flow of mineralising fluids. Still more strange is the fact that the central part of the Middleton Tyas–Sleightholme Anticline (Wells, 1957), lying to the north and north-east, is virtually unmineralised. The remaining prospects are, as before, described from south to north.

Marrick Barf South Vein Lead ore

SE 09 NE; Yorkshire 52 NE, 53 NW
Direction N60°W, turning to E–W, throw 20 ft (6 m) NE

Marrick Barf North Vein Lead ore

SE 09 NE; Yorkshire 52 NE, 53 NW
Direction N60°W to N70°W, throw variable, NE

Marrick Great Vein

SE 09 NE; Yorkshire 52 NE, 53 NW
Direction N30–60°W, displacement at least 230 ft (70 m) NE

Bents Vein Lead ore

SE 09 NE; Yorkshire 53 NW
Direction N60–80°E, throw 70 ft SE to over 150 ft (46 m) NW

Bents North Vein

SE 09 NE; Yorkshire 52 NE, 53 NW
Direction E–W, throw N

Marrick Great Vein is a major fault which extends from beneath the Swale alluvium (where it may link with Downholme Vein, p. 152) passing under Marrick village, and bounding the Marrick Barf outlier of Main Limestone to the north-east. It is only known to be mineralised at its north end, where it splits giving off the two Marrick Barf veins to the WNW, and the Bents Vein to the ENE. The Great Vein itself terminates against Bents North Vein, north of which the faulted-in outlier of coal-bearing Ten Fathom Grit is undisturbed along the Great Vein direction. Geological mapping towards the west requires North and South veins to be crossed by faults trending ENE, making angles of about 40° with them, and also calls for a parallel WNW fault south of South Vein. However, in spite of the numerous fractures, mineralisation at Marrick was feeble; a little calcite and baryte are present, but such galena as was obtained seems to have come from flots in the Main Limestone, on the south side of North Vein and the west side of Great Vein. Bents Vein was tried in 1843 by the Wild Sike Level [0763 9885] but only shale is to be seen on the dump; a cross-measures level in Red Beds nearby was also unsuccessful. Bents Vein, after Bents North Vein joins it, changes its throw to NW and continues to the vicinity of Marske Hall as a main fault, there impinging on the Marske Fault. The ground traversed is mainly drift-covered, and no prospecting has been done north-east of the Bents workings. Marrick was an important lead smelting centre, there being a mill here as early as 1592 (Raistrick, 1975); the site of Cupola Mill, north of Ince Wood is known, and the impressive ruins of two others remain on the south bank of Ellers Beck (Clough, 1962). The Marrick veins are certainly not such that these mills could have been kept supplied with ore from them, and it appears likely that much of the production from Hurst was in fact smelted here.

Cleaburn Pasture Vein Lead ore

NZ 00 SE, SE 09 NE; Yorkshire 38 SW
Direction N38°E, throw 125 ft (38 m) SE

Cleaburn Pasture String Lead ore

NZ 00 SE; Yorkshire 38 SW
Parallel and 475 ft (145 m) to SE, throw transferred in part

Skelton = Wallnook Vein Lead ore

NZ 00 SE; Yorkshire 38 SW
Direction N60–75°W, throw N

The Wallnook Main Vein of Hurst was for the most part a barren fault where tested there (pp. 147, 149). Shafts sunk in the vicinity of the Reeth – Hurst road are reputed to have found associated NNE scrins from which a little production was obtained. However in a stretch starting 3000 ft (914 m) ESE of the road and continuing for 2700 (822 m) very numerous old shafts into Richmond Cherts found lead ore with calcite and baryte. The vein appears to die out by virgation with the Cleaburn Pasture Vein and String, the former also having been exploited from numerous shallow shafts continuing along strike for 2800 ft (853 m) to the SSW. At the end of this stretch, a barren E – W fault seems to have terminated the oreshoot; the mapping suggests that this fault, turning towards the NW, almost links up with the eastern end of the Copperthwaite workings. Cleaburn Pasture Vein was tested about 1870 by means of a level [0770 9983] driven on the base of the Underset Limestone from the upthrow side of the fault at 920 ft (281 m) OD, but only shale gouge appears to have been found. An older level, 50 ft (15 m) above this, from the downthrow side, which must have intersected the vein beneath the Main Limestone, was also unsuccessful and may have found that the old shaft-workers had already taken the ore from that limestone.

North of the Skelton Vein workings, shafts were sunk to test the Marske Fault, supposedly the south-eastward continuation of Wallnook North Vein, though here and for the rest of its 2-mile (3.2 km) course to the River Swale, the direction of downthrow is in the opposite sense to that proved at Prys Mine. Nothing appears to have been found, and no further trial is known to have been made between here and the Swale. On the east side of the river the fault passes north of the large roadstone quarry in Main Limestone and Chert situated beside the Richmond – Reeth main road 4 miles (6.4 km) from Richmond; here the beds turn down south into the fault and there are traces of galena in calcite. The Marske Fault dies out a short distance to the south-east.

Moss Mire Vein Lead ore

NZ 00 SE; Yorkshire 38 SE
Direction N50°E, throw about 15 ft (5 m) NW

Remotely situated south of Munn End [0777 0210] and north of the Wallnook North Vein, a belt of shallow shafts 1500 ft (457 m) long marks the course of this vein through the upper beds of the Richmond Cherts. Calcite and galena are present on the tips. The workings do not extend to the outcrop of the Main Limestone in the Helwith valley to the north-east, and no test of this vein by adit has been made. Around the hamlet of Helwith a few isolated shafts have tested the Helwith Fault and showings in the cherts, but without success.

Telfit Vein

NZ 00 SE; Yorkshire 38 NW, SW
Direction N72°E, W – N 78°E, throw exceeding 150 ft N

Telfit Branch Vein

NZ 00 SE; Yorkshire 38 NW, SW
Direction N78°E, throw about 100 ft (30 m) N

Clints Vein Lead ore

NZ 00 SE; Yorkshire 38 SW
Direction N65°W, throw 15 ft (5 m) NE

Clints East Vein

NZ 00 SE; Yorkshire 38 SW
Direction N25°W, throw 20 ft (6 m) W

Orgate Vein Lead ore

NZ 00 SE, 10 SW; Yorkshire 38 SW
Direction N65°E, throw 100 ft (30 m) SE

This group of veins occur on the east side of the valley of Marske Beck between Telfit Farm and Orgate Bridge. As already noted, Telfit Vein forms a link with Shaw Vein and the Helwith Fault by way of an untested stretch varying in strike between ENE and E – W. East of the farm, an ESE trend is established, and the fault brings Main Limestone on the south side against Ten Fathom Grit to the north. Telfit High Level, at about 780 ft (238 m) OD tested this stretch, probably with the Twentyseven Fathom Grit on the footwall and Black Beds on the hangingwall, but without finding much ore. An old record states that two parallel strings were cut, one down 10 ft (3 m) S, the other down 30 ft (9 m) N. East of this working the Branch Vein comes in from the south, augmenting the throw to the minimum figure cited above. Clints Vein may also be regarded as a flyer from the Telfit vein-fault; ore was worked from it on the high ground, in Main Limestone, where the East Vein and other strings were found, probably with associated flots. These deposits were worked from small levels high on Clints, but a test of the veins was also made from Telfit Low Level [0890 0249] at 700 ft (213 m) OD in the Twentyseven Fathom Grit, the first part of the level being driven SSE to reach the vein. The large dump contains sandstone and shale, but little evidence of mineralisation. Orgate Vein gives rise to a prominent nick in Clints Scar, where it brings Main Limestone against Red Beds to the south. A level was driven below the Scar at 880 ft (268 m) OD along the vein, with Underset Limestone on the footwall and Main Limestone on the hangingwall. Calcite and a little coarsely crystalline baryte are present, but there is no indication that considerable quantities of ore were worked. Orgate Level [0934 0165] driven about 1870 from the east bank of Marske Beck at 625 ft (191 m) OD was driven 480 ft (146 m) N to cut the vein, in which a rise of 120 ft (37 m) was made, probably reaching the Five Yard Limestone on the hangingwall. The dump contains sandstone, shale, a little calcite and galena, but there is no evidence of ore dressing. In the past it is said that there were two smelt mills (Clough, 1962), one of which may have been working at the end of the 17th century. It is difficult to believe that the Telfit – Orgate group of veins could have maintained smelters for any long period, and Clough mentions that ore was brought here from Arkengarthdale to smelt. The various fractures unite into a main fault which runs SE to Clapgate Spring, in the next valley. Here the Main Limestone can be seen to be faulted down to the north nearly 100 ft (30 m), but no mineralisation has been found.

Clapgate Scar Veins

NZ 10 SW; Yorkshire 38 SW
Direction N80°E, throw N

A complex disturbance which crosses Clapgate Gill ¾-mile (1.2 km) N of Clapgate Spring and bridge, is also mapped as branching from the Telfit Vein. The fault gives rise to a feature on the west bank, bringing down the Main Limestone from 825 ft (251 m) OD on the south side to the stream bed at 700 ft (213 m) on the north; this is probably accomplished partly by folding. Branching fractures within the monocline were worked from two levels on the downthrow side, where calcite and baryte were associated with some galena; further ore may have been obtained from shafts at higher levels. On the east side of the gill nothing appears to have been found. Northwards the base of the Main Limestone rises again to 775 to 800 ft (236–243 m) OD before encountering another fault downthrowing north, but this one appears to be unmineralised.

Thringill Hill Vein

NZ 00 SE; Yorkshire 38 NW, 37 NE
Direction N75°W, changing to N85°E, throw 25 to 75 ft (8–24 m) N

Holgate[1] Pasture Vein

NZ 00 SE; Yorkshire 38 NW, 37 NE
Direction N70°E, changing to N80°W, throw about 100 ft (30 m) S

High Waitgate Vein Lead ore

NZ 00 SE, NE, NZ 10 SW; Yorkshire 38 NW
Direction N55–70°W, small S throw

West Feldom Vein Lead ore

NZ 10 SW; Yorkshire 38 NW
Direction N20–30°W, small E downthrow

The structurally interesting but poorly mineralised Moresdale Ridge disturbance (p. 144) splits into two main branches east of the Rispey Wood exposures. The southern one becomes the Thringill Hill Vein crossing Throstle Gill (Waitgate Gill) at [0846 0405], while the northern branch crosses Holgate Pasture, and Throstle Gill at [0840 0468]. Both branches exhibit the same abrupt changes in direction that are such a remarkable feature of the disturbance farther west. Evidently they looked interesting to the old-time prospector, as there are shallow shafts into them at wide intervals; but mineralisation proved to be sparse. Shafts on the Thringill structure [0786 0399], found a little galena, with calcite and baryte. As exposed on the east side of Throstle Gill valley, a broad belt of ferruginous dolomite accompanies the faulting, but that is all. The Holgate Pasture structure eventually converges on the High Waitgate Vein, which serves as a link with the Feldom Vein (p. 157), ½-mile (0.8 km) to the south-east. According to Wells' (1957) revision of W. Gunn's mapping, Feldom Vein lines up exactly with the Thringill Hill disturbance, and it is possible that the two join beneath the obscured ground north of the abandoned Cordilleras Farm. High Waitgate Vein has a remarkable belt of scrins associated with it, covering an area roughly 1000 × 500 ft (300 × 150 m), a feature which shows up strikingly on the air photograph, the closely-spaced fractures, branching nearly eastward from the vein in Main Limestone. Another similar belt of scrins was found associated with West Feldom Vein, which cuts through the same wide outcrop of the limestone, but here the direction of fracturing is NE. Closely-spaced shafts associated with High Waitgate Vein near the farm of that name suggest that flots were also exploited. All these workings appear to be very ancient, and almost completely grassed over. The scale is very small compared with the mines in the central and western part of the North Swaledale Mineral Belt.

Downholme Vein Lead ore

SE 19 NW, Yorkshire 53 NW
Direction N76°E, displacement 60 ft (18 m) NW

White Earth Vein Lead ore

SE 19 NW; Yorkshire 53 NW
Direction N86°E, displacement 3 ft (1 m) S

High Spring Vein = Marske Fault

NZ 10 SW; Yorkshire 38 SW
Direction N75°W, displacement 85 ft (26 m) S

[1]Spelt Hallgate on some older six-inch maps

Historical records show that a mine was being opened at Downholme in 1396 (Raistrick, 1975, p. 19). The vein runs through the village and there are remains of shafts, now largely overgrown, at the eastern end of it, both on the line of the fault and to the south. A little galena can still be found, with calcite, but the deposit cannot have been substantial. There may also have been a level into the vein at the Richmond Cherts horizon from the hangingwall. About ⅓-mile (0.5 km) N of the village, more is to be seen at White Earth in the steep west-facing feature formed by beds from the Main up to the Crow Limestone. Here a level has been driven into White Earth Vein at the top of the Main Limestone; the absence of workings below this suggest that flots were being sought. The level, at 690 ft (210 m) OD yielded calcite, aragonite, and minor amounts of baryte. Small trials were made in the vein at higher levels. The Marske Fault is considered to be the prolongation of Wallnook North Vein from Prys Mine. Mapping requires a substantial southward downthrow south of Marske village, and though no mineralisation is known to have been found in this neighbourhood, there are remains of trials in the wood north of the Council Quarry in the Main Limestone and Chert at High Spring [118 000] on and north of the line of the fault. Shafts marked 'Old Pits' or 'Old Coal Pits' on the six-inch Ordnance Map probably explored flots with calcite and galena on the footwall side of the fault, while farther north in West Wood, small faults with only a few inches of throw affect the limestone and top of the underlying sandstone, and contain calcite and galena.

Thorpe Under Stone Veins Lead ore

NZ 10 SW; Yorkshire 38 SE
Direction near E–W, and N15°W, the former downthrowing N

Whitcliffe Scar Vein

NZ 10 SW; Yorkshire 38 SE
Direction N38°W

Deep Dale Vein

NZ 10 SW; Yorkshire 38 SE
Direction N40°E

The E–W vein at Thorpe Under Stone [126 005] brings down Main Limestone and overlying cherts on the south side. A few shafts have tested the vein itself, but the chief interest lies in limestones, particularly the pale crinoidal ('Red Beds') limestone in the Richmond Cherts to the south, where the weak SSE fracture has probably acted as a leader for flots. The flots were worked from at least seven short levels and accompanying shallow shafts, the mineralisation consisting of galena in a calcite matrix. An old quarry on the south side of the fault reveals the character of the ground; there are many cavities and joint planes in the limestone, which has a decomposed appearance, coated with calcite and in places with galena. The length of ground exploited may have extended as far as the 'Old Pits' shown on the map, ¼-mile (0.4 km) S of the E–W vein and the distribution of old levels suggests a maximum width up to 700 ft (0.2 km), but in the nature of the deposits, only limited parts would be workable.

On the small vein near the west end of Whitcliffe Scar, a level at 780 ft (238 m) OD was driven at the base of the Main Limestone, while another, above the top gave command of the cherts and intercalated limestones. Calcite including scalenohedra from cavities, a little pink baryte and small amounts of galena occur on the tips. The small vein, nearly at right angles to the direction of Whitcliffe Scar Vein, tested by a level in Deep Dale contained a little calcite and baryte in flaggy sandstone below the Main Limestone.

Near Hudswell there are two occurrences in Crow Beds, one on the NE-trending fault crossing Sand Beck [0022 9957] (see p. 161), and the other on a possible E–W vein near the church. Both carried calcite and galena, but cannot have yielded much ore. KCD, JRE

OUTPUT AND RESOURCES

Where production figures are available for individual mines, these have already been stated in terms of long tons of galena concentrates. Additional general figures summarised from company and estate papers in the North Yorkshire County Records Office by Fieldhouse and Jennings (1978, app. 5) and elsewhere, taken in conjunction with official statistics from 1845 on, enable the following record (Table 21) to be compiled for the Belt as a whole. The A.D. Mines include all those in Swaledale west of Reeth and north of the Stockdale Disturbance, together with Surrender Mine which crossed the watershed into Arkengarthdale. The C.B. Mines cover all the remainder in the Arkengarthdale drainage. Hurst, and the mines lying east and south-east of it are listed separately:

Table 21 Recorded production, lead metal and concentrates, Area 11

Group	Years	Lead metal long tons	Concen- trates long tons	Lead %
A.D. Old Gang–Lownathwaite	1696–1700	2 200*	3 750	59‡
A.D. Mines	1786–1794, 1796–1799	12 005†	20 008	60‡
	1800–1809, 1811	21 766†	36 639	60‡
	1817–1844	50 499†	72 141	70‡
	1845–1867	39 702†	56 235	70.6§
	1868–1913	24 279§	35 230	69§
C.B. Mines	1783–1791, 1794–1799	10 095†	16 825	60‡
	1845–1867	29 462‖	41 731	70.6§
	1868–1912	26 167§	38 330	68.3§
Hurst	1852–1890	9 340§	13 375	69.8§
		225 515	334 264	67.4¶

Sources
* Raistrick and Jennings, 1965, p.155.
† Fieldhouse and Jennings, 1978, appendix 5, p.492.
‡ Assumed as basis for calculation of concentrates production.
§ Official statistics, *Mem. Geol. Surv.*, 1848–1881; Home Office 1882–1913.
‖ Calculated by deducting A.D. Mines production (including South Swaledale) as given by Fieldhouse and Jennings from the total figure for Swaledale (excluding Hurst) in official statistics for this period.
¶ Average value.

Gaps in the record (Table 21) may be filled in tentatively on the following basis. To cover all mining in the belt prior to 1696, a total of 20 000 tons of concentrates is adopted. By that year Gunnerside Gill was already yielding 440 tons lead (or say 750 tons concentrates) per year. Smelt mills at Orgate and Marrick had already been working for a hundred years, replacing crude bole-hill smelting; the Bathurst family had acquired exclusive rights in Slei Gill mill in 1629 and Lord Wharton's Hartford mill at Gilling had been in work since 1671 (Raistrick, 1975). The figure may be conservative, but this was the period when the industry was growing from small beginnings. For the A.D. Mines, the gap between 1701 and 1786 unfortunately covers a highly active period, when the shallower oreshoots were largely worked out from hushes and shafts. A figure for this period may be obtained by averaging the lead outputs at the beginning and end of the period (respectively 440 and 632 tons) suggesting 45 560 tons lead as the total, or 76 000 tons of 60 per cent Pb concentrates. The gap from 1812 to 1816 coincides with Easterby, Hall and Co's reconstruction at Old Gang, and is perhaps best left unfilled. The very successful work of this concern at the C. B. mines had however increased output there to such an extent as to call successively for the replacement of Low Moulds Old Mill by the Octagon four ore-hearth plant in 1803–5, and the erection of the Langthwaite New Mill, with three ore hearths, in 1821. No doubt need be felt, even though the records no longer apparently exist, that the years from 1800 to 1844 were productive years in Arkengarthdale and having regard to the mining evidence, a figure of 1400 tons concentrates per year is adopted, giving 63 000 tons for the period. The hushes and some of the shafts in Arkengarthdale certainly predate 1783 when the earliest surviving records begin, and for the earlier part of the 18th century, a figure of 25 000 tons concentrates is included. Finally, following experience in other districts, it is assumed that two-thirds of the production predated the beginning of records for Hurst in 1851, and 26 750 tons concentrates is assumed to include also the smaller mines between Hurst and Richmond. The estimated addition to the recorded output of concentrates is thus taken at a rounded figure of 215 000 tonnes, making the total output of the North Swaledale Belt slightly over 550 000 tonnes of galena concentrates.

If it is asked whether adequate smelting capacity existed to deal with this ore, Raistrick's account (1975) makes it clear that it did. At least 18 different mills operated within the belt between 1589 and 1913. Working from the dates given by him leads to a rough figure of 1738 mill-years; and accepting his figure of a little over 1 ton lead smelted per ore-hearth per 12-hour shift, or about 300 tons per year suggests that there could have been capacity to smelt 780 000 tonnes of galena, allowing only one hearth per mill. In fact, the larger mills such as Marrick, Octagon, Langthwaite, Old Gang and Surrender had more than one hearth each. It is not, however, surprising that the smelt mills were not used to maximum capacity, for as Raistrick points out, the advantage of the ore-hearth was its flexibility; it could easily be shut down and opened up, according to the supply of ore from the mines. The reverberatory furnace, demanding more continuous operation, was used only at the Cupola Mill at Marrick, and then only in the later half of the 19th century.

No proved or even probable reserves of lead ore remain. The root beneath the Lownathwaite – Old Gang lode could be pursued farther in depth, and searches could be made for others beneath Surrender and the junction-area of Blackside Vein with Stodart, Dam Rigg, and Justice veins (pp. 134–139). Ore could possibly exist in the bearing beds concealed beneath younger cover-rocks in the north-western part of the belt. The only tangible resource, however lies in unrecovered lead and zinc in mineral wastes, and in the associated spar minerals, particularly baryte and fluorspar. The latter are slowly being recovered in Hard Level Gill, while a bulk sampling programme has been launched in Arkengarthdale and at Hurst preparatory to treatment of the dumps in a modern plant yet to be erected. The results of this programme indicate, in general, somewhat lower values than those listed in Table 15, where the samples are of tailings and did not include waste rock, but the implications for the relative distribution of minerals are unchanged. Since bulk sampling of a substantial part of the Belt, including the stretch from Beldi Hill to Old Gang, has not been undertaken, no accurate figure for the tonnage of residues can be given, but about 750 000 tonnes may be taken as a very rough figure. Comparing this with the production of lead (Table 21 and associated discussion) would suggest a very high grade of ore, over 30 per cent Pb. This figure is high even by comparison with the richest deposit in the northern Pennines for which accurate data are available (Boltsburn; see Dunham, 1944) and is difficult to accept. There are two probable explanations: (i) considerable tonnages of tailings have been carried away by streams and rivers; (ii) careful sorting of the ore underground was practiced, leaving behind the spar minerals as far as possible. The second possibility was probably correct prior to the driving of the horse levels, when all the ore had to be hauled up narrow shafts, often by hand. A stage may therefore be reached when the reworking of some parts of the mines for spar minerals and unrecovered sulphides may be worth considering.

The lead mines of the North Swaledale Belt provided a living, often in conjunction with small-holding, for many generations of miners, smelters and associated workers (Raistrick, 1955, 1975; Cooper, 1948, 1960; Jennings, 1959; Fieldhouse and Jennings, 1978; Barker, 1972). Has this village industry any future apart from the recovery of wastes? In its old form, almost certainly not; but if substantial new oreshoots could be found, central treatment of the ore produced might make a modern operation possible. Although the contention of Carruthers and Strahan (1923), that erosion has cut through the zone of mineralisation, is true for a substantial part of the Belt, in that the bearing beds from the Underset up to the Crow Limestone lie above the valley bottoms, it has previously been maintained (Dunham, 1958, 1959, 1974) that targets worth exploring still remain. The present detailed study has shown that in addition to the root zones mentioned above, the recognised bearing beds are partially or wholly untested on the north side of the Hurst structure, and west of the Blakethwaite workings, where Rowell and Scanlon (1957) confirmed the existence of many faults in higher Namurian strata. The whole question of whether beds below the Middle Limestone may have become mineralised also remains open. KCD

REFERENCES

Barker, J. L. 1972. The lead miners of Swaledale and Arkengarthdale in 1851. *Mem. North. Cavern Mine Res. Soc.*, Vol. 2, pp. 89–97.

Bradley, L. 1862. *An inquiry into the deposition of lead ore in the mineral veins of Swaledale, Yorkshire.* (London: Edward Stanford.) 40 pp.

Brook, D., Davies, G. M., Long, M. H. and Ryder, P. F. 1977. *Northern Caves.* Vol. 5, 2nd edit. (Clapham: Dalesman) 160 pp.

Carruthers, R. G. and Strahan, A. 1923. Lead and zinc ores of Durham, Yorkshire and Derbyshire, with notes on the Isle of Man. *Spec. Rep. Miner. Resour. Mem. Geol. Surv. G. B.*, Vol. 26.

Clough, R. T. 1962. *The lead smelting mills of the Yorkshire Dales and Northern Pennines.* 2nd Edit. 1980. (Keighley: Clough.) 332 pp.

Cooper, E. 1948. *Muker. the story of a Yorkshire parish.* Chapt. 10. (Clapham: Dalesman.)

— 1960. *Men of Swaledale: an account of Yorkshire farmers and miners.* (Clapham: Dalesman.)

Crabtree, P. and Foster, R. 1963. Sir Francis Mine. *Cave Sci.*, Vol. 5, pp. 1–24.

— 1965. The Kisdon Mining Company. *Bull. Peak Dist. Mines Hist. Soc.*, Vol. 2, pp. 303–306; Vol. 3, pp. 63–67, 119–124.

Dakyns, J. R., Tiddeman, R. H., Russell, R., Clough, C. T. and Strahan, A. 1891. The geology of the country around Mallerstang, with parts of Wensleydale, Swaledale and Arkendale. *Mem. Geol. Surv. G. B.*, 213 pp.

Dunham, K. C. 1944. The production of galena and associated minerals in the Northern Pennines; with comparative statistics for Great Britain. *Trans. Inst. Min. Metall.*, Vol. 53, pp. 181–252.

— 1948. Geology of the Northern Pennine Orefield. Vol. I—Tyne to Stainmore. 1st Ed. *Mem. Geol. Surv. G. B.*, 357 pp.

— 1952. Fluorspar. 4th Ed. *Spec. Rep. Miner. Resour. Mem. Geol. Surv. G. B.*, Vol. 4., 143 pp.

— 1958. Non-ferrous mining potentialities of the Northern Pennines. Proc. Symposium on the future of non-ferrous mining in Great Britain, *Inst. Min. Metall.*, London, pp. 115–147.

— 1959. Epigenetic mineralisation in Yorkshire. *Proc. Yorkshire Geol. Soc.*, Vol. 32, pp. 1–29.

— 1974. Epigenetic minerals. Pp. 294–308 in *The geology and mineral resources of Yorkshire.* Rayner, D. H. and Hemingway, J. E. (Editors). (Leeds: Yorkshire Geological Society.) 405 pp.

— and Dines, H. G. 1945. Barium minerals in England and Wales. *Geol. Surv. Wartime Pamphlet*, No. 46.

Fieldhouse, R. and Jennings, B. 1978. *A history of Richmond and Swaledale.* (Chichester: Phillimore.) 520 pp.

Jennings, B. 1959. *Leadmining in Swaledale.* M.A. thesis, University of Leeds.

Lodge, P. D. 1966. Hydraulic pumping and mining machinery, Sir Francis Level. *Mem. North. Cavern Mine Res. Soc.*, Vol. 2, pp. 151–160.

Mitchell, W. R. 1977. Old Gang and beyond. *Dalesman*, Vol. 38, pp. 801–803.

Raistrick, A. 1927. Lead mining and smelting in West Yorkshire. *Trans. Newcomen Soc.*, Vol. 7, pp. 81–96.

— 1955. *Mines and miners of Swaledale.* (Clapham: Dalesman.) 92 pp.

— 1975. *The lead industry of Swaledale and Wensleydale.* Vol. 1 – The mines; Vol. 2 – The smelting mills. (Buxton: Moorland.) 120 pp. each volume.

— and Jennings, B. 1965. *A history of lead mining in the Pennines.* (London: Longmans.) 347 pp.

Rowell, A. J. and Scanlon, J. E. 1957. The Namurian of the north-west quarter of the Askrigg Block. *Proc. Yorkshire Geol. Soc.*, Vol. 31, pp. 1–38.

Small, A. T. 1978. Zonation of Pb-Zn-Cu-F-Ba mineralisation in part of the North Yorkshire Pennines[1]. *Trans. Inst. Min. Metall.*, Vol. 87, pp. B 10–13.

Wells, A. J. 1955. The development of chert between the Main and Crow limestones in North Yorkshire. *Proc. Yorkshire Geol. Soc.*, Vol. 30, pp. 177–195.

— 1957. The stratigraphy and structure of the Middleton Tyas – Sleightholme anticline. *Proc. Geol. Assoc.*, Vol. 68, pp. 231–254.

[1] Dr Small's representative collection (1978) of minerals and ores illustrating the zonal pattern has been placed in the Geological Survey rock-store (1980).

CHAPTER 10
Mineral deposits

Details, Area 12 THE RICHMOND COPPER MINES

Towards the east, the outer zone of the North Swaledale Mineral Belt gives place to a smaller mineralised area from which the production, albeit only small in tonnage, has been dominated by ores of copper. Since Area 12 (shown in Figure 28) contains veins at Feldom, Forcett and Merrybent that are plainly transitional in type from the familiar calcite-baryte-galena veins of the eastern part of Area 11, it is proper to include the deposits as part of the Northern Pennine Orefield, but in addition, some of the mineralisation, particularly around Middleton Tyas, is unique to this area. As a result of the Mineral Reconnaissance Programme of the Survey which has included a new investigation of this area since 1973, it is possible in this chapter to give special consideration to the genetic problems of these deposits.

The earliest reference to the Richmond Copper Mines occurs in a charter of Edward IV, 1454 but according to Raistrick (1975, p. 96) it is impossible to identify the exact site. The Feldom Vein, structurally linked with the North Swaledale Belt (see p. 71), may have been exploited as early as 1675, by which date the high mill at Whashton, close to Copper Mill Bridge, was already in existence. This mill was being worked in 1728 by 'the lessees of the copper mines of Feldom' (Raistrick, *op. cit.*). A lease of 1718 of the Town Pasture of Whitecliff, north-east of Richmond, gives royalty rate for copper as well as lead, and this may indicate that the Gingerfield workings (p. 157) were active at that time; they were already old works by 1763. The Middleton Tyas mines, the most important of Area 12, have been the subject of short accounts by Raistrick (1936, 1975) and of a very full historical and economic investigation by T. R. Hornshaw (1975), who shows that the first discovery here was made about 1733 in a limestone quarry worked by Leonard Hartley. Trials under leases of 1736 and 1738 from Hartley were unsuccessful, but by 1742 profitable workings had been established on property belonging to Lady d'Arcy by a partnership with which she was associated. By 1750 copper had been discovered on glebe lands and leased by the Rector, Dr Mawer, to G. Tissington, who had mined ore worth £40 000 by the time of the Rector's death in 1763. Hartley had meanwhile developed properties that seem to have been at least equally valuable nearby. By 1763 the recoverable ore was becoming exhausted, and mining seems to have ceased about 1779. Contemporary accounts of the Middleton Tyas operations remain from two foreign visitors, R. R. Angerstein (1755) from Sweden, and G. Jars (1765) from France. Among the papers at the North Yorkshire County Record Office that formed the basis for Hornshaw's interesting study there is a sketch-map by R. Richardson dated 1754 (ZDG (A) IV 5) and a crude sketch showing some shafts and underground workings by Tissington (ZAW 117); however, no detailed underground plans of the Middleton Tyas workings are known, so that their extent can be judged only from indications at surface, fast disappearing in arable land.

Little further interest was taken in the deposits until 1856, when discoveries were made leading to the establishment of a mine on a vein-complex at Merrybent, 1¼ miles (2 km) NW of Middleton Tyas. The mine was developed on three levels and for a few years produced both copper and lead ores, but it was unsuccessful financially and was closed by 1879. The operation is well documented by Mining Records plan 536 (Department of Energy) and T. R. Hornshaw's account (*op. cit.*, pp. 118–138). A small amount of work was done here shortly before the first World War, but with little result; however, the last phase of copper mining was by then proceeding at Billy Bank, on the south side of the Swale opposite Richmond, where a small mine had been in operation from 1905 in Hudswell parish on a vein that had probably been tried before the making of the first six-inch Ordnance Survey in 1850, which shows a copper mine here. The primary geological survey, by W. Gunn, carried out in the 1870s recorded the veins so far mentioned and established that a gentle anticline, plunging south-east, controls the bedrock geology of Area 12. This is now known as Middleton Tyas – Sleightholme Anticline; the mapping was revised by A. J. Wells (1957) and has been further revised as a result of the recent boring programme. The solid formations at surface in the area of present interest range downward from the Richmond Cherts to the sandstone below the Underset Limestone, and lower members of the Brigantian sequence have been penetrated in the shafts at Merrybent. Around the nose of the anticline, the Main Limestone is unusually thin, reaching only 25 ft (8 m), whereas the Underset Limestone at 75 ft (23 m) is exceptionally thick in comparison with the normal Swaledale sequence. The Main Limestone resumes its normal thickness at Richmond and at Feldom, while the Underset makes a wide crop within the fold. With the top part of the underlying sandstone it is the principal bearing horizon at Middleton Tyas.

No memoir for Sheet 41 (Richmond) reached the stage of publication, and W. Gunn's manuscript does not cover the present area. Raistrick (1936) published the first geological account of Middleton Tyas, basing it on the writings of Angerstein and Jars, and on correspondence between Hartley and William Brown, builder of the Newcomen atmospheric steam engine used for pumping. This was installed by Tissington in agreement with Hartley in 1754–55. This account correctly identifies the horizons of the copper mineralisation as the Underset and Main limestones, but Raistrick's later (1975) description states that the Middleton Tyas deposits occur in the Magnesian Limestone. The Upper Permian does in fact onlap unconformably on to the Carboniferous immediately to the east of the mines, but the mineralisation, apart from traces of malachite staining, is in Brigantian and (elsewhere in Area 12) in Pendleian strata. The first detailed study of the mineralogy was by Dr T. Deans[1] following his work on the

[1]Deans demonstrated his material to the Mineralogical Society in 1951 and took one of us (KCD) over his localities in 1952, but did not publish his results. Minerals collected at the time are in the Russell Collection of the British Museum (Natural History) and in the mineral collection of the University of Durham. Dr Deans kindly made available his notes to the Mineral Reconnaissance Group of the Institute in 1973.

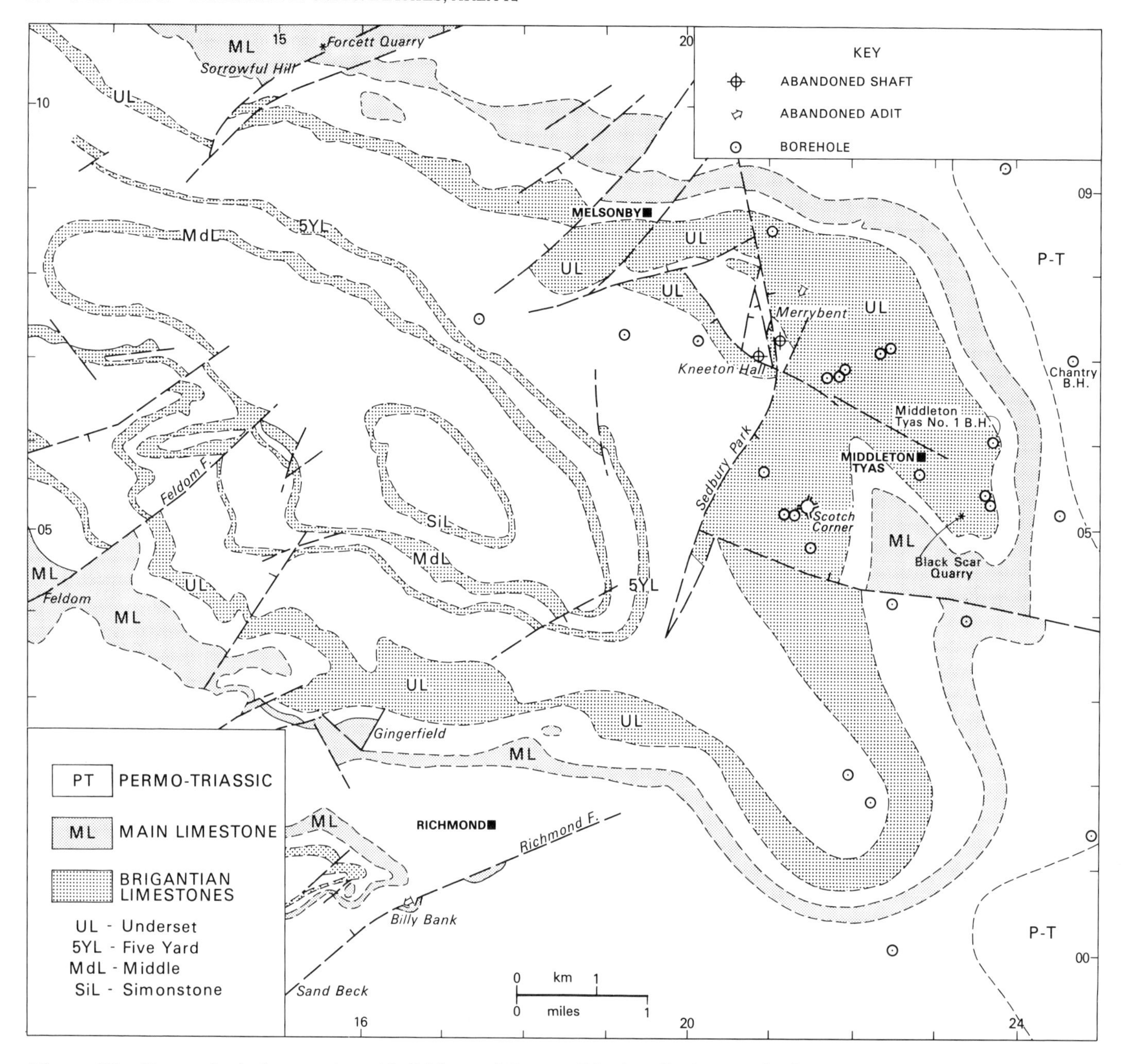

Figure 28 Key geological map, Area 12, Richmond Copper District. Geology revised by A. J. Wadge, 1972–78

Kupferschiefer – Marl Slate (1950). At Forcett Quarry (p. 90) the north-east vein carries chalcopyrite in a gangue of calcite, unaltered witherite and baryte, the sulphide occurring in plates and coarse crystals. Small veins with bornite, covelline and chalcocite were also found. The shale at the base of the Main Limestone carried flattened sulphide nodules for an exposed distance of 450 ft (137 m) along the vein, and to a distance of 60 ft (18.3 m) away from it. The nodules are flattened, ranging up to 2 in (5 cm) diameter; the centres consist of bornite which has probably replaced pyrite of diagenetic origin, surrounded by bornite enclosing chalcocite and chalcopyrite plus covelline. Assay of a large sample showed 45.9 per cent Cu, 13.2 per cent Fe and 12 per cent insoluble. Oxidation products including malachite, azurite and limonite were also noted. At Black Scar Quarry, Middleton Tyas (Figure 28), now considered to be in Underset Limestone, Dr Deans again found nodules of similar type at the decomposed bottom of the limestone; some friable carbonate and baryte surrounds the nodules. In sand pockets in the limestone, relict limestone, with chalcocite, covelline, malachite, baryte and tiny crystals of native copper occur with limonite. At Merrybent Dr Deans found chalcocite and covelline replacing galena. Mr P. R. Simpson who has examined this material and additional specimens collected by the Mineral Reconnaissance Group, considers that the primary copper phases were chalcopyrite and djurleite, and that secondary enrichment introduced digenite, bornite and covelline. Later alteration in the oxidation zone produced widespread malachite and azurite. To these two products may be added native copper and goethite. Genetic aspects will be considered further below.

The North Yorkshire Records (Hornshaw, *op. cit.*) show that the ore won from Middleton Tyas was of extraordinarily high grade. Angerstein thought it the richest in Europe, and figures given by Hartley to Brown indicate a recovery of 66 per cent Cu; an assay made at Jermyn Street a century later (p. 33) showed 65.8 per cent Cu. Some of Hartley's ore was sold for £52 per ton, while the production by the d'Arcy partnership fetched an average of £28 10s. 0d. over a long period. The range of copper content in ore with only very crude beneficiation probably was from 30 to 66 per cent, and this is understandable if the main source was material like the nodules and sand-fillings described above.

On the other hand, the tonnage produced was small. Lady d'Arcy's partnership produced, from 1842 to 1867, according to the histogram in Hornshaw (p. 63) altogether 400 tons (of 21 cwt) which were sold at the average price cited above. If this same price is applied to the sum on which Tissington paid tithe over much the same period, £40 483, his output may have been 1434 tons. No data are available from which the output by Hartley, Shuttleworth and any other operators can be gauged; but it appears difficult to envisage a total for Middleton Tyas exceeding 3500 tonnes of ore, perhaps averaging 45 per cent Cu, and thus representing about 1575 tonnes of metal. Official statistics which give the production of copper ore from Merrybent and Billy Bank, show copper recoveries amounting respectively to 13 and 6.3 per cent, indicating that enrichment at these mines was probably of less consequence than at Middleton Tyas (p. 162).

The recent mineral reconnaissance survey included geochemical and geophysical studies covering a significant part of Area 12, the results of which are available on open file at the Survey (Wadge and others, 1981). Soil sampling was carried out along traverse lines more or less at right angles to strike, covering the eastern and south-western flanks of the anticline and totalling 46.5 km in length. Samples of the bottom boulder clay were taken by power auger along parts of some lines, where drift was more than a few meters thick. The survey revealed anomalous copper values over an area exceeding 6 km^2 covering (1) the Underset Limestone along the east flank, where peak values of 2300 ppm were recorded (ii) the Main Limestone from Moulton [235 040] to Aske Moor [140 035], west of Richmond, here reaching 5000 ppm. Bedrock was tested by churn drilling in three areas, Middleton Lodge [220 069], Southfields Farm, west of Black Scar Quarry [227 048] and Gingerfield [155 026]. Limestone and the underlying sandstone and mudstone tested in numerous borings not only showed anomalously high copper, but also lead and zinc in all three lithologies, though with some exceptions. Regarding the eastern soil and drift anomaly, it can be argued that dispersion by ice of cupriferous drift from Middleton Tyas towards the south-south-east in accordance with the drumlin trend could explain it, while the western anomalous area perhaps results from glacial transport east-south-east from the neighbourhood of Gingerfield. However it plainly emerges that Area 12 characteristically contains fairly widespread copper in both rocks and soils, and it cannot be said with certainty that no undiscovered deposits exist.

The geophysical method adopted was Induced Potential, as the only one capable of detecting metal concentrations in the selected area east of Middleton Tyas, where the drift cover is between 20 and 40 m thick. Unfortunately several strong artificial conductors masked effects from bedrock, and only one likely target was identified. Boring on this failed to find mineralisation.

Feldom Vein **Copper ore, lead ore**

NZ 10 SW; Yorkshire 38 NE, NW
Direction N53–45°E, displacement about 30 ft (9 m) SE increasing towards the centre of the Middleton Tyas–Sleightholme Anticline

The linkage of this vein through the High Waitgate and possibly through the Thringill Hill Vein with the North Swaledale complex has already been described (p. 152). The Feldom workings consist of many shafts covering a stretch 3000 ft (914 m) NE from [1170 0412] with Main Limestone on both walls of the vein. The shafts have been described as bell-pits, but the scale is much greater than the typical bell-pit of the district, and they are more comparable with the shafts on Friarfold Vein believed to have been worked in the 16th and 17th centuries. The period of working at Feldom was probably late 17th and early 18th century, but although the ground is suitable, the stage of driving levels from surface never seems to have been reached. The minerals present on the dumps include baryte both in coarse plates and in the form that may be secondary after witherite; plentiful calcite, galena, chalcopyrite, a little covelline and the oxidation products of the sulphides. The wallrocks are extensively dolomitised and disseminated sulphides, including chalcopyrite, occur in the pale brown dolomite. The impression gained is that the sulphides were finer-grained than usual. A farm nearby is called Buddle House, suggesting that dressing floors once existed, but it is not obvious where. It is known that there was some copper production, and blebs of the metal can be seen in the slag at the site of High Whashton Mill, near Copper Mill Bridge, but no record of the output of copper ore or lead ore remains. The scale of the workings would however justify minimum estimates respectively of 400 and 1000 tons, and these are tentatively adopted on p. 162. They are likely to be conservative. After the sandstone and shale beneath the Main Limestone appears at surface on the footwall side, the vein has been tested by shallow workings for ½ mile (0.8 km) to the NE, but with little result. The fault continues for almost 2 miles (3.2 km) farther, to Whashton village, where Middle Limestone is on both walls, but nothing workable has been found in this stretch.

The last work on the Feldom Vein may have been in 1754, when Leonard Hartley mentioned in a letter to William Brown that he had been unsuccessful in finding anything there (Hornshaw, 1975, p. 108).

Sorrowful Hill Vein

NZ 11 SE; Yorkshire 24 NE
Direction N65°E, small NW downthrow

The trials at Sorrowful Hill [151 105] probably explored a weak vein which was subsequently exposed as a fault crossing the south-east part of Forcett Quarry, East Layton. As seen in 1951–52, patches of primary mineralisation up to 9 in (22 cm) wide could be found in which plates of chalcopyrite were set in unaltered witherite, calcite and baryte. Clearly no acid solution had attacked these, but elsewhere along the fault, limonite-stained secondary baryte, probably the alteration product of the witherite, could be found. Copper sulphides typically formed during secondary enrichment—bornite, covelline and chalcocite—could also be found in veins within the fault, but more strikingly as replacements of pyritic nodules in the thin shale separating the Main Limestone from the underlying Tuft Sandstone. The dimensions of the deposit as exposed in 1951 have already been mentioned, and possibly it was this secondarily enriched layer that was reached by the trial pits, though evidently it was not regarded as workable in spite of the high grade of the nodules. The underlying sandstone also carried small amounts of the copper sulphides. The bottom 2 ft (60 cm) of the limestone here has been dolomitised, possibly in connexion with the mineralisation, though elsewhere the bottom post of the limestone may be dolomitised far remote from mineral veins. The Forcett occurrence, though without economic interest, is of some importance in confirming the linkage of the veins in Area 12 with the northern Pennine field, and in confirming the existence of secondary copper enrichment affecting both the chalcopyrite of the vein, and the diagenetic pyrite in the shaly layer at the bottom of the limestone. In 1972 the following section was exposed in the quarry bottom [1556 1066]:

	ft	in	m
MAIN LIMESTONE, grey, fine-grained, dolomitic, with azurite and malachite traces	6	6	2
Siltstone, dark grey, with sulphide nodules		3–4	0.1
TUFT SANDSTONE, grey, fine-grained, siliceous, with carbonaceous streaks and malachite staining	2	0	0.6
Mudstone, dark grey, silty, pyritic, micaceous with malachite staining	6	6+	2+

Analysis of malachite-stained sulphide nodules from the siltstone showed: 42.0 per cent Cu, 20.0 per cent Fe, 31.5 per cent S. The green-stained sandstone assayed 2 to 5 per cent Cu. The fault, with which dolomitisation of the limestone is associated continues to the north-east, but the mineralisation appears to die out.

High Langdale Trial

NZ 10 NE; Yorkshire 25 NW

At [1912 0927] a shallow shaft was sunk in green-stained limestone encountered in a quarry in the Main Limestone. Probably this dates from the period of active prospecting in the 18th century, but the lack of tips suggests that little was found.

Middleton Tyas Veins, Flats and Pipes — Copper ore

NZ 20 NW; Yorkshire 25 SE, 39 NE

Flats related to vein-fractures but owing to the inaccessibility of the workings and lack of records, the exact position of the fractures is uncertain. W. Gunn's maps in 1889 showed two veins trending N60°W, and one N26°E (see also Hornshaw, 1975, map 3). Recent revision has suggested the version given in Figure 29, with two or three fractures trending N65°E, and a curviform NNW fracturing linking them. Neither version can be proved.

In a dozen fields surrounding St Michael's Church [235 056] and extending ¼-mile (0.4 km) to the north and west, there are numerous remains of the shallow shafts sunk for copper during the 30 years in the mid-18th century when the village was the scene of active mining, dressing and smelting. Since the pumps stopped at the end of this period the workings have been wholly inaccessible, and the exact shape and form of the deposits remain, in the absence of plans, somewhat conjectural. However the contemporary observer Angerstein (1755 and English translation in Hornshaw, 1975, p. 139) has recorded that there were two vertical veins called pipes (a term familiar to Tissington's Derbyshire miners who worked them) and a flat vein ('floetz'—an archaic German term equivalent to flat, flot or float in the Northern Pennine Orefield). William Brown gives the width of one such flat as 90 ft and the thickness as 1¼ ft (27.4 × 0.38 m) and Hornshaw points out that if this was the flat running from Church Field to Goosehill, the length was about 900 ft (say 275 m). The pipes contain black, green and blue 'platt' (presumably gouge with malachite and azurite) while the flat deposits occur at two levels, an upper one resting on clay, then, about 12 ft (3.6 m) below, a second level also referred to as the 'underbed', richer than the first and resting on sandstone. G. Jars (1765) writing towards the end of the most active period, records that 'kidneys' of ore are found, in cavities of different sizes, 'filled with iron-stained sand', and goes on to list in poetic language, copper minerals that can easily be recognised as covelline, chalcocite, cuprite and bornite, mixed with malachite and occasionally chalcopyrite, 'but not often'. It is now clear that the bearing bed throughout the Middleton Tyas mines is the Underset Limestone, and Dr Deans' observations at Black Scar Quarry in this limestone now fall into place as representing a lean but nevertheless typical example of the deposits; the fissures filled with gouge impregnated with copper carbonates were recorded, together with cavities filled with limonite-coated secondary copper sulphides, and the 'underbed' at the base of the limestone with its enriched pyrite nodules, resting on sandstone into which copper mineralisation had also penetrated. One further significant fact must be added: the deposits, for the most part, lay below present water-level and were reached by shafts 48 to 51 ft (14.6–15.5 m) deep in the vicinity of the church, from which large quantities of water had to be pumped. Dolomitisation is the characteristic wallrock alteration, as at Forcett. The uncertainties regarding the Middleton Tyas deposits have led to at least four suggestions as to the origin of the mineralisation:

1. syngenetic copper sulphide mineralisation close to the bases of the Underset and Main limestones;
2. derivation from Kupferschiefer-type sedimentation, for example the Marl Slate; Deans at his Mineralogical Society demonstration canvassed this possibility but did not publish it;
3. brine-derived sulphide mineralisation. This has been advocated by Small (1978) for his C-zone (essentially our Area 12), the Stainmore Trough supplying low Na:K ratio metal-rich brines which mixed around Middleton Tyas with high Na:K ratio formation waters. On this view, veins of transitional type at Feldom and Billy Bank represent mixing of the two sorts of brine.
4. strong supergene enrichment in copper of previously-existing lean sulphide deposits, probably in an arid environment.

With respect to the first hypothesis, drilling of several cored and rotary percussion boreholes at both Middleton Tyas and Forcett Quarry has failed to locate any sediments containing primary syngenetic copper sulphides though, of course, pyrite is widespread in this form. Background levels of Cu, and to a lesser extent, of Pb and Zn are higher hereabouts in the Underset and Main cyclothems than elsewhere, but in the light of other evidence, this is ascribed to permeation by metalliferous brines.

Where the Marl Slate member of the Upper Permian succession is well developed in north-east England, it is enriched in Pb over a wide area (Hirst and Dunham, 1963) as Deans (1950) also found to be the case in north Germany; the area workable for copper in Germany is relatively small, and no such conditions have been found in this country. The mineralised ground lies close to the base of the Permian rocks, which dip eastward at 1–2°. Two boreholes were drilled to test the hypothesis that the mineralisation was connected with the Permian, immediately to the east of Middleton Tyas at Low Chantry [2469 0705] and Halnaby [2608 0724] through the lower part of the Permian sequence, but neither Marl Slate nor cupriferous sediments were found. At Low Chantry, attenuated Lower Magnesian Limestone is separated by thin breccia from the underlying Richmond Cherts, while at Halnaby a basal Permian breccia is overlain by Middle Permian Marls; thus the Permian on-laps the Middleton Tyas – Sleightholme Anticline, which must have stood up as an island in the Zechstein Sea. A small

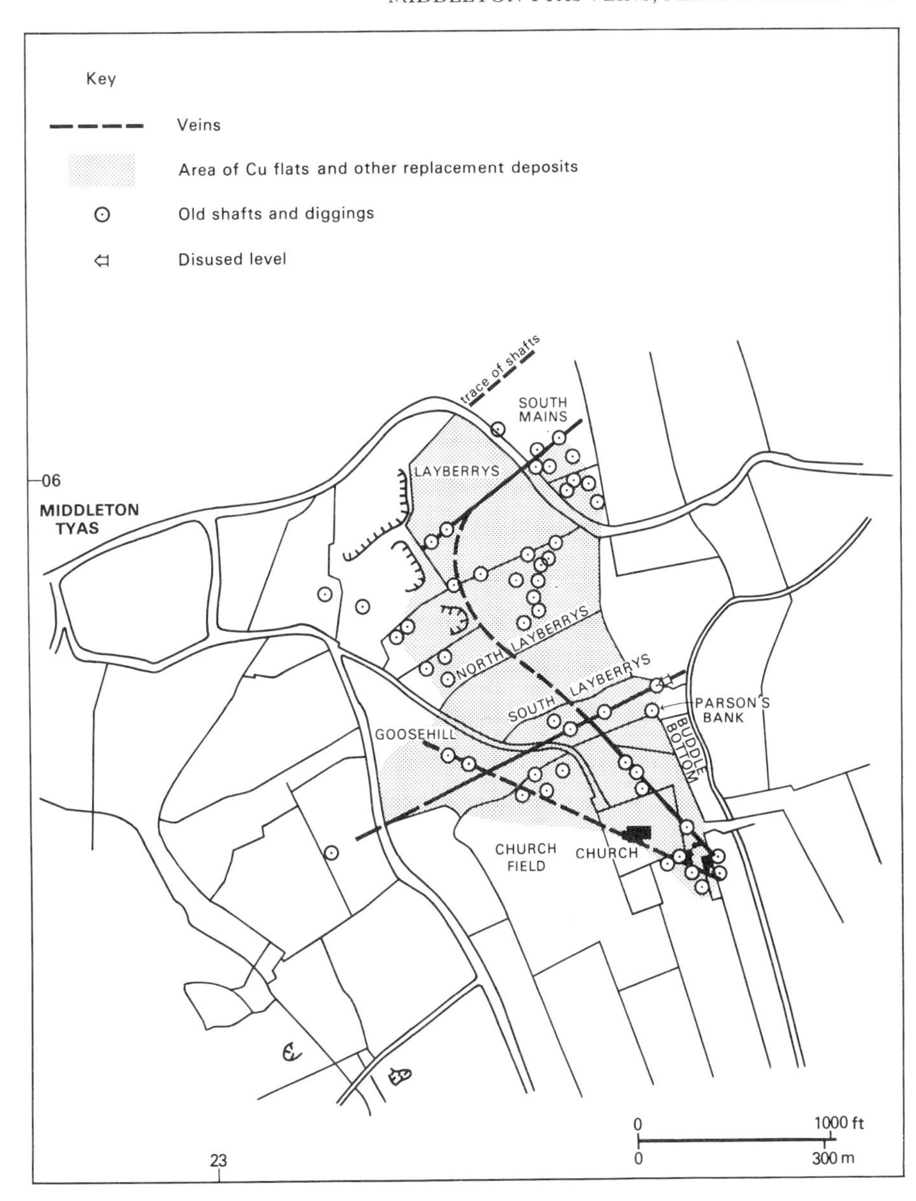

Figure 29 The Middleton Tyas Copper Mines, showing the probable vein pattern

mineral vein with clear baryte and pearly low-iron ankerite was found in the Main Limestone at Low Chantry, but there is no reason to connect this with the Permian. Thus hypothesis (2) must be discounted.

There is clear evidence of a transition from the North Swaledale Mineral Belt to the deposits of Area 12; and as discussed in Chapter 7, fluid inclusion studies identify the mineralising fluids as hypersaline brines. The same conclusion may thus be held to apply to the primary mineralisation at Middleton Tyas, though material suitable for inclusion investigation is difficult to obtain in the present condition of the workings. Small (*op. cit.*, p. 13) found only monophase inclusions in calcite with chalcopyrite from Kneeton Hall, indicating a formation temperature below 70°C, containing brine with K/Na = 0.3[1]. Further insight into the temperature of primary mineralisation is provided by the presence of djurleite ($Cu_{1.96}S$) which according to Roseboom (1966) is stable only up to 93 ± 2°C. It may be regarded as probable that the original copper mineralisation was of low grade, consisting of chalcopyrite and djurleite in veins and disseminated in dolomitised limestone. Possibly the Billy Bank veins near Richmond, represent such a stage.

[1] Figure quoted from A. T. Small, Durham, PhD thesis, 1977, by permission.

Supergene copper enrichment has been a well-known process since the numerous investigations in western North America in the early part of this century, summarised by W. H. Emmons (1917). It consists of the leaching of copper sulphides from the upper parts of the deposits, transport of copper in highly acid sulphate solutions downwards, and precipitation by reaction with existing sulphides near the water table. It is at its most effective in arid regions where the water table is deep. We suggest that the Middleton Tyas Anticline was such a region in Lower Permian times, and that the deposits owe their much enhanced grade to precipitation of copper not only on primary sulphides including chalcopyrite, pyrite and djurleite, but also on diagenetic pyritic nodules in the calcareous shales beneath the cleaner limestones. The mineral suite bornite – covelline – digenite – chalcocite is characteristic of supergene enrichment sequences. Formation of cavities in the limestone, even though partly dolomitised, would be likely to accompany this process. The enrichment is taken to predate the present hydrological cycle, since the enriched deposits are below the water table. During the present cycle, however, oxidation of the sulphides to cuprite, malachite and azurite, and the local reduction to native copper in the presence of organic matter has no doubt continued. We conclude therefore that a combination of hypotheses (3) and (4) offers the more likely explanation of these small but rich deposits.

The coincidence of a number of favourable conditions as seen at Middleton Tyas might be repeated in other places, for example on strike of the Underset Limestone to the north or south; from this point of view the geochemical results around Middleton Lodge are interesting. They might recur to the north-east or east under the Permian but in the absence of geophysical or geochemical methods capable of penetrating through thick drift, marl or evaporites and dolomite, targets are very hard to identify.

Merrybent West Fault

NZ 20 NW; Yorkshire 25 SW
Direction N15°E, downthrow SE

Black, Robert Raw's, Smithson's and Lowe's Veins

Parallel to West Fault

Kneeton Vein = **Merrybent Main Vein** Copper ore
Fairly's Vein = **Sedbury Park Fault** Lead ore

Direction N20°W, swinging S of North-West Cross Vein to S30°W, downthrow 105 ft (32 m) E

North-West Cross Vein Lead ore

Direction N50°W, changing westward to N38°W, downthrow S

Merryfield Main Vein and North-West Cross Vein limit a structural block exposing arenaceous beds beneath the Underset Limestone (the Nattrass Gill Hazle of the orefield north of Stainmore), and bring in the limestone to the east and south respectively. The West Fault links the two a short distance west of the mine workings, and these fractures define the area of present interest worked from the *Merrybent Mine* (abandoned). This was the scene of a brief revival of copper mining during the third quarter of last century. Work started about 1856, perhaps following up one of Hartley's prospects west of Kneeton Hall of the previous century; an old adit here is referred to in some reports as Hartley's Level, and there is an ancient trial shaft. The mine was in production from 1863 to 1874; the history of the two companies concerned has been written by T. R. Hornshaw (1975, pp. 118–138), largely based on the papers of the Havelock-Allan family, now in North Yorkshire County Records. His account also includes a reduction of the abandonment plan (Health and Safety Executive No. 536) and reproduces a report by Charles Bawden of St Day which, though difficult to reconcile with the plan, gives some geological information about the underground workings. A geological section of the 300 ft (91 m) deep Engine Shaft [2108 0738] has survived in the Radcliffe Records and in BGS files.

The initial development by the Merrybent and Middleton Tyas Mining and Smelting Company was probably the driving of the Upper Adit, starting close to the Engine Shaft site at 455 ft (138.7 m) OD. This adit gave access to workings in Kneeton Vein, the western part of Main Vein, about 1000 ft (305 m) long, and at the south end a crosscut to the east linked with what may have been Hartley's entrance in Kneeton Hall Quarry, passing through a NNE fault downthrowing west before reaching the quarry. The average depth of the Upper Level workings was no more than 30 ft (9 m) below surface, but Bawden records that on Kneeton Vein, good bunches of lead ore were mined from the back, and that these continue into the sole. This led to decisions to sink the Engine Shaft in 1866, and to drive a new adit system, Bussey's Level, starting from Waterfall Beck at [2125 0893] near Trundall's House, running directly to the shaft at 1850 ft (564 m) from the portal, at 565 ft (111 m) OD. A branch from this by-passed the shaft and continued SW for 850 ft (259 m) to give access to a southward drive on Lowe's Vein some 900 ft (274 m) long, leading to workings which followed North-West Cross Vein to the WNW. It is uncertain whether any production was obtained from Lowe's Vein, but the Cross Vein appears to have been mineralised, since six men were reported working there in 1871. Bussey's Level should just have given full control of the Underset Limestone on the hangingwall of North-West Cross Vein, with Nattrass Gill Hazle on the footwall, and it must be assumed that similar conditions hosted the oreshoot on Kneeton or Main Vein. It is possible to use the recently-drilled BGS borehole Middleton Tyas No. 1 [2371 0603] to interpret the geological record of Merrybent Engine Shaft, by postulating that two faults must pass through the shaft. These are identified with the two parts of Main Vein, the western (Kneeton) vein throwing about 45 ft (13.7 m) E, the eastern (Fairley's), 60 ft (18.3 m) E. Probably these merge before reaching the Cross Vein. Before 1870 the Upper Level had been driven 840 ft (256 m) SW and then 610 ft (186 m) WNW, the latter stretch passing through the four small veins listed in the arenaceous beds below the Underset Limestone. Bawden's report of 1867 is worth reproducing: '... up to the present nothing of consequence has been found, and in such a stratified country I have never known payable mineral to occur'. According to Hornshaw (Map 6) this drive was continued into the adjacent royalty, but without finding anything worth working.

The Engine Shaft had been well sited to exploit the Main Vein in depth; from Bawden the assumption appears to have been that a limestone formation would be reached. As is now clear from geological mapping, this could not be the Main Limestone, but mainly limestone strata comprising the Five Yard and Middle limestones at least 66 ft (20 m) thick were available on the footwall of the western fracture above the 50 Fathom Level. A 23 Fathom Level was driven for 425 ft (130 m) along the eastern (Fairley's) fracture, and the western fracture was tested from a crosscut, but not developed. On the former, the sandstone or Nattrass Gill Hazle would be on the hangingwall against a footwall of shale with sandstone bands; on the latter, the mixed beds would predominate on both sides. When the mine closed, probably for reasons other than the state of the workings, only about 150 ft (46 m) of driving had been done on the 50 Fathom Level, certainly not enough to test the possibilities of the various structures in the Five Yard – Middle limestones. It is not even known whether any mineralisation was seen at this level. The poor production records (p. 162) give little encouragement, however, to reopen this marginal mine. An attempt to do so in 1913–15 led to the production of less than 2 tons of copper ore.

The Merrybent West Fault was tested by an adit from Waterfall Beck about 1000 ft (305 m) long in a SSE direction, with two air shafts. Nothing workable appears to have been found.

On W. Gunn's interpretation of the geology of Middleton Tyas the North-West Cross Vein of Merrybent is mapped as continuing into the two WNW veins he inserted through the flat ground around the church. Although a different interpretation for Middleton Tyas has now been proposed, Gunn's conception of this linkage has been retained on Figures 28 and 29.

Near Kneeton Hall, in addition to the working from the quarry already mentioned, there are remains of shafts on either side of the road to Sedbury Home Farm, 650 ft (200 m) WSW of the Hall, with galena and copper minerals in crinoidal limestone, some of it reddened, on the heaps. The northern one possibly formed part of Leonard Hartley's trial in 1754; the southern one may have been included in work done by Tissington on Dovecot Hill about the same time. Both must have given access to the North-West Cross Vein, and it may be conjectured that the possibility of a link with Middleton Tyas may have been considered then, particularly in view of shafts sunk between Kneeton Hall and the Great North Road to the east.

Mineral specimens representative of the Merrybent mineralisation are not easy to find; galena replaced by covelline has already been mentioned. A. T. Small (1977), who made the most thorough recent examination, found calcite and galena; galena marginally replaced by chalcopyrite and bornite; these minerals replaced by covelline, digenite and chalcocite and the whole assemblage of sulphides partly altered to malachite, azurite and limonite. The presence of primary chalcopyrite here does not clearly emerge, though it might be regarded as very probable. A typical enrichment sequence of supergene origin is in evidence, as might be expected from the fairly high grade of the ore (Table 22). Oxidation effects are normal.

Melsonby Trial

NZ 10 NE; Yorkshire 25 SW

South of Melsonby village, west of the road from Richmond at [1978 0815] a small quarry exposes Underset Limestone with calcite and malachite on joints. On the south side of the quarry, a trial shaft is believed to be the one sunk by Hartley about 1754. Dr Earp was unable to find any evidence of the proximity of a fault.

Gingerfield Vein Copper ore

NZ 10 SE; Yorkshire 38 NE, SE, 39 SW
Direction N25°E, downthrow E

West of Gingerfield Farm, this small vein brings up the Main Limestone to the west about [1607 0261] and farther south near the northern edge of the former racecourse, where trial shafts have tested the vein, and perhaps produced a small tonnage of copper ore. Activity here dates from about 1762 (Hornshaw, 1975, p. 108). Another trial was made in Main Limestone ½-mile (0.8 km) to the west-north-west at Rasp Bank while to the south-east of Gingerfield there was at least one more near Bend Hagg [172 022].

Covering this ground along the outcrop of the Main Limestone on the southern flank of the Middleton Tyas – Sleightholme Anticline, six lines with an aggregate length of 2 miles (3.2 km) were sampled geochemically (Wadge and others, 1981), the soils showing strongly anomalous copper values over maximum widths up to ¼-mile (0.4 km), and reaching maximum values of 3500 ppm at Gingerfield, and 4280 on Pilmoor Hill north of Richmond. The anomalous area might be explained by ice-borne debris from a single exposed copper deposit, but this cannot be the Gingerfield Vein since the anomaly begins considerably west of this. A churn drilling pattern at Gingerfield showed highly anomalous Cu in the limestone as well as in underlying sandstone and mudstone; Pb values were slightly high, Zn not above background. In the present state of knowledge it is not easy to decide whether the 2¼-mile (3.6 km) long copper anomaly indicates the existence of an undiscovered substantial copper oreshoot, a number of small bodies, or a general dissemination of copper minerals in small quantities through the Main Limestone and the beds immediately below. The southern margin of the limestone appears to be terminated east of Pilmoor Hill, by a fault downthrowing south. This was mapped in a nearly E – W direction through the limestone by the primary geological surveyors as a mineral vein linking with the southern end of the Gingerfield Vein, but this interpretation has not been included on the Richmond (41) Sheet, revised in 1970.

Richmond Fault Copper ore, lead ore

NZ 10 SE, SE 19 NE; Yorkshire 38 SE, 39SW
Direction N60–72°E, maximum downthrow about 100 ft (30 m) N

Billy Bank Veins Copper ore

NZ 10 SE; Yorkshire 38 SE
Direction N60–72°E (two), E – W (one).

The Richmond Fault crosses the River Swale at the north end of Billy Bank [1649 0065], bringing Main Limestone to the south against the base of the shale above the Richmond Cherts to the north. The throw is thus equal to the thickness of the cherts, which appears to be less than the maximum thickness of 130 ft (40 m) observed by Wells (1955) 2 miles (3.2 km) to the west. The fault was known to be mineralised at the time of the primary survey (published 1889) and probably there were old workings to be seen then; indeed these workings, or others on the same fault where it passes through Richmond Borough, north of The Green, now obscured, may have been the 15th century Richmond Copper Mines to which reference has already been made (T. R. Hornshaw, 1976; Anon, 1978). This fault dies out in the vicinity of Low Wath Cote, about 1 mile (1.6 km) ENE of Richmond Market Place. The Billy Bank Mine, which post-dates the primary geological survey, worked from 1906 to 1915, and was the last producer in the Richmond copper area. Two adits not far above river level gave access to workings on the west bank, a northern one (now collapsed) where the fault crosses the river, and a southern one at [1650 0057]. The latter was entered by a party from the Moldywarps Speleological Group consisting of L. Beevers, A. Holmes, J. Knight, J. Longstaff and P. Ryder in 1969 and a survey of the workings was made. The following summary of the underground workings is based upon this survey, and an account of a more recent visit (Anon, *op. cit.*). The southern adit gives access to narrow natural caverns following ENE joints, perhaps somewhat extended by mining. At 175 ft (53 m) a NNW crosscut 45 ft (14 m) long leads to the main Billy Bank Vein which has been followed about 600 ft (183 m) to the WSW; of this stretch the presence of rises in a length a little over 400 ft (122 m) suggests stoped ground on an oreshoot, the drive here reaching a maximum width of only 5 ft (1.5 m). At the west end of this stretch the E – W vein crosses, this having been examined over a length of 270 ft (82 m) including three rises. North of the main vein, another run of workings or natural caverns is reached by crosscuts 75 ft (22 m) and 55 ft (17 m) long respectively. These workings probably connected with the northern adit, and are believed to be in the immediate footwall of the Richmond Fault, probably following minor sympathetic fractures. Minerals recorded from the underground workings include pink calcite, chalcopyrite and oxidation-zone copper minerals including chrysocolla, malachite and azurite. The chalcopyrite occurs interbanded with and disseminated in calcite and must be assumed to be of contemporaneous origin with this mineral, at least in part. No record of supergene enrichment sulphides has been found here, and the relatively low grade ore (Table 22) is taken to be typical of vein-mineralisation in Area 12 in its primary condition. On the east bank of the Swale, tips from ancient workings in the Temple grounds are reported as containing calcite, chalcopyrite and malachite also. R. T. Clough (1962) reports that a smelt mill for copper was built in Billy Bank Wood in 1585; evidently there is a long history of small-scale copper mining here. Very little spoil remains, but it appears that dolomitisation accompanied the copper veins in the Main Limestone, and that the veins extended up into the Richmond Cherts. A brief note on the mine by R. G. Carruthers appears in Dewey and Eastwood (1925).

The Richmond Fault continues WSW and where it cuts through Sand Beck, approximately 1 mile (1.6 km) from Billy Bank, some galena was obtained at the horizon of the Crow Limestone and Ten Fathom Grit, but there was no substantial working.

OUTPUT AND RESOURCES

Since the more important period of copper mining preceded, by nearly a century, the issue of official statistics, it has been necessary to include some estimates, based as far as possible on geology and history, in Table 22.

The production, amounting to a little less than 2000 tonnes of copper, is unimpressive when compared with the lead production of the orefield. At the same time, indications of copper are persistent around Richmond, and if it is asked whether a substantial copper orebody could be found in Carboniferous limestone, the discovery at Gortdrum, Co. Tipperary (I. S. Thompson, 1967) of an epigenetic chalcopyrite deposit, enriched with bornite and chalcocite, containing nearly 50 000 tonnes of copper, supplies the answer. A decisive explanation of the copper anomalies in Area 12 has not yet been obtained, and it is unfortunate that the local conditions make geophysics difficult or unworkable, for this would be required to identify targets for drilling; prospecting would otherwise be too chancy. AJW, KCD

Table 22 Recorded and estimated production of copper and lead concentrates, Area 12

Mine or Group	Years	Copper concentrates		Lead concentrates	
		long tons	%	long tons	%
Feldom	17th–18th century	400*	?12	1000	?60
Middleton Tyas	1742–1779	3500†	?45		
Merrybent	1863–1874 1910–1912	1042‡	13.1	1284	73
Kneeton Hall	1892–1896	32	18		
Billy Bank	1905–1912	1515	6		
		6489	28	2284	67§

* Estimate, perhaps conservative, based on scale of workings.
† 800 tons, d'Arcy and 1 year Hartley; Tissington, calculated from tithes, 1434; the rest estimated.
‡ This and subsequent figures from Official Statistics.
§ Average value.

REFERENCES

ANGERSTEIN, R. R. 1755. *Journal of a journey through England (1753–1755).* Vol. 2, pp. 159–165. English translation *in* Hornshaw, T. R. (*op. cit. infra*, 1975), p. 129.

ANON. 1978. *The Billy Bank copper mine in Richmond.* Darlington and Stockton Times for August 5th., p. 13.

CLOUGH, R. T. 1962. *The lead smelt-mills of the Yorkshire Dales and Northern Pennines.* 2nd Edit. 1980. (Keighley: Clough.) 325 pp.

DEANS, T. 1950. The Kupferschiefer and the associated lead-zinc mineralisation in Silesia, Germany and England. *Rep. 18th Int. Geol. Congr., London, 1948.*, Part 7, pp. 340–351.

— 1951. *Notes on the copper deposits of Middleton Tyas and Richmond.* Abstract in *Mineral. Soc. Not.* No. 74 for meeting of June 7th.

DEWEY, H. and EASTWOOD, T. 1925. Copper ores of the Midlands, Wales, the Lake District and Isle of Man. *Spec. Rep. Miner. Resour., Mem. Geol. Surv. G. B.*, Vol. 30.

EMMONS, W. H. 1917. The enrichment of ore deposits. *Bull. U. S. Geol. Surv.*, No. 625.

HIRST, D. M. and DUNHAM, K. C. 1963. Chemistry and petrography of the Marl Slate of S. E. Durham, England. *Econ. Geol.*, Vol. 58, pp. 912–940.

HORNSHAW, T. R. 1975. Copper mining in Middleton Tyas. *Publ. North Yorkshire. Rec. Off.*, No. 6. Northallerton. 153 pp.

— 1976. The Richmond copper mine. *J. North Yorkshire County Rec. Off.*, No. 3, pp. 77–85.

JARS, G. 1765. *Voyages metallurgiques.* Vol. 3, pp. 72–75. (Paris.)

RAISTRICK, A. 1936. The copper deposits of Middleton Tyas, N. Yorks. *Naturalist* (for 1936), May 1, pp. 111–115.

— 1975. *The lead industry of Wensleydale and Swaledale.* Vol. 1: The mines. (Buxton: Moorland.) 120 pp.

ROSEBOOM, E. H. 1966. An investigation of the system Cu-S and some natural copper sulphides between 25° and 700°. *Econ. Geol.*, Vol. 61, pp. 641–672.

SMALL, A. T. 1977. Mineralisation of the Stainmore Depression and northern part of the Askrigg Block. Unpublished PhD thesis, University of Durham.

— 1978. Zonation of Pb-Zn-Cu-F-Ba mineralisation in part of the North Yorkshire Pennines. *Trans. Inst. Min. Metall.*, Vol. 87, pp. B9–14.

THOMPSON, I. S. 1967. The discovery of the Gortdrum deposit, Co. Tipperary. *Trans. Can. Inst. Min. Metall.*, Vol. 70, pp. 85–92.

WADGE, A. J., HUDSON, J. M., PATRICK, D. J., SMITH, I. F., EVANS, A. D., APPLETON, J. D. and BATESON, J. H. 1981. Copper mineralisation near Middleton Tyas. *Miner. Reconnaissance Programme Rep. Inst. Geol. Sci.*, No. 54.

WELLS, A. J. 1955. The development of chert between the Main and Crow limestones in North Yorkshire. *Proc. Yorkshire Geol. Soc.*, Vol. 30, pp. 177–196.

— 1957. The stratigraphy and structure of the Middleton Tyas – Sleightholme anticline, North Yorkshire. *Proc. Geol. Assoc.*, Vol. 68, pp. 231–254.

CHAPTER 11
Mineral deposits

Details, Area 13 SOUTH SWALEDALE AND WENSLEYDALE

The northern limit of Area 13 is defined by the line of the Stockdale Disturbance and its supposed continuation, the Great Stork Vein (Figure 30). The junction of these faults, neither of which has been proved to carry payable mineralisation, occurs in unexposed ground, in part alluvial, in the vicinity of Healaugh and the simplest interpretation is that the generally E – W Stockdale 'Vein' swings east-south-east to become the Great Stork structure, the substantial downthrow of both being to the north. The structural culmination of the present area lies on the south side of the Stockdale 'Vein', east of Gunnerside, where the equivalent position of the base of the Main Limestone reaches 1850 to 1950 ft (564–594 m) OD. Fairly numerous fissure veins have been worked on the southern slope of the Swale valley, from opposite Muker eastwards to Grinton, but only one of any importance lies on the northern slope, the Friar's Intake Vein, terminated to the north by the Stockdale fault system. A new feature of the present area as compared with the North Swaledale Mineral Belt is the presence of a series of NNW-trending fracture belts, at least six being recognisable, cutting through the Swaledale – Wensleydale watershed between the Buttertubs Pass in the west and Bellerby Moor in the east. Each of these is structurally continuous for several miles, but the oreshoots appear to have been small and patchy except perhaps in the easternmost belt, at Keldheads and Cranehow Bottom. More continuous mineralisation has been found on west-north-west veins like Summer Lodge and Whitaside – Apedale (Figure 30). Several important east-north-east veins have been worked, and one major E – W zone. The stratigraphical horizons favourable to mineralisation show a wider range in Area 13 than in Area 11, perhaps because lower beds in the Brigantian and the top part of the Asbian are exposed on the valley sides and bottoms; nevertheless, an important part of the lead production has come from strata between the base of the Main Limestone and the Crow Chert, exposed high on the valley sides on either side of the watershed. (Bradley, 1862 gives details for South Swaledale.) The orefield does not effectively extend into the higher part of Wensleydale, west of Buttertubs, for although the side valleys that provide a more or less radial drainage around Great Shunner Fell (2340 ft, 713 m) expose all the potentially favourable strata, no ore deposits have been found. On the south side of upper Wensleydale, the side valleys of Widdale, Snaizeholme, Sleddale and Raydale are likewise curiously lacking in mineralisation. On the east side of the area, however, mineral veins continue close to the region where the favourable beds disappear beneath the mid-Pendleian unconformity on Bellerby Moor, and in the vicinity of Leyburn and Middleham.

Many of the important mines of Area 13 lie within Lord Bolton's estate, between Carperby, Apedale Head and Wensley, but mining ceased early in the present century after having been strongly in decline during the second half of the previous century. The only recent activity has been washing of tailings, chiefly at Wet Grooves, for fluorspar (Dunham, 1952) and barytes. Only a few minor veins lie within the military reservations associated with Catterick Camp, where the favourable beds are largely concealed by late Pendleian and later strata.

Description of the deposits starts in the south-west part of the mineralised area and as far as possible the veins are described in order from west to east.

Sargill – Providence Vein Lead ore

SD 89 SE, NE Yorkshire 51 SW, NW
Direction N10–15°W, throw 70 ft (21 m) E

Stags Fell – Lover Gill Vein Lead ore

SD 89 SE, NE Yorkshire 51 SW, NE
Direction N10–40°W, throw 50 to 100 ft (15–30 m) W

From the Wensleydale side, the Sargill – Providence Vein was worked from the remotely-situated *Sargill Mine* (Figure 30, loc. 1). The main level [8953 9295] was driven at the base of the Main Limestone on the upthrow side of the fault at 1620 ft (494 m) OD. At 475 ft (145 m) from the portal an air shaft (Dakyns and others, 1891, p. 169) passed through a coal 4 in (10 cm) thick within grit, perhaps the coal in the Ten Fathom Grit, before passing into the upfaulted Main Limestone. At 175 ft (53 m) N of the shaft, the drive was turned to NNE for 200 ft (61 m) perhaps to cross through the fault-zone before resuming its NNW course on the vein. At 1500 ft (457 m) from the level mouth a string in the footwall nearly at right angles to the vein was discovered and followed for 1060 ft (323 m) WSW until a fault parallel to Sargill Vein was cut, but as no workings were developed on it, it must be presumed to have been barren. The main forehead on Sargill Vein was driven on for a further 2650 ft (808 m) prior to abandonment shortly before the primary geological survey by C. E. de Rance in 1870. He records that rich ore was left in a sump near the forehead; this would presumably give access to the Main Limestone on the hangingwall of the fault. We have been unable to find a section of the stoped ground but it appears likely that this was mainly against a footwall of Main Limestone. At the mouth of the level there is a large dead heap, but the quantity of material remaining from the dressing of lead ore is small, consisting of brown dolomitic carbonate and limonitised limestone. Enough galena was apparently being obtained in 1840 to justify the erection of the smelt mill situated ¾-mile (1.2 km) SSE of the mine; according to Raistrick (1975b) ore was carried over the hills to be smelted at Summer Lodge before that date. The output recorded in the official statistics is small: 147 tons concentrates in 1856–58; 11 tons in 1865; 129 tons 1881–85; 194 tons 1888–93. From 1860 to 1868 the mine is stated to have been in the possession of the Sargill Mines Company, with Lord Wharncliffe as lessor; thereafter it passed to Henry Pease Esq., and on his death in 1881 to Mr F. H. Pease. The operation was clearly intermittent and not successful during this period. At an earlier stage, according to Raistrick (1975a, b) Sargill was included in a mining area held by the London Lead Company under a lease in 1734 from Edward Wortley. Stags Fell Mine is mentioned in their minutes in 1738, but this is not the same as Sargill Mine (see below), nor is Sargill on the Glover Gill Vein if this is the same as the vein here called Lover Gill, following the Ordnance Survey.

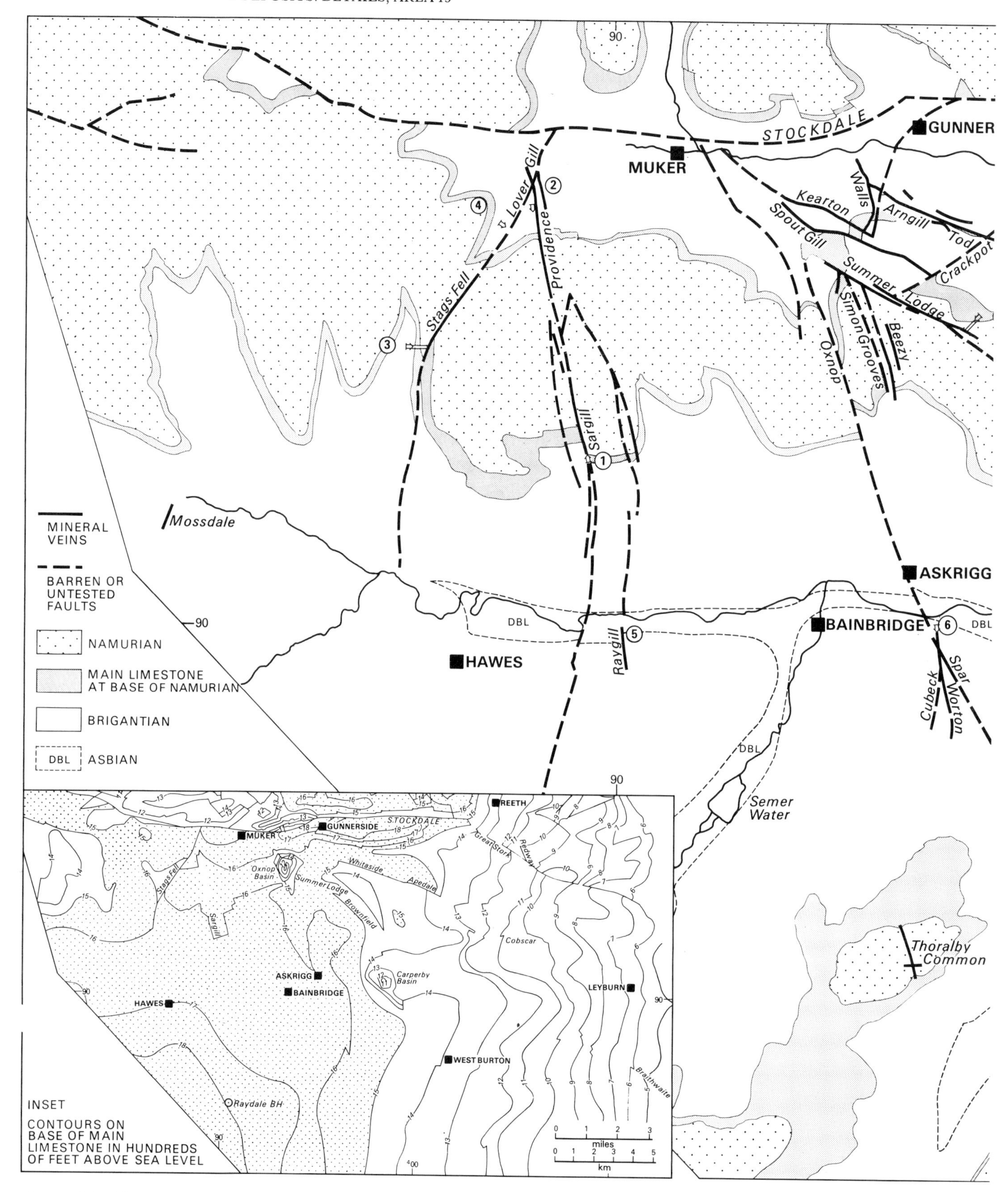

Figure 30 Key and structure map, Area 13, South Swaledale and Wensleydale. Contours on the base of the Main Limestone in feet above Ordnance Datum

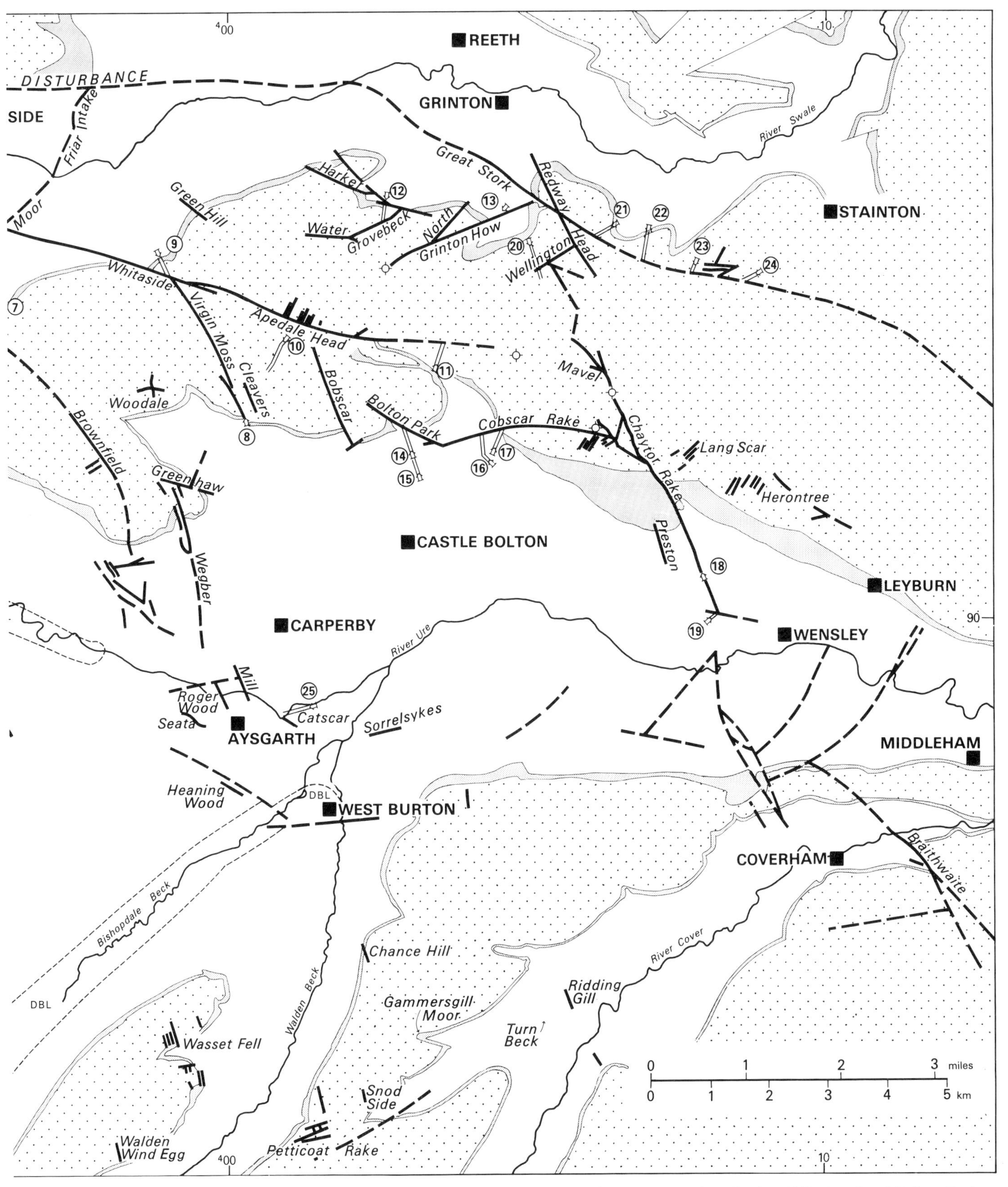

Key to numbered levels and mines **1** Sargill Level; **2** Swaledale Providence Level; **3** Stags Fell Groove Adit; **4** Lover Gill Level; **5** Raygill Level; **6** Warton Hall Level; **7** Summer Lodge Level; **8** Virgin Moss Level; **9** Whitaside Low Level; **10** Wood's Pot Level; **11** Cat Scar Level; **12** Grovebeck Level; **13** California Level, Grinton; **14** Bolton Park Middle Level; **15** Bolton Park Low Level; **16** Bolton Gill Level; **17** Bolton Gill East Level; **18** Keld Heads Level; **19** Wensley Park Level; **20** Devis Hole Level; **21** Hags Gill Level; **22** East Level; **23** Derbyshire Level; **24** Dagget's Level; **25** Aysgarth Level

Southward from Sargill Level, the vein apparently splits into two faults with the displacement divided between them, shifting the outcrops of the Middle and Simonstone limestones north of the hamlet of Litherskew, but there is no indication that either fracture has been prospected here. South of the River Ure, a fault on the same line continues to displace the beds eastward across Burtersett High Pasture and the east flank of Wether Fell where, however, the direction has changed to N15°E. Taking into account the northward extension of the gently curving fracture system, it has been traced for altogether 7.5 miles (12 km) to its junction with the Stockdale Fault.

North of the northern forehead of Sargill Level there is an untested gap 5300 ft (1612 m) to the southern forehead of a level from Providence Mine in Swaledale, the line of the fault being traceable by the displacement of late Pendleian and Arnsbergian strata, including the Tan Hill Coal. The portal of the level from Providence [8874 9653] (Figure 30, loc. 2) is now collapsed, but the drive was close to 1650 ft (502 m) OD and was at a geological horizon corresponding with that of Sargill, thus at the base of the Main Limestone on the footwall side of the vein. The total length driven was 2800 ft (852 m) including a number of changes in direction where the level was driven across the fracture zone. In the northernmost 600 ft (182 m) of workings in the Main Limestone, the mineralised zone was as much as 200 ft (60 m) wide, including flot deposits as well as subparallel fractures. The ground was exploited not only from the level already mentioned, but from three additional levels at lower horizons, the second of which, 380 ft (115 m) long was driven a short distance above the base of the Main Limestone on the hangingwall side, at about 1570 ft (477 m) OD. The final development, 'New' Level, started in 1843, which continued for 1880 ft (572 m) in beds beneath the Main Limestone, communicated with flot workings above by means of rises including Modesty Rise, 760 ft (231 m) S of the adit entrance. The dumps from these workings show highly limonitised limestone, typical of flot workings within the zone of oxidation, with cerussite-fringed galena, calcite, aragonite and very small amounts of baryte. Production from Providence or Mukerside Mine for the most part preceded the official statistics, but 461 tons of concentrates are recorded for 1867–68. Plans (ZLD 41/28, 30, 31) indicate that the mine was under re-examination in the 1870s, but without positive result, even though Modesty Flot was reopened.

The Lover Gill Vein, though in Wensleydale it lies west of Sargill – Providence Vein, crosses the latter north of Providence Mine and is east of it when it reaches the Stockdale Fault. Probably it was first discovered where it shifts the outcrop of the Main Limestone near the head of the Buttertubs Pass and where there is evidence of opencasting. However the southernmost working is Stags Fell Groove [8649 9471] high up in Fossdale, which seems to have been a level (Figure 30, loc. 3) driven beneath the Main Limestone on the downthrow side of the fault, the position of which is indicated by surface mapping and the presence of a shaft near the Buttertubs road 1050 ft (319 m) E of the entrance. Some pale fluorite is present on the heaps, but not much galena appears to have been found. North of the watershed, an oreshoot was worked opencast at Lover Gill Hush on the eastern slope of the Cliff Beck valley, against a footwall of Five Yard Limestone. In 1868–71 a level [8830 9689] (Figure 30, loc. 4) was driven SE 75 ft (23 m), cutting a fracture parallel to the line of Lover Gill Vein as indicated by the hush, but about 200 ft (60 m) farther NW. This was followed for about 600 ft (183 m), before an oblique drive cut across to the vein, reaching it near Ralph Geordy's Shaft. The beds explored here on the downthrow side of Lover Gill Vein must have been those beneath the Three Yard Limestone. Dump material at Lover Gill includes amber-coloured fluorite, baryte, witherite and calcite in addition to galena and traces of chalcopyrite. No separate production is recorded in the official statistics beginning 1848.

Lover Gill Vein is considered to cross through Sargill – Providence Vein about ¼-mile (0.4 km) NE of the northern end of the hush, but there is no evidence of prospecting here or in the stretch up to the exposure in Thwaite Beck on the west side of Thwaite village, where C. T. Clough found fluoritic veinstuff in beds near the Hardrow Scar Limestone. Some evidence of mineralisation thus occurs over a wide stratigraphical range on this vein, yet no substantial oreshoot has yet been found. Immediately north of Thwaite, it is cut off by the Stockdale Fault.

Mossdale Vein

SD 89 SW; Yorkshire 65 NW
Direction N20°E, throw E

Exposures in Mossdale Beck around [823 917] show some mineralisation in a fault affecting Middle Limestone, shale and associated thin limestones. Fluorite, carbonates and traces of yellow sulphide are present. This minor occurrence is interesting as the westernmost of a belt in the bottom of Wensleydale, (Small, 1978) extending into the headwater region which appears otherwise to be unmineralised.

Raygill Vein

SD 99 SW, 98 NW; Yorkshire 66 NW, SW
Direction near N – S, throw 15 ft (4.6 m) W

In Sargill Beck a fault in this direction brings in the Five Yard Limestone at a waterfall, while near Sargill Gate, leading to the fell north-east of Litherskew, the same fault affects the feature of the Middle Limestone. On the same line it was exposed in the River Ure at the time of the primary survey. In the course of operations by the Leeds Mining Co. in 1862–71, a level was driven southward in it, starting north of the Bainbridge – Hawes road at [9017 8987] (Figure 30, loc.5). The first 820 ft (244m) of this adit appears to have followed a string diverging E of S, but a crosscut 60 ft (18 m) W picked up the main vein, and it was followed 825 ft (251 m) S; a rise 44½ ft (13.6 m) near the south forehead gave access to a drift 300 ft (91 m) long running south. In 1979–80 Mr G. C. Clarkson and Mr M. Scarr gained access to the workings and a survey was made by Mr J. D. Carlisle for Earby Mines Research Group. This shows that the deposit worked was a flat, extending 550 ft (168 m) on strike, 15 ft (4.5 m) up, reaching maximum widths of 40 ft (12 m) W and 70 ft (21 m) E of the vein. The thickness of ore observed was about 3 in (10 cm). The vein appeared to be barren, and little seems to have been found in an intersecting fracture running N85°W. The record of 52 tons of lead ore by Wensleydale Mining Co. in 1862 (Hunt, 1863) is possibly from this mine. The exposure in the Ure was in Thorny Force Sandstone, so the level must command Hawes Limestone. Minerals on the dump include amber fluorite, baryte, some sphalerite and galena, malachite and pearly ankerite. The disturbance can be followed southward as far as the outcrop of the Hardrow Scar Limestone.

Worton Spar Vein

SD 99 SW, SE, 98 NE; Yorkshire 66 NE, SE
Direction N35°W, throw perhaps 90 ft (27 m) W

Worton Old Vein

SD 99 SW, SE; Yorkshire 66 NE, SE
Direction N5°W

Cubeck Vein

SD 99 SW, SE; Yorkshire 66 NE, SE
Direction near N – S, throw 24 ft (7.3 m) E

In the vicinity of Cubeck, south of Worton village, there are remains of ancient mining operations, probably in the Gayle Limestone, connected with a fracture belt which displaces the sharp feature of the Hardrow Scar Limestone. We were informed that serious subsidence at the farm at Cubeck in 1847 caused it to be rebuilt on a new site farther north. The farm is close to the mapped line of Cubeck vein, which can be traced up the hillside to the south, through at least four old shafts and a place where the stone wall persistently subsides. It was said that when an excavation was made about 700 ft (213 m) S of the farm to bury a horse, workings or caverns were revealed below. It is not known whether the subsidence at the farm was due to cavernisation or mine-workings or both; mining has however been carried on, as attested by two poor-quality plans in the collection of the Health and Safety Executive (1672) dated 1884. North of Cubeck an old adit (9547 8971) may be tentatively identified with a drive shown running SSE on the plan marked 'High South Workings'. This adit, on the Spar Vein, reached the area of intersection with Cubeck Vein at 165 ft (50 m). Thence the Cubeck Vein workings, lying west of the Spar Vein continue southward for 780 ft (238 m) to the forehead. From the intersection area a crosscut leads ENE to a shaft, the remains of which can be seen in the angle between the road from Worton, and the road following the top of the Hardrow Limestone towards Thornton; it is not, however, certain that this is at the same horizon as the adit. Southward from the intersection area, the workings believed to be on the Spar Vein diverge only slowly from Cubeck Vein for 280 ft (85 m) before the former assumes its S35°E course. J. R. Dakyns, who carried out the primary survey, considered that the Spar Vein continues SSE for 4400 ft (1.34 km) to the old Thornton Lead Mine, where old opencast workings and shafts mark the last 900 ft(274 m) of this extension. Here the Middle Limestone to the east is brought into contact with the Five Yard Limestone and underlying beds. A little purple fluorite, baryte and galena remain in these overgrown workings.

Worton Old Vein passes beneath Worton village and runs alongside the road to Cubeck; presumably the shaft mentioned above is on or near it. At 280 ft (85 m) S of the Cubeck/Spar Vein intersection, Percival's crosscut runs E to reach a shaft 180 ft (55 m) from Cubeck Vein which may also be on Old Vein. The principal prospect on this vein, however, was Warton Hall Level [9550 9009] (Figure 30, loc. 6) at a few feet above 700 ft (213 m) OD. This drive was probably undertaken not later than 1884, the dates on the plans mentioned above; unfortunately no plan of Warton Hall Level has been found. The second Health and Safety Executive Plan, marked 'Southern part of Warton Mine' shows a drift 285 ft (87 m) long but the direction (SSE) does not agree with Old Vein. The fairly extensive dump from Warton Hall Level shows that it penetrated the Thorny Force Sandstone, and probably reached the uppermost part of the Great Scar Limestone. The minerals include purple and amber fluorite, in part replacing limestone, ankerite, calcite, sphalerite and traces of chalcopyrite. The fluoritised limestone closely resembles that of Seata (pp. 180–181) and should any considerable tonnage be present, this would be of economic interest. The official statistics credit Worton Mine with 363 tons of lead concentrates between 1875 and 1883. Thornton Moor is mentioned from 1862 to 1874 the operator being Percival & Co. and the lessor Lord Bolton; but no production is recorded. The crosscut mentioned above suggests that Percival & Co. was also at Worton.

The Worton Spar Vein is believed to continue NNW as a fault passing beneath Askrigg village east of St Oswald's Church; an attempt was made to test it by means of two shallow shafts ½-mile (0.8 km) farther north, near Straits Lane. Traces of mineralisation were found, probably close to the footwall position of the Simonstone Limestone, but nothing workable was obtained. From here the fault continues, with a slightly more northerly trend, into the Oxnop fault system, described below.

Oxnop Faults

SD 99 SW, NW; Yorkshire 51 SE, NE
Main fault trending N10–15°W, variable W downthrow; accompanied by subparallel and diverging faults on the W side, downthrowing E

Satron Tarn String Lead ore

SD 99 SW; Yorkshire 51 SE
Direction N20°W

Simon Grooves Strings Lead ore

SD 99 SW; Yorkshire 51 SE
Direction N8–18°W, small W downthrow

Beezy String Lead ore

SD 99 SW; Yorkshire 51 SE
Direction N10–15°W

Oxnop Beck head makes a col in the watershed followed by the road from Askrigg to Muker in Swaledale. The course of the Oxnop main fault is evident on the Wensleydale side from the shift of the base of the Main Limestone from Stackhill House [9421 9309] to Kittle Rigg [9410 9347], but no other major disturbance of the run of the limestone crop is evident. The cliffs and other features around Oxnop Kirk make it evident that other faults have developed west of the main fault in a belt about ¼-mile (0.4 km) wide (Figure 31). Between Stotter Gill and Castle How [933 961] the base of the Main Limestone dips down the eastern hillside from 1550 to 1050 ft (471–319 m) towards the main fault before rising again in a series of steps to the west. The faulted synclinal structure, plunging gently north, continues beyond the Spout Gill Vein (p. 169) but probably it closes into a basin before reaching the Stockdale Fault to the north. This remarkable faulted basin, already mentioned in Chapter 6, is only paralleled in the northern Pennines at Carperby. A number of veins diverge from the structure on the east side; these are considered in the present and the two succeeding sections. South-west of the structure a prominent electromagnetic anomaly has been detected.

The main fault was tested by means of a hush about 400 ft (122 m) long at Oxnop Beck Head [938 944], where limestones and shales dip into it from the west at 15–18°, but without much success, though alteration of the limestone is apparent. Trials were also made at the foot of Oxnop Kirk. On the higher ground east of the valley, where the Little and Red Beds limestones, with cherts, are the surface formations, the Satron Tarn, Simon Grooves and Beezy strings were worked, their courses being marked by lines of shallow shafts. On the first of these the line of shafts suggests mineralisation over a length of 1250 ft (380 m) near Beezy Bottom, SSE of the Tarn, and again over a shorter stretch NNW of the tarn. This string converges southward upon the second, probably a double fracture with perhaps two mineralised stretches in the 2000 ft (610 m) length of Simon Grooves Mine [944 952], while a line of shafts 1800 ft (547 m) long apparently on the same strings was worked at the Beezy mines. Beezy String, the eastern feature, was worked from a level [9475 9422] and from widely-spaced shafts along the ½-mile (0.8 km) of ground to the NNW (Figure 31). The debris from all these workings is essentially similar. Limonitised limestone with brown, partly oxidised carbonate encloses cerussite-coated crystals of galena, suggesting that the ore was obtained from flots in limestone. It is not certain, however, that the Main Limestone was everywhere penetrated, the lowest working where the horizon is known being Beezy Level, at the top of the limestone. Some ore came from chert of the Richmond Cherts, on the evidence of mineralised fragments in the heaps. On Simon Grooves String, about 1050 ft (320 m) NNW of Beezy Level mouth, there is an unusual dense massive goethite which Raistrick (1975a p. 85) ascribes to replacement of massive iron carbonate, but which could also have resulted from oxidation of pyrite. The goethite also encloses cerussite-sheathed galena. Raistrick states that these mines were working before 1765, and that they continued into the 19th century, when they may have contributed to the production from Satron Moor mentioned below. They were abandoned before official statistics began, in 1861, to show individual mine outputs in Yorkshire. Estate records however show 2 tons of lead raised from Oxnop Gill Head between 1909 and 1912 by G. Calvert &

Company, representing the last activity in the area. To the north, the four strings intersect the Summer Lodge veins, while southward it seems a reasonable hypothesis that they combine to form the Worton Vein (p. 166).

On the west side of the Oxnop Pass, very little prospecting seems to have been done. There is, however, evidence that a series of faults pass through the ground; for example a small outlier of presumed Lower Howgate Edge Grit which occurs at Red Braes [932 949] with its base at 1700 ft (517 m) OD, nearly 100 ft (518 m) below the base of the main outlier capping Oxnop Common to the west implies, as De Rance's primary survey indicates, the presence of a N – S fault. This he considered must change direction near Lealamb Pot [9310 9467] to SE. However, an airborne electromagnetic survey made on behalf of the Department of Industry in 1975 (Evans and others, 1983) revealed a well-defined anomaly running S10–15°E from near Lealamb Pot to the watershed [9347 9395], a distance of over ½-mile (0.8 km). It is a possible interpretation that this is related to a continuation of the direct line of the fault. Although two boreholes were drilled (Figure 31), one of them reaching Main Limestone, the cause of this anomaly has not yet been established; but it could well be related to a substantial concentration of sulphides.

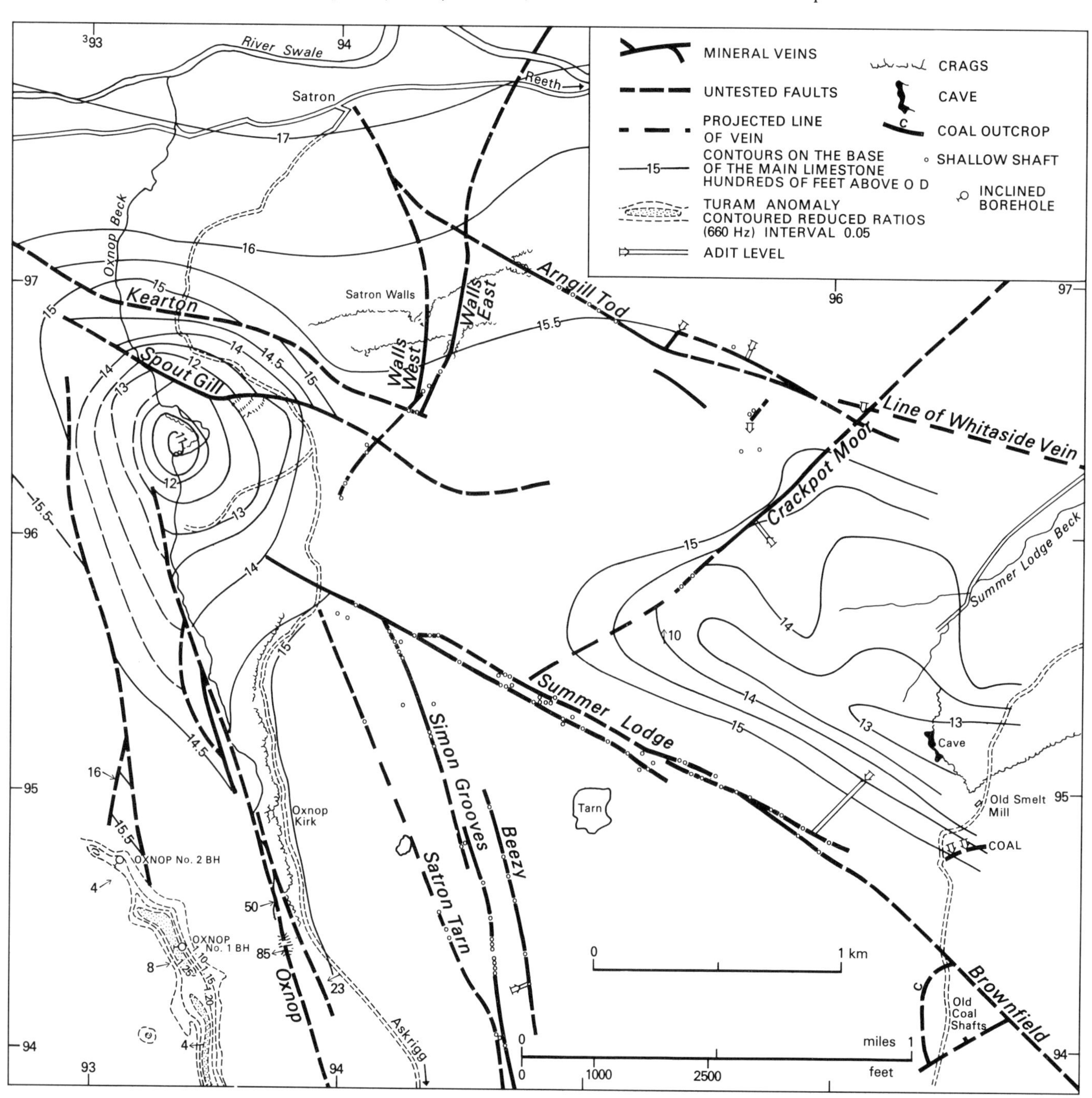

Figure 31 The Oxnop Basin, Oxnop and Summer Lodge Mines. Contours on the base of the Main Limestone in feet above Ordnance Datum.

Stotter Gill – Summer Lodge Veins Lead ore

SD 99 NW, NE, SE; Yorkshire 51 NE, SE
Direction N65°W

Some 1½-miles (2.4 km) of continuous workings mark the course of this mineralised belt across Satron and Summer Lodge moors, north of Summer Lodge Tarn (Figure 31). Altogether the remains of 72 shafts can be identified from air photographs and from surface mapping. As already noted, below Stotter Gill, at the western end, the Main Limestone dips rapidly west into the Oxnop structure, but there is an almost equally unusual monoclinal dip north of the western stretch of the belt, where the base of the Main Limestone dips NE from 1580 ft (480 m) OD at Hog Gill Hole [9569 9510] to almost 1250 ft (380 m) in Summer Lodge Beck. The veins form a belt up to 300 ft (91 m) wide in which several subparallel or diverging fractures are involved. Stotter Gill has been widened and deepened by hushing, but throughout the remainder of the belt shafts have been sunk through Richmond Cherts, probably into the Main Limestone. White chert forms prominent knoll-like features on Satron Moor (see also p. 56), suggesting lenticular deposits, interbedded with coarse limestone of the Red Beds, which probably carried at least some of the deposits. At the eastern end, Summer Lodge Level [9611 9505] (Figure 30, loc. 7) at 1520 ft (462 m) OD started near the top of the Main Limestone and reached the first vein 950 ft (289 m) SW of the portal, near or below the base of the limestone. The tailings from ore dressing here, and indeed the debris throughout the belt are dominated by brown dolomitic carbonate; only small amounts of baryte, aragonite and calcite can be found (analyses, Table 16), with some chert and limonitised limestone. The appearance of the gangue material closely resembles that from flots worked elsewhere, for example at Windegg, and that such deposits occurred here is indicated by a note by the Agent of the A.D. Company on a visit in 1864: 'G. Calvert & Co. have sunk about 4 Fms in to the Main Lime in search of a float—there are strong indications but not sufficiently firm as there is no cover on the limestone. I have very little hope of success. There is a better chance farther W, where the Main Lime is capped by chert.' Most of the workings are ancient; Raistrick (1975a) states that they were fairly rich in the 18th century, with an important smelt mill where the road crosses Summer Lodge Gill. Records formerly in the A.D. Mine Office showed that there was still activity at the Satron Moor end during the period 1835–50, when the lessees' share of the lead produced was 739 pieces (pigs) or about 62 tonnes; allowing for royalty, the quantity of concentrates produced must have been of the order of 100 tonnes. Fieldhouse and Jennings (1978) quote the production of the south Swaledale mines of the A.D. Group as 1087 tons lead, 1817–1880. This would represent about 1550 tons of concentrates, mainly from Oxnop – Summer Lodge.

Mineralisation on the Summer Lodge run ceased a short distance SE of the Summer Lodge Level head, but in this vicinity a fault is given off, running SE to become the Brownfield Vein of Beldon Bottom (pp. 170–171).

Spout Gill Vein Lead ore

SD 99 NW; Yorkshire 51 NE
Direction N65°W – N85°E, throw 150 ft (46 m) S

Kearton Vein Lead ore

SD 99 NW; Yorkshire 51 NE
Direction N42–80°W, small throw S

Whereas the Stotter Gill vein appears to swing into and unite with the Oxnop main fault, these two sinuous veins apparently cut right across the Oxnop structure (Figure 31). Spout Gill Hush [936 965], was not a large one by comparison with many in the district, but Raistrick (1975a, p. 32) cites a record of 1732 that the mine was 'exceeding rich'. The deposits are said to have been flots and probably these were in the bottom part of the Main Limestone, where the vein brings this into contact with the Underset Limestone. The hush has worked through into beds beneath the Main Limestone, and remarkably little evidence of mineralisation remains. Levels and shafts near the head of the opencut have also left very little debris. According to Raistrick and Jennings (1965) the area was first leased by Lord Wharton to Thompson and Partners of Richmond, but early in the 1730s the Company of Mine Adventurers took over the property and improved both hydraulic mining methods and smelting practice. Raistrick (1975a, p. 33) quotes a statement that in one year the mine produced £40 000 worth of ore; at the mid-18th century price of lead (around £15 per ton) this might represent at least 3000 tons of concentrates, and if it is correct the puzzling question of the whereabouts of the tailings from dressing such a quantity needs to be answered. Raistrick adds that this may be a miner's tale, but maintains that it indicates that some rich ore was mined. Ground easily accessible above water level is said to have been exhausted by 1768, when levels were driven, but without much success. Nevertheless the mine was still being operated in a small way by James Brown & Co. from 1816 to 1836 with a lessee's share of the output, according to A.D. Mining Office records, of 845 pieces of lead, representing, after allowing for ¼ royalty, about 85 tonnes of lead, or say 120 tonnes of concentrates. Very small-scale activity continued until 1850 under T. Calvert.

Kearton Vein, so-called here because the disturbance is exposed crossing Oxnop Beck in Kearton's Wood, is a sinuous minor fault with very little mineralisation. It was tested by a shallow shaft close to where it shifts the outcrop of the Main Limestone at the west end of Satron High Walls, dolomitic limestone being present.

Walls West Vein Lead ore

SD 99 NW; Yorkshire 51 NE
Direction N35°E turning to N – S

Walls East Vein Lead ore

SD 99 NW; Yorkshire 51 NE
Direction N 0–20°E, small E throw

Both veins (Figure 31) have been opened up where they cross the Underset Limestone outcrop of Satron Low Walls and the Main Limestone crags of Satron High Walls. Surprisingly the East Vein was tested by a line of 12 shallow shafts over a length of 600 ft (182 m) where it crosses the sandstone beneath the Underset Limestone. The principal ore-bearing workings were, however, on the East Vein in the Main Limestone and Main Chert, a length of some 900 ft (274 m) ending at the intersection with Kearton Vein. Tailings from hand dressing are chiefly brown carbonate, in part cavernous suggesting the presence of flots, with some baryte. The West Vein can be traced by widely-spaced shafts almost to its intersection with Stotter Gill Vein. T. Calvert was working at Walls in 1823 with Edmund Metcalf and some others; in a statement written at Calvert's house they state 'We have cut a vein above Spout Gill that appears to have a great slip of beds'. In 1828 Calvert produced 60 pigs of lead from Walls and minor, perhaps one-man, activity continued to 1850.

Crackpot Moor Vein = Friar Intake Vein Lead ore

SD 99 NE; Yorkshire 51 NE
Direction N35–15°E, throw 20 ft (60 m) NW

A feebly-mineralised but persistent fracture crossing below Blea Barf on Crackpot Moor in the cherts and limestones above the Main Limestone was investigated by means of a crosscut level at [9574 9601], a line of shallow shafts 375 ft (114 m) long north of this, and a short level [9612 9651], a little galena and brown carbonate being found (Figure 31). This work was being done by J. Davies in 1818–23, when some 65 pieces (about 5 tonnes) of lead were produced. Production of a similar order continued as late as 1850. The fracture if projected towards the SW would intersect the

Summer Lodge veins in the neighbourhood of Dry Gill, but there is little evidence of it here. The NE projection of the vein or string links up with the position of Friar Intake Vein, which was worked north of the River Swale between Gunnerside and Low Row, and it is a reasonable assumption that the two are continuous, though at Friar Intake the direction changes to N15°E. The Friar Intake deposit is unique in Swaledale in being productive in the Simonstone Limestone. Surface workings and shafts mark its course up the hillside through this limestone and the overlying Middle Limestone. The main level [9740 9755], enters as a crosscut driven WNW in beds above the Hardrow Scar Limestone, reaching the vein at 400 ft (122 m) and continuing northward in it for 1500 ft (456 m) to the forehead reached in 1876. Underground workings at a higher level had reached 510 ft (155 m) farther north by 1871. The spoil suggests that the vein was banded with galena, amber fluorite, baryte and traces of chalcopyrite. Some sphalerite may also have been present, since green smithsonite also occurs. About 1880, attempts were made to find this vein on Old Gang property, north of the Stockdale Fault (ZLB 41/1–61), several short levels being driven on Brownsey, but it was not located and must be presumed to have cut out against the fault.

No separate production figures are available, but Raistrick states that the mine was already at work in the 17th century, that Abraham Fryer and partners worked it in the 18th century, and that in 1740 it was the subject of litigation by the Wharton trustees (1975a, p. 66); thus it sustained operations for two centuries, though never on any but a small scale.

Woodhall Vein Lead ore

SD 99 SE; Yorkshire 67 NW
Direction N20°W, may have small NE throw

Disher Force Vein Lead ore Fluorspar

SD 99 SE; Yorkshire 67 NW
Direction N30–55°W, throw NE

Oxclose Vein

SD 99 SE; Yorkshire 67 NW
Direction N60°W, branch of Disher Force Vein

Wet Grooves Knot Vein and Strings Lead ore Fluorspar

SD 99 SE; Yorkshire 67 NW
Direction N – S

Wet Grooves Cross Vein

SD 99 SE; Yorkshire 67 NW
Direction N30–45°W, throw NE

Wet Grooves Vein Lead ore

SD 99 SE; Yorkshire 67 NW
Direction E – W, turning to ENE

Thackthwaite vein Lead ore

SD 99 SE; Yorkshire 67 NW
Direction E – W, throw N

Blue Scar = Brownfield Vein Lead ore

SD 99 SE; Yorkshire 67 NW, 52 SW, 51 SE
Direction N – S, thinning northwards to N45°W; throw, at Blue Scar 1 ft (0.3 m) E, at Swinhaw Bottom Crags, 6 ft (1.8 m) E, at Brownfield Mine 42 ft (13 m) NE, throws SW at Windgates Colliery, before uniting with Summer Lodge Vein.

Brownfield Strings Lead ore

SD 99 SE; Yorkshire 52 SW, 67 NW
Direction ENE

North-west of the village of Carperby in Wensleydale lies another area of abnormal structure. Looking north from the Aysgarth – Bainbridge main road, a virtually flat series of low scarps containing beds from Hardrow Scar up to Middle Limestone appear in the foreground, with Ivy Scar, of Underset Limestone, displaying a dip of 15°W behind, evidently detached from the foreground formations. On further investigation the Ivy Scar dip proves to be part of a closed faulted basin, about ¾-mile (1.2 km) in diameter, and with an amplitude of about 250 ft (76 m) (Figure 32). On the north-east side of the structure the Brownfield Vein dies out; this fault, of changeable throw, may be regarded as linking the structure with the two other unusual ones at Summer Lodge and Oxnop respectively (pp. 167 – 169). Another exceptional feature of the Carperby area is the high proportion of limestone in the Underset cyclothem; of the total thickness of 90 ft (27 m), over 50 ft (15 m) are in the Underset Limestone (p. 48). The basin structure produces unusually wide outcrops of this limestone on the hillside, but near the north of Thackthwaite Beck, erosion has cut through the limestone to reveal a closed area of beds down to the Three Yard Limestone, and the sandstone underlying it. The veins listed above are the principal fractures associated with the structure, though by no means the only ones. Lead ore has been won from an unusually wide range of beds, from the Simonstone Limestone up to and including the Richmond Cherts, a total range of some 800 ft (243 m), but the oreshoots found have been small and patchy. Probably the largest concentration was in The Knot [9865 9020] which we regard as a landslipped mass of Underset Limestone from Ivy Scar, but now resting against a face of Preston Grit and associated beds. The limestone of The Knot has been extensively mineralised, both by fracture-filling and by replacement, and it is now honeycombed with workings. It stands opposite, and is considered to have formed part of a N – S vein that continues across the slope above the scar, where it was worked in shallow opencuts; it also seems likely that at least one of the levels in The Knot continued northward into it. This vein dies out to the north, perhaps by virgation with a NNE string. South of The Knot, the plateau or shelf crossed by the Oxclose road was the main site of dressing operations, probably dealing with ore from most of the veins, not merely with Knot Vein. The extensive dumps have been reworked since World War II, yielding fluorspar. An estimated figure of 5000 tonnes has been included in Table 2 for Wet Grooves and Keld Heads. Oxclose Vein is structurally the southern boundary of the basin; it seems to have been reached by a deep shaft near the foot of The Knot, which may have linked with an adit starting near the Askrigg – Carperby road at Haw Bank [9859 8997] close to 750 ft (229 m) OD beneath the Simonstone Limestone. There is evidence of much veinstuff, with fluorite, baryte, galena, sphalerite and calcite; it is not certain, in the absence of plans, whether this came from Oxclose Vein, from an extension of the level beneath The Knot, or from an oblique drive into Disher Force Vein. The same suite of minerals, though in smaller quantities, is found on the heaps from Disher Force Level [9808 9035] which gave access to the vein of that name in the Middle Limestone. Above the Force, recent excavations for a fish pond revealed good quality fluorite in the vein. The workings of *Wet Grooves Mine* include The Knot and extend for about ¼-mile (0.4 km) north-westward to where deeply-eroded gorges, probably in part the result of hushing, expose the Underset Limestone in the headwaters of Eller Beck. Here the Wet Grooves Vein and Cross Vein and associated replacement of the Underset Limestone was worked; there is a level and shafts on the vein and a few shafts on the cross vein. Farther north, Thackthwaite Vein has been worked eastward from its origin at the cross vein along the course of Thackthwaite Beck by means of shallow shafts. Brought in on the north wall of this vein, the Preston Grit has been opened up by an adit probably aiming for Blue Scar Vein, but with little success. From Thackthwaite Vein northwards, fluorite is no longer found in the veins, but brown dolomitic carbonate or calcite is commonly present. The Blue Scar workings in Underset Limestone, and on the same vein the level in the Main Limestone seems to have yielded little ore; the Brownfield Vein was however apparently a

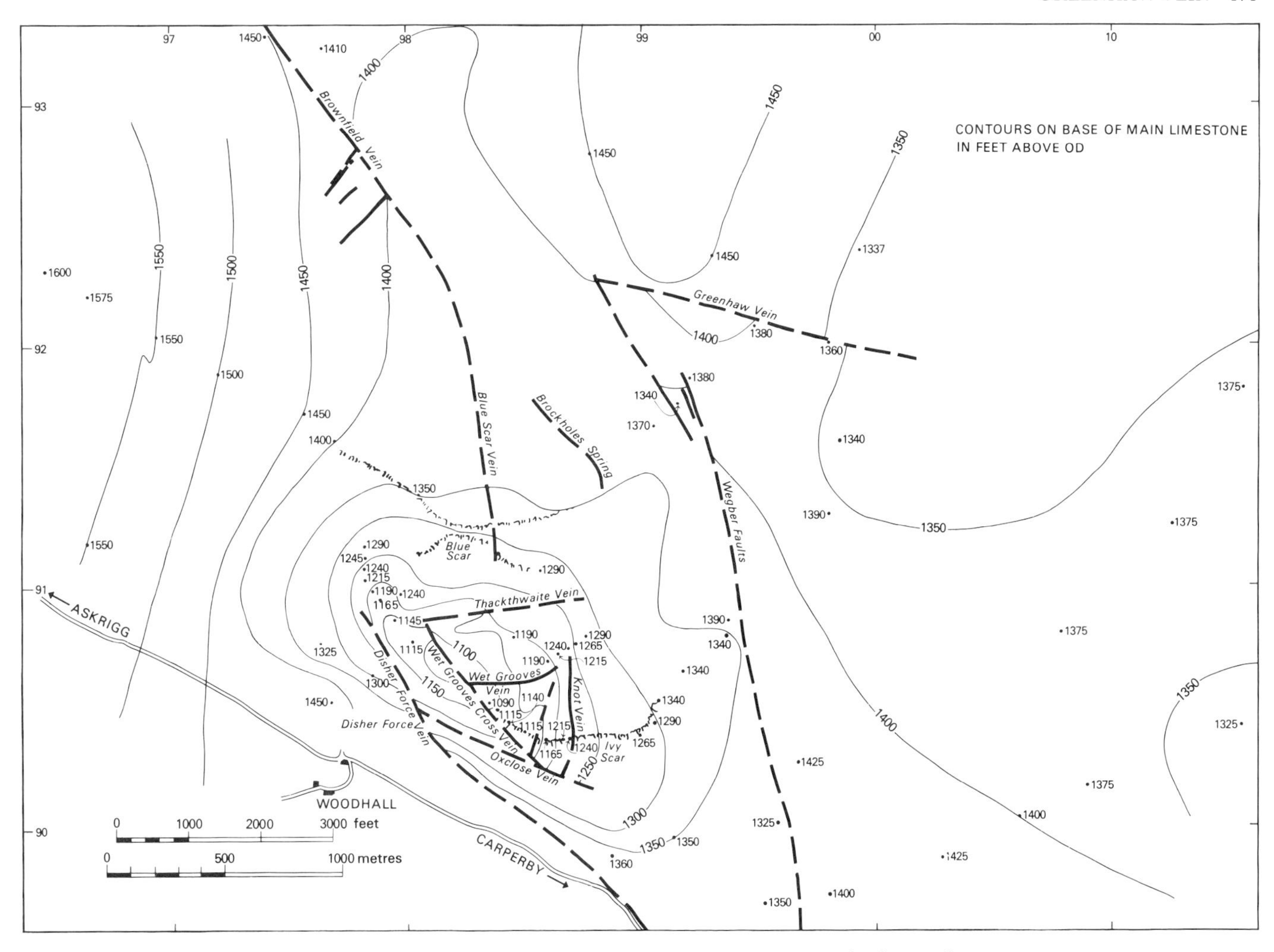

Figure 32 The Caperby Basin, showing the relationship of veins. Contours on the base of the Main Limestone in feet above Ordnance Datum.

significant producer in Main Limestone at the *Brownfield Mine*, nearly 1 mile (1.6 km) farther north, where a very wide spread of shafts worked strings, and probably flots in Red Beds Limestone as well as Main Limestone (Figure 32). Watson's Shaft [9770 9244], near the fell wall, is said to have been sunk 150 ft (46 m), passing through shale 40 ft (12.1 m), 'Red Beds' 45 ft (13.7 m), limestone (the Red Beds Limestone) 10 ft (3 m), shale 5 ft (1.5 m) and so to the base of the Main Limestone, on one of the strings west of the main vein. There are numerous small tailings heaps, containing baryte, witherite, aragonite, calcite, limonite and malachite as well as galena. This remotely situated mine, worked exclusively from shallow shafts probably belongs to the 17th century (Raistrick, 1975a). During the BGS mineral reconnaissance survey (Evans, Patrick, Wadge and Hudson, 1983) airborne EM measurements showing a series of disjointed anomalies along the Brownfield Vein were followed up by a ground survey using Turam and VLF-EM. These revealed a belt of anomalies 1.4 km long, centred on Brownfield Mine, but extending beyond the limits of working. The form of the anomaly is stated to indicate a conductor at least 130 ft (40 m) deep, which could be a downward extension of sulphide mineralisation. This anomaly has not been drilled owing to the inaccessibility of the area. From [9758 9316 to 9732 9350], in the headwaters of Beldon Beck, Main Limestone is exposed along a strip on the west wall of Brownfield Vein, and hereabouts W. Gunn records that the throw of the fault was established as 42 ft (12.8 m) NE, presumably in underground workings. This strip marks the end of the active lead mining ground related to the vein, but it can be followed for a further 1½ miles (2.4 km) to the NW, partly through the high grits, where it forms the eastern boundary of the Windgates workings on a faulted-in outlier of what is considered to be the Tan Hill Coal. Here its throw must have reversed. As already noted, it unites with Summer Lodge Vein near Summer Lodge Level head (p. 169). A long stretch on this vein, where the Main Limestone and Richmond Cherts/Red Beds Limestone are covered by beds above the mid-Pendleian unconformity, remains untested.

Wegber Faults

SD 99 SE; Yorkshire 67 NW
Direction near N10°W, throw 20 to 30 ft (6–9 m) W, decreasing to N

Brock Holes String — Lead ore

SD 99 SE; Yorkshire 67 NW
Direction N10–30°W, little or no throw

Greenhaw Vein — Lead ore

SD 99 SE; Yorkshire 67 NW
Direction N75°W, throw 90 ft (27 m) S

The Brock Holes String cuts the Swinehaw Hill cliff in Main Limestone 1650 ft (502 m) E of Blue Scar Vein, and the associated deposits, probably at least in part flots, can be followed 1500 ft (456 m) to the NNW. The workings though numerous are all small and debris includes only brown altered limestone and calcite. The strings may have influenced the northward course of Thackthwaite Cave in the Underset Limestone (plan: Brook and others, 1977.) Farther east, a small fault-trough, forming part of the Wegber faults, cuts through the same cliff-line at Keld Heads [992 918], dropping the base of the Main Limestone between the outer faults, which are about 275 ft (84 m) apart, by about 20 ft (6 m). There is a little mineralisation, chiefly in the form of limonitic alteration of the limestone. The eastern member is considered to continue south and become the fault which shifts the outcrops of the Middle and Simonstone limestones east of Wet Grooves, but it appears never to have been prospected. In the vicinity of the outcrop of the top of the Main Limestone, 1200 ft (365 m) N of Keld Heads there is a group of shafts suggesting former working of flot deposits. The Wegber faults north of these workings encounter the Greenhaw Vein, which lifts the Main Limestone by more than its thickness to the north. Flots were worked in association with this vein and accompanying strings, tips from the shafts showing cavernous limestone with galena and calcite. Pits on the line of Wegber east fault continue on the north side of Greenhaw Vein, but seem to have yielded little.

Smithy Gill Strings Lead ore

SD 99 SE; Yorkshire 52 SW
Direction N62°E

Woodale Vein Lead ore

SD 99 SE; Yorkshire 52 SW
Direction E – W

Woodale Cross Vein Lead ore

SD 99 SE; Yorkshire 52 SW
Direction N – S

Woodale Head Vein Lead ore

SD 99 SE; Yorkshire 52 SW
Direction N32°W

West Bolton Moor, forming the northern slope of the headwaters of Beldon Beck, contains the remains of many small workings, together making up the *Woodale Mine*. All are on the outcrop of the Richmond Cherts and it seems likely that the bearing beds were limestones, particularly the Red Beds Limestone, in that formation, for not a single working occurs on the outcrop of the underlying Main Limestone. The two principal Woodale veins cross at [9871 9381]; shafts along the E – W vein have been sunk at intervals of about ½-mile (0.8 km) while there are closely-spaced shafts along the cross vein for 1300 ft (395 m) mainly south of the crossing. The cross vein has been regarded as sufficiently interesting to test by means of a shaft [9876 9457] through the grits just within the boundary of Whitaside Mine (p. 173) but it is not known whether this reached the vein in bearing beds. Three pits [985 943] aligned NE, also at a high stratigraphical level, suggest the discovery of a fracture in grits but there is no evidence that it was mineralised. In the veinstuff at Woodale, baryte and galena occur in chert and limestone, but the quantity of tailings is very small, this applying equally to the pits along Woodale Head Vein. A shallow but long trench, trending N20°E, marked Woodale Strings on the six-inch map shows no evidence of having discovered workable mineralisation. Taking the Brownfield, Greenhaw and Woodale workings together, lead ore was evidently widespread around the head of the Beldon valley, largely in limestone and chert above the Main Limestone; but there is no evidence that substantial, concentrated oreshoots were present.

Virgin Moss Vein = Virgin String Lead ore (barytes, witherite)

SD 99 SE, NE, SE 09 SW; Yorkshire 52 SW
Direction N25°W, throw 2 to 20 ft (0.6–6 m) E

Cleaver's Rake Lead ore

SD 09 SW; Yorkshire 52 SW
Direction N22°W, probably no throw

Virgin Moss Vein is the fifth of the NNW lodes that cross the watershed from Wensleydale and can be identified in Swaledale. However, the productive ground on this one lay on the Wensleydale side, probably largely or wholly in the cherts and limestones above the Main Limestone and below the Ten Fathom Grit. The main level [0024 9335] (Figure 30, loc. 8) is driven on the top of the footwall position of the Main Limestone at 1480 ft (450 m) OD. No plan has been found, but the dump suggests that the level could be ½-mile (0.8 km) or more long. There are shafts at intervals up the rising fell side for 5300 ft (1.6 km) within the Bolton Royalty, and at least two of these, Dressing Shaft [9989 9423], at 1770 ft (538 m) and Hill Top Shaft [9984 9438] at 1826 ft (555 m) OD were deep shafts used for drawing ore, for they are surrounded with substantial tailings heaps. At 800 ft (243 m) farther NNW there is another deep shaft, sunk from much the same elevation, still indicating strong mineralisation, but the final shaft in the Bolton Royalty, 1575 ft (480 m) NNW of Hill Top Shaft shows only shale and sandstone on the tips; it may have failed to reach the bearing beds, or the oreshoot may have pinched. All these shafts must have been sunk through the mid-Pendleian unconformity, but there is no evidence on the heaps of mineralisation of grits or shales, only of pale limestone and platy chert. Almost certainly the bearing beds were in the Richmond Cherts, and the shafts must have been sunk 250 to 300 ft (76–91 m) to command these. The fact that much of the ore was not extracted through Virgin Moss Level points to an early period of shaft working, prior to the driving of the adit. It is regrettable that no sections of the strata cut in these shafts have been preserved, but this is consistent with their antiquity. Nevertheless the mine worked until the later part of the 19th century, and towards the end of its life some shipments of barytes are said to have been made. A flat containing pure witherite was also cut, and picked material from this can be seen near the level mouth (analysis, Table 16). Witherite is in fact persistently present not only here but in the heaps at higher levels. The other minerals present include, beside galena and its oxidation products, hemimorphite and calcite, with limonitised crinoidal limestone. Strangely, no attempt seems to have been made to follow the vein across the Main Limestone outcrop, or into lower strata at outcrop; however, a sump is said to have been sunk 210 ft (64 m) NNW of the adit portal to the Underset Limestone, without success.

Cleaver's Rake is subparallel to Virgin Moss Vein, about 1050 ft (319 m) farther east; there are a few shafts into the Richmond Cherts over a length of more than ½-mile (0.8 km), but it was not worked across the outcrop of the Main Limestone.

Virgin Moss Vein continues into Whitaside ground, and was cut at that mine in Main Limestone; its dimensions had probably decreased, for here it was called Virgin String. A short drive was made on it (p. 173) and a shaft was sunk 2800 ft (851 m) S of the Whitaside workings; this appears only to have found calcite in the pale limestones.

Arngill Tod Nick = Whitaside = Apedale Head Vein Lead ore

SD 99 NW, NE, SE; SE 09 NW, SW; Yorkshire 51 NE, 52 NW, SW
Direction varies between N70°W and E – W, maximum throw, 90 ft (27 m) S

Harker Vein Lead ore

SD 99 NE; Yorkshire 52 SW
Branch running N70°E

Apedale Head Cross Strings Lead ore

SD 09 SW; Yorkshire 52 SW
Subparallel strings on N side of Apedale Head Vein, N10–20°E

This 6 mile (9.6 km) long vein system (Figure 30) is one of the two most important in Area 13, judging from the spoil remaining from operations along it; mining must have begun at least by the 17th century for Raistrick and Jennings (1965, p. 124) record the purchase of Whitaside Mine by the London Lead Company. It is strange that the mine is so poorly recorded under these circumstances for only an inadequate plan (R27F) dated 1858 with no sections remains, probably from a period subsequent to the Company's ownership. In Lord Bolton's Royalty, there are plans of Apedale Shaft and Harker's Level, both belonging to the later part of the 19th century, but the more interesting stretch at Apedale Head is poorly recorded, without sections.

The vein in its westernmost exposure was found at Arngill Tod Nick [9468 9707] where opencut workings have created a gash through the Underset Limestone of Satron Low Walls. North-west of the Nick, it was tested by pits into the Twentyseven Fathom Grit where it crosses Walls East Vein (p. 169), and was mapped by C. T. Clough as uniting with Walls West Vein south of Satron, where a northward downthrow of the Middle and Simonstone limestones is apparent, but no further trials have been made. East-south-east from Arngill Tod Nick, the opencut and pits revealed mineralised ground at least 1600 ft (487 m) long, with a branch vein to the north-west. The veinstuff includes brown carbonate and a little baryte but the vein must have been narrow in this stretch. A short crosscut level ½-mile (0.8 km) ESE of the Nick proved a northern branch, throwing 6 ft (1.8 m) N, and also probably tested the vein below the Underset Limestone, but found no ore. From here the course is obscured by drift cover on the western slope of the Crackpot Beck valley, and no attempt seems to have been made to investigate the crossing through the Crackpot Moor - Friar Intake line. The disturbance due to Whitaside Vein can be detected in Crackpot Beck, where the Middle Limestone on the east slope of the valley is faulted down to the south; thus if there is a continuous fracture from Satron Low Walls to Whitaside, as seems likely, the direction of displacement has reversed. Two levels investigated the vein in the Middle Limestone and adjacent strata, but found nothing here. Strong mineralisation begins at *Whitaside Mine* almost a mile (1.6 km) farther ESE, but it was limited to the Main Limestone and overlying cherts. A shallow hush, at least four adits and numerous bell-pits mark the course of the workings over the next mile, up to the boundary of the Bolton Estate. Low Level [9823 9616] at 1360 ft (414 m) OD (Figure 30, loc. 9) and Middle Level [9827 9600] at 1430 ft (435 m) OD were crosscut adits from the north (footwall side) driven beneath sandstone features in the Twentyseven Fathom Grit. They produced no ore, even though the upper one must have had Underset Limestone on the hangingwall of the vein. A level from the north at about 1490 ft (454 m) seems to have been more successful, but the main working level, with portal at [9862 9578] was driven at 1510 ft (459 m) OD, a little above the hangingwall base of the Main Limestone. At about 300 ft (91 m) from the portal what is now regarded as the main vein turned towards the east, but the adit continued along an unnamed string in the previous direction of the vein for 1150 ft (350 m) and a belt of flots was exploited from rise workings for some 600 ft (182 m), a shaft (Morley's Shaft) communicating with the surface. The adit was continued as a crosscut to the NE, cutting at 270 ft (62 m) a second branch which became known as Harker's Vein. The Virgin String (see p. 172) was found cutting through this 510 ft (155 m) to the ESE and was tested for a short distance, apparently without encouragement. The drive was continued 900 ft (274 m) further in Harkers Vein and in this stretch several crosscuts were made, including one to the NE which eventually linked with the main Whitaside - Apedale Vein. It is evident from the irregularly-distributed surface shafts, and the ramifying workings between Whitaside and Harker veins that considerable areas of flot ground were exploited, the western part of this being known as Murton's Float. The spoil heaps show that much hand dressing was done *in situ*, leaving behind dolomitic carbonate, limonitised limestone, a little baryte and hemimorphite as well as partly-oxidised galena (analysis, Table 16)[1]. Surprisingly, no attempt seems to have been made to prospect the Main Limestone on the north (footwall) side of this productive area. At the shaft called Morley's Folly [9958 9555] the main vein begins to resume its ESE course, but it appears that in the stretch where it was trending nearly E - W, a series of ESE strings were given off which acted as feeders for the limestone replacement deposits. Some 500 ft (150 m) W of the estate boundary there are remains of workings on the south side of the main vein in a coal which must lie approximately 200 ft (61 m) above the top of the Ten Fathom Grit. The belt of flots continued through the boundary into Apedale ground but perhaps for only a short distance, for only three shafts—Round Shaft, [9993 9530] collared at 1806 ft (488 m) OD, and two others—lie on this trend, far to the south of the main vein. The shafts here must reach depths approaching 300 ft (91 m), but the flot workings must have been in Main or perhaps Red Beds limestone. Apart from these, the shafts related to *Apedale Head Mine* are along the course of the vein, the deep ones being respectively 250, 1350 and 1800 ft (76, 395 and 547 m) from the boundary fence. It is noticeable that the character of the tailings associated with these is different from that around the shafts on Whitaside; here the tailings are dominated by chert fragments, as at Hurst, and the conclusion may be drawn that vein oreshoots in chert wallrocks were exploited here rather than flots in limestone. The main entrance to the mine, Apedale Head Level [0126 9479] at 1388 ft (422 m) OD, enters on the base of the Main Limestone from the hangingwall side and probably extends as far as the second of the deep shafts mentioned above, though no plan has been found to substantiate this. The extensive excavations where the Main Limestone is faulted up more than its own thickness to the north in places show the limestone dipping steeply to the south. The ground north of the footwall of Apedale Head Vein is remarkable for the presence of many mineralised strings. One of these, running N45°W has been prospected for 1200 ft (364 m) by an adit driven on top of the Main Limestone at (480 m) OD known as Russell's Level. There are at least six more, running NNE, that have been wrought opencut in the Main Limestone; Raistrick's map (1975a, p. 88) illustrates this very well. On the other hand, the hangingwall (S) side of the Apedale Head vein has revealed no strings or flots and it is not even certain that Bob Scar Vein (p. 175) actually reaches its projected junction with this vein. The ground south-west of the vein was tested by a drive known as Woods Pot Level [0088 9476] (Figure 30, loc. 10) running SW for 1975 ft (600 m) on top of the Main Limestone at 1560 ft (474 m) OD, but no vein or string was cut; this drive reached almost half-way to Virgin Moss Vein. Down the valley from Apedale Head, the vein passes into beds below the Main Limestone, but at Apedale Shaft [0186 9456] sunk in 1868 an oreshoot 650 ft (198 m) long in Underset Chert and Limestone was successfully worked, the shaft being 200 ft (61 m) deep and thus penetrating beds of the Twentyseven Fathom Grit. West of this operation the vein again crosses Main Limestone and higher strata, but becomes progressively feebler, after giving off a branch to the NE. A brave attempt was made to prospect virgin ground a mile (1.6 km) to the east of Apedale Shaft by driving Harker's Level from Cat Scar [0346 9407] (Figure 30, loc. 11). The level, on the base of the Main Limestone at 1190 ft (362 m) OD runs 1375 ft (418 m) N24°E to cut a vein on line with Apedale Head Vein. This was opened up for 700 ft (213 m) W and 2800 ft (852 m) E of the level head, presumably in Main Limestone between 1880 and 1894, a period of considerable discouragement owing to the low price of lead. The small dump contains calcite, baryte and galena and nearby there are remains of a dressing floor, but few tailings, suggesting that what was found was not economic at the ruling prices. A plan (ZLB 41/174) indicates that other local

[1]The tailings suggest that the flots were rich but oxidised. The analysis can be recast: PbS, 5.5; $PbCO_3$, 9.7; $ZnCO_3$, 15.4; $BaSO_4$, 43.4; $CaCO_3.MgCO_3$, 15.9; $CaCO_3$, 1.8; FeO.OH, 8.0.

prospecting, from Harker's Shaft immediately west of the west forehead, and Anderson's Level [0300 9437], 1000 ft (305 m) farther west, driven on the top of the Main Limestone, was thought to cast doubt on the identity of the vein tried from Harker's Level with the main Apedale Head Vein. At 1100 ft (335 m) E of the head of Harker's Level a vein came from the NW into the vein being followed, but a drive 220 ft (66 m) long in this did not cut another vein where Apedale Head Vein was expected to be. An *en échelon* shift of the vein, east side south, as shown on Raistrick's sketch map, is the most acceptable interpretation. The farthest east trial for the Apedale Head vein system was at Middle Moss Shaft [0522 9410] sunk through the coal of Preston Moor and strata equivalent to the Grassington Grit Formation to reach the Crow Limestone at 240 ft (73 m) depth below the collar at 1440 ft (440 m) OD. A section derived from the records of the A. D. Company is given on Figure 13. This shaft is said to have been sunk in search of Mavel Vein, a branch leaving Chaytor Rake towards the NW, 335 ft (102 m) N of Cranehow Bottom Shaft (p. 178). The positioning of Middle Moss Shaft indicates that the expectation was that Mavel Vein would swing to link up with the Apedale Head vein-system, but although a crosscut was driven N for 300 ft (91 m) no vein appears to have been cut, and no other fracture that might have linked with the Apedale Vein was noticed on the west wall of Chaytor Rake or its northern branch, Cranehow New Vein. A small group of shafts ¼-mile (0.4 km) E of Harker's Level portal known as Golden Grooves lie considerably south of the Apedale Head Vein and no structure is known to pass through them; the spoil contains sandstone from the Ten Fathom Grit, shale, limestone and chert but no veinstuff. Other shafts or bell-pits scattered alongside the Apedale road prospected the upper part of the Main Limestone and there are also numerous sink holes. Lander's Level, on the base of the Main Limestone 600 ft (182 m) SE of Harker's Level, driven in 1902 found a NW string; it was the last prospect in this part of the Bolton Royalty.

Whitaside Mine may have contributed to the 98 177 tons of lead concentrates produced in Swaledale between 1845 and 1867, but it is separately credited with only 47 tons in 1862 and 12 tons in 1871; the workings must have been exhausted by the mid-century. The Wensleydale mines produced a total of 28 306 tons from 1848 to 1865 to which Apedale Head Mine may well have been a contributor. The Underset oreshoot, worked from Apedale Engine Shaft (p. 173) yielded 526 tons from 1876 to 1883. Clearly, however, the main period of output from Whitaside and the Apedale mines preceded the middle of the 19th century, and probably preceded the beginning of it.

Green Hill Ends String Lead ore

SD 99 NE; Yorkshire 52 NW
Direction N37°W

Harker South Vein Lead ore

SD 09 NW; Yorkshire 52 NW, NE
Direction N70°W, throw 20 to 25 ft (6–7.5 m) SW

Harker North Vein Lead ore

SD 09 NW; Yorkshire 52 NW, NE
Direction N45°W, small SW throw

Grovebeck Vein Lead ore

SD 09 NW; Yorkshire 52 NW, NE
Direction N60°E, throw 5 ft (1.5 m) NW

Water String Lead ore

SD 09 NW; Yorkshire 52 NW
Direction near E – W

North of Whitaside Mine there are small workings exploiting flots in the Main Limestone at Green Hill Ends [995 965] probably related to a fracture that has also yielded a little ore near the Three Yard and Five Yard limestones. The remaining veins in this group have been worked on a larger scale by the *Harkerside Mines*. Extensive hushes have been made on the south side of High Harker Hill to open up the South Vein in the Main Limestone and overlying cherts over a length of ½-mile (0.8 km). The fault terminates the outcrop of the Underset Chert and Limestone on both sides of the Grovebeck valley, and the presence of numerous bell-pits on the fault below the Main Limestone indicates that productive ground was found at the Underset level. An adit [0134 9745] at 1300 ft (395 m) OD from the north end of the hush system would also give command of the Underset beds, but no record of its course remains. Much of the production here, however, came from flots in the Main Limestone, and in this connexion a document, probably by Mr F. Kendale, passed on to W. Gunn in December 1873 by Mr R. Metcalf of the Hurst mines may be quoted: 'It (Harker Vein) is attended by Floats in the 12 Fas. Limestone which has somewhat impoverished it. We never have a good float coming exactly up to a vein. When floats are poorish the veins are generally rich and vice versa. This is a general rule in Grinton Moor Field.' The tailings and debris along the Harker South hushes include a great deal of cavernous altered limestone such as might be expected in flots near the surface, together with calcite, baryte in small quantity, much limonite, partly oxidised galena and hemimorphite. Harker North Vein is a weak branch from the north side of the South Vein, while Grovebeck Vein appears to originate from the hangingwall side of the South Vein. Grovebeck Level [0275 9671] (Figure 30, loc. 12) at 1310 ft (398 m) OD runs slightly W of S as a crosscut in the bottom beds of the Main Limestone for 820 ft (250 m), cutting the vein at 700 ft (213 m) from the portal, then following it SW for 1890 ft (576 m) through ground mainly productive in flots where it apparently terminated against a NW-trending cross string, from which was given off an E – W string. Known as Water String it may be surmised that this was a natural cavern feature rather than a vein. It was shown on the plan as extending almost ½-mile (0.8 km) from the point of discovery in a westward direction, but how much of this was actually explored is not clear. A gin shaft sunk 1270 ft (386 m) SW of the point where Grovebeck Vein was lost appears to have found nothing. Attempts to explore Water String from surface are shown on the six-inch Ordnance Survey map as Wildgoose Trials.

Much of the ore obtained on Harkerside thus seems to have come from replacement ground in limestone. Flots occur widespread and not all can be correlated with known fractures. For example, the *Guy Mine* midway between Green Hill and High Harker consists of shallow shafts through the cherts scattered irregularly over an area approximately 500 ft (152 m) square, south of the Main Limestone feature of Beldow Hill, and another similar group of pits occurs nearer to South Hush. Excavations at the head of Grovebeck valley are far wider than would be necessary to exploit the veins. The Kendale MS referred to above mentions a Brownagill String trending N – S and throwing 1½-ft (0.45 m) W '. . . worth very little of itself but the floats flying out of it were good'. Although no exact site for this is known, Browna Gill lies between the Guy Mine and the unnamed flot ground to the east, and Raistrick (1975a, p. 81) states that the Browna Mine was working in 1700; in fact, that the Harkerside mines have a long history, going back at least to the dissolution of the monasteries, when they were granted along with other mines in Grinton parish to Metcalf of Nappa, subsequently passing to Lord Wharton.

Grinton How Vein Lead ore

SD 09 NW; Yorkshire 52 NE, SE
Direction N60–67°E, throw 6 ft (1.8 m) NW

Grinton North Vein Lead ore

SD 09 NW; Yorkshire 52 NE
Direction N45°E

The *Grinton Mines* lie athwart the Reeth – Redmire road and afford easily accessible examples of hushes and of the debris remaining from hand dressing of lead ore. How Vein has been proved over a total length of 6750 ft (2.05 km) but the surface evidence suggests that ore was found mainly in the topmost beds of the Main Limestone, where there was replacement ground, and in the overlying cherts, which make up much of the extensive debris. The mineralised ground continued south-west under cover of Ten Fathom Grit and the overlying measures up to and including the massive grit of Greets Quarry, but there is no evidence that any of these beds were ore-bearing. Two shafts, How Greets [0294 9575] and High Greets [0282 9567] were sunk through the grit outlier, the second of these being just within the Bolton Royalty, though it does not correspond with the shaft of that name shown on the six-inch map. The ultimate trial in this direction, Low Greets Shaft [0253 9548] was collared just below the base of the grit at 1625 ft (494 m) OD and the following section was recorded by W. Gunn and J. R. Dakyns (see also Figure 11, section 2):

	ft	m
Measures	66½	20.2
Crow Chert	29	8.8
Crow Limestone	19	5.8
Ten Fathom Grit	24	7.3
Plate	51	15.5
Black Chert	7½	2.3
Red Beds	41	12.5
Thin Limestone	5	1.5
Main Limestone		

There can be little doubt that the bearing ground ranged from the top 10 ft (3 m) of the Main Limestone up to the Black Chert, a thickness of little more than 60 ft (18 m). This may well be representative of the situation at many of the mines along the Swale – Ure Watershed (see Figure 12), and once again, much of the ore had probably been extracted from hushes and shafts before the driving of deep levels began. The principal adit, How Level [0426 9629] at 1310 ft (369 m) OD enters as a short crosscut in the middle part of the Main Limestone, and is known to continue as far as the head of the hush. At 1420 ft (432 m) from the portal, a NNW crosscut 700 ft (213 m) long gave access to the North Vein, which proved to be a branch given off from How Vein; it is understood to have yielded ore, but to have died out before reaching the Main Limestone outcrop. It is noticeable that no very great quantity of tailings is associated with How Level, but the material present, chiefly limonitised limestone, calcite, a little marcasite and baryte, in addition to partly oxidised galena, is typical of flot ground. Below the outcrop of the Main Limestone, a few bell-pits and a cross hush, Swinston Hush, must have given some encouragement, for an adit, California Level [0467 9657] (Figure 30, loc. 13) was driven beneath the Underset Limestone for 1410 ft (429 m) reaching beneath the lower part of the hush; in this a NW branch or cross vein was cut at 840 ft (256 m) from the portal, but the tips show little evidence that workable ore was got from either vein. A trial to the bottom of the Main Limestone was made at a shaft on the vein 1510 ft (459 m) ENE of How Greets Shaft, the depth of the shaft being 180 ft (55 m) but the result is not known. The only production credited to Grinton in the official statistics is 5 tons lead concentrates in 1872–73, but Fieldhouse and Jennings (*op. cit.*) record lead equivalent to 3833 tons concentrates between 1775 and 1801.

Bob Scar Vein Lead ore

SE 09 SW; Yorkshire 52 SW
Direction N22°W, throw 24 ft (7.3 m) E

Walker Wife Rake and Strings Lead ore

SE 09 SW; Yorkshire 52 SW
Direction N47°E

Bolton Park Vein Lead ore

SE 09 SW; Yorkshire 52 SE, 67 NE
Direction N50–70°W, throw 20 ft (6 m) SW

Bob Scar Vein lies roughly parallel to, and a mile east of Virgin Moss Vein (Figure 30). It is exposed in the prominent Main Limestone cliff of Bob Scar, which Raistrick (1975a, p. 86) regards as the place probably referred to by Leland (1546) as the great rock 2 miles (3.2 km) from Castle Bolton 'where my Lord Scrope seketh for Lede'. Bob Scar Level [0200 9302], in the upper part of the limestone at 1370 ft (417 m) OD, and the associated shafts to the north-west, worked flots, with the usual gangue of limonitised brown limestone and calcite. The workings continued in strength for at least 3000 ft (912 m) to the NNW, and in this stretch there appear to have been four or five independent shaft mines, Craddock (which may have connected with the adit), Bob Scar, Silver Bottom, Round and Moorcock, respectively 990, 1900, 2300, 2825 and 3010 ft (301, 578, 699, 859 and 915 m) from the adit, each with the remains of hand dressing associated. Not much appears to have been found by the Mark cross hush, but another group of old shafts clusters 2000 ft (610 m) ahead of Moorcock Shaft. Beyond this the vein, if it exists, was barren; Bobscar Drift beneath the Main Limestone from the south-west bank of Apedale Beck may have cut a fracture, but got no ore, and it is not recorded where Bob Scar Vein impinges on Apedale Head Vein if indeed it reaches so far. South of Bob Scar cliff, a short level tried for the vein in the beds below the Underset Limestone; this passed through an ENE fault throwing SE which might be regarded as part of the Walker Wife group, and which may terminate Bobscar Vein to the south. The three Walker Wife fractures are feeble strings only, but with some flot ground associated. They form a link towards the more important Bolton Park Vein, also found at outcrop in the Main Limestone feature east of Rowantree Scar. The *Bolton Park Mine* worked levels: No. 4 near the base of the Main Limestone on the footwall side [0287 9314] at 1320 ft (401 m) OD; No. 3 below the limestone on the downthrow side, at 1280 ft (389 m) OD; No. 2 [0298 9279] beneath the Underset Limestone, at 1130 ft (344 m) OD, and No. 1 or Dents, at [0303 9235] 955 ft (290 m) OD in the Preston Grit between the Three Yard and Five Yard limestones. Nos. 1 and 2 were crosscut adits driven a few degrees W of N, the lowest one being nearly 2000 ft (610 m) long, but probably a branch fracture system was followed. A moderate-sized heap of cuttings is associated with the Main Limestone workings, the minerals present being brown carbonate, calcite, aragonite and slightly oxidised galena. Siliceous flaggy sandstone dominates the tip from the third level (Figure 30, loc. 14), but some veinstuff with baryte, calcite and galena is present; however, no dressing appears to have been done here. The lowest level (Figure 30, loc. 15) seems only to have found calcite. The two upper levels followed Bolton Park Vein and a southern branch or loop, No. 4 level forehead standing 2110 ft (643 m) from the portal. No. 3 level covered only a third of this distance but near the north-west end a sump was sunk to the Underset Limestone. Probably the prospect in depth on this vein was not encouraging, for Nos. 1 and 2 levels were not driven under the productive ground in the Main Limestone, but stopped at what must be regarded as the westward continuation of Cobscar Rake, described below. This vein hades south, being cut by No. 2 Level 440 ft (134 m) from the portal and by Dent's Level 1911 ft (581 m) inbye. No. 2 Level was just entering Bolton Park Vein when it was suspended. All these workings are shown on a plan dated 1866, at NYCRO. Bolton Park Mine was coming to the end of its useful life in that year, the first for which separate statistics are available. The production was 475 tons of lead concentrates; up to 1870, 101 tons more were mined, but no production is recorded after that year.

Cobscar Rake Lead ore ?zinc ore

SE 09 SW, SE; Yorkshire 52 SE, 67 NE
Direction generally E – W, variable; throw 16 ft (4.9 m) S

This strong vein has been worked more or less continuously from its origin in Bolton Park for 2 miles (3.2 km) to its junction with the Chaytor Rake vein system to the east (Figure 30). It was no doubt discovered at outcrop in the Main Limestone south of Thorny Bank Hill [0472 9305] where the northernmost section of Redmire Limestone Quarry has now cut through it. The mile of ground to the west of this outcrop, where Brigantian strata are at the surface, was exploited by means of two adits driven from the deep valley of Bolton Gill; their objective was the Preston Grit and it is probable that they were inspired by the discovery of rich oreshoots at this horizon in Keld Heads Mine. Bolton Gill Level [0428 9252] at 800 ft (243 m) OD (Figure 30, loc. 16) runs NNW as a crosscut in measures between the Five Yard and Middle limestones for 1470 ft (447 m) to cut the vein adjacent to an air shaft [0405 9290]. All the development was to the west, covering a total length of ½-mile (0.8 km). Rise workings begin after a short drive at the adit horizon; the western part of these is known to have been in Preston Grit. Like other developments in this district, the drives take a series of turns towards the north-west which are taken to indicate that the vein was in several subparallel fractures. At 1360 ft (414 m) W of the level head, there is a major turn in this direction which lines up with Bolton Park Vein and is reasonably regarded as its confluence with Cobscar Rake; the Rake, however, resumes its western course after 260 ft (79 m). It also gives off a string to the SSE 220 ft (67 m) in this direction. On the opposite side of Bolton Gill, a second adit [0432 9263] at 820 ft (249 m) OD (Figure 30, loc. 17) in sandstone beneath the Five Yard Limestone runs NNE as a crosscut 1215 ft (369 m) long. No information is available as to the extent of driving from this, if indeed any was done; it would be surprising if this had not been carried under the rich ground in the Main Limestone to the east, but the modest hushing between the base of this limestone and the bottom of the Gill may indicate that the vein was poor in the later Brigantian strata. A small dressing floor is represented by remains, now much overgrown in the Gill south of these adits, at least 50 tons of tailings being visible. It must be concluded that while the Rake was not barren in the Preston Grit, these developments failed to expose large oreshoots. Some light may be thrown on this by a section preserved by W. Gunn and J. G. Dakyns of 'the first surface rise' in Bolton Park Mine. Since this section includes the whole thickness of the Five Yard Limestone, it cannot refer to the air shaft above Dent's Level at Bolton Park, but it could be a surface rise above Bolton Gill Level on the west side of the Gill. The following is the section:

	ft	m
Boulder Clay	60	18.2
Preston Grit	21	6.4
Plate	21	6.4
Grit	18	5.5
Girdles	5	1.5
Fossil Lime (Five Yard)	57	17.3

It thus appears that the grit is here split by shale, whereas at Keldheads it had an unbroken thickness of 45 ft (13.7 m).

The position in the Main Limestone and overlying cherts and limestones in the mile of ground eastward from Redmire Quarry was more favourable. Here the numerous surface cuts and shafts, and the wide spread of tailings left from the hand dressing of ore mined from them, give evidence of extensive mineralisation with galena in a matrix in which baryte and calcite were abundant, with traces of fluorite. Sphalerite is present, with hemimorphite; one of the buildings near the workings is still called Calamine House, and the possibility that some 'dry bone' oxidised zinc ore was sold is suggested. Where the Rake is cut through by the quarry there is a gap 60 ft (18 m) or more wide, indicating the extent of mineralised and altered limestone associated with the vein, but no indication can be found of the actual width of the vein. Near the ruins of Cobscar Smelt Mill the direction of the vein changes to ESE, and at least 20 strings, better appreciated from air photographs than on the ground, come into it from the south-western side in a stretch 1500 ft (456 m) long up to its junction with a branch of Chaytor Rake. Midway in this stretch is the remains of the most important shaft [0618 9295] on Cobscar Rake, and probably, from the concrete foundations associated with it, the most recently worked. No separate output figures are available.

Chaytor Rake = Cranehow Bottom New Vein Lead ore (fluorspar, barytes)

SE 09 SE, NE
Direction N22–26°W, throw 44 ft (13.3 m) W, decreasing to N

Preston Vein Lead ore

SE 09 SE, NE; Yorkshire 52 SE, 67 NE
Direction N22°W, small E throw, dies out N

Thorntree Vein Lead ore

SE 09 SE; Yorkshire 67 NE, 68 NW
Direction N40°W, small E throw

Stopmore Rake Lead ore

SE 09 SE; Yorkshire 52 SE
Direction N25°E

Mavel Vein

SE 09 SE; Yorkshire 52 SE
Direction N55°W

Cranehow Bottom Old Vein Lead ore

SE 09 NE; Yorkshire 52 SE
Direction N20°W

Chaytor Rake, the seventh and easternmost of the NNW veins crossing the Ure – Swale watershed exists as far as is known, only between two fractures; a nearly E – W 'dowky' (gouge-filled) vein cut in the south at [0826 9008] and the ENE Wellington Vein which it meets at [0538 9575] in the north (Figure 30). The fracture system is continuous through the 4 mile (6.4 km) stretch between these points, giving off a number of branches, but it appears not to be continuously mineralised, nor did not carry oreshoots throughout. *Keldheads Mine,* which worked the southern stretch of Chaytor Rake, was the most important mine in Wensleydale. A strata section given to W. Gunn and J. R. Dakyns (see Figure 7, section 7 and Figure 8, section 15) during the primary survey, was repeated in the A.D. Company records. The main working level [0795 9076], at 600 ft (183 m) OD (Figure 30, loc. 18) was driven with the unusually thick Five Yard Limestone on the hangingwall of the vein, and the underlying measures on the footwall. This level extends a total length of 8380 ft (2.55 km) to reach 440 ft (134 m) beyond Tattersall's Shaft. The principal oreshoots were in Three Yard Limestone and the underlying Preston Grit, with a second run, beneath a partial gap caused by shale, in the 'Top Limestone' of Keldheads, equivalent of the Five Yard Limestone. Only a crude section dated 1828 (NYCRO ZBO(L) 8) shows the first 1700 ft (518 m) of the workings related to the Main level, but it is probable that ore was already being won above the level from 1000 ft inbye. Details of stoping remain from 2450 ft (747 m) to the N end of these oreshoots near Blue Hillock Shaft, 6250 ft (1.9 km) N of the level mouth. The maximum height stoped was 150 ft (46 m) as shown on

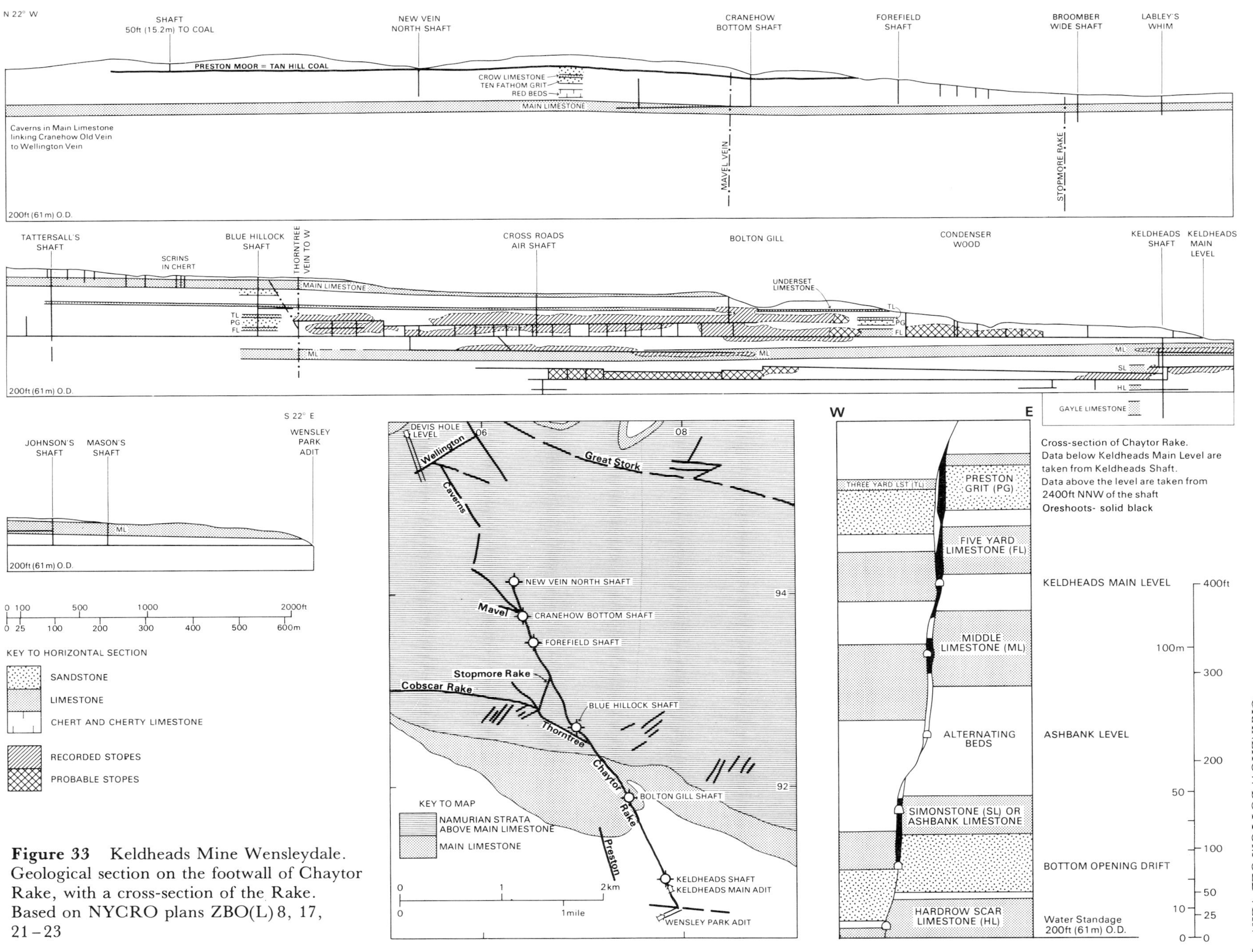

Figure 33 Keldheads Mine Wensleydale. Geological section on the footwall of Chaytor Rake, with a cross-section of the Rake. Based on NYCRO plans ZBO(L) 8, 17, 21–23

Figure 33 and it may be estimated that the recorded stopes have yielded about 150 000 tons crude ore. Near Bolton Gill there was a small oreshoot in Underset Limestone, but the long stretch of Main Limestone exposed north of Bolton Gill was apparently barren above the Main Level oreshoots. Near the portal of the level, a shaft was sunk to 472 ft (144 m) depth. Raistrick (1975a) figures the massive chimney of the boiler house which supplied steam for the pumps in this shaft and mentioned that pump rods were formerly visible. There were levels at 78 ft (24 m), 216 ft (66 m) (Ashbank Level), 264 ft (80 m) and 319 ft (97 m) (the Bottom Opening Drift). Figure 33 shows that in the Middle Limestone, ore was worked virtually only when the vein, here throwing 35 ft (10.7 m) W brought limestone into contact with limestone. On the other hand, the thinner Ashbank Limestone evidently had enough hard sandstone near it to admit mineralisation over a greater height. The 319 ft level was driven 4400 ft (1.34 km) N and though the only recorded stoping was near the shaft, the driving of 4800 ft (1.49 m) of stope drifts above it strongly suggests the presence of a lengthy oreshoot. The water stand from the shaft at 389 ft (118 m) depth was in Hardrow Scar Limestone, but no workable ore seems to have been found; and though the shaft penetrated what would now be called the Gayle Limestone, there was no development at this horizon.

South of Wensley station, a crosscut adit was driven ENE from Wensley Park (Figure 30, loc. 19). Besides finding the dowky vein referred to above, a drift north from the level head, known as Ashbank Level, became the main drainage for Keldheads Shaft. The adit was in beds below the Middle Limestone, but the upper part of the Ashbank (= Simonstone) Limestone appeared on the footwall of the vein. However, this had dropped below the level before the shaft was reached, and the drive was not, therefore, continued northward. Mineralisation above Ashbank Level was found only near the shaft; elsewhere, only calcite was found. The mine was one of very few in Wensleydale equipped with power (steam)-driven dressing facilities. The once-large tailings heaps have partly been used as the base of a railway siding, partly worked for fluorspar. The tailings (analysis, Table 16) contain colourless fluorite with yellow sulphide inclusions, probably chalcopyrite; barium minerals; dolomite and calcite, with sphalerite, galena and their oxidation products.

North of the shaft in the Underset Limestone, where pink baryte is present on the heap, there is a stretch 3750 ft (1.14 km) long with virtually no surface evidence of mineralisation, though this includes the whole outcrop on both sides of the vein, of the Main Limestone. Then, in the Richmond Cherts N of Blue Hillock Shaft, numerous shafts appear along the line of Chaytor Rake and also along the branch known as Thorntree Vein which provides the link with Cobscar Rake (p. 176).At the horizon of the Keldheads Main Level, Thorntree Vein branches from Chaytor Rake to the NW, 6128 ft (1.87 km) N of the portal, and though this is not connected through to the Thorntree workings, a strong impression is given that as the Keldheads orebody in the Preston Grit ended, the main strength of the mineralisation went NW into the Main Limestone. Labley's Whim [0680 9294], Wide Shaft [0674 9313] and Forefield [0652 9348] were the principal working shafts in a well-mineralised stretch nearly ¾-mile (1.2 km) long, in which coarse white baryte, unusual for this district, is conspicuous at Labley's Whim, and elsewhere the type of baryte here regarded as probably secondary after witherite is present, with brown carbonates, but not fluorite: the upward and outward change from fluorine to barium to CaMg carbonate mineralisation is shown as strikingly as anywhere in the Askrigg Pennines by the Chaytor Rake system. Strongly limonitised limestone is also present. Near Wide Shaft, Stopmore Rake, worked south-south-west of Chaytor Rake, reaches the latter. The surface workings show up conspicuously on air photographs, and at one point there is a shift SW side W, perhaps effected by a cross fracture. Stopmore Rake itself may be shifted by Chaytor Rake, for Wide Shaft and three others to the NNE probably sunk to it line up to suggest an E side S movement of about 100 ft (30 m).

The next section of Chaytor Rake was worked in the midst of the Preston Moor – Grinton Moor colliery area, where there are hundreds of closely-spaced bell-pits sunk to extract the coal in the Grassington Grit (see Middle Moss Shaft section, Figure 13). In Cranehow or Cryna or Crina Bottom, a glacial meltwater drainage channel crossing the coalfield, Cranehow Bottom Shaft [0640 9375] was sunk on the rake. Starting in medium-grained sandstone above the coal at about 1225 ft (373 m) OD, the shaft was about 220 ft (67 m) deep to reach the top of the Main Limestone; a section (NYCRO ZBO(L)23) shows a level in the upper part of the limestone driven N, with a rise to the Crow Limestone, but beyond this, no information remains as to the horizon of the drift which must have connected the several shafts. The spoil suggests that at least part of the workings were in limestones of the Richmond Cherts, but it is not known whether Main Limestone was worked, or whether any ore was found in the Crow Beds, the workings were probably mainly north of the shaft. The Old Vein fracture (here regarded as a branch), which left the Rake 200 ft (61 m) N of the shaft, was worked at least as far as a shaft 1375 ft (418 m) NNW of Cranehow Bottom Shaft, and yielded some galena in a matrix of carbonates, and baryte with a little sphalerite present. It is said, however, to have been less productive than New Vein which is here taken to represent the continuation of the main fracture. This was mined for a distance of a mile (1.6 km) up to the boundary with the Ellerton Moor mines at Lunton Hill, the last 1500 ft (457 m) of this stretch turning to a sinuous but average N – S course. Near the boundary, natural caverns seem to have been cut which formed a link with the workings considered to be on the same vein from Devis Hole Level (p. 179) where there was a drive on it running 1020 ft (370 m) S from Wellington Vein, with flot workings in communication.

Mention must also be made of the weak parallel vein to the west of Chaytor Rake north of Preston village, downthrowing east. Baryte, calcite and galena were found in this by shafts into the Underset Limestone, but although it may have reached the southern margin of the Main Limestone outcrop, it seems not to have been discovered in the wide workings of Preston Quarry in this limestone.

Keldheads Mine is reputed to have produced about 1000 tons galena per year during the first half of the 19th century, and the official statistics credit it with an output of 9518 tons from 1866 to 1888, when it closed. The Preston Smelt Mill, nearby, was already at work before 1700, but during the 19th century it was replaced by a mill higher up the valley at Condenser Wood, with stone flues running nearly 2 miles (3.2 km) up the hillside to the chimney of the old Cobscar smelter. According to Raistrick (1975b) it shared with the London Lead Company's mill at Egglehope in Teesdale the distinction of being the most advanced ore hearth mill in the country, its processes, equipment and experimental work being quoted in Percy's standard 19th century text on the metallurgy of lead (1870). R. T. Clough (1962) states that Keldheads Mine, known to have been worked as early as the 12th century, is the oldest mine of which there is any record in Wensleydale, and was the richest in the county, employing in its heyday 250 men and boys according to the Victoria County History (Backhouse, 1912). Using arguments similar to those employed for the area in general on p. 183, it can safely be credited with 30 000 tons of lead concentrates, but the true figure may be substantially greater.

Wellington Vein Lead ore

SE 09 NE; Yorkshire 52 NE, SE
Direction N65°E, throw 6 ft (1.8 m) N

Robinson's Vein Lead ore

SE 09 NE; Yorkshire 52 SE
Direction N58°W, throw 3 ft (0.9 m)

Redway Head = Old Stork Vein Lead ore

SE 09 NE; Yorkshire 52 NE, SE, 53 SW
Direction N 16°W, throw 50 ft (15 m) E, north of Great Stork Vein 3 ft (0.9 m) W

East Vein and Strings Lead ore

SE 09 NE; Yorkshire 52 SE, 53 SW
Direction N75°E

With the exception of Great Stork Vein and its associated fractures, discussed in the next section, these are the principal veins of the *Ellerton Moor Mines*, on the southern slope of Swaledale east of the Grinton mines (p. 175). Cranehow Bottom (New) Vein was also worked from this side. Wellington Vein was probably the chief producer, probably mainly during the early years of last century. It was worked for a total length of 3750 ft (1.14 km), this stretch including eleven large shafts, with associated tailings heaps. These shafts doubtless represented the first operations on the vein. Subsequently an adit, Devis Hole Level [0516 9605] (Figure 30, loc. 20) was driven from Lemon Gill at 1155 ft (351 m) OD running SSE for 1500 ft (457 m) to cut the vein, in the upper part of the Main Limestone. Wellington Vein proved to be followed by a linear natural cavern. The mine has been entered and explored in recent years by members of the Earby Mines Research Group and the Moldywarps Speleological Group (Brook and others, 1977, pp. 58–60). They record the existence of a drift, Pearson's Level, running nearly due E from the crosscut adit, 440 ft (132 m) from the entrance, the forehead of which is 800 ft (244 m) from the point where it leaves the adit; it is not clear what structure was being followed. At 150 ft (46 m) ahead of this junction, begins the remarkable maze of phreatic caverns, already mentioned on pp. 108–110 containing over a mile (1.6 km) of passages in an area only 400 × 150 ft (122 × 46 m). The dominant trend follows the major joint-set of the northern Pennines, close to N20°W. The conjugate set is present, but less well developed, and another repeated trend is N65°W. On reaching the vein, it was found that the drives both to the ENE and the WSW had been made by only slightly modifying the linear cavern, and it may be supposed from this that the vein fracture was not strongly mineralised in the upper part of the Main Limestone (see also Raistrick, 1938, 1973). Examination of the shaft dumps, and the fact that the lead workings lay considerably above the adit, suggest that the productive horizon was the Red Beds Limestone in the Richmond Cherts, and that the ore came mainly from flots. The mine plan shows that the Wellington Vein was opened up for 730 ft (222 m) WSW and 2200 ft (669 m) ENE of the crosscut adit, which cut the vein 1190 ft (362 m) from the portal. The continuation of the adit beyond the vein was in limestone, but found no additional vein; the cave explorers, however, report a rift near the forehead dropping into a short series of natural caverns. The WSW level along Wellington Vein was found in places to be almost entirely a natural cavern, and this was followed more than 600 ft (183 m) beyond the end of the workings shown on the mine-plan into two series (the Northern Occidental and Southern Occidental) of natural caverns mainly aligned parallel to the vein; 'good formations' (presumably of calcite stalactites) were reported, but no ore. To the ENE the development at the level of Devis Hole Adit on the other hand extends considerably beyond the limit shown on the cave explorer's plan though they had penetrated beyond the points, 300 and 330 ft (91 and 100 m) from the crosscut junction, where Cranehow Bottom and Robinson veins respectively leave the south wall of Wellington Vein,. noting that sumps had been sunk into the floor of the 1000 ft (305 m)-long drive in the latter vein. The objectives of both developments are believed, however, to have been flots at a higher level, associated with a joint-system more or less parallel and perpendicular to the direction of Cranehow Bottom New Vein. Ahead of these workings, Whim Shaft [0574 9554] was sunk on Robinson's Vein; collared at 1390 ft (423 m) OD, a little below the sandstone (of the Grassington Grit) underlying the Preston Moor Coal, this shaft must have been at least 160 ft (49 m) deep to the top of the Main Limestone. Some 400 ft (122 m) of driving followed the vein, but the abundance of brown carbonate on the dumps suggests that flot working was also undertaken. A crosscut, at what horizon is not known, linked this shaft with Cooper Shaft, 615 ft (187 m) to the NE, where dolomitic gangue is again present; the crosscut was carried on in the direction of Redway Head Vein (below) but did not reach it.

Returning to the main ENE drive on Wellington Vein, at 2200 ft (669 m) from the level head Redway Head Vein crosses Wellington, shifting it E side S approximately 90 ft (27 m), a shift which is also plainly shown by the alignment of the surface shafts. The vein seams or associated flots seem to have been equally productive on either side of this intersection. Redway Head Vein (Figure 30) has been worked by a series of shallow shafts to the NNW over a distance of 1800 ft (549 m) to its intersection with Great Stork Vein, while to the SSE a deeper shaft 650 ft (198 m) from the intersection shows much chert on the dump, with a little dolomite. The intersection of Wellington with Great Stork Vein is dealt with in the next section. Throughout the Ellerton Moor deposits, the matrix of the galena was calcite and dolomite, with a little sphalerite and hemimorphite derived from it. Baryte is only sparingly and locally present. Official statistics record only 215 tons of concentrates for 1872–75, so that it must be concluded that the mines were exhausted before 1848 (see Chapter 1).

Great Stork Vein

SE 09 NW, NE; Yorkshire 52 NE, SE, 53 SW
Direction N60–70°W, throw 162 ft (49 m) NE at Ellerton Moor

James Raw Rake Lead ore

SE 09 SE; Yorkshire 53 NW, SW
Direction N50°W, throw 5 ft (1.5 m) N

Hags Gill = Heggs Gill Vein

SE 09 NE; Yorkshire 53 SW
Direction N – S, approx.

Stainton North and South Veins Lead ore

SE 09 NE; Yorkshire 53 SW
Direction N85°W

The reasons for regarding Great Stork Vein as the eastward continuation of the Stockdale 'Vein', and the interpretation of the concealed ground around Healaugh have already been discussed (p. 119). Great Stork is, in fact, a powerful fault line, which can be traced continuously from south of Healaugh, through the Ellerton and Stainton mines across the Bellerby Ranges to beyond Constable Burton [166 908], a distance of at least 11 miles (17.6 km). No investigation seem to have been made on it in the Grinton mines area, where it cuts through early Brigantian strata, but in connexion with the Ellerton mines, many attempts have been made to prospect it. In the MS notes believed to have been given by Mr R. Metcalf of the Hurst mines to W. Gunn in 1873, the following statement appears: 'This is partly a cross vein. Though this vein bears ore in all beds where seen, it never would work by bing tale at any price that could be given'. To put this into current language, Great Stork Vein, though mineralised wherever it has been cut, was uneconomic even at contract prices ruling in the 19th century. The westernmost trial was a crosscut level [0493 9703] driven S from Cogden Gill starting at the base of the Underset Limestone on the downthrow side and probably entering Three Yard Limestone on the opposite side of the fault; the small dump consists of limestone and calcite. Not far away, bell-pits were sunk on Old Stork Vein (the equivalent of Redway Head Vein north of Great Stork) in the Main Limestone on the downthrow side, but little appears to have been found. The junction of these two veins was explored by a level from the north side of Black Hill, east of Heugh Nick, but without success. Hags Gill Level [0658 9632] at 1050 ft (319 m) OD (Figure 30, loc. 21) was driven WSW in Main Limestone from the Gill along the course of Wellington Vein, cutting Great Stork at 2200 ft (669 m) inbye; the dump contains galena, carbonates and baryte, probably from productive ground on Wellington Vein south of Great Stork, the throw of which must have brought Twentyseven Fathom Grit on the footwall into contact with Main Limestone. Heggs or Hags Gill Vein, mentioned in the Metcalf MS, is difficult to identify; the northern part of Hags Gill runs due south, and is perhaps a hush on a weak vein. Bradley (1862) states that this vein was productive in the Red Beds. In Hags Gill Level another vein was cut 780 ft (237 m) from the portal, known as James Raw Rake. North of the level a few bell-pits were sunk along this vein, but it became more important in the other direction for as it converged on Great Stork Vein, numbers of such pits are associated with it and cover the ground between the two veins, indicating the presence of extensive flots. East Level [0709 9635] (Figure 30, loc. 22) at 990 ft (301 m) is a crosscut adit running a little W of S, reaching Great Stork Vein 1940 ft (590 m) from the portal, east of the intersection of Raw and Stork veins. This adit starts about 60 ft (18 m) below the base of the Main Limestone; if the section of Brigantian strata is similar to that at Keld Heads Mine, the Preston Grit should have been on the footwall of the Stork Vein here. Mineralisation was found; calcite, baryte, limonite and a little galena occur with limestone, sandstone and shale on the tips; but little or no ore dressing was done. Bell-pits north of Stork Vein suggest the presence of ENE and WNW flyers off the vein here, and East End Vein with its associated strings (p. 179) lies south of Stork Vein, though East Level is not known to have been driven to it. The easternmost workings related to Great Stork Vein are those of the *Stainton Mines*, extending about ½-mile (0.8 km) from the head of Juniper Gill to The White Bog, south-west of the village. Here the bearing beds are the Crow Chert and Limestone, exposed by hushing and with many bell-pits sunk into them; the outcrop is terminated to the east by a NE flyer from Great Stork Vein, bringing in grit (presumably the Grassington Grit), and here the workings cease. There were three adits: Derbyshire Level (Figure 30, loc. 23), from Juniper Gill, running S beneath the Ten Fathom Grit, probably passing through the main fault; an adit in the hush on Stainton North Vein, and Dagget's Level [0890 9553] (Figure 30, loc. 24) at 1010 ft (307 m) OD probably driven SW under the grit to reach the vein. Some galena, baryte, calcite and limonite is present on the tips from each of these, and these minerals occur also in the debris from the hushes. Stratigraphically, this must represent the highest position reached by mineralisation in Swaledale. The beginnings of two adits probably designed to reach the veins at lower levels were made from the headwaters of Stainton Moor Beck, Wyvill's Level [0948 9581], of which no details remain, and another starting 610 ft (186 m) farther west, driven 675 ft (205 m) WSW to an air shaft. Neither level seems to have reached the vein, and no exploration is known to have been attempted farther east, where the bearing beds are concealed at increasing depths owing to the general eastward dip of the measures. Summarising, the evidence suggests that Great Stork Vein may have functioned as a channel for mineralising solutions, particularly in early Brigantian strata; but the workable oreshoots were all in associated minor fractures or flots. No information on output is available; all that can be said is that no large tailings heaps remain.

Lang Scar Strings Lead ore

SE 09 SE; Yorkshire 68 NW
N35–46°E

Herontree Allotment Strings Lead ore

SE 09 SE; Yorkshire 68 NW
N35–40°E

Two remarkable areas where closely spaced mineralised joints (strings or scrins) have been worked lie NNE and NE of Preston Under Scar, both on the military reservation. The country rock in both areas is the Richmond Cherts, and the strings do not continue either into the underlying Main Limestone or the overlying Ten Fathom Grit. The Lang Scar area [077 927], has a maximum strike length of nearly ½-mile (0.8 km) and a width of 1200 ft (365 m) in which at least 22 subparallel lines of shallow workings can be distinguished on the air photograph. The Herontree area lies on both sides of the Leyburn – Preston road [0861 9225], the corresponding dimensions being 1500 by 1400 ft (456 by 426 m) (Figure 30). As both areas lie on rising ground, a range of strata must be involved, probably representing half the thickness of the Richmond Cherts. According to notes by W. Gunn, flots were also worked and during the primary survey he obtained the following section of a shaft sunk through the Ten Fathom Grit, which from his description is on the west side of the Leyburn – Reeth road [084 930]:

	ft	m
Plate	36	10.9
Crow Chert	18	5.5
Crow Limestone	18	5.5
Ten Fathom Grit	10	3.0
Plate	21	6.4
Black Chert	15	4.6
Main Chert, 6 in shale at base	78½	23.8
Main Limestone		

Gunn adds that the term Main Chert as used in Wensleydale, includes equivalents of the Red Beds, Black Beds and Main Chert of Swaledale, and thus it covers in part the Richmond Cherts of the classification employed in this memoir (following Wells, 1955). This shaft is situated about 1700 ft (516 m) NE of the highest surface working in the Langscar Scrin belt, and is one of several shafts, marked 'Old Coal Pits' on some editions of the six-inch Ordnance map, but merely 'Old Pits' on the current edition, that are almost certainly trials for and workings on concealed string or flot deposits of similar type to those worked from surface farther SW. From Herontree Shaft there was a drift running SW, while from an adjacent shaft, 103 ft (31 m) deep, a crosscut was driven due E. Along the west side of the Leyburn – Reeth road hereabouts, a few shafts investigated a NNE-trending minor fault or vein, but no information is available about what was found. Virtually this line marks the end of attempts to work deposits concealed beneath the Ten Fathom Grit. No similar attempt seems to have been made to follow the Herontree strings under cover. The minor fault just mentioned crosses the Leyburn – Preston road south-east of Whipperdale Quarry and swings to SW to shift the outcrop of the Ten Fathom Grit immediately south of Yarker Bank Farm [105 915]. This fault is marked as a vein on the one-inch map and is in fact the easternmost one in Wensleydale north of the Ure; but it yielded negligible amounts of ore. Raistrick (1975a) records that Longscar Rake (Lang Scar) was worked for a time from 1760 by Lord Scrope, but never became an important mine. Now it is hard to find any mineral, extensive as both sets of workings are, and it must be concluded that the deposits were narrow fillings of joints in the brittle Richmond Cherts, with some related replacement, presumably in limestone beds.

KCD, JRE

Seata veins Lead ore (fluorspar)

SD 98 NE; Yorkshire 67 SW
Direction sinuous, average N45°W

A small number of deposits south of the river remain to be described in Wensleydale. The Seata veins (Figure 30) lie about 1 mile (1.6 km) SSE of the southernmost workings on the Carperby complex of veins (p. 170), but there is no evidence of a connexion between them, though in the intermediate ground three pits near Ballowfield [9936 8930] investigated a NW-trending fracture that might be the continuation of Disher Force Vein. On the same line a small flot with fluorite and galena was formerly seen in the railway cutting. The Seata workings, situated east of the quarry of the same name [9925 8850] are largely in the Hardrow Scar Limestone. The pattern suggests two sinuous veins, followed by shallow pits, with associated replacement ground, into which at least four short adits have been driven and three shafts sunk. The most striking feature is the presence of dense fluoritised limestone similar to that seen at Worton Hall Level (p. 167). The associated minerals include galena, sphalerite, traces of chalcopyrite, brown carbonate and white calcite. The workings extend some 1500 ft (456 m) along strike, but there is insufficient evidence to show the width of the flot; however, the 25¾ ft (7.85 m) of limestone exposed in the quarry is unmineralised.

Roger Wood Vein Lead ore

SE 08 NW; Yorkshire 67 SW
Direction N36°W, small E throw

Aysgarth Mill Vein

SE 08 NW; Yorkshire 67 SW
Direction N25°W, throw NE

Catscar Vein Lead ore

SE 08 NW; Yorkshire 67 SW
Direction N58°W

At the west of Aysgarth village in the fork between the two roads there are much overgrown remains of shafts which, according to Raistrick (1975a), worked flots; a little purple fluorite has occasionally been found. These were presumably associated with the small fault crossing the river to the north, here called the Roger Wood Vein. The direction and throw of this does not fit in with any of the veins that might have been expected to come through from Haw Bank, Carperby. The mineralised fault here called Aysgarth Mill Vein (Figure 30) crosses the Ure at [0038 8890] and is associated with dips up to 18° in the adjacent strata; it includes fractures throwing both upstream and down. Several loose blocks of baryte have been found close to the fault plane, but there is no evidence of working. The vein could possibly be regarded as a prolongation of the Wegber faults (p. 171). Catscar Vein [0101 8844] at Aysgarth High Force intersects without displacement the Hardrow Scar Limestone, 18 ft (5.5 m) of which are exposed in the cliff above the force. The vein, no longer exposed, is represented by a muddy opening in the cliff, in which there may formerly have been an adit. Aysgarth Level [0132 8866] (Figure 30, loc. 25) appears to have been driven from below Middle Force as a 1200 ft (365 m) long crosscut to this vein. Raistrick (1975a, map 8) reproduces a plan from the early part of the 19th century for an elaborate dressing floor related to this level, but there is no certainty that it was ever constructed. No shafts were sunk on either side of the river in connexion with Catscar Vein and no dumps from this activity have been found.

Heaning Wood Fault

SE 08 NW; Yorkshire 67 SW
Direction N50°W

A fault, possibly with some mineralisation (Figure 30), intersects the Gayle and Hardrow limestones with intervening measures. There are no exposures or dumps, but there is an adit driven into Hardrow Scar Limestone at [0002 8736] and a brick-lined shaft in the valley bottom [0014 8728].

Sorrelsykes Vein

SE 08 NW; Yorkshire 67 NE
Direction N80°E

Levels [0257 8815] below the Hardrow Scar Limestone, and in the limestone [0260 8816] mark the position of this vein. The lower one found only mudstone, but there is abundant veinstuff on the upper one, consisting of calcite, baryte and some galena. In the Simonstone Limestone the position of the vein seems to be marked by closely spaced joints with the same trend. The trend is almost at right angles to that of the small veins at Aysgarth and it could hardly represent any of these.

West Burton Vein Lead ore

SE 08 NW; Yorkshire 67 SW, SE
Direction N83°E, throw 23 ft (7 m) S

This vein was discovered at outcrop at West Burton Force in Walden (Figure 30) and traces of mineralisation, including colourless fluorite, can still be seen *in situ* [0190 8675]. The exposed strata include the Gayle and Hawes limestones, and a level driven east in the latter gave access to workings which included a sump 192 ft (59 m) deep. This must have penetrated below the Thorny Force Sandstone into Great Scar Limestone. A plan dated 1884 (Health and Safety Excutive, 1339) shows that the sump, 210 ft (64 m) E of the adit portal, gave access to a drift 55 ft (17 m) below the adit, which ran at least 335 ft (102 m) to the west, and 420 ft (128 m) to the east. The vein was also tested at 114 and 162 ft (35 and 49 m) below the adit. Down to the 162 ft (49 m) crosscut, the inclination of the vein was 83°S, but it changed to 50° between here and the bottom drift from the sump. Presumably the condition of the vein was not encouraging in depth, for the plan shows very little drivage below the 55 ft (17 m) drift. A block 45 x 30 ft (14 x 9 m) is shown as stoped under the adit east of the sump; and some shallower under-stoping was done nearer the portal. There was an air shaft to surface east of the sump, and a drift 64 ft (20 m) above the adit, 125 ft (38 m) long. Veinstuff dumped in the field north of West Burton Force includes pale fluorite, more coarsely crystallised than is usual in this region, with calcite, galena and sphalerite. Between 1864 and 1881, under the successive ownerships of John Tattersall & Co., and Mr Henry Pease, the mine produced a total of 733 tons of lead concentrates, from which silver was also recovered. The West Burton Vein loses throw in the Middle Limestone to the east of the workings, and a single shaft [0240 8679] shows calcite as the only vein mineral. Still farther east, scattered shafts have been sunk into the Main Limestone and Richmond Cherts on Dovescar Plain. One such shaft [0346 8687], on the projected line of West Burton Vein, found traces of galena; but the others, none of them successful, are off this line. A row of shallow shafts [0409 8710] defined a small local vein.

Braithwaite Vein Lead ore

SE 18 NW; Yorkshire 84 NW
Direction N30–40°W, throw 40 ft (12 m) NE

Braithwaite Branch Vein Lead ore

SE 18 NW; Yorkshire 84 NW
Direction N70°W

Braithwaite West Vein Lead ore

SE 18 NW; Yorkshire 68 SW, 84 NW
Direction N43°W

These veins are the easternmost deposits in the present area (Figure 30); they were worked in Coverdale, a southern tributary of the Ure, south of Middleham, close to Braithwaite Hall. The gentle dip of about 2°E in Coverdale is interrupted in the lower reaches of the river by a roughly N – S anticline with dips of 3 to 6° on its flanks. Braithwaite Vein breaks the crest of this fold. The principal workings were from shafts north of the road opposite Braithwaite Hall, and the oreshoots appear to have been on the main and branch veins within a few hundred feet of their intersection. The Main Limestone on the footwall of the main vein is here faulted against Richmond Cherts and Ten Fathom Grit on the hangingwall. A shaft [1197 8582] was still open in 1977, and flaggy sandstone of the Ten Fathom Grit could be seen in it beneath 8 ft (2.9 m) of stony clay. From the positions of shafts it appears that the extent of worked ground on the main vein was at least 1000 ft (304 m) long, and up to 600 ft (182 m) long on the Branch Vein. Work was carried below water level, necessitating pumping in this area of gentle topography; Raistrick (1975a, p. 95) refers to the operation of the pumps by a large water wheel, situated on the river about 1500 ft (456 m) away, through wooden rods suspended on timber triangles. All that remains now is a stone-walled leet, the mine having closed in 1866; there is no certain evidence of a drainage adit. The minerals present on the heaps round the shafts include pale amber fluorite, baryte, calcite, sphalerite, traces of chalcopyrite and galena, but no considerable quantity of any of these is left. The mine had a long history; Raistrick and Jennings (1965) record a grant of the mines in the lordship of Middleham by King Henry VIII to Sir James Metcalf in 1531, and of a lease held by James Ward in 1603, though neither identify the Braithwaite deposit. It was however certainly working by 1734, when the London Lead Company improved the smelt mill at West Burton, which for a time became known as Braithwaite Mill. During the 19th century concentrates were sent to the Bollihope Mill in Weardale, owned by the London Lead Company, for desilverising. The official records give the output in the last years of production, 1854 to 1866, as 586 tons of lead concentrates, yielding 4.6 oz. silver per ton lead metal. Some production was also obtained from the West Vein, nearer the river, where shafts along a strike length of about 400 ft (122 m) were sunk in the Middle Limestone and overlying sandstone, obtaining galena in a calcite matrix. AAW, KCD

Chance Hill Vein

SE 08 SW; Yorkshire 83 NW
Direction N15°W

Snod Side Vein

SE 08 SW; Yorkshire 83 SW, SE
Direction N18°W

Petticoat Rake Lead ore

SE 08 SW; Yorkshire 83 SW
Small complex, 3 veins N70–75°E, intersected by 4 scrins running N5°W, 1 vein N15°W

Wasset Fell Veins Lead ore

SD 98 SE; Yorkshire 83 NW, SW
7 small veins trending N10–15°W, 1 N80°E

Walden Windegg Vein Lead ore

SD 98 SE; Yorkshire 83 SW
N10°W

These veins are remotely situated in the high ground (about 1850 ft 564 m OD) around the headwaters of Coverdale and Walden (Figure 30). At Windegg [9775 8105] the following succession was measured; below the estimated position of the Cockhill Marine Band:

	ft	m
Grassington Grit Formation		
Sandstone	15	4.6
Mudstone	25	7.7
Sandstone, medium and coarse grained, in part pebbly	140	42.7
Richmond Cherts		
Limestone, grey, biosparite; many chert layers	42	12.8
Main Limestone, grey	65½	20
Sandstone	2	0.6

At Petticoat Rake, 2 miles (3.2 km) to the east, the pebbly sandstone at the bottom of the Grassington Grit Formation has given place to mudstone, 130 ft (40 m) thick with three coal seams, while the Richmond Cherts, still containing crinoidal grainstones-interbedded with blocky cherts, have thickened to 75 ft (23 m). The Petticoat Rake workings consist of shafts, so numerous around [015 817] as to suggest that flats were worked in the crinoidal grainstone and perhaps in the upper part of the Main Limestone. No shaft appears from the spoil to have penetrated below this limestone. The minerals present in addition to galena include calcite, a little aragonite, smithsonite and limonite. At [0152 8169] early six-inch Ordnance Survey maps record a 'washing machine' and there is evidence that some dressing was done. Nearby adits appear to have worked coal seams. To the south-east at Tomlin Haw, an ENE fault throwing 30 ft (9 m) SE has been tested, but appears to have proved to be barren. One shaft [0166 8137] which has been sunk to about 150 ft (46 m) is open and brick-lined, but the dump contains only limestone and cherty limestone. The shafts on the single vein at Chance Hill [023 846] penetrated Richmond Cherts, Main Limestone and beds to 50 ft (15 m) below the limestone; a little galena, accompanied by dolomitised limestone, appears to have been found. Shafts and an adit at Gammersgill Moor [032 833] may have been intended to test the south-eastward continuation of this same vein but no vein is recorded on the primary survey. It is nevertheless possible that this was the locality described by Raistrick (1975a, p. 95) as the Gammersgill Mine, where a good flot was found and worked up to 1845. Two other possible sites where almost all signs of old workings have been since obliterated were seen by Dalton in the Middle Limestone. Flats were noted in Turn Beck [0535 8344] and a vein in Ridding Gill [0575 8380] (Figure 30).

Turning now to Wasset Fell (Figure 30), the closely-spaced veins here cut through spurs of Main Limestone and Richmond Cherts at Baxton Edge [991 833] and at Wildgath End [995 825]. The workings, chiefly shallow shafts, show that dolomitised limestone accompanied all the veins, but the amount of galena discovered seems to have been small. Finely fibrous aragonite (p. 94) with minor amounts of witherite is present here. The westernmost vein at Baxton Edge was followed up into the Grassington Grit, and at Wildgath also, the westerly veins were tested at this level, but very little mineralisation appears to have been found. A short vein, separate from the main group, was tried and yielded dolomite [996 835]. On Walden Windegg vein [982 814] (Figure 30), seven shafts were sunk through Grassington Grit, but the abundance of dolomitised limestone on the tips shows that the worked ground was either in limestone of the Richmond Cherts, or in Main Limestone. In addition to galena, baryte and calcite were found.

Thus although the important bearing beds were readily accessible, all the trials around Coverhead and Waldenhead failed to reveal strong mineralisation. AAW

OUTPUT AND RESOURCES

Outputs from individual mines have already been given, but part of the official statistics refer to Wensleydale mines as a whole, and the complete statement of recorded production is given in Table 23. For this area, very little data is available prior to the commencement of official records in 1851. The scanty production from South Swaledale recorded by Fieldhouse and Jennings (*op. cit.*, app. 5) strongly suggests that this sub-area was approaching exhaustion by the end of the 18th century, but it remains possible that small tonnages may have been included in the official figures for Swaledale for 1845–67; these would therefore have found their way into Table 1, p. 5, but any such error will have only a small effect on the general picture. The mines on the southern slope of Swaledale were among the first discovered in the valley; a royal mandate had been extended to the miners of Grinton as early as 1219 (Raistrick, 1975a, p. 18). The deposits worked were predominantly flots in the Red Beds Limestone, and the upper beds of the Main, for the most part readily accessible from surface; indeed some must have cropped out. When levels were driven in the 18th and early 19th centuries at deeper levels, some of them, as at Devis Hole, found mainly barren phreatic caverns following the feeding veins. It is not possible to arrive at a satisfactory separate figure for the output of these mines. However, for the area, including Wensleydale and the southern tributaries of the Ure, the following argument may tentatively be applied. Experience in other parts of the Pennines (see for example Dunham, 1944, p. 1, and this volume, Table 1) suggests that at least half of the ultimate production had been achieved before the commencement of the official records in the mid-19th century. After 1851, however, Wensleydale had only twenty-five more years of substantial production left, and for this reason, and to cover the earlier exhaustion of South Swaledale, it is suggested that a figure implying that two-thirds of the ultimate output was achieved before 1850 would not be too high. The figure used in Table 1 would imply a total output of 128 000 long tons of concentrates in Area 13.

Table 23 Recorded production, lead metal and concentrates, Area 13

Group	Years	Lead metal long tons	Concentrates long tons	Lead %
Grinton	1775–1801	2 300*	3 833	60‡
	1872–73, 1890–93	131†	198	66
Oxnop-Summer Lodge	1817–1872	1 081*	1 550	70‡
Mukerside, Whitaside, Ellerton	1861–1896	577†	791	73
Wensleydale mines	1851–1863	16 815†	26 356	64
Keldheads	1864–1888	6 600†§	9 520	69
Other Wensleydale mines‖	1864–1896	1 277†	1 899	67
West Burton, Braithwaite	1854–1881	841†	1 318	64
		29 622	45 465	65¶

* Fieldhouse and Jennings, 1978, appendix 5, p.492.
† Official statistics, *Mem. Geol. Surv.*, 1848–1881; Home Office 1882–1897.
‡ Assumed as basis for calculation of concentrates production.
§ Including How Bank.
‖ Including Apedale, Askrigg, Bolton Park, Gayle, Sargill, Virgin, Wet Grooves, Whitaside, Woodhall and Worton.
¶ Average value.

No proved reserves of lead ore remain in the mines of Area 13. Regarding Keld Heads Mine, Clough (*op. cit.*, p. 101) remarks that towards the end of the 19th century, it was found that the River Ure was penetrating the lowest levels of the mine, flooding the still rich workings. If this statement is correct, there must have been shaft workings beneath Wensley Park Level (p. 178) and ore must have been continuing into the Simonstone cyclothem. It is very much to be regretted that confirmation has not been obtained, owing to the lack of records. The trial level driven by the Leeds Mining Company at Raygill (p. 166) in 1862–71 showed flot mineralisation in the Hawes Limestone, while that of the Worton Mining Company (p. 167) produced some ore from the upper part of the Kingsdale Limestone in 1877–83. Perhaps the latter trial would have been carried further but for the sharp fall in lead prices; but neither of these explorations found substantial oreshoots. The Wet Grooves and adjacent mines yielded some production from Simonstone and Middle Limestones, but the ore mainly came from the Underset level; all underground activity ceased in 1879. The Bolton Park Company, succeeding Storey and Company, thoroughly tested this mine down to the Preston Grit, with little success, and ceased operations in 1871. The continuation of the Apedale Vein (pp. 172–174) yielded some production at this horizon, but evidently there was no inducement to sink deeper, and the final output from this vein came from the Underset Limestone, all operations ceasing in 1902. West Burton Mine (p. 181) working in the Gayle and Hawes limestones, had ceased operations by 1881. Thus while there is some evidence that mineralisation continues into early Brigantian and even late Asbian strata, the prospect for large oreshoots in depth is not attractive on the existing evidence and these deposits have attracted virtually no attention during the present century.

Apart from Keld Heads, it can be argued that the bulk of the output from Area 13 came from flots in the Red Beds and upper Main limestones. Here, following the formation of the Swaledale Mining Association under Mr J. C. D. Charlesworth's influence in 1887, the Grinton Mining and Smelting Company may have carried on some prospecting at Ellerton Moor, Grinton, Summer Lodge and Whitaside until 1897, but the last small production was achieved in 1893. No doubt, undiscovered ore in flots still remains, but the cost of searching for this under cover is prohibitive under present conditions. At Walden Head and Wasset Fell (p. 182) prospecting ceased in 1881.

The tailings heaps at Wet Grooves and Keld Heads may not yet be exhausted for fluorspar and barytes. Baryte and witherite may remain underground and the former mineral in surface heaps at Virgin (p. 172) but the very remote situation is against any modern exploitation; the last working, by the Virgin Moss Mining Company was in 1897. KCD

REFERENCES

BACKHOUSE, J. 1912. Modern mining. In *Victoria History of the County of York*. (London.) 373 pp.

BRADLEY, L. 1862. *An inquiry into the deposition of lead ore in the mineral veins of Swaledale, Yorkshire*. (London: Stanford) 40 pp.

BROOK, D., DAVIES, G. M., LONG, M. H. and RYDER, P. F. 1977. *Northern caves*. Vol. 5 – The northern dales. (Clapham: Dalesman.) 158 pp.

CLOUGH, R. T. 1962. *The lead smelting mills of the Yorkshire Dales and Northern Pennines*. 2nd Edit. 1980. (Keighley: Clough.) 323 pp.

DAKYNS, J. R., TIDDEMAN, R. H., RUSSELL, R., CLOUGH, C. T. and STRAHAN, A. 1890. The geology of the country around Mallerstang, with parts of Wensleydale, Swaledale and Arkendale. *Mem. Geol. Surv. G. B.,* 213 pp.

DUNHAM, K. C. 1944. The production of galena and associated minerals in the northern Pennines; with comparative statistics for Great Britain. *Trans. Inst. Min. Metall.*, Vol. 53, pp. 181–253.

— 1952. Fluorspar. 4th Edit. *Spec. Rep. Miner. Resour. Mem. Geol. Surv. G. B.*, Vol. 4, 143 pp.

EVANS, A. D., PATRICK, D. J., WADGE, A. J. and HUDSON, J. M. 1983. Geophysical investigations in Swaledale. *Miner. Reconnaissance Programme Rep., Inst. Geol. Sci.*, No. 65.

FIELDHOUSE, R. and JENNINGS, B. 1978. *A history of Richmond and Swaledale*. (Chichester: Phillimore.) 520 pp.

HUNT, R. 1863. Produce of lead ore and lead in the U.K. for the year 1862. *Mem. Geol. Surv. G. B.*, 40 pp.

LELAND, J. *c.* 1543. *Itinerary*. Vol. 1. 1907. TOMLIN SMITH, L. (Editor). (London: George Bell.) 352 pp.

PERCY, J. 1870. *Metallurgy of lead, including desilverization and cupellation*. (London: John Murray.) 561 pp.

RAISTRICK, A. 1938. *Two centuries of industrial welfare. The London (Quaker) Lead Company, 1692–1905*. (London: Friends Historical Society.) 168 pp.

— 1973. The London (Quaker) Company mines in Yorkshire. *Mem. North. Cavern Mine Res. Soc.*, Vol. 2, pp. 127–132.

— 1975a. *The lead industry of Swaledale and Wensleydale*. Vol. 1 – The mines. (Buxton: Moorland.) 120 pp.

— 1975b. *The lead industry of Swaledale and Wensleydale*. Vol.2 – The smelting mills. (Buxton: Moorland.) 120 pp.

— and JENNINGS, B. 1965. *A history of lead mining in the Pennines*. (London: Longmans.) 347 pp.

SMALL, A. T. 1978. Zonation of Pb-Zn-Cu-F-Ba mineralization in part of the North Yorkshire Pennines. *Trans. Inst. Min. Metall.*, Vol. 87, pp. B10–13.

WELLS, A. J. 1955. The development of chert between the Main and Crow limestones in North Yorkshire. *Proc. Yorkshire Geol. Soc.*, Vol. 30, pp. 177–195.

CHAPTER 12
Mineral deposits

Details, Area 14 WHARFEDALE, HIGH MARK, GRASSINGTON MOOR

The River Wharfe is fed by two headwaters, Oughtershaw Beck and Langstrothdale (Greenfield Beck) that sweep around the northern flank of the half-dome of the Askrigg Block after rising almost centrally in this structure. Where Cray Gill joins from the direction of Bishopdale, the river turns almost due south and flows along the eastern flank of the dome. As a result of the south-westward rise of the strata, wide areas of early Brigantian and Asbian beds are exposed in the country west of Upper Wharfedale, though even here, some outliers of higher strata remain. It is a striking fact that the western flank and centre of the half-dome contain very few known mineral deposits, even though strata that are good carriers of ore elsewhere are present. The Hawes (sheet 50) one-inch geological map embraces not only the head-water streams of Wharfedale, but also the major western valley of Littondale and beyond that, Ribblesdale, Widdale and the Greta, and must be regarded as one of the largest areas of exposed Lower Carboniferous limestones in the country to contain no significant ore deposits. Along Upper Wharfedale, however, the position is different. On the western slopes of the valley, generally high up above the bottom, nests of small veins that have yielded lead ore from very numerous bell-pits and shallow shafts begin to appear at Upper Fosse Gill, opposite Starbotton, and continue southward with five extensive clusters lying in the interfluve between Arncliffe and Hawkswick on the Skirfare and Kettlewell in the main valley. Here the mineralised limestones range from the upper part of the Great Scar Limestone up to the Simonstone Limestone, but although very numerous shallow shafts have been sunk, there are very few adits and no substantial mine has been developed. In the wide area of Great Scar and Hawes limestone exposure south of Littondale and west of Wharfedale, here called High Mark, very numerous small veins, mostly striking NE, have been worked, but here too no substantial deposit has been found.

The major mining districts of Area 14 lie on the east side of Wharfedale, beginning in the north with Buckden Pike and to the south occurring at Starbotton, Kettlewell Providence, Conistone, Mossdale and finally at Grassington Moor. Production of lead ore from the Grassington Moor mines has far exceeded that from all the rest of the district put together, and here the dumps have been worked over more than once for fluorspar and barytes. There is no current activity. Although some ore has been raised from Brigantian limestones in all these districts, a new factor enters into the geological situation in Upper Wharfedale, in that the mid-Pendleian unconformity brings the basal Millstone Grit (the Grassington Grit) into contact with the Main Limestone at Buckden. Just north of Kettlewell, this limestone is cut out by the unconformity which thereafter transgresses across the Underset, Three Yard and Five Yard limestones until, throughout Grassington Moor, it causes the Grit to rest on an eroded surface of Middle Limestone. Compared with the North Swaledale Belt, where thick early Pendleian shales everywhere prevented access by the mineralising fluids to the Howgate Edge grits and Mirk Fell Ganister, the cutting out of these shales by the unconformity in Upper Wharfedale permitted the Grassington Grit to be extensively mineralised, and here it is the principal bearing bed.

Grassington Moor and an area on the Skirfare–Wharfe interfluve north-west of Kettlewell belong to the estates of the Duke of Devonshire; between these two areas the ground is divided between the Trust Lords of Kettlewell Manor and the Freeholders of Conistone. No doubt there was very early mining of the small veins exposed in the bare limestone ground, but according to Raistrick (1973, p. 90) although Will'mo Coksen de garsyngton had sold some lead to Fountains Abbey between 1446 and 1458, there is no evidence of mining on Grassington Moor until 1603 when the Yarnbury deposits were discovered. The Earl of Burlington, to whom the mines were now reserved, brought miners from Derbyshire and until 1620 attempted to work the mines, but in that year he abandoned the attempt and put the mines out to lease. A great period of mining by small working partnerships now followed, and by 1640 Laws and Customs similar to those in Derbyshire had been introduced with a Barmaster and Barmote Court. Raistrick (*op. cit.*) gives interesting details of the vicissitudes of the system, culminating in revised laws in 1737, but the field continued to be torn by conflicting claims over disputed boundaries. During the 17th century it is likely that exploitation of the Grassington Grit oreshoots spread northward to Mossdale and perhaps Buckden, and it is reasonable to suppose that the elements of the geological situation were beginning to be recognised. In the 18th century, local capitalists began to become interested in Grassington. Their initials or names—Wardle, Overend, Whickham, Bagshaw, Dr King and many others—appear on contemporary surveys and in some cases on meer stones, marking lease boundaries, that can still be seen. The influx of capital enabled horse whims to be installed for winding and pumping and some shafts to be deepened, but Raistrick's study shows that towards the end of the century there was a serious and continuing fall in lead output. To remedy this, the Cavendish Estate began to take a controlling interest in the mines of the whole field, providing funds for the driving of a major deep adit system, known as Duke's Level, under the direction of two successive chief mineral agents, Cornelius Flint from 1779 and John Taylor from 1818. This ambitious scheme, very well conceived in relation to the geological situation, was begun in 1790 and gave the Grassington mines a new lease of life which carried them on until 1876. The veins of the Grassington complex head E and SE into the property of the Freeholders of Hebden where, however, the Grassington Grit Formation disappears under the Thirtyfive Fathom or Nidderdale Shale and the overlying Red Scar Grit. One successful though short-lived mine was developed at Bolton Gill on the continuation of a vein from Yarnbury, but although some miles of driving were undertaken, no other workable deposit has been found. Synthesis of the whole geological picture nevertheless suggests that the critical part of Hebden Moor has not yet been prospected. In view of its relevance to Grassington Moor, Hebden Moor forms part of the present chapter.

In this chapter, description of the mining areas proceeds from the Ure – Wharfe watershed where several minor veins are briefly described, including some high in Walden, southward through the districts on the west side of Wharfedale, including those in Littondale and around High Mark. The eastern mines, from Buckden Pike to Grassington and Hebden moors then follow. It should be noted that Wharfedale where it turns east and south-east after crossing the Craven faults is not included in the present chapter, but in Chapter 13.

The Grassington lead mining activities during the 19th century were documented in a series of papers by S. Eddy (1844, 1845, 1859a, 1859b) who managed the Duke of Devonshire's mines, and by his son and successor J. R. Eddy (1883). The laws and customs of this field were put on record by Bean (1737) and discussed by Raistrick (1936) and by Raistrick and Jennings (1965). Raistrick has also given two more general views, setting the mining in its local context (1936, 1973). The first Geological Survey account of the mines, apart from the brief mention by Dakyns and others (1890) was that of Carruthers and Strahan (1923) while the 4th edition of the Special Report on Fluorspar (Dunham, 1952) contained the first results of the present resurvey. Dickenson (1902) and Raistrick's three accounts of the mineral deposits (1938a and b, 1953) have also been drawn upon, with a number of local publications mentioned in connexion with individual mines. Foster Smith (1968) included data from the Mining Journal between 1850 and 1865 in his account of Buckden, Starbotton and Dowber Gill. Clough (1980) has dealt comprehensively with the smelt mills.

Cam West Vein

SD 88 SW; Yorkshire 81 NW, SW
Direction N45°E turning to N60°E

Cam East Vein

SD 88 SW; Yorkshire 81 NE
Direction N25°E

Bardale Head Vein

SD 88 SE; Yorkshire 81 NE
Direction N25°E

Wether Fell Veins

SD 88 NE; Yorkshire 65 SE
Direction N86°E and N10°W

On the north side of Cam Hill High Road, a former Roman road running from Ribble Head to Bainbridge in Wensleydale, there is an extensive outlier of Main Limestone. A shaft [8197 8301] on Cam West Vein (Figure 15) passes through the limestone into sandstone, and another shaft 575 ft (175 m) to the SW may have done so. The shallow workings between show slight dolomitisation of the limestone, but no other veinstuff was found. The total length of ground tested was 1100 ft (335 m). The main shaft on the East Vein [8433 8356] started above the top of the Main Limestone and shows a little galena and calcite on the tip. Here 800 ft (244 m) of ground was tested from four additional shafts to the SW, but little appears to have been found. This vein is mapped as continuing across the shoulder of the Dodd Fell Hill outlier of Lower Howgate Edge Grit to Bank Gill. Paralleling the road from Gayle into Wharfedale, a feeble vein has been tried by a shaft [8631 8441] and about 300 ft (91 m) of openwork, now grassed over, to the SSW. No veinstuff is to be seen. Trials on the Wether Fell veins [8766 8679 and 8748 8738] were equally unpromising (Figure 15). All these veins may be little more than slightly mineralised joints. They are very remote from the main centres of mineralisation.

Oughtershaw Hall and Beckermonds Deposits

SD 88 SW; Yorkshire 81 SE

In the stream east of the road opposite Oughtershaw Hall, [c. 874 812] two small veins carrying traces of copper minerals were formerly exposed (Figure 15). At Beckermonds Farm the primary survey records a vein [8733 8024] striking NNE. At this point a narrow zone containing a little pale mauve fluorite with traces of chalcopyrite, and with associated dolomitisation and silicification, is seen in the Great Scar Limestone, but it is difficult to be sure of the direction of strike. There is also a probable mineral flat [8720 8021] on the north side of Greenfield Beck near Beckermonds Farm, indicated by a gap in exposure, with blocks of dolomitised limestone. The BGS Beckermonds Scar Borehole [8635 8016] (Wilson and Cornwell, 1982 and p. 8), after passing through silicified limestone, with a 2-cm baryte-chalcopyrite-malachite vein, from 88 ft 3 in to 98 ft 3 in (26.90–29.95 m), and cavernous ground from this depth to 117 ft 2 in (35.70 m), passed through a mineralised zone from 119 ft to 166 ft 2 in (36.25 – 50.64 m) which, from the prevalence of low-dipping textures, we interpret to be a flat. The minerals present, in order of abundance, include calcite, dolomite, baryte, colourless, pale yellow and deep purple fluorite, pyrite and chalcopyrite. The best stretch, from 153 ft 7 in to 160 ft 6 in (46.82–48.92 m) averaged 1 per cent Cu, while from 129 ft 11 in to 142 ft 10 in (39.60–43.55 m) the average was 0.1 per cent Cu. Lead values ranged between 20 and 40 ppm, little above normal background and no more than 10 ppm Zn was found. The limestone remaining in the zone had been partly replaced with authigenic silica and in places with pyrite. In the zone described, the borehole was inclined at 14° from vertical, and apparent dips varied between 9° and 19° in the core, except at 165 ft (50.29 m), where the apparent dip was 44°. The base of the flat lies 165 ft 7 in (50.47 m) above the base of the Danny Bridge Limestone and 712 ft 9 in (217.25 m) above the unconformity at the base of the Carboniferous (borehole depths in both cases). Disseminated pyrite with a little chalcopyrite was noted close above the unconformity and for some 26 ft (7.90 m) below it. The distance between the borehole site and the exposed mineralisation at Beckermonds Farm is 3850 ft (1.17 km) and the probable flat near Beckermonds Farm is at about the same stratigraphic horizon as the flats in the borehole. An induced polarisation survey with dipole-dipole array was undertaken to try to establish the connexion (A. D. Evans, personal communication) but the results were inconclusive due to the effects of borehole casing and forestry fences. At the same time it may be said that the occurrence in the borehole was unexpected, and this type of ground may be more extensive in the Great Scar Limestone than has hitherto been supposed. A substantial increase in the introduced mineral content would, however, be needed to make it of economic interest.

Fosse Gill Veins Lead ore

SD 97 SW; Yorkshire 98 SE
Direction near E – W

Around Starbotton as elsewhere in Upper Wharfedale, the steep valley sides expose the Great Scar Limestone, while the early

Brigantian cycles make gentler slopes above. The two veins, 90 ft (27 m) apart have been worked on the north side of Upper Fosse Gill [9431 7428], the streams in which rise from springs at the base of the Simonstone and Hardrow Scar limestones (Figure 15). These streams disappear into a sink hole, but the water reappears in Fosse Gill proper, in the Great Scar Limestone. Four bell-pits mark the course of the northern vein, ten and a short opencut the southern. Baryte, calcite and galena are present and the range of the workings is from the lower part of the Simonstone to the upper part of the Gayle Limestone, beneath which the mineralisation evidently dies out.

Black Rock Vein Lead ore

SE 97 SW; Yorkshire 115 NE
Direction N28°W

A line of 12 bell-pits remotely situated on Old Cote Moor marks the course of this vein for a distance of 1900 ft (579 m) (Figure 15). What appears to have been the main shaft [9395 7266] was sunk at the top of the Hardrow Scar Limestone and ore was probably won from this limestone. Some may also have come from the sandstone beneath the Simonstone Limestone, but the vein does not reach this limestone. Some dressing was done near the main shaft, calcite, baryte and galena being present.

Charlton's Vein Lead ore

SD 97 SE; Yorkshire 115 NW
Direction N75–90°E

Wiseman's Vein Lead ore

SD 97 SE; Yorkshire 115 NW
Direction N80°E

Sunter's Vein Lead ore

SD 97 SE; Yorkshire 115 NW
Direction near E–W

The workings on these veins constitute *Starbotton Moorend Mine* (Figure 15), also known in the early part of the 19th century as New Providence, and after 1869 as Wharfedale Mine (Raistrick, 1973, pp. 135–137). Charlton's Vein was probably found with the others in the limestones above the Great Scar, within the liberty of the Trust Lords of Kettlewell, at least as early as the 18th century. Prospecting to the west, in the liberty of the Duke of Devonshire early in the 19th century revealed promising ground in the Simonstone Limestone and in 1850–58 Moorend Shaft [9505 7276], collared at 1500 ft (472 m) OD was sunk to 111 ft (33.8 m) and a '20-Fathom' Level was driven westward from the bottom[1]. The level was driven in the Dirt Pot Grit, here 27 ft (8.2 m) thick. Workable ore was found in the sandstone, but except in a stretch 150 ft (46 m) long near the shaft, it is not shown as persisting up into the Simonstone Limestone; however, 'old man' workings may well have extracted anything there at an earlier stage. At 360 ft (104 m) W of the shaft a crosscut was driven N 40 ft (12 m), finding North Vein, either a branch from Charlton's Vein or the same vein shifted W side N by a cross fracture. At 120 and 220 ft (36–67 m) W of the shaft, sumps were put down 90 ft (27 m), and from the second a '30-Fathom' Level was developed westward. From the crosscut the 20-Fathom Level was driven 600 ft (183 m) W19°S reaching by 1862 the point where it would have re-entered Kettlewell liberty. Two more sumps, on North Vein, yielded ore almost down to the 30-Fathom Level, but trouble was experienced with water either in these or in two sumps started below the 30-Fathom Level. Meanwhile on the Kettlewell ground to the east, mining had continued in all three veins, and an adit [9549 7279] was started on Charlton's Vein. This was taken over when the Wharfedale Mining Company was formed in 1868 and by 1870 it had crossed the boundary, and a rise had been put up in promising ground (said to have yielded sufficient ore to pay expenses in that year and to pay off some of the debt on the mine) in the Devonshire royalty. A crosscut was put out to Wiseman's Vein and some ore was got, but not enough to make a mine. The intention was to join up with Moorend Shaft and with the 30-Fathom Level, thus relieving the water problem but in 1877 before this had been done the mine was abandoned. The stratigraphical range of the patchy oreshoots here is indicated by the fact that the adit is at about 1260 ft (384 m); ore was thus found over 240 ft (73 m), from the Simonstone down to limestones equivalent to the Hawes. Apparently the vein did not extend up to the Middle Limestone, and attempts to follow the veins down into the Great Scar Limestone *sensu stricto* on the well exposed side of the valley were not successful. All three veins, and the small Sun Vein south of Sunter's Vein, carried, in addition to galena, baryte, pale yellow fluorite, calcite and a little sphalerite. Dolomitisation of the limestone is conspicuous. Production of lead ore 1860–66 was 300 tons, 1869–77, 275 tons; the total output may have been of the order of 1000 tons. This was the only mine on the west side of Wharfedale to achieve even a modest degree of development and equipment and it serves to place in perspective the numerous other small workings (see also Dickinson and May, 1968; Foster-Smith, 1968).

[1]Tonks examined a plan and section of this mine in the possession of the Trust Lords of Kettlewell in 1940, by kind permission of Mr Cutcliffe Hyne.

Coldstreak Vein Complex Lead ore

SD 97 SW, SE; Yorkshire 115 SE

Veins	Direction	Length explored		Pits
7	E–W	5250 ft	(1.6 km)	36
3	NNW	1600	(0.49)	13
3	WNW	2500	(0.76)	12
3	ENE	1500	(0.46)	11

These extensive shallow workings are situated on Hawkswick Moor, on the gentle slope formed by the equivalents of the Hawes to Hardrow limestones, above the steep side of Littondale where the Great Scar Limestone forms the lower part of the slope (Figure 15). The complex is one of several where, if individual names were given to the veins, as they may have been when they were leased to free miners, the record of these no longer exists, though Raistrick mentions Blea Scar Vein as the southernmost of the E–W set, and Great Spar Vein as the northernmost. There is however insufficient data to merit individual descriptions for the veins and only the dominant and subordinate directions, and the number of bell-pits and shallow shafts mapped can be stated. One level [9453 7174] at 1100 ft (335 m) OD was driven, starting near the west end of the strongest vein, at almost the lowest level at which any indication of mineralisation was found. It does not appear from the dump, mainly of limestone, to have been carried far. The shaft heaps and shallow opencuts reveal that baryte and a little pale fluorite accompanied the galena, which may have been sufficient to keep a few working partnerships going at the time when the raising of a few tons of lead ore a year was enough to provide a living, but this complex was evidently not regarded as worthy of serious development.

Middlesmoor Vein Complex Lead ore

SD 97 SE; Yorkshire 115 NE

Veins	Direction	Length explored		Pits
11	N80°W to N80°E	10650 ft	(3.25 km)	38
5	N65°E	3500	(1.06)	20
3	N10–30°W	2450	0.95	18

Situated on the spur between Littondale and Wharfedale, south-east of the Coldstreak complex, the stratigraphical position of these veins or mineralised joints is similar. Directions near E – W again predominate, but here in addition to the numerous shallow shafts, small opencuts also remain, with calcite, pale fluorite, baryte and a little galena on the heaps, though in small quantities only. The vein directions listed above do not correspond with the master joints in the Great Scar Limestone, for these run N30°W and more or less at right angles to this. There are however signs of a bisectrix corresponding with the dominant near E – W direction among the veins. A fault shifting the limestone scars on the east side of Wharfedale and throwing south, apparently springing from the Mossdale veins, crosses the dale in a WNW direction and seems to terminate scars below the Middlesmoor Complex. It might join with the NNW vein which marks the eastern boundary of the complex but its presence is not otherwise discernible. Its course, however, if projected westward, would also pass through the Coldstreak complex.

There are no separate records of production from Middlesmoor, but it was probably small.

Davey Rake Lead ore

SD 97 SW; Yorkshire 115 SE, NW
Direction N80°E, N75°W, N60°W

Situated about 1 mile (1.6 km) S of Arncliffe village, this vein was probably discovered at outcrop [9323 7023] in the equivalent of the Hawes Limestone and developed from shafts for 1200 ft (366 m) to the W and nearly 2000 ft (604 m) to the ESE, the work commencing in 1756. These workings became waterlogged, but in 1813 two levels, the lower 1300 ft (336 m) SE of the original shaft, were driven to dewater them. The range of ground above the lower level was over 150 ft (46 m), and the easternmost shafts start 50 ft (15 m) down the hillside below the horizon of this level. There is a suggestion here that mineralisation continued somewhat deeper into the Great Scar Limestone than has proved to be the case elsewhere in the district. The mineralisation does not however appear to have been strong. Farther west, Blue Scar flanking Cowside Beck exposes a weak string more or less on the line of the N75°W portion of Davey Rake, while 1500 ft (457 m) N of the Davey workings another small E – W vein has been tried; this is perhaps the one referred to by Raistrick (1973, p. 144) as Snotty Bessy Rake, though it is some distance from Lineseed Head, where the only veins run NE.

High Mark Veins Lead ore

SD 96 NW, NE; Yorkshire 115 SW, SE

Veins	Direction	Length explored
14	N60–65°E	15 600 ft (4.75 km)
25	N40–50°E	29 100 ft (8.86 km)

Within a triangular area defined by spotheight 1650 ft (503 m) [9148 6770] near Clapham High Mark, a point [9347 6942] near Scar Bank and with its SE point cut off in Barstow's Kilnsey Moor [between 9538 6738 and 9442 6688], numerous small veins exposed in the outcropping scars and pavements have been prospected, principally by shallow opencut, with some shafts and a few adits (Figure 15). Wager (1931) and Raistrick (1938b, 1973) have commented on the persistent NE direction of these veins, the former maintaining that many of the veins were deposited in widened joint planes in view of the exact parallelism between their directions and that of one of the joint sets. The High Mark area contrasts remarkably with the other complexes of small veins so far described, and with Starbotton Cam Pasture (p. 189) in the complete absence of veins paralleling the NNW joint direction, which is dominant here as throughout the rest of the district. On High Mark, analysis suggests that two groups of NE veins should be distinguished: the N60–65°E set appears at the western and eastern parts of the triangle, the N40–50°E set in the centre, and at ½-mile (0.8 km) E of Proctor, the two sets intersect without displacement. The spacing between veins varies from about 100 ft (30 m) to 800 ft (244 m); thus though the veins are numerous they are not so closely spaced as in the scrin belts of Wensleydale (p. 163). Some of them are faults with a few feet of measurable displacement but many show no throw. Vein-widths, though they can no longer be measured, were clearly small from the amount of unmineralised limestone thrown out from the trials, perhaps no more than a few inches. The principal introduced minerals were pink baryte and galena, accompanied by some dolomitic alteration. Pale yellow or colourless fluorite is found in smaller quantities, and only in the easternmost veins. Sphalerite is sporadically present and is fairly abundant in the vein at Lineseed Head [924 681]. The primary surveyors, who mapped the veins with great care, recorded prehnite from a shaft [9380 6752], but we have neither been able to find the material in the Geological Survey collections, nor to confirm the record. Chalcopyrite occurs in trace amounts enclosed in the baryte. Stratigraphically, the veins occur above the position of the Girvanella Band as mapped by Garwood and Goodyear (1924) and recently remapped in more detail during the resurvey of Geological Sheet 60 (Settle). They therefore cut not only the upper part of the limestone equivalent to the Hawes Limestone, but may reach beds equivalent to the Gayle or Hardrow. Below the Girvanella Band, only one has been worked down into the Great Scar Limestone as far as the Davidsonina septosa Band, about 115 ft (35 m) below the Girvanella Band. This is the vein at Barstow's Kilnsey Moor which cuts across the eastern corner of the triangle; here a level [9538 6738] enters about 100 ft (30 m) below the probable position of the D. septosa Band, but seems not to have been successful. Much of the ore from this area seems to have been carried to washing places near the spring feeding Howgill Beck, and about ½-mile (0.8 km) to the south-east, but the quantity of tailings remaining, only about 1000 tons, suggests that no considerable production was got from High Mark.

During the resurvey, Mrs Linda Jones was able to map a persistent bedding-plane 26 ft (8 m) below the Girvanella Band, which approximates to the base of the Hawes Limestone. This falls gently north from about 1670 ft (509 m) OD at Clapham High Mark and 1650 ft (503 m) OD at Proctor High Mark to 1470 ft (448 m) OD near the northern apex of the triangular area of veins as defined above. At Buckden the base of the Main Limestone lies 636 ft (194 m) above the corresponding bedding plane, while at Fountains Fell the separation is 505 ft (154 m). The High Mark area lies in the crestal region of the half-dome of the Askrigg Block and if the Main Limestone were present, its base would be at around 2250 ft (686 m) OD, dipping gently north at approximately 100 ft (30 m) per mile; otherwise, apart from the minor shear fractures that have been mineralised, the region has suffered little structural disturbance. Thus although it may be the case that these small veins were feeders of more substantial deposits at higher stratigraphical levels, now eroded, it would hardly be expected that large concentrations of ore formerly existed. The strong mineralisation begins on the eastern flank of the half-dome, where powerful fracturing produced good channelways. To this area, east of the Wharfe, the remainder of the chapter is devoted.

Buckden Veins Lead ore

SD 97 NE; Yorkshire 98 NE, 82 SE
Subparallel veins and strings forming a belt up to 375 ft (114 m) wide with average trend N7°W (Figure 15). Displacement probably small, to the W

The centre of the belt lies 500 ft (152 m) W of the summit of Buckden or Ramsden Pike (2300 ft; 701 m). The zone which can be followed for 1¾-miles (2.8 km) has been worked in three separate royalites, at *Bishopdale Gavel Mine* [main shaft: 9581 7965] in the north, at *Buckden Gavel Level* [portal: 9555 7815] and from *Walden Road Level* in Cam Gill [9601 7706], the altitudes of the two levels both being 1750 ft (533 m) OD. It is not unreasonable to consider that the Buckden veins lie in the same near N – S zone as the small veins worked on the summit of Thoralby Common, the Worton veins (p. 166) and the Oxnop main fault (p. 167). The discovery in the Buckden area was probably made in Bishopdale Gavel, where the west vein must have cropped out in the Grassington Grit (here directly resting on the Main Limestone) and in the limestone below. Working here was entirely by shafts and opencut, except for a short level driven in limestone from the north. The shaft dumps show grit veined by baryte, galena and calcite, and limestone much altered to brown carbonate. Flats are said to have been found, and the spoil resembles that from the flot workings in Swaledale/Arkengarthdale. The mineralisation dies out soon after the zone enters the exposed Main Limestone. To the south, the bell-pits continue into the Buckden royalty of the Duke of Devonshire, the southernmost shaft, 1200 ft (366 m) S of the boundary, having been sunk through the Thirty-five Fathom Shale into grit and limestone. Here the surface shafts cease, evidently because the increasing cover of shale, eventually capped by Red Scar Grit, was becoming too great. The underground workings from the shafts may well have penetrated farther south. Raistrick (1973, p. 140) states that the Gavel mines developed rapidly so that by 1710 some 30 to 40 tons of ore per year were being sent to the smelt-mill, leading to the building, about 1735, of a new mill at Birks, south-west of Buckden, with sufficient capacity to deal with it. This production continued through the 18th century, and perhaps 3500 tons of concentrates should be credited to this phase of mining. The tailings, from hand dressing of the ore on the fellside above 1800 ft (549 m) consist of the minerals mentioned above together with a little fluorite, smithsonite and anglesite.

In 1803 the bold step was taken of starting the Higgs or Buckden Gavel Level at the base of the Main Limestone in the head of Buckden Beck, to search for the southward continuation of the mineralised zone, more than ½-mile (0.8 km) S of the existing workings. After passing through at least two sets of N – S strings, the level entered the west vein 1525 ft (465 m) ENE of the portal and by 1822 had been driven a further 1550 ft (472 m) N in the vein. Levels were developed at 20, 45 and 60 ft (6, 13.5 and 18 m) above the adit-level and crosscuts from these revealed that the zone contains at least four veins or leaders. Productive flats were found at the top of the limestone and at a lower level, perhaps 15 ft (4.6 m) below the top. The 'upper drifts' shown on a small-scale plan preserved in the official records (Yorks. 6) possibly refer to workings in the Grassington Grit, mineralised samples of which can be seen on the dump. An analysis of tailings is quoted in Table 17. Subsequently the main adit was also extended south along the west vein, reaching 930 ft (283 m) from the level head by 1842. These new developments, according to Raistrick (*op. cit.*) increased the demands on the Birks smelt-mill beyond its capacity and after tests at both Birks and the Cupola Mill at Grassington Moor in 1814, a new smelt-mill was erected at Starbotton. At about the same time, the Buckden ore zone was being explored still farther south, from the Walden Road Level high on the side of Cam Gill Beck, also driven on the base of the Main Limestone. The direction lines up exactly with that of the west vein at Buckden Gavel, but here it was known as Wiseman's or Newbould's Pipe, suggesting that ore was got from a linear cavernous deposit. In the first 300 ft (91 m) of the Cam Gill level there is evidence from bell-pits of intersecting strings runnings E – W and NW and local tradition has it that flats were found. The length driven in Walden Road Level was 1935 ft (584 m), leaving about 1500 ft (457 m) of ground untested between the forehead, which had reached the royalty boundary, and the south drive from Higgs Level.

Although the figures are not complete, some idea of the production from the Buckden zone can be gained from data given by Raistrick for Starbotton Low Mill. In its first five years, 1814–18, 716 tons of lead were made from Buckden Gavel ore, and 49 tons from Cam Gill ore. After a short recession, 100 tons lead per year was made until 1865, while up to the closing of the mines in 1877 another 225 tons was perhaps produced. For 1814–77 lead metal output was thus not less than 5000 tons, to produce which not less than 6600 tons of ore would be required. Taking into account the earlier working at the northern end of the zone, a minimum output of 10 000 tons of concentrates is indicated (see also Foster-Smith, 1968).

Cam Pasture Veins — Lead ore

SD 97 NE, SE; Yorkshire 98 SE, 115 NE

Veins	Direction	Length explored		Pits
4	N12–18°W	8100 ft	(2.47 km)	36
6	N25°W	4300	(1.3)	numerous
3	N60–65°W	4950	(1.5)	18
9	N75–90°E	10700	(3.26)	51

From the hillside north-east of Starbotton Village, a crescent-shaped area with very numerous shallow workings extends south to High Side and swings east to the Cam Gill Road at Scabbate Gate Wood (Figure 15). Stratigraphically these extend from the lower part of the Simonstone Limestone (in the south only) through the Hardrow Scar into beds probably equivalent to the Gayle Limestone. The four subparallel NNW veins occupy the northern part of the crescent, where they have been prospected or worked from shallow pits, but not very continuously. They appear to terminate against one of the WNW veins that has been worked opencast [962 744] in the Hardrow Scar Limestone south-east of which a 'horsetail' of closely-spaced N25°W veins diverge from the south side. This appears to have been perhaps the most productive part of the complex, and substantial heaps of white tailings (analysis, Table 17) remain from hand dressing here. The structure is comparable with that at Hurst (p. 147) though on a much smaller scale, for the close-set veins, each perhaps 600 ft (183 m) long probably terminate against a second fairly strong WNW vein to the south. A series of ENE and E – W veins come in farther south, and the NW-trends disappear. The southernmost ENE vein lines up exactly with Charlton's Vein at Starbotton Moorside Mine on the opposite side of Wharfedale. As with the vein complexes described from the west side of the dale, Starbotton Cam Pasture bears all the signs of having been worked, mere by mere, by small partnerships of miners, never penetrating far, and producing only a few tons of ore per year. From the Wharfedale side, an attempt was made to explore in depth when Spout or Springs Wood Level [9587 7435] was driven at 1025 ft (312 m) OD (Richardson, 1966). This reached the ground beneath the opencut mentioned above but the lowest workings on the hillside are above 1200 ft (366 m) OD and though calcite, baryte, a little fluorite and traces of copper occur on the dump, these probably came from rises, for no mine was developed. There were also short levels on two of the ENE veins and, from the SE, on the WNW fracture that seems to cut off the 'horsetail', but none of these found much mineral. Spout Gill Level, driven in 1853, was presumably the last attempt to work here. No details of production are available, but mining may well go back to early times.

The Cam Pasture complex lies athwart the projected course of the Buckden zone, had it continued south from Walden Road Level. As there is no sign of the N7°W trend among the veins here, it must be assumed to have died out in this direction. Comparing the two districts, pale yellow fluorite is noticeably more abundant at Cam Pasture; baryte is probably the commonest gangue mineral at both; dolomite is less in evidence at Cam Pasture; calcite is common to both areas. Small amounts of malachite, smithsonite and cerussite can be found.

Hooksbank Vein Lead ore

SD 97 SE; Yorkshire 116 NW, 99 SW
Direction N75°E, probably throws N

Dowber Gill Veins

SD 97 SE; Yorkshire 116 NW
Three small veins trending N70°E, one turns to N45°E; and N – S crosscourse

Smith's Vein, North Vein

SD 97 SE; Yorkshire 116 NW
Direction N60°W, turning (as North Vein) to E – W and ENE, throw 30 ft (9 m) SW

Old Providence Vein Lead ore

SD 97 SE; Yorkshire 116 NW
Direction varies between N85°W and N80°E, throw 14 ft (4.3 m) N

Exhibition Vein Lead ore

SD 97 SE; Yorkshire 116 NW
Direction N75°E, throw 12 ft (3.6 m) S

Australia Vein Lead ore

SD 97 SE; Yorkshire 116 NW
Direction N75°E, throw 30 ft (9 m) N

Rain Slack Veins Lead ore

SD 97 SE; Yorkshire 116 NW
Three weak veins and associated strings, trending N73–87°E, three N70°W

This group of E – W to ENE veins are those of the *Old Providence Mine*, the most important operation in the liberty of the Trust Lords of Kettlewell. Raistrick (1973, p. 133) states that the Kettlewell smelt mill worked nearly continuously from 1660 to 1877, and this mine was its chief, though not its only, source of ore; his account includes a map of the veins and principal adits. Hooksbank Vein, on the north side of Dowber Gill may have been one of the earliest worked; it lies only ½-mile (0.8 km) E of the southernmost vein of the Cam Pasture area and was exploited from shallow shafts. Brackmint Level [9860 7312] at 1250 ft (381 m) OD, driven in limestone equivalent to the upper part of the Hawes Limestone, reached the vein as a NNW crosscut 1080 ft (329 m) long; the dump contains amber fluorite, calcite, baryte, galena and traces of malachite. Little is known about the Dowber Gill veins save that bell-pits on the north side of the gill show where they were worked; amber fluorite, in cubes up to 1½-in on the side, could formerly be found here. The Old Providence and the several Rain Slack veins were almost certainly found at outcrop in the Simonstone and Middle limestones; this ground is interesting as the northernmost in Wharfedale in which the latter limestone becomes productive. Rain Pit is a large pothole near the top of the Middle Limestone, through which an ENE vein runs, throwing 10 ft (3 m) S. No doubt, underground water offered a serious obstacle to the mining of these veins, so much so that in the 18th century an ambitious drainage project was undertaken. Rain Slack Low Level was driven S from the bank of Dowber Gill at [9854 7295], starting at 1150 ft (350 m) OD and following a crosscourse for altogether 2575 ft (785 m) in a direction S5°E. The horizon of this drive is close to that of the Girvanella Band, which is exposed near the adit mouth; it is, therefore, low down in the Hawes Limestone. It is not known whether the crosscourse was mineralised, but at least ten veins were intersected, including Old Providence and the principal Rain Slack veins. Ventilation was probably got from a shaft on Old Providence Vein, collared just below the outcrop of the Dirt Pot Grit. No information remains to show whether rises—which would be a considerable operation—were put up in the other veins; perhaps the Low Level fulfilled its purpose by enabling shaft working to continue above. There is no evidence of drives from the level on the various veins cut, other than a short drift on Old Providence, and it must be assumed that here as elsewhere, productive ground did not extend down to the Girvanella Band, a conclusion supported by the absence of tailings around Low Level mouth. In the 19th century, Old Providence became the most important producer. Middle Level [9920 7280] was driven from the base of a cliff of Middle Limestone at close to 1400 ft (427 m) OD, probably with the intention of reaching the vein east of the old workings from surface shafts in the Simonstone and Middle limestones. The level followed a crosscourse running SSE, and found Old Providence Vein 675 ft (206 m) inbye on the west side of the crosscourse; on the east side it diverged from the crosscourse and resumed its E direction 230 ft (70 m) farther in. Another apparent shift E side S was found 250 ft (76 m) ahead, and a third 700 ft (213 m) beyond this. A crosscut was driven S on this latter crosscourse, and after driving 225 ft (59 m) farther on the Old Providence Vein, crosscuts were driven N and S. Here there is some conflict between the available plans, but a geological section preserved in the Health and Safety Executive Collection (R.312 D) shows the general situation at this end of the mine. Two new veins, Exhibition and Australia, were found south of Old Providence Vein, and a connection was made to North or Smith's Vein which was already known from a surface level. The coal shaft, sunk to test the 2 ft (0.6 m) coal (Figure 12, Section 12) above the Bearing Grit of the Grassington Grit Formation, can be identified on surface [9995 7248], serving to fix the position of Australia Vein. There is no evidence that any of the veins were worked east of this shaft. The section, borne out by recent mapping (Figure 13) indicates that the mid-Pendleian unconformity is here separated by about 120 ft (36 m) of shale with thin limestone beds from the top of the massive Middle Limestone. Under these circumstances the Bearing Grit may well not have been mineralised, and no evidence that ore was worked from the grit was found on the dumps from Middle Level, nor from a level 150 ft (46 m) from the stream, which would reach the vein in shale beneath the grit, and which shows litle sign that it was successful. No attempt was made to follow the veins by opencut across the outcrop of the grit. The evidence appears to indicate that the oreshoots here were in Middle and Simonstone limestones. Surface evidence further shows that the Old Providence Vein split up and died out at about the base of the Hardrow Limestone, east of the line of Low Level.

Satisfactory production data for Old Providence and Rain Slack mines have unfortunately not survived, but it is possible to make a guess in the following way. Through the courtesy of Dr Raistrick we have been enabled to see the Lead Book and Day Book of Kettlewell Smelt mill for the final period, 1859–86, during the early part of which the mill was reconstructed, improving the ore hearth and adding a slag hearth and roasting furnace. From 1859 to 1874, ore from Providence yielded 860 tons of lead production, compared with a total production by the mill of 1418 tons. Comparison with Hunt's figures confirms that all the Providence ore, 1302 tons, in this period was smelted at Kettlewell. The best year's production from the smelter was 1867, when 207 tons lead were made, but the average capacity seems to have been about 140 tons. Prior to reconstruction, the average capacity might be assumed to have been between 50 and 60 tons annually. If the Dowber Gill – Langcliff group of mines, certainly the best-developed in the Trust Lords royalty, supplied 60 per cent of the ore over the previous 200 years (the mill having been built in 1660), the metal produced would have been 6000 tons; if only half of this, a possible minimum yield of ore for the life of these mines of 5800 tons is obtained; if the full 60 per cent, the figure for concentrates would exceed 10 000 tons.

Among the associated minerals, fluorite is more prominent here than anywhere in the present district north of Grassington, but it remains subordinate in amount to baryte and calcite, and there is no indication that the spar minerals are present in workable quantities.

Summer Haw North and South Veins

SD 97 SE; Yorkshire 116 NW
Direction N80°E av.

Cocklake = Moorhead = Silver Rake = Fearnought South Vein Lead ore

SD 97 SE, SE 07 SW; Yorkshire 116 NW
Direction variable between N80°E and N70°W, but general trend E – W, throw 2 ft S

Old Fearnought Vein Lead ore

SE 07 SW; Yorkshire 116 NW
Direction N68°W, throw 3 ft (0.9 m) S

Bell and Dragon Vein Lead ore

SD 97 SE; Yorkshire 116 NW
Direction N74°E, throw about 15 ft (4.6 m) S

Corrigan String = Hearty Lass String

SE 07 SW; Yorkshire 116 NW
Direction near E – W

Park's = Odd Mere = Twenty Meers = Mossdale Vein Lead ore

SD 97 SE, SE 07 SW; Yorkshire 116 NW
Direction near E – W, throw 6 ft (2 m) S

Jeffries Vein

SE 07 SW; Yorkshire 116 NW
Direction E – W

Sarginson Vein = Sarginson Rake = Dreadnought Vein Lead ore

SE 07 SW; Yorkshire 116 SW
Direction E – W, throw perhaps 10 ft (3 m) S

Peck O'Malt Vein Lead ore

SE 07 SW; Yorkshire 116 NW, SW
Direction N43°W

Astonishment Vein

SE 07 SW; Yorkshire 116 NW, SW
Direction N60°E

Butcher's Rake Lead ore

SE 07 SW, 06 NW; Yorkshire 116 NW, SW
Direction N70°W

Royal Exchange Vein Lead ore

SE 07 SW; Yorkshire 116 NW, SW
Direction N40°W

Mossdale North Vein Lead ore

SE 07 SW, 06 NW; Yorkshire 116 NW, SW, SE
Direction N30°W

Glover's Rake and String

SE 07 SW; Yorkshire 116 SW
Direction N77°E and N70°E respectively

The veins listed above are those of the *Conistone Moor or Mossdale mines* (Figure 23). Thanks to Dr Raistrick's work on ancient plans, it is possible to identify most of the veins, whereas the primary surveyors were not able to do this. The usage here follows that of Raistrick (1966). The western ends of the veins can be seen in the limestones exposed on the side of the valley about 1 mile (1.6 km) E of Conistone village. The complex is dominated by the generally E – W-trending veins, and workings continue for 2 miles (3.2 km) to the east, the maximum width of the worked belt being about 1000 ft (305 m). The first ground leased here was in 1686 and as Raistrick (1966, 1973) in his interesting historical accounts shows, the royalties belonged to the Freeholders of Conistone and throughout more than 200 years of working, the operators were small partnerships of miners, in some cases, particularly in the 19th century, with some capital behind them. The field was never unified into a single operation as Grassington was. At one time there were as many as fifty separate small workings. The veins found in the limestones included the Summer Haw and Bell and Dragon veins, and Cocklake Vein, the west end of the northern run of workings. The amount of spoil is small and the veins were evidently lean, but some amber fluorite remains at the level of the Hardrow Scar Limestone. There were closely spaced pits on Bell and Dragon Vein in the Simonstone Limestone and some baryte is present in the Middle Limestone, where Park's Vein runs parallel before turning away to initiate the second main run of workings to the east. In beds that may be the equivalents of the Gayle Limestone, Bell and Dragon Vein gives off a fault trending N70°W which can be followed across Wharfedale, where several limestone scars clearly terminate against it, to join up with the easternmost vein of the Middlemoor vein complex; this downthrows about 30 ft (9 m) S but does not appear to have been prospected for mineralisation. The veins north of Bell and Dragon probably merge into it, but at an unpromisingly low stratigraphical level.

The principal workings at Conistone Out Moor are in the Bearing Grit of the Grassington Grit Formation, here (according to the record preserved in Geological Survey Vertical Sections Sheet 28, 1874), 56 ft (17 m) thick, compared with 91 ft (28 m) at Providence (Figure 13, section 12). Small shafts are very numerous along the Cocklake – Fearnought South and Park's – Mossdale veins, and Corrigan – Hearty Lass String between these main veins can also be followed at surface. The positions of the remaining veins are less obvious and reliance has to be placed on plans in which there are some inconsistencies; in particular the exact positions of Jeffries and Astonishment veins are in some doubt. The E – W veins throw S, the stepping down of the grit-shale contact being visible in both Gill House Beck and Swarth Gill. Main shafts on the northern run were at Moorhead [0068 7033], Silver Rake [0112 7035], near Old Wife Pot Hole, New Fearnought [0150 7038] and Old Fearnought, 800 ft (244 m) to the east. The Cocklake – Fearnought Vein is reminiscent in its trend of the 'Quarter-point' veins in Weardale, which swing from WNW to E – W, the productive parts in general lying on the E – W portions (Dunham and others, 1965), but it is not clear that a similar mechanism worked here. In the Fearnought stretch, where a sharp turn is made from E – W to ESE, two subparallel veins gradually diverging W to nearly 100 ft (30 m) apart were worked. East of Mossdale Beck, the Fearnought Vein joins Mossdale Vein near North Mossdale Shaft [0228 7023], the main operations from which were on Mossdale North Vein, running NNW, Royal Exchange Vein in a similar direction and Mossdale Vein east of Mossdale Beck. The Dreadnought workings appear to have been entered by a level (Figure 22) on the west bank. Other shafts of importance were at Twenty Meers [0177 7024] where Butcher's Rake leaves the southern main vein, at Odd Meres on the same vein, 4700 ft (1.43 km) to the west, at the junction with Peck O'Malt Vein; and Peck O'Malt Shaft [0045 7009] at the junction of the latter vein with Sarginson's Rake. From all these workings and from the numerous bell-pits and shallow shafts, the quantity of mineralised spoil is small. Baryte and calcite are the commonest associated minerals, but amber fluorite can be found between New and Old Fearnought shafts. At North Mossdale the shafts evidently penetrated limestone; they were equipped with horse whims for winding and it is probably safe to conclude that small oreshoots existed in the limestone here. However, by 1871 Royal Exchange and Mossdale North veins had been worked out to the boundary but were proving impossible to follow in depth without expensive pumping; this brought mining to an end east of Mossdale Beck. Meanwhile in 1848 the Conistone Moorhead Company had been formed with a complicated history of ownership (for which see Raistrick, 1973, p. 128) with holdings farther west, including the Moorhead stretch of the main northern vein, Odd Mere to Astonishment on the southern run, and they drove a deep adit [0129 6966] from Mossdale Beck, ¾-mile SW of North Mossdale Mine, at about 1380 ft (420 m) in the Simonstone Limestone. The direction as indicated by Geological Survey records is N28°E, presumably chosen to cut the Sarginson Rake – Dreadnought run east of Sarginson's holdings, and near the junction with Astonishment Vein. The distance to the run would be 2440 ft (743 m), but Raistrick states that the level was 'over 1000 yards' (914 m) long and its course after entering the vein complex is not clear. If it had been driven straight ahead it would just have reached Old Fearnought Shaft (collar: 1570 ft (478 m) OD), but possibly the additional footage was used in opening up towards the west where the company's holdings lay. Since there is sandstone (as well as much limestone and some chert) on the dump, it may safely be concluded that the level passed through at least one of the S-throwing faults and entered the Dirt Pot Grit, but it is equally clear that very little payable ground was found. The level may be referred to as the Leeds Level since a new company was formed by John Gledhill of Leeds in 1863 to work it; but after another reconstruction, the mine was abandoned in 1870.

Raistrick's conclusion after much study is worth reproducing here: 'All records show that none of the Coniston Moor veins were ever rich. All were thin with pockety ore. The veins were suited to the sporadic ventures of the Freeholders who were all farmers to whom the mining was never more than a subsidiary part of their living' (1973, p. 110). Most of the ore came from the Grassington Grit, and not from the 186 ft (56.7 m) of Middle plus Simonstone Limestone made accessible by the shafts at North Mossdale and the Leeds Level.

The dying phases are shown by the lead book of the Kettlewell Mill, 1859–74; Silver Rake produced 205 tons of lead, Dreadnought 6, Fearnought 12, Sarginson 7, North Mossdale 77, the total representing about 500 tons of ore. Official statistics give production of ore for Kettlewell and Conistone Liberties, as 1667 tons, but part, perhaps the larger part, of this would come from Old Providence Mine. The generalised figures given by Raistrick (1973) for royalty payments received by the Freeholders of Conistone between 1805 and 1876, representing one-thirteenth of the value of the lead produced, may be used to indicate that ore (assumed to yield 66 per cent lead metal) production from Conistone Out Moor as a whole must have been of the order of 4000 tons over this period; perhaps a like or somewhat greater amount was won over the previous 150 years.

South-east of North Mossdale Mine lies the extensive cavern system surveyed many years ago by the British Speleological Association, in the Middle Limestone (Figure 23). No veins were found in it, but its existence almost reaching to Mossdale North Vein South may have enabled the shafts at the mine to penetrate the limestone. The next mineralised ground begins on Grassington High Moor, almost 2 miles (3.2 km) to the south.

Bycliffe Vein Lead ore Fluorspar Barytes

This and the veins listed below are on SE 06 NW; Yorkshire 116 SW, SE, 134 NW, NE
Direction N60–90°W, throw 25 ft (7.6 m) S near Richards Shaft

Legrim's Vein

Direction N80°W to N80°E, throw 170 to 198 ft (52–60 m) S

Richards Vein

Direction N85°W, throw 120 ft (37 m) S at Richards Shaft

Alexandra Vein Lead ore

Direction N87°E, southern element of fault trough with throws 30 ft (9 m) S and 45 ft (13.7 m) N

Grimes Groove North Vein = Grassington North Vein Lead ore

Direction N85°W, throw 35 ft (10.6 m) S at Turf Pits crosscut

Standfast Vein Lead ore

Direction N87–75°E

Chatsworth Vein? = New Moss New Vein Lead ore Fluorspar

Direction near E – W

Eddy's Vein Lead ore

Direction N85°W

Coalgrove[1] Head Vein Lead ore Barytes Fluorspar

Direction N85–90°W

Grassington Middle Vein = Friendship Vein

Direction N80°E to N80°W, throw 30 ft (9 m) S at Cottingham Shaft

Lees Vein Lead ore

Direction E – W

Old Rippon Vein Lead ore

Direction N80°W

Grassington South Vein; Hartington Vein Lead ore

Direction N80°W, Branch S70°E

New Rippon Vein = Ripon Vein Lead ore

Direction N64°W

Burnt Ling Vein Lead ore

Direction N65°W, with E – W branches on Green Hill; throw 120 ft (37 m) NE

Stool Veins Lead ore

Many subparallel strings, N70–85°W

South Grimes Groove = Ringleton and Piper Plat Veins = Moss Vein Lead ore Fluorspar

Direction N75–55°W, throw 48 ft (15 m) N at Ringleton Shaft

Castaway = Ripley = Coalgrove[1] Beck Vein Lead ore

Direction N50–55°W, small NE throw

Sixtynine Vein Lead ore

Direction N60°W

[1] Called Coalgroove on older six-inch maps.

Galloping Vein

Direction N50°W

Palfrey = Pawfrey = Porphyry Vein Lead ore

Direction N65°W, small SW throw

North Cavern = Slanter Vein Lead ore

Direction N64–45°W, throw 45 ft (13.7 m) SW beyond Moss New Shaft

Cavendish Vein Lead ore Fluorspar

Direction N57°W, probable small NE throw

Blow Beck Vein Lead ore

Direction N70°W, turning to N55°W

The *Grassington mining area*, the most important in Area 14, lies in the moorland country centred about 1½-miles (2.4 km) NE of the town (Figure 34). It is divided into two parts by the barren Grassington Grit outlier running from Downs Pasture to the site of the former Cupola smelting works; the veins listed above belong to the part north-east of the outlier, known as Grassington High Moor, and the south-western or Yarnbury portion will be described in the next section. Both parts fall within the royalty of the Duke of Devonshire, and after a long period of exploitation by free miners, followed by small companies, from about 1790 the whole field was consolidated under the Duke's agent, and operated as a unified industry until the mines closed in the 1880s. The records, now preserved at Chatsworth House, are extensive and though not complete in every detail (there is, of course, not much information on the free miner's operation) a fairly complete picture can be obtained of the geological and mining situation. The writings of Raistrick (1936, 1955, 1973) and of Raistrick and Jennings (1965) give a fascinating picture of the discovery, laws, mining history and 19th century mechanisation of these mines. The geology is briefly described in Carruthers and Strahan (1923), and the sampling of the dumps during 1941 by Dunham (1952). Some spar minerals were worked from the dumps in the 1930s, and between 1955 and 1964 Dales Chemicals Ltd operated a flotation mill for the production of fluorspar and barytes (Dickinson, 1965). The latter mineral is being produced from time to time from a gravity plant at Yarnbury, but the field is otherwise now abandoned.

In the extensive area of exposed Great Scar Limestone below the level of the Girvanella Band near Ghaistrill's Strid [992 645] a small E–W vein carries pink baryte and a little chalcopyrite. At Lea Green, about 1 mile (1.6 km) north of Grassington, some trials have been made on NNE and ENE fractures which have yielded a little galena, with amber fluorite and baryte: these workings date from the 18th century, but there is one that Raistrick (1973, p. 90) identifies as much older. Farther north, the continuation of Grey Vein of Yarnbury runs NNW across the limestone scars of Conistone Old Pasture, and a mile (1.6 km) beyond this a fault produced by the combinations of Burnt Ling, Legrim's and a branch from Bycliffe Vein forms a feature crossing the Great Scar Limestone to join up with the Grey Fault near Throstles Nest Barn [9791 6946]. Hardly a trace of mineralisation has been found on these faults from which, nevertheless, the elaborate fracture-pattern of the Grassington field springs.

The oreshoots at the High Moor were found on the thirteen WNW to E–W and eleven NW-trending veins listed above (Figure 34), many of which persist for a mile (1.6 km) or more. The westernmost shaft on Bycliffe Vein, called Green Bycliffe [0185 6844] is centrally situated in the outcrop of the Middle Limestone. A party including R. D. Leakey and others from the British Speleological Association surveyed this shaft and an adjacent one in 1941 and entered levels at 39 and 80 ft (11.8 and 24.4 m) below surface, with sump workings 12 ft (3.6 m) below and a second sump going down 60 ft (18 m) further into water. There was stoped ground in the Middle Limestone and the deepest sump may have gone into Simonstone Limestone. The dumps contain baryte, fluorite, a little witherite, partly oxidised galena and some copper staining. At 1530 ft (466 m) farther to the SE, How Gill Shaft (Figure 34, loc. 1) also worked the Middle Limestone, but only a short distance beyond this the vein begins to cross the outcrop of the Grassington Grit, here separated from the limestone by only a thin layer of shale, or in places resting directly on the limestone. This is the stratigraphical situation throughout the field, some sections exposed during stoping showing an irregular surface on top of the limestone (Figure 34). Some small production also appears to have been got in the western part of the area from Porphyry (Palfrey) Shaft [0159 6804] (Figure 34, loc. 2), sunk into the Simonstone Limestone, from the scrins in Middle Limestone at Stool Mine, and from flyers associated with Burnt Ling Vein on Green Hill (where, incidentally, the irregular top surface of the limestone is well exposed). Analyses of the tailings at Green Bycliffe, How Gill and Porphyry are given in Table 17, p. 100, nos. 48–52. The high lead-content is no doubt due to the presence of cerussite or anglesite which was not recovered during dressing.

As the veins are followed into the Grassington Grit, it is quite clear from the great increase in the number of closely-spaced shallow shafts, pits and opencuts that most of the ore was found where the veins cross the lower thick sandstone, the Bearing Grit. This is amply confirmed by the stope-sections of the deeper workings to the east, but here, too, some workable ground was found in the Middle Limestone. Even where such sections are unavailable, as in the ancient workings in the western half of the field, it is not difficult to identify the places where oreshoots were found, on strike, from the clustering of superficial and shallow workings, accompanied by well-mineralised spoil. This has been done and the deeper oreshoots have been included in calculating the following figures: the aggregate lengths of the veins add up to 17.9 miles (28.6 km) while the length of workable ground was about 5.6 miles (8.9 km). The vertical dimension of the oreshoots was determined by the thickness of the hard part of the Grassington Grit Formation – usually the lower sandstone known as the Bearing Grit – less some part of the throw of the vein. In a few cases, Slanter Vein for example, a higher sandstone, the Top Grit, was also mineralised. The stratigraphy of the formation varies considerably even within the small compass of the mining field (Figure 13, sections 15, 17, 18). The grit was at its most favourable development at *Coalgrove Beck Mine*, where in places the thickness of hard sandstone was over 200 ft (61 m), with only two thin shale partings; this gave rise to the most substantial oreshoot at Grassington (Figure 13, section 15), for the ore also continued down 60 ft (18 m) into the Middle Limestone. A thickness of 90 to 100 ft (27–30 m) for the Bearing Grit was more generally worked, and even in this, wedge-shaped incursions of shale were found, for example on Slanter and Hartington veins. The greatest vertical dimension of an orebody between limestone walls was at Sarah Shaft on Cavendish Vein, 110 ft (34 m), but even here it did not prove to be workable to the bottom of the Middle Limestone. The evidence remaining as to the widths of the veins is scanty, but there is nothing on the plans to suggest that they normally exceeded the ordinary width of a drift, 4 ft (1.2 m).

The structural picture is summarised in Figure 34. The largest displacements are on Legrim's and Richards veins (S) and Burnt Ling Vein (NE). As there has been some confusion about nomenclature, it should be said that the vein (and fault) here regarded as Legrim's comes from Seeds Hill and Middle Pasture in the west, and after passing through Palfrey Vein, without displacement, it suffers an apparent shift E side SE of above 400 ft (122 m) on North Cavern Vein. From here a course ENE takes it through Bycliffe Vein, but in the meantime Richards Vein has branched off to the E, taking with it a major part of the throw. A small fault, apparently the continuation of Legrim's, was mapped north of Bycliffe Vein running NE and shifting the outcrop of the Red Scar

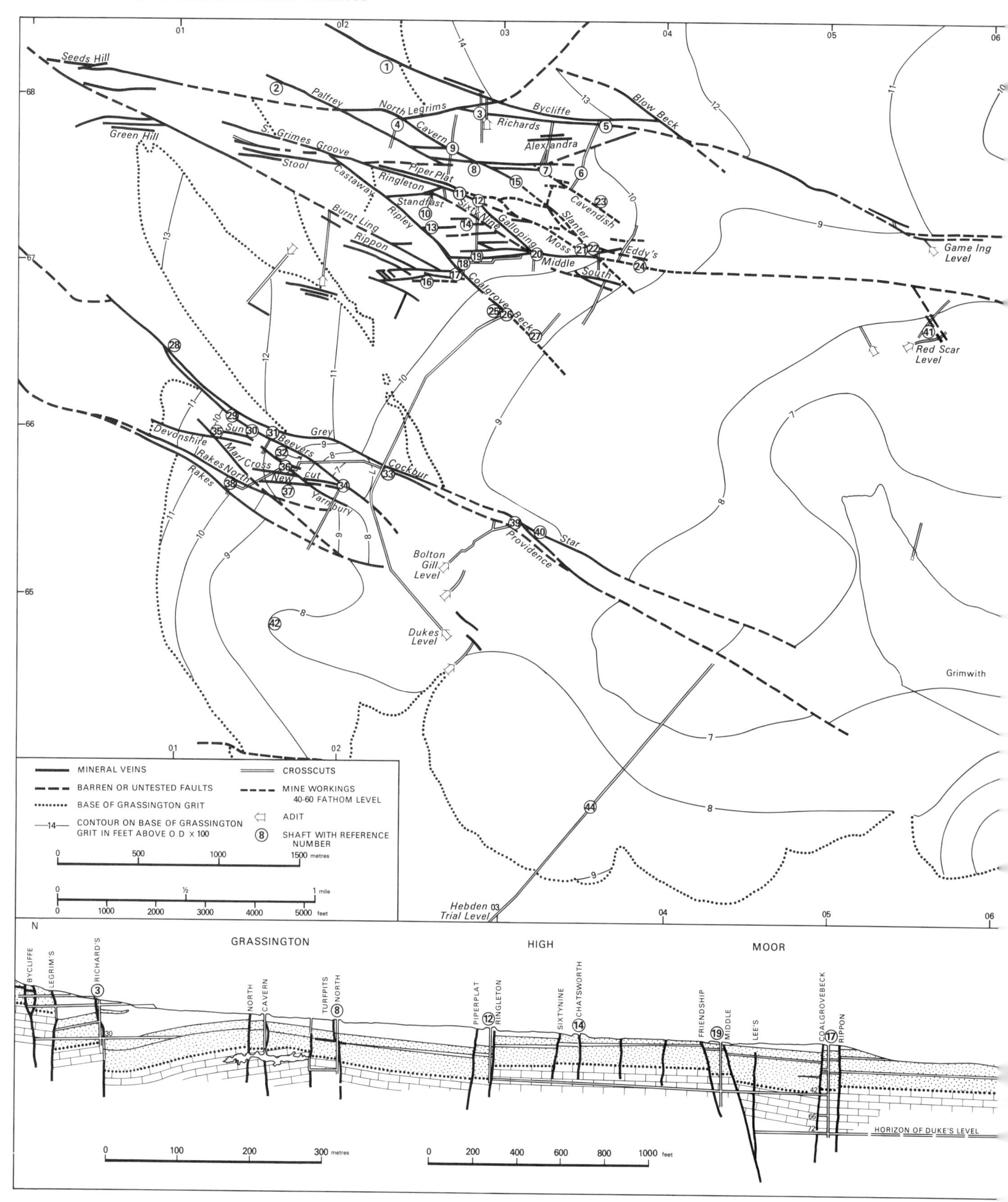

Figure 34 Structure map and sections of principal mine workings, Grassington and Hebden moors. Structure contours on the base of the Grassington Grit in feet above Ordnance Datum. Mining information from the plans preserved at the estate office of the Trustees of the Chatsworth Settlement, Bolton Abbey

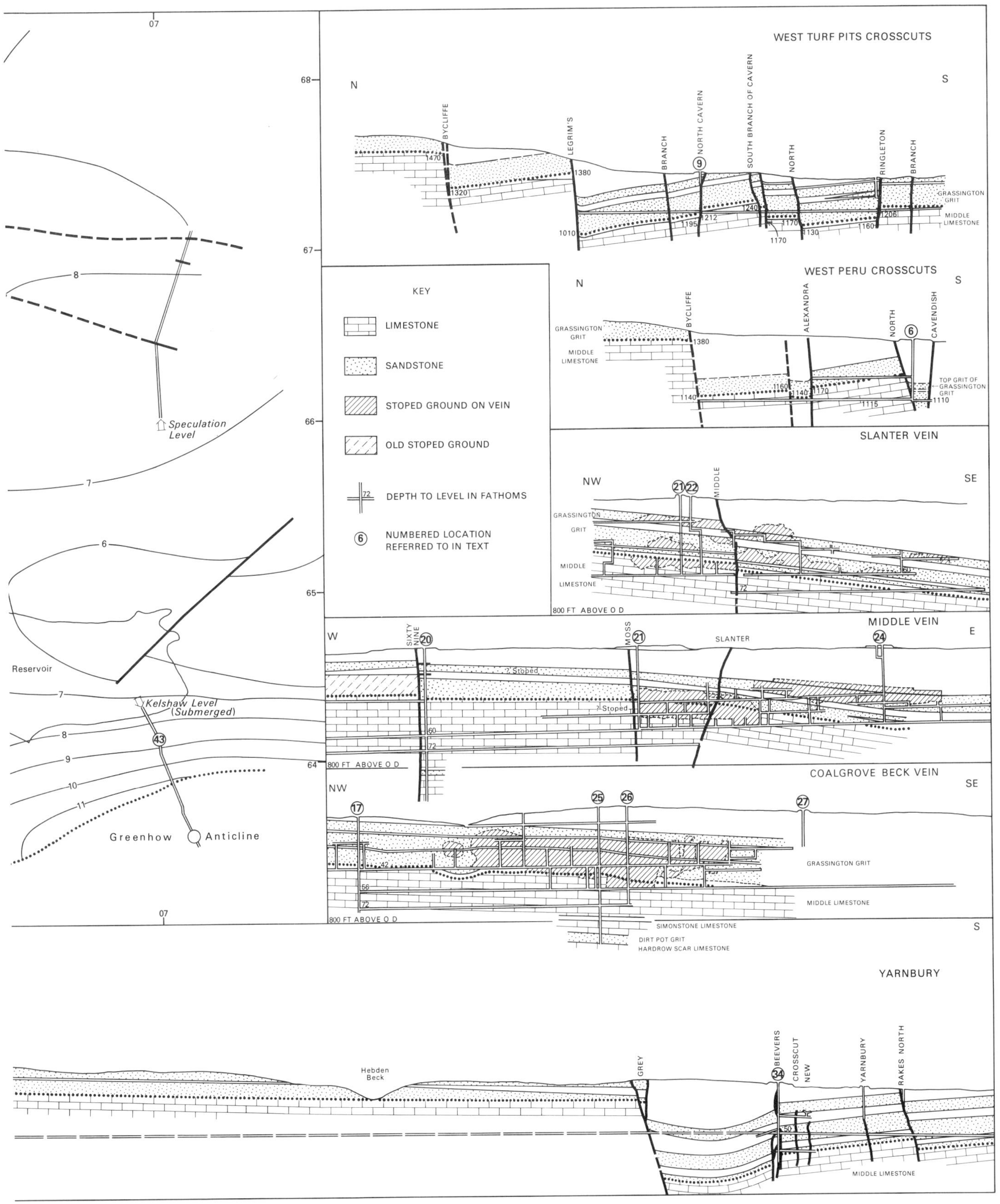

Key to numbered shafts **1** How Gill; **2** Porphyry; **3** Richard's; **5** Legrim's; **5** Peru; **6** Low Peru; **7** West Peru; **8** Old Turf Pits; **9** West Turf Pits; **10** Turf Pits Air; **11** High Ringleton; **12** Low Ringleton; **13** Chatsworth; **14** Glory; **15** Henry's; **16** Worsley; **17** Taylor's; **18** Lee's; **19** Friendship; **20** Coalgrove Head; **21** Old Moss; **22** New Moss; **23** Sarah; **24** Cottingham; **25** Brunt's; **26** Engine; **27** White Hillock; **28** Good Hope; **29** Tomkin; **30** Barratt; **31** Mason; **32** Nixon; **33** Cockbur; **34** Beevers and Union; **35** Devonshire; **36** Bowden; **37** Davis; **38** Rake; **39** Hebden Moor; **40** Cook's; **41** Red Scar; **42** Mire; **43** Kelshaw Aket Air; **44** Hebden Trial Air

Grit. This is the only NE-trending fault in the area for which there is reasonable evidence; such fractures are remarkably lacking in the mines. Burnt Ling Vein was a poor producer, probably because its throw is similar to the thickness of the Bearing Grit which could, therefore appear on only one wall of the vein at any given place. The remaining displacements are all less than 50 ft (15 m), most of them much less, so that good channels could be created with Bearing Grit on both walls. Even so, the oreshoots are not laterally continuous and it is not clear in many cases why they terminated sideways. Vein intersections might explain some cases, but certainly not all. There is no consistent relationship in the apparent shifts; as noted above, North Cavern seems to displace Legrim's Vein; Ripley – Coalgrove Beck Vein is supposed to be displaced by Middle Vein, but a case can also be made for a W side S shift of Middle Vein on the general line of Ripley – Coalgrove Beck. There appear to be sinistral movements on Slanter and Caunter veins (Figure 35). Several veins cross without displacement. It is most satisfactory to conclude that the fracture-systems came into existence simultaneously, but there may have been some local adjustments later, possibly after the minerals had been introduced.

The progress of exploitation of the field can be followed from Figure 34. The early phase of working is represented by the lines of numerous bell-pits and shallow shafts, especially in the areas of outcrop of the Bearing Grit. There is a general ESE dip, illustrated by the structure contours, which carries the Bearing Grit under cover of the Thirtyfive Fathom Shale towards the east, and by the later years of the 18th century, although some deeper shafts were equipped with winding and even with simple pumping equipment—Brunt's Shaft [0296 6670] (Figure 34, loc. 25) and the adjacent Engine Shaft are cases in point; so too are Old Turf Pits [0278 6754] (Figure 34, loc. 8), Peru Shaft [0361 6780] (Figure 34, loc. 5) on Bycliffe Vein and at least one of the Ringleton shafts—a stage was reached when, as Raistrick convincingly shows, output from the field was falling off rapidly because water was preventing development in depth and down dip. The problem was solved by the driving of the deep drainage level known as Duke's Level, some account of which is given in the succeeding section on Yarnbury. For the present sub-area, it may be noted that the level connected with Engine Shaft [0302 6669] (Figure 34, loc. 26) on Coalgrove Beck Vein in about 1830, at 864 ft (263 m) OD, about 50 ft (15 m) above the base of the Middle Limestone and 72 fathoms (128 m) below surface. Thence it was linked to Brunt's Shaft and continued NW to Taylor's and Lee's shafts (Figure 34, locs. 17, 18), thereafter following successively Lee's, Friendship and Coalgrove Head veins eastward to beyond New Moss Shaft (Figure 34, loc. 22). It proved everywhere to be below the bottom of the mineralised ground, but it enabled a series of higher levels to be developed, from which the eastern area of the field was exploited, adding fifty years to its life. These levels, indicated by heavy dashed lines on Figure 34, included:

1. The 56-Fathom on Coalgrove Beck Vein (935 ft, 285 m OD) in Middle Limestone from Taylor's to Engine Shaft, but passing through the unconformity 400 ft (122 m) SE of the latter; no ore was got below this nor beyond 470 ft (143 m) SE of Engine Shaft, but the drive was continued 1230 ft (375 m) beyond this and a crosscut was put out 800 ft (244 m) almost below White Hillock shaft (Figure 34, loc. 27) (which had already proved the upper part of the grit to be barren) without success.

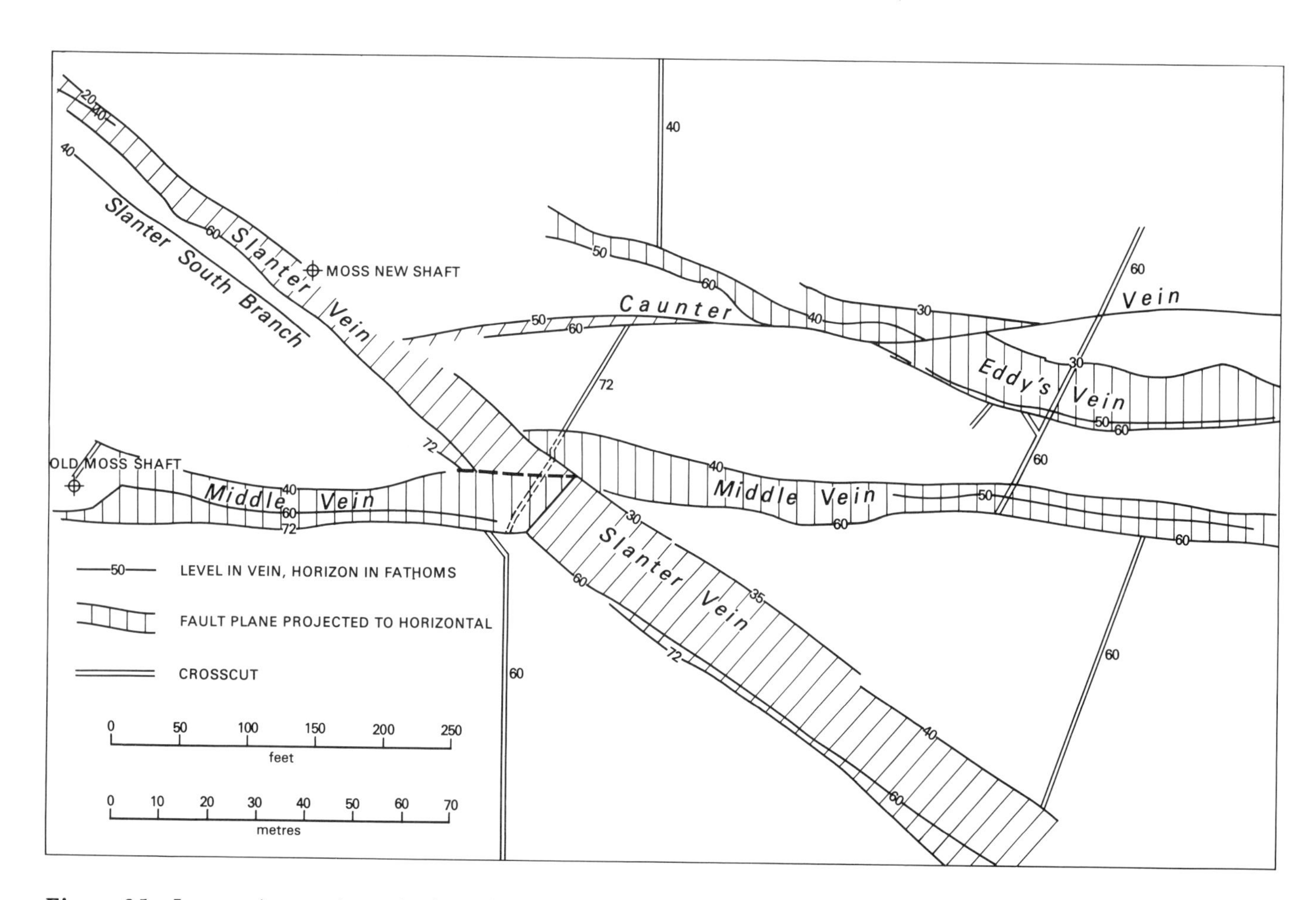

Figure 35 Intersecting oreshoots in the neighbourhood of the Moss shafts, Grassington Moor, to illustrate apparent sinistral movements on Slanter and Caunter veins

2. The 42-Fathom (starting at 1032 ft 315 m OD) followed Coalgrove Beck Vein back to Taylor's Shaft, whence there were westward connexions to Old Rippon and Burnt Ling veins, NW to Chatsworth Shaft (Figure 34, loc. 13) and N to Lee's, Friendship and eventually by means of a long crosscut to Low Ringleton Shaft (Figure 34, loc. 12); beyond Taylor's Shaft much of this drive must have been in limestone, the level reached at the bottom of Ringleton Shaft being 1080 ft (329 m) OD showing either that the crosscut had a higher gradient than normal, or that there were steps up in its level not indicated on the plan; three small veins including North Gregory were found between Friendship Shaft (Figure 34, loc. 19) and Chatsworth Vein and were explored for a few hundred feet, and a connexion was made to Glory Shaft (Figure 34, loc. 14) on the latter vein.
3. North and east of Ringleton, the 42-Fathom linked by means of a rise with a higher level, the 50-Fathom of Henry's and New Moss shafts (Figure 34, loc. 15, 22), at a little above 1100 ft (335 m) OD. This provided a link by way of the W – E portion of Piper Plat Vein to Slanter Vein, where there was a level above designated the 40-Fathom at about 1155 ft (352 m). This continued NW beyond Henry's Shaft to the beginning of the cavern ground (described below) and to North or New Vein; and SE to rises close to Moss New Shaft (note that this is not the same as New Moss Shaft, Figure 34, loc. 22) coming up from the Moss 60-Fathom Level at 1010 ft (308 m) OD and thence draining to the 72-Fathom.
4. There was a linked development area including West Peru, Low Peru and Sarah Shafts, with the main level, called 40-Fathom, at 1100 to 1110 ft (335–338 m) OD. Crosscuts N gave access to Alexandra and Bycliffe veins (the latter either barren or worked out), and around Sarah Shaft the important Cavendish oreshoot. This was followed down to Sarah 56-Fathom Level at 1000 ft (305 m) OD so that pumps were certainly installed at this shaft. What is not clear is whether the 40-Fathom Level drained back to Slanter or Ringleton; there is a crosscut N from Slanter towards No. 2 Caunter Vein of Low Peru but neither version of the plan shows this connected through, nor is it clear whether it comes from the 40- or 50-Fathom level of Henry's Shaft. Perhaps the driving of the 60-Fathom N crosscut from Cottingham Shaft (described below) which crosses the line of Cavendish Vein without finding an extension of the orebody, provided drainage to the 72-Fathom by that route; or perhaps the pumps at Sarah were able to deal with the flow.
5. Finally the Middle Vein was developed by means of the 42-Fathom from Engine and Taylor's Shaft as far east as Coalgrove Head Shaft; this was linked to Old Moss Shaft by a 50-Fathom Level, but beyond this the main level was the 60-Fathom at 960 ft (293 m) OD, giving access to the Cottingham, Eddy's, South and Hartington oreshoots, and draining back to the important system of rises from the 72-Fathom at [0359 6701].

In none of these bottom levels is there any evidence of a substantial oreshoot continuing beneath the level. Towards the south-east, the Bearing Grit dips below these horizons, but in every case the worked oreshoots have terminated against a more or less vertical east or south-east boundary. This in no way precludes the possibility that undiscovered oreshoots exist farther down dip, but it is important to make the point that the evidence, reasonably well documented, points to the fact that these mines were exhausted, and that the bottoms of the known oreshoots lay substantially above Duke's (72-Fathom) Level.

Two trials were nevertheless made below Duke's Level, in the hope that limestones lower in the sequence might prove to have trapped ore, even though the Middle Limestone had already turned barren. The evidence elsewhere in the Northern Pennine Orefield shows that this was a reasonable possibility to test. Brunt's Shaft (Figure 34, loc. 25) at Coalgrove Beck was sunk to 694 ft (212 m) OD, 170 ft (52 m) below Duke's Level, passing through the Simonstone Limestone and Dirt Pot Grit into the upper part of the Hardrow Scar beds. Coalgrove Head Shaft (Figure 34, loc. 20) was continued to 217 ft (66 m) below Duke's Level, bottoming at 656 ft (200 m) OD, 109 ft (33 m) below the base of the Dirt Pot Grit, and thus perhaps in limestone equivalent to the Gayle or lower part of the Hardrow Scar. Neither trial is adequately documented, but the fact that no attempt was made to drive off at Coalgrove Beck, while at Coalgrove Head the deepest level was a short drive called the 60-Fathom at about 1050 ft (320 m) OD, considerably above Duke's Level shows that neither investigation was successful. J. R. Dakyns examined samples from the Coalgrove Head sinking during the primary survey in the late 1860s, so that it must have been in progress or recently completed at that time. Veinstuff was seen in dark limestone at 559½ ft (170.5 m), 564 ft (178 m), 607 ft (185 m) and 637½ ft (194 m) depth below surface, but calcite is the only mineral mentioned.

West of the area of deep levels linked with Duke's Level, there were two other significant trials during the 19th century. From West Turf Pits Shaft [0267 6767] (Figure 34, loc. 9), crosscuts were driven N and S close to 1200 ft (366 m) OD, exploring the ground from Legrim's Vein in the north to the Sixtynine Vein in the south (Figure 34 illustrates the geological results). The S crosscut cut across the western end of the cavern system associated with North Cavern Vein, which has already been illustrated in Figure 23. Very large numbers of bell-pits not obviously related to veins in this area suggest that some may have been sunk into the caverns to collect residual galena from the cave-fill, or there may have also been some flats. There is however no mention of this form of deposit in Grassington records. It is difficult to judge how much new production came from this crosscutting; Legrim's was evidently not producing at the depth penetrated, though, with the Grassington Grit on the footwall, it was well mineralised near surface. The limestone on the footwall of Legrim's was penetrated by the crosscut 180 ft (55 m) below the unconformity and thus probably near the base of the Middle Limestone. The other important development was the sinking or reopening of Richard's Shaft, near Grassington No. 2 Level, with crosscutting S at the 30-Fathom horizon, about 1285 ft (392 m) OD, a section of which appears in Figure 34. Drives were carried through Legrim's and Bycliffe veins, the latter already known from shallower adits to include a complex horse or loop here. The 30-Fathom Level was then extended along Bycliffe Vein nearly ½-mile (0.8 km) W to How Gill Shaft. No stope section remains, but it is possible that much of this ground had already been extracted by the free miners.

The pattern of long crosscuts at the High Moor, ranging from 1280 ft (390 m) OD down to 935 ft (285 m) (or 864 ft (263 m) if Duke's Level is added), mostly commanding the bearing bed in the Grassington Grit and for long stretches exposing the underlying Middle Limestone, is far more comprehensive than anywhere else south of the North Swaledale Mineral Belt; but as will be shown later, it does not rule out the possibility of an extension of the field under cover.

Production data will be discussed after the Yarnbury mining area has been described.

Yarnbury North = Grey = Cockbur Vein Lead ore
SE 06 NW; Yorkshire 134 NW, NE
Direction N45°W, N87°W, throw 175 to 350 ft (53–106 m) SW

Contre = Beevers Old Vein Lead ore Barytes
Direction N45–65°W, throw 60 ft (18 m) NE near Beevers Shaft

Sun Vein Lead ore Barytes
Direction N80°W

Devonshire Vein
Direction N80°W, throw about 20 ft (6 m)S

Yarnbury Vein Lead ore Barytes
Direction N55–65°W, throw 35 ft (11 m) SW

Crosscut Vein Lead ore

Direction N80°W, little displacement

Beevers New Vein

Direction N85°W

Marl Vein Lead ore

Direction N40°W, throw 24 ft (7 m) NE

Rakes North Vein Lead ore

Direction N65°W turning to E–W, throw 30 ft (9 m) N changing to 40 ft (12 m) S

Rakes Vein

Direction N65°–45°W, throw 30 ft (9 m) N changing to 43 ft (13 m) S

Although the earliest mining in the Grassington field, following the arrival of miners from Derbyshire about 1603 was on the Low Moor, the *Yarnbury mines* lying SW of the grit outlier between Downs Pasture and Cupola assumed their most productive period after the driving of Duke's Level. Evidence of mining to shallow depths from opencuts and pits testifies to the period prior to 1790, but some large oreshoots, not far below the surface, were not found until development in depth and the sinking of new shafts to the Duke's Level horizon took place. As on the High Moor, the displacement of the Grassington Grit – Middle Limestone boundary by the veins as they cross the western outcrop is easy to follow. The only productive workings west of this boundary in the limestone, were at the Good Hope shafts [0093 6645], where calcite is the main constituent of the tailings. Some 1350 ft (411 m) to the NW, the displacement on Grey Vein has almost died out, but it is taken up by an *en échelon* fracture lying about 750 ft (229 m) to the SW, joined to the expiring Grey Vein by an E – W link. Trials here found only a little galena and baryte. The Rakes Vein fractures can also be traced as they cross the Grassington Grit – Middle Limestone boundary ½-mile (0.8 km) S of Good Hope shafts, but here, pits into the Middle Limestone and the top of the Simonstone found little. Upon entering the grit, Rakes North Vein appears to have carried a shallow oreshoot extending E towards its junction with Marl Vein; baryte and hemimorphite occur in the spoil. Less extensive workings exploited Rakes Vein. The powerful Grey Vein can be followed SE, separating limestone from Grassington Grit to the south by shallow cuts and pits as far as Cockbur plantation. Immediately south of the fracture there is evidence of many subparallel breaks in the grit, some of which have yielded ore; Contre Vein is the most important of these, uniting with others towards the SE and diverging as Beevers Old Vein, in the extensively disturbed ground around Yarnbury House. A few pits also mark the site of Devonshire Vein. Like the High Moor mines, those on the Low Moor were in difficulties with water in the later years of the 18th century and Raistrick (1973) has shown how the rapid fall in output was threatening the future of lead mining in both parts of the Grassington field.

Duke's Level driven at 850 ft (259 m) OD from the west side of Hebden Beck [0264 6479] was planned by Cornelius Flint from Longstone in Derbyshire and started in 1792. Originally conceived as a boat level, the large cross section was abandoned after John Taylor succeeded Flint as the Duke's chief agent in 1818, by which time it was nearing the Yarnbury veins; the remainder of the drive and crosscuts were the normal 7 × 4 ft (2.1 × 1.2 m) size (Raistrick, *op. cit.*). The course of the level is shown on Figure 34. Unfortunately no details of the geology have come to light among the records, but the general situation is not difficult to deduce. The portal is in the Bearing Grit and the first 1550 ft (472 m), to the point where Rakes Vein was cut, crosses obliquely the very gentle anticline in this formation also penetrated by the south crosscut from Beevers Shaft (p. 199 and Figure 34, loc. 34). There is no record of either rising or driving on Rakes Vein, which must have been barren here; some surface trials on this vein short distances west of the line of the tunnel were also unsuccessful. Surface mapping suggests that an upthrow of the beds to the north was found on Rakes Vein, as in Beevers crosscut. Shortly the drive was turned to NNW and after 1275 ft (389 m) Beevers Vein was cut. The northward dip of the measures may well have carried the Bearing Grit below the level at this point and since this vein throws NE still higher measures would be entered north of it. A rise on the vein failed to find workable ore. After driving close to this vein, another turn was made, this time to NNE and after 836 ft (255 m) Cockbur Vein, the extension of Grey Vein, was encountered. The ground crossed between Beevers and Cockbur was the east side of the small but deep basin lying north of Beevers Shaft, and shale above the Grassington Grit Formation may have come into the level here. The Cockbur structure was found to consist of a number of subparallel fractures, one of which was explored at tunnel level for 850 ft (259 m) to the ESE and 400 ft (122 m) to the WNW. Some ore was produced here, and a connexion was made with the surface by means of Cockbur Shaft. From Cockbur the main tunnel continued on the NNE line for 1925 ft (587 m) to an air shaft near Cupola Mill. Two further changes of direction, to NE and then to ENE were made before Engine Shaft on Coalgrove Beck Vein was reached after 2090 ft (637 m) of additional driving. The effect of the Grey – Cockbur Fault was to bring Middle Limestone into the tunnel on the N side, and surface and underground data are consistent with the view that the drive followed the strike of this formation around the curving margin of the basin, as indicated on Figure 34, about 50 ft (15 m) above the base of the Middle Limestone. The length of Duke's Level to Engine Shaft is 1.45 miles (2.34 km) and the additional drivage on Coalgrove Beck, and the various elements of Middle Vein to the end beyond New Moss Shaft totals 4350 ft (1.33 km).

Returning to Yarnbury, an extensive system of drives at Duke's Level horizon was made to open up this area in depth in spite of the disappointing showings where the tunnel cut through the principal veins. No doubt ore had been left in the soles of a number of workings farther west as water became more difficult to deal with prior to the driving of Duke's Level. From the WNW forehead on Cockbur Vein a crosscut known as Black Drift was driven W to Beevers – Contre Vein, and one branch linked up successively Nixon, Mason, Barrett and Tomkin shafts (Figure 34, locs. 32, 31, 30, 29) as the 50-Fathom Level, probably ending 150 ft (46 m) below the intra E_1 unconformity, in Simonstone Limestone on the footwall of this vein. The oreshoot here was confined to the Bearing Grit. East of Nixon Shaft another branch led to Bowden Shaft and thence to Rakes Shaft (Figure 34, locs. 36, 38), with a further branch following Crosscut Vein back to Beevers Shaft, from which a crosscut at the same horizon had been driven 1450 ft (442 m) SSW towards the newly-sunk prospecting shaft known as Mire Shaft (see below) by the time the mine was closed down. Drivage from Duke's Level to develop the Yarnbury area thus totalled 1.04 miles (1.66 km) making the whole length of the Duke's Level system 3.31 miles (5.33 km).

At Yarnbury the succession in the Grassington Grit Formation differs from that found at the High Moor, the available sections being summarised as follows:

	ft	m
Exact position of top of the formation uncertain		
Top Grit	48–60	15–18
Shale with thin coal in centre	42–60	13–18
Bearing Grit	60–75	18–23
Shale	19–28	6– 9
Hard Grit	16–20	5– 6
Shale	9–22	3– 7
Limestone		
Mean thickness	225	69

At Mire Shaft [0158 6482] (Figure 34, loc. 42), not taken into account in the summary above, the Bearing Grit has thickened to

96 ft (29 m) and the coal rests almost directly on it, indicating a return to conditions comparable with High Moor.

The stope sections preserved in the Cavendish Estate records show that all three sandstones in the Grassington Grit Formation carried ore but that none was found in limestone. The largest oreshoot was on Contre Vein, from Mason Shaft crosscut NW to 150 ft (46 m) beyond Tomkins Shaft, with stope area 1045 × 100 ft (319 × 30 m), less a small block between Barrett and Mason shafts. Although drained by Duke's Level, 150 ft (46 m) below, the working level at the bottom of the oreshoot was the 20-Fathom Level at about 1015 ft (309 m) OD. The Contre Vein was tested also with a 35-Fathom level, but no stoping is shown related to this. The ore from the Contre oreshoot was brought to surface by means of an incline to the 20-Fathom Level, the portal of which bears the date 1828. To the north, Grey or North Vein was tested not only at the 50-Fathom (= Duke's) Level, but also at the 20-Fathom between Nixon and Mason shafts, but very little stoping ground was found and here the old miners had virtually exhausted the vein. The footwall (N) of Grey Vein where penetrated by the NE crosscut from Mason Shaft at 855 ft (261 m) OD was 325 ft (99 m) below the base of the Grassington Grit, and must have been in limestone equivalent to the Hardrow Scar. This represents a third deep underground test of the limestone, but it was no more successful than at Coalgrove Head and Coalgrove Beck. Beevers Vein, to the south-east, may be regarded as the continuation of Contre Vein, but no doubt it is also linked with Grey Vein by oblique fractures. Two separate oreshoots were worked on it after the early 19th century development. One, lying mainly north-west of Beevers Shaft was 450 ft long, with a maximum height of nearly 100 ft (137×30 m) and must have been related to the Top Grit on both sides of the vein. The other, 250 × 60 ft (76 × 15 m) was influenced by the Bearing Grit, and continued below the 50-Fathom Level. The shaft was accordingly sunk to 770 ft (235 m) and equipped with pumps, and a winze was carried down 60 ft (18 m) further, testing the Middle Limestone on the footwall side of Beevers Vein. A crosscut was driven back to Grey or Cockbur Vein from the shaft bottom and linked with Duke's Level by means of a rise on or near this vein. This crosscut follows an erratic course, no doubt because it was driven across the small but deep basin-like structure indicated by the cross section of Beevers Vein, which carries the base of the Grassington Grit Formation down to 620 ft (189 m) from above 1000 ft (310 m) OD at Nixon Shaft, before it rises again to about 780 ft (238 m) OD (calculated from the top of the formation) east of the line of Duke's Level. The disturbance on Cockbur or Grey Vein may have been too great here, perhaps plugging up the fissure with gouge, to provide good ground for mineralisation. Also near the bottom of Beevers Shaft, a ribbon oreshoot on Crosscut Vein rises to the west with the Bearing Grit, matching exactly the thickness of this bed. This was worked 270 ft (82 m) E of the shaft, and 700 ft (213 m) W, by which stage the base of the grit had passed above the 50-Fathom Level. A second oreshoot, 290 × 60 ft maximum (88 × 18 m) was worked on this vein near Bowden Shaft. Devonshire and Sun veins carried oreshoots respectively 300 and 700 ft (91 and 213 m) long opposite the Bearing Grit, and after these veins united to form Yarnbury Vein, an oreshoot 530 ft (162 m) was found, probably at the Top Grit horizon. Yarnbury Vein cuts through Crosscut Vein, shifting it E side S, but only a short distance to the south it imparts an E side N shift to New Vein. Marl Vein, about which no details are available, appears to shift Devonshire Vein. The trial on the two Rakes veins from the 50-Fathom Level seems to have found no workable ground. The driving of Duke's Level confirmed earlier results from lines of pits into the Grassington Grit, that Beevers and Rakes veins become unproductive before they reach the Hebden Beck; only Cockbur gave promise of continuity (see Hebden Moor, p. 200).

Baryte, generally pink and close grained, was the principal gangue mineral in the oreshoots worked for their galena, but calcite and fluorite were also present. Tailings from Good Hope, the former dressing floor near Mason Shaft, and from Beevers, gave $BaSO_4/CaF_2$ ratios respectively of 4.2, 6.7 and 0.7 respectively, the lower ratio at Beevers corresponding with the deeper penetration. The zinc figures were respectively 2.0, 1.1 and 1.2, while loss of lead, which ranges from 2.1 to 3.6 per cent in the tailings, points to considerable oxidation or ineffective dressing. The same gangue minerals are found throughout the High Moor, and the $BaSO_4/CaF_2$ ratios form an interesting comparison with those for Yarnbury (see Table 17 for analytical data); the tailings around the north, west and south margins of the High Moor area show baryte dominant: Green Bycliffe, 3.4; Howgill, 3.7; Richards and Bycliffe Levels, 3.3; Palfrey, 2.5; Coalgrove Head, 2.4; Coalgrove Beck, 4.0, whereas there is an area, east of centre, where fluorite predominates, as represented by Turf Pits 0.6, Ringletons 0.1, West Peru 0.5 and Sarah 0.6. Zinc varies from 0.2 per cent at the Ringletons to 1.2 at Coalgrove Beck mill site, but no correlation with the baryte/fluorite ratio is evident. The presence of mercury minerals at Grassington was established by the late A. W. Kingsbury (1968), who found cinnabar, metacinnabarite and calomel at Turf Pits together with the rarer oxidation zone minerals minium (after cerussite), rosasite, jarosite and plumbojarosite, and the zinc phosphates parahopeite and spencerite. At West Turf Pits he recorded argentojarosite, with yellow jarosite, while the Ringleton heaps yielded this mineral as well as cinnabar. Raistrick (1973, p. 89) notes the presence of aurichalcite and smithsonite all along Castaway – Ripley vein and the abundance of hemimorphite on Rakes Vein; he also found minium coating baryte plates among the debris from Ripley Vein. There is no doubt that the oreshoots in the Bearing Grit worked in the western half both of High Moor and Low Moor were well within the oxidation zone, but it is difficult to say where the lower limit of this may have been. The fluorite at Grassington, where not coated with oxidation minerals of iron and manganese, is clear and colourless, or pale amber; the dark purple variety, according to Raistrick, was found at one locality along Middle Vein. Determinations of yttrium-contents are cited in Table 11.

A more complete picture of lead ore and lead production from Grassington can be obtained than for any other mining area south of the area described by Dunham (1948), thanks to the records preserved by the Cavendish Estates. From 1735 on there are fairly continous records of lead smelted, with only a few years missing, and for these, average figures have been inserted. Indications given by Raistrick of occasional receipts and of smelt mill capacity in the earliest days suggest that an output of 50 tons per year during the period 1629–1700 might be a reasonable assumption, rising to 118 tons average in the years 1701–34. From 1735 to 1800 the records plus some insertions give the following total:

Production of lead at Grassington

	Tons metal
1620–1734 (estimated)	8 000
1735–1790 (1754–63 estimated)	23 257
1791–1844 (1835–44 estimated)	25 424
1845–1881 (Hunt's statistics)	23 258
	79 939

If recovery of metal from the ore is taken at 60 per cent prior to 1844, this being the actual recovery in 1845–49, the output of concentrates from 1620 to 1844, totals 93 134 tons, to which must be added 33 322 (Hunt's figure) for 1845 to 1881, making a grand total of 126 456 tons. Silver was extraced from the ore only during the last four years' life of the mines, when the recovery was 3.72 oz. per ton lead metal, a figure previously regarded as too low to justify desilverisation of the lead.

From the Estate records, it is of some historical interest to note that Castaway Vein produced very little after 1753, and Piper Plat is last mentioned in 1739. Bycliffe Vein yielded 2179 tons metal between 1739 and 1820, but only 168 during 1821–33; the performance of Burnt Ling Vein was similar, 957 tons and 43 tons respectively. Coalgrove Heads Mine on Middle Vein was the leading producer of the period 1752 to 1819 with an output of 5277 tons metal, but production was only 120 tons for 1821–33, while Coalgrove Beck increased from 671 to 946 tons, though in 1833 the main period of extraction of the Coalgrove Beck Oreshoot, the largest in the area, still lay ahead. On Middle Vein, interest shifted from Coalgrovehead eastward to the Moss shafts and Cottingham; the former are first mentioned in 1764 and up to 1820, 1038 tons metal had been produced, rising to 1533 in 1821–33. In these records, however, the effect of driving Duke's Level can best be appreciated from the rise in output at Yarnbury (Low Moor) from only 711 tons metal for 1735 to 1820, when the level connected with Bowden Shaft, to 3007 tons in 1821–33.

Since 1881, if any lead has been produced, it has been as a product of processing tailings to extract barytes and fluorspar.

Hebden Moor Bycliffe Vein Lead ore

SE 06 NW, NE; Yorkshire 116 SE
Direction N70°W, turning to E – W, throw about 200 ft (61 m) S

Coal Hole = Middle Vein

SE 06 NW, NE; Yorkshire 134 NE
Direction E – W to N70°W, throw about 35 ft (11 m) N

Red Scar Vein

SE 06 NE; Yorkshire 134 NE
Direction N35°W, throw said to be 96 ft (29 m) SW

Star Vein Lead ore

SE 06 NW; Yorkshire 134 NE
Direction N64°W, throw 15 ft (4.6 m) S

Cockbur Vein

SE 06 NW; Yorkshire 134 NE
Direction N47–65°W, throw 130 ft (40 m) S

The Grassington royalty of the Duke of Devonshire marches to the east with that of the Freeholders of Hebden Moor, the boundary crossing Coalgrove Beck north of Duke's Level (the first part of which was driven through Hebden property) and crossing the Bycliffe Vein a short distance east of its intersection with Blow Beck Vein. There is evidence that Bycliffe and at least three other veins from Grassington Moor continue into Hebden Moor, but all course through areas where the Grassington Grit Formation is concealed beneath the Thirtyfive Fathom Shale (Nidderdale Shales) or younger formations such as the Red Scar Grit and Lower Follifoot Grit. The only successful mine in Hebden Moor, to which the present section is devoted, was on a continuation of Cockbur Vein from Yarnbury, where it was possible to follow the Bearing Grit under cover by means of an adit from Coalgrove Beck. However, it became obvious once the stratigraphical relations of the Grassington Moor oreshoots were worked out more than a century ago that the area north of Hebden offers important targets for prospecting through the cover of younger rocks and considerable efforts were made during the 19th century, though without decisive results. Modern prospecting technology, using geochemistry, geophysics and the diamond drill have not been applied to the area, probably because the main streams draining it, Gate Up Gill and Trunla Gill are both catchments for Grimwith Reservoir, situated in the south-east part of the Moor.

As indicated on Figure 34, Bycliffe and Richards veins impinge on Blow Beck Vein, but east of this intersection only a single fracture is mapped across Groove Gill Rigg, carrying the considerable combined displacements of both. The fault can be mapped for more than 2 miles (3.2 km) across Hebden and Appletreewick moors before it joins up with the Merryfield – Providence structure of Ashfold Side Beck, described under the Greenhow district (p. 206). At 1250 ft (381 m) ESE of the intersection, a line of pits was sunk into the Grassington Grit on the footwall side, but the debris contains no veinstuff for the next 750 ft (228 m) where baryte and amber fluorite with a little galena begin to appear. Ore was mined from two opencuts as indicated on the north-east side of Groove Gill, and the Game Ing Level [0562 6711] runs west at 1150 ft (351 m) OD from the side of Gate Up Gill, following first a southern branch of Bycliffe Vein, then a northern one, for over 1500 ft (457 m). The footwall in the opencuts and level is the Nidderdale or Thirtyfive Fathom Shale, with thin beds of flaggy sandstone, and the hangingwall is Red Scar Grit. Even at this horizon, some lead ore production was evidently obtained, but how much is not known. Although some small shafts are associated with the level, there is no evidence that the Bearing Grit on the hangingwall was reached by any of these. Baryte is present on the dump of the level, with grit and much shale. Apart from a few pits on the east side of Gate Up Gill, there is no further evidence of prospecting along the vein for the next mile to the east, until the intersection in Speculation Level is reached. This level, driven prior to the middle of the 19th century, starts in a small outlier of beds above the Red Scar Grit near Trunla Hill; fragments of the Colsterdale Marine Beds can be found on the heaps. It is documented by a section which accompanied a prospectus issued by the Grimwith Mining Company in 1863 (Dickinson, 1966 and information supplied to Mr Tonks by Dr Raistrick in 1940). Starting at 1144 ft (349 m) OD, it runs 1425 ft (435 m) N4°W to a shaft on Hard Hill. The hill rising ahead of the level is effectively a dip slope on Red Scar Grit, so that before Hard Hill is reached, the level passes through the grit into shales beneath. At Hard Hill a pair of faults were cut, the southern one throwing N, and the northern one, identified from its trend as the Middle Vein of Grassington Moor, throwing S. Several shafts were sunk to test this vein on strike, finding baryte in the grit, but not much galena. Although it is difficult to be certain, it does not appear that any of the shafts were sunk to the Grassington Grit, which would have required some 400 ft (121 m) of sinking. From this point the Speculation Level turns to N20°E and continues for a further 2300 ft (701 m). North of Middle Vein, the dip of the strata changes abruptly to 15°N and the level again passed through the Red Scar Grit. In this formation, Coal Hole Vein was cut 975 ft (297 m) N of Middle Vein, throwing a few feet S; from its trend this could be a branch from Middle Vein leaving it farther west. No work seems to have been done on it. Bycliffe Vein was cut where expected, 2175 ft (663 m) from Middle Vein, but with Nidderdale Shale on both sides of the fault, it proved to be unmineralised, and nothing further was found to the north end of the level. Experts including Nathan Newbold of Greenhow, who contributed to the Grimwith Company's prospectus, maintained that Speculation Level was driven 'too high in the beds' a conclusion that experience at Grassington, though not at Providence, would endorse[1]. The geological interpretation of observations made in this level for the issue of the prospectus agrees well with J. R. Dakyns' later findings, when he carried out the primary six-inch geological survey about 1870. However another geological section, on the line of Gate Up Gill, accompanying the prospectus printed by J. F. Mosser of Leeds, is less satisfactory in that the grit exposed at Bullfront Waterfall [058 674] is identified in effect with the Grassington Grit Formation, whereas in fact it is the Red Scar Grit, with the characteristic Colsterdale Marine Beds exposed on top of it; the same is true of a section from the Grimwith Company dated 1876

[1] Turam and VLF observations, following up airborne EM measurements made in connexion with the BGS Mineral Reconnaissance Programme for the Department of Industry have shown a belt of anomalies, in part double, following the south side of Bycliffe Vein and its north branch, between [060 672] and [080 672], 1¼ miles (2 km) long. These could indicate mineralisation in the Grassington Grit at depth, but they have not yet been drilled (Wadge, Bateson and Evans, 1983).

along the line of Gate Up Gill, preserved in the Health and Safety Executive Collection, no. R.121 F.

The Company's main effort was the driving of *Red Scar Level* at 1060 ft (323 m) from Gate Up Gill (Figure 34), the original object of which seems to have been to prove the Middle Vein. A careful section of this (Health and Safety Executive R.121 F) shows that it was driven in shale below the Red Scar Grit. Numerous minor fractures were cut, but from 505 to 565 ft (153–172 m) from the portal it passed through a strong disturbance trending NNW which was believed to downthrow 16 fathoms (29 m) SW, although a sump sunk 80 ft (24 m) to a sandstone believed to be the Top Grit of the Grassington Grit Formation showed only 8 ft (2.4 m) throw in this direction. It is possible that the supposed large throw was worked out from the misleading section along the Gill mentioned above, for the Red Scar Vein fails to produce any noticeable shift in the Red Scar Grit features on either side of the valley. The level was continued 415 ft (126 m) beyond this vein, but did not reach the position of Middle Vein. A shaft was then sunk on the vein [0558 6661] proving the following section:

	ft	m
Drift (called 'gulf')	66½	20.3
Shale with girdles (sandstone)	100¾	30.7
Coal	¾	0.2
Sandstone ('Top Grit')	32¼	9.8

From close to the bottom of the shaft, drifts were driven 300 ft (91 m) SSW and 700 ft (213 m) NNE, the latter passing through Red Scar Vein, which was evidently mineralised since baryte can be found on the dump. The northern drift was about 300 ft (91 m) from the position of Middle Vein, which is considered to pass through Red Brow Sike, when it was abandoned, and evidently Red Scar Vein was not sufficiently encouraging for any driving to have been done along it. It is nevertheless interesting that it appears to be another of the NNW set of veins, paralleling Coalgrove Beck, Slanter and Cavendish at Grassington Moor. The geophysical surveys mentioned on the previous page also showed an anomaly corresponding with and extending beyond the known extent of Red Scar Vein (Wadge, Bateson and Evans, 1983, p. 35). A start was also made with a projected long adit at 925 ft (282 m) OD, the Yorke Level, from Hollin Kell [0552 6528], near the western end of Grimwith Reservoir, but this did not get very far. The company was wound up in 1876, having produced only a few tons of ore.

Meanwhile from 1853 on, in that part of Hebden Moor adjacent to the Yarnbury mines, an intensive search for the continuations of the veins had been in progress under W. S. Winn (Raistrick, 1973, p. 82). Numerous pits across the line of Beever Vein on the east side of Coalgrove Beck had failed to reveal it, but in Bolton Gill, pits and opencuts revealed the continuation of Cockbur Vein. The Hebden Moor Mining Company, formed in 1856, sank and equipped a shaft high up on the side of the gill [0307 6543] (Figure 34, loc. 39); a small vein called Chance Vein was investigated near the shaft, but a drive to the E opened up Cockbur Vein 250 ft (76 m) from the shaft. This was identified from its substantial S downthrow; the direction implied a turn towards the SE from the average direction of Grey – Cockbur Vein (Figure 34). It is not clear how much ore was mined from this vein, but it proved to have another vein, Star Vein, running ESE, to the north of it. There was now sufficient encouragement for the driving of a surface adit, and Bolton Gill or Bottle Level was driven at a little above 850 ft (259 m) OD. The mine as it eventually developed is illustrated in plan on Figure 34. A mine section, copied in BGS records from an original in Dr Raistrick's possession in 1980 shows, in cross-section the Star and Cockbur veins both hading S, with displacements as cited above. The strata on the hangingwall of Star Vein are shown on a longitudinal section which suggests that this vein was the principal producer from the Bearing Grit. In the western part of the section an understope is shown below Bottle Level but it is not possible to be certain whether this was on Cockbur Vein (in which case it would be in Bearing Grit) or on Star Vein (where it would be in Middle Limestone on the footwall side) but it is certain that the workings must have entered Middle Limestone north of Cockbur Vein. The numerous short crosscuts from the Cockbur Vein workings suggest a complex disturbance, consistent with the large throw. It is unlikely that the Bearing Grit on the downthrow side of Cockbur Vein was penetrated, and if only the Bearing Grit was mineralised, this had passed below Bottle Level on Star Vein before work was finally abandoned nearly ½-mile (1.6 km) ESE of Engine Shaft. The disturbed strata which prevented the completion of Cook's Shaft might be interpreted as the result of a cross fault, which could also have caused the temporary shift in Cockbur Vein, but this can only be regarded as a speculation. The minerals present in the *Bolton Gill Mine* were dominated by baryte, though a little fluorite was present, the $BaSO_4/CaF_2$ ratio being appreciably higher than at Yarnbury. Production of concentrates from 1856 to 1872 amounted to 2989 tons, but it had been evident from 1867 on that the extent of workable ground had been proved. Two further attempts were made to find Beever Vein, even though it had been barren where cut a short distance inside Bottle Level. An adit, Charger Level, was driven NE from the dressing floor in Grassington Grit, failed to find it, and Lanshaw Level, driven after 1877 and therefore the last work done here, cut two fractures in the right direction but both were barren.

The continuations of both Star and Cockbur veins can be followed to the ESE of Bolton Gill Mine without difficulty since they fault down the Red Scar Grit in steps on Great Black Hill and Bolton Haw. The final attempt to open up Hebden Moor for mining was the driving of the *Hebden Trial Level* [0283 6297], from the side of Hebden Beck, downstream from the bridge on the Grassington – Pateley Bridge road, starting at 595 ft (181 m) OD with the object of intersecting Cockbur Vein and others to the north. It is unfortunate that no plan or geological record of this tunnel has been found. Raistrick gives its direction as NE (1973, p. 85) and its length as 7830 ft (2.39 km), in which case it should have reached Cockbur Vein. According to Raistrick's account, both Beever Vein and Cockbur were found to be barren. The level starts in dark shales with thin limestones, belonging to the Bowland Shales, some 300 ft (91 m) below the base of the Grassington Grit to the north. In 1942 the first 1040 ft (317 m) were accessible, and in an unwalled section from 318 to 422 ft (100–129 m) inbye, it was possible to examine the northern elements of the North Craven Fault, through which it passed. The drive was straight up to this point, the direction being N49°E, but beyond the fall blocking it a small turn must have been made to enable the level to connect with the air shaft [0357 6374], now collapsed. The level had already entered limestones of the Yoredale facies before reaching Cockbur Vein.

As Figure 34 illustrates, the Grassington Moor – Hebden Moor area may be considered structurally as a gentle semi-basin, open to the east, with the Grassington Grit faulted by steps into it from the north, and dipping into it from the west and south. The structure contours on the intra-Pendleian unconformity shown on this diagram can be no more than approximate except where they are based on levelled positions, ascertained at outcrop or in the extensive underground workings of the mines; beyond the workings it has been necessary to rely on calculations based on the top of the Grassington Grit, where local stratigraphical information has been taken into account, and towards the east, the base of the Red Scar Grit for which a separation of 425 ft (130 m), justified by the data at Bolton Gill and Red Scar mines, has been applied uniformly. These contours, admittedly subject to an error of perhaps ± 50 ft (15 m) serve to give a picture of the gentle structure, rimmed to the west and north with mineralised veins, and unexplored—save for the Trial Level and one sinking at Red Scar—in the centre. It emerges that the Trial Level was very well sited to command the upper beds of the limestone beneath the unconformity, and therefore the Grassington Grit, over the whole area, and it may be regarded as a misfortune that the results of the first phase—albeit involving 1½ miles (2.4 km) of tunnelling—were so discouraging as to prevent the execution of the larger scheme which might well have opened up the whole concealed area. It is nevertheless appreciated that under 1980 conditions, the rewards from the discovery of a repetition of Grassington Moor mineralisation could hardly meet the cost of restoring the first 1½ miles (2.4 km) of the drive, and adding to it 2 additional miles (3.2 km), plus crosscutting, rising and sinking. Further, the precautions necessary to safeguard Grimwith Reservoir would inevitably affect any such development.

OUTPUT AND RESOURCES

The production of lead in Wharfedale came principally from the Grassington Moor mines, active from the 17th century but closed before the end of the 19th. Records of lead smelted at Grassington are available from 1736, by which year output was approaching 150 tons per year; but the continuity of the record is broken in 1753–64, and 1833–44. Few of the Grassington veins came to outcrop, and it is likely that exploitation of veins exposed high on the limestone sides of the valley, farther north began earlier, but for most of these no records remain prior to 1860; by which time most of them had been exhausted.

Table 24 Recorded production, lead metal and concentrates, Area 14

Mine or Group	Years	Lead metal long tons	Concentrates long tons	Lead %
Buckden Gavel	1859–1884	708*	1 175	60.2
Starbotton Cam and Moorend	1860–1878	413*	743	55.6
Kettlewell and Old Providence	1849–1875	1 991*	3 129	63.6
Conistone Mossdale, Moorhead, Outmoor, Fearnought, Silver Rake	1860–1872	760*	1 131	67.1
Grassington Moor	1736–1753	5 027†	8 378	60‡
	1765–1833	31 664†	52 758	60‡
	1845–1881	23 258*	33 322	69.7
Grassington royalties	1863–1874	382*	685	55.7
Hebden Moor and Grimwith	1856–1872	2 008*	3 000	66.8
		66 209	104 321	63.5‖

* Official statistics, *Mem. Geol. Surv.* 1848–1881; Home Office 1882–1885.
† Cavendish Estate records; the figures are more complete than those published by Raistrick (1973, pp.107, 116).
‡ Assumed on basis of recovery during the first five years of the official records, 1845–1849.
‖ Average value.

Production at Grassington prior to 1736 is considered, taking into account all the evidence, to have been at least 12 000 tons of concentrates (pp. 199–200). The two gaps in the subsequent record may reasonably be filled in by averaging, adding 20 000 tons more. Reasons have been given (p. 189) for regarding the output at the Gavel mines (Bishopdale, Buckden and Walden Road) as no less than 10 000 tons of ore, so an additional 9000 tons may here be taken into account. To this may be added 7000 tons for the Kettlewell mines (p. 190) and 6000 for the Conistone group (p. 191). For the vein-complexes west of the dale there is no basis of estimation other than the shallow but extensive ancient workings; for these 2000 tons may well be an underestimate. The estimated additional figure to add to the recorded figure is thus 56 000 tons making total concentrate production, ascertained and estimated, 160 000 tons.

No proved reserve of lead ore is known to remain in any mine in this area and the evidence clearly indicates that the developed ore at the Grassington mines was exhausted by the time they closed in 1881. Had the important prospect known as Hebden Trial Level (p. 201) been pursued in spite of the disappointing early results, a considerable extension of the Grassington Mining Field under cover might have been found (Dunham, 1959), but for reasons already discussed the project is not now attractive, except insofar as geophysical evidence points to possible mineralisation, probably in Grassington Grit, north of Wigstones [060/080 672] on Bycliffe Vein. The Mossdale and Old Providence veins (pp. 190–192) might carry further oreshoots under Namurian cover to the east, but access to this ground is very difficult. The remaining veins of the area clearly die out in depth in exposed limestone and there are few if any targets for prospecting, unless concealed oreshoots exist in the lowest beds of the Carboniferous succession.

Since attention was drawn to the existence of useful amounts of fluorite and baryte in the tailings at Grassington (Dunham, 1952) these have been treated (Dickinson, 1965), with the recovery of 10 545 tones. The dumps are by no means exhausted of these minerals, but the remaining grade is low due to admixture with rock 'deads'. KCD

REFERENCES

Bean, S. 1737. *Rara Avis in Terris: or, The laws and customs of the lead mines within the mineral liberty of Grassington cum Membris...* (Leeds.)

Carruthers, R. G. and Strahan, A. 1923. Lead and zinc ores of Durham, Yorkshire and Derbyshire, with notes on the Isle of Man. *Spec. Rep. Miner. Resour. Mem. Geol. Surv. G. B.*, Vol. 26.

Clough, R. T. 1980. *The lead smelting mills of the Yorkshire Dales and Northern Pennines.* 2nd edition. (Keighley: Clough.) 332 pp.

Dakyns, J. R., Tiddeman, R. M., Gunn, W. and Strahan, A. 1890. The geology of the country around Ingleborough, with parts of Wensleydale and Wharfedale. *Mem. Geol. Surv. G. B.*, 103 pp.

Dickenson, J. 1902. Lead Mining districts of the North of England and Derbyshire. *Trans. Manchester Geol. Soc.*, Vol. 27, pp. 218–265.

Dickinson, J. M. 1965. The Dales Chemical Company. *Mem. North. Cavern Mine Res. Soc.*, pp. 1–6.

— 1966. The Grimwith Mining Company Limited. *Mem. North. Cavern Mine Res. Soc.*, pp. 1–13.

— and May, M. C. 1968. Wharfedale Mine. *Mem. North. Cavern Mine Res. Soc.*, pp. 9–12.

Dunham, K. C. 1948. Geology of the Northern Pennine Orefield: Vol. I, Tyne to Stainmore. *Mem. Geol. Surv. G. B.*, 357 pp.

— 1952. Fluorspar. 4th edit. *Spec. Rep. Miner. Resour. Mem. Geol. Surv. G. B.*, 143 pp.

DUNHAM, K. C. 1959. Non-ferrous mining potentialities of the northern Pennines. *Non-ferrous Mining in Great Britain and Ireland. Inst. Min. Metall., London*, pp. 115–148.

— DUNHAM, A. C., HODGE, B. L. and JOHNSON, G. A. L. 1965. Granite beneath Viséan sediments with mineralisation at Rookhope, northern Pennines. *Q. J. Geol. Soc. London*, Vol. 121, pp. 384–417.

EDDY, J. R. 1883. On the lead veins in the neighbourhood of Skipton. *Proc. Yorkshire Geol. Polytec. Soc.*, Vol. 8, pp. 63–69.

EDDY, S. 1844. On the geology of the Grassington mines, near Skipton, Yorkshire. *Trans. R. Geol. Soc. Cornwall*, Vol. 6, pp. 186–189.

— 1845. An account of the Grassington lead mines, illustrating a model of the mine. *Rep. Br. Assoc. Adv. Sci. for 1844*, pp. 52, 53.

— 1859a. On the lead mining districts of Yorkshire. *Rep. Br. Assoc. Adv. Sci. for 1858*, pp. 167–174.

— 1859b. On the lead mining districts of Yorkshire. *Proc. Yorkshire Geol. Polytec. Soc.*, Vol. 3, pp. 657–663.

FOSTER-SMITH, J. R. 1968. Notes on the lead mines on the east side of Wharfedale between Buckden and Kettlewell. *Mem. North. Cavern Mine Res. Soc.*, pp. 14–21.

GARWOOD, E. J. and GOODYEAR, E. 1924. The Lower Carboniferous succession in the Settle district and along the line of the Craven faults. *Q. J. Geol. Soc. London*, Vol. 80, pp. 184–273.

KINGSBURY, A. W. G. 1968. *Demonstrations to the Mineralogical Society in March and May*. (Proceedings not published.)

RAISTRICK, A. 1936. Rara Avis in Terris: or The laws and customs of lead mines in Yorkshire. *Proc. Univ. Durham Philos. Soc.*, Vol. 9, pp. 180–190.

— 1938a. The mineral deposits. Pp. 343–349 *in* The geology of the country around Harrogate. HUDSON, R. G. S. (Editor). *Proc. Geol. Assoc.*, Vol. 49.

— 1938b. Mineral deposits in the Settle - Malham district, Yorkshire. *Naturalist* (for 1938), pp. 119–125.

— 1948. *Grassington and Upper Wharfedale.* (Clapham: Dalesman.)

— 1953. The lead mines of Upper Wharfedale. *Yorkshire Bull. Econ. Soc. Res.*, Vol. 5, pp. 1–16

— 1955. The mechanisation of the Grassington Moor mines, Yorkshire. *Trans. Newcomen Soc.*, Vol. 29, pp. 179–193.

— 1966. Conistone Moor Mines, Wharfedale, Yorkshire. *Mem. North. Cavern Mine Res. Soc.*, pp. 27–32.

— 1968. Fourteen and Ten Meers mine. *Mem. North. Cavern Mine Res. Soc.*, pp. 1–7.

— 1973. *Lead mining in the Mid-Pennines.* (Truro: Bradford Barton.) 172 pp.

— and JENNINGS, B. 1965. *A history of lead mining in the Pennines.* (London: Longmans.) 347 pp.

RICHARDSON, D. T. (Editor). 1966. Spring Wood Level, Starbotton. *North. Cavern Mine Res. Soc.*, Individ. Surv. Ser. No. 1.

WADGE, A. J., BATESON, J. H. AND EVANS, A. D. 1983. Mineral reconnaissance surveys in the Craven Basin. *Mineral Reconnaissance Programme Rep. Inst. Geol. Sci.*, No. 66.

WAGER, L. R. 1931. Jointing in the Great Scar Limestone of Craven, and its relation to the tectonics of the area. *Q. J. Geol. Soc. London*, Vol. 87, pp. 392–424.

WILSON, A. A. and CORNWELL, J. D. 1982. The IGS borehole at Beckermonds Scar, north Yorkshire. *Proc. Yorkshire Geol. Soc.*, Vol. 44, pp. 59–88.

CHAPTER 13
Mineral deposits

Details, Area 15 NIDDERDALE, GREENHOW HILL, SKYREHOLME AND THE CRAVEN REEF BELT

The final area to be described forms a belt up to 5 miles (8 km) wide extending from Pateley Bridge in the east to Settle in the west, a total length of 21 miles (33.6 km), with the North Craven Fault forming a generally W–WNW diagonal across the rectangle so formed. For convenience, a few small deposits in Upper Nidderdale, north of Pateley Bridge are included first; these might be regarded as giving some indication that mineralisation may continue beneath the late Pendleian cover north of the Bycliffe Vein. In this chapter however the main concern is with the supposed continuation of the Bycliffe veins into Ashfold Side Beck and their approach to the North Craven Fault; with the anticline of Greenhow Hill, showing three culminations and two deep basins along the north side of the fault; and with the Skyreholme structure mirroring the folding *en échelon* to the west, on the south side of the fault. The belt embraces the transition from the basin facies of Bowland to the block facies of the Craven Uplands, with the North Craven Fault somewhere near but not exactly along the transition. The hinge line of Lower Carboniferous times is considered to have been situated a short distance south of what eventually became the site of the North Craven Fault. Associated with the hinge are the well-known limestone reef-knolls, from the doubtful case of Whithill near Appletreewick, to the well-displayed knolls of Thorpe Kail, Elbolton, Stebden, Butter Haw, Carden and Langerton (Figure 6). The Greenhow and the Skyreholme anticlines are highly mineralised with numerous veins; the reef-knolls contain only a few. Finally, a number of veins and some oxidised zinc deposits in limestone occur in the Settle area, mainly between the North and Middle Craven faults.

The area is a classical one in British geology and has an extensive literature (see pp. 12–66 for stratigraphy and pp. 71–82 for structure). Greenhow Hill is noteworthy for the clear evidence of intra-Pendleian earth movements, as a result of which the unconformity at the base of the Grassington Grit Formation transgresses across and cuts out beds from the Five Yard Limestone equivalent down to the lower part of the Great Scar Limestone, involving at least 450 ft (137 m) of strata. Where beds of Yoredale facies are present beneath the unconformity, there is evidence of thickening of the limestone members at the expense of the clastic sediments, while beneath the Girvanella Band the massive limestones of the Asbian and earlier stages thicken appreciably from west to east. The North and Middle Craven fault systems defining the southern boundary of the Askrigg Block structurally define the southern limit of the Northern Pennine Orefield. It is however taken to include mineral deposits along the hinge line, but not to extend into the Craven Basin to the south, where the stratigraphy has changed completely.

The varied structure and stratigraphy of the present area has provided more diverse environments for introduced mineralisation than elsewhere in the region. Along Ashfold Side Beck, host rocks for vein oreshoots extend as high as the Cayton Gill Beds, and include the Follifoot, Red Scar and potentially, the Grassington Grit. On Greenhow Hill the Grassington Grit and higher beds are not mineralised except adjacent to the North Craven Fault, and the oreshoots occurred at stratigraphical levels in limestone ranging from beds above the Girvanella Band on Coldstones, down to and including the lowest exposed limestone at Blackhill, 1250 ft (381 m) lower in the sequence. The greatest vertical range of any oreshoot so far proved is, however, a little over 500 ft (152 m). Beneath the shaly beds associated with the unconformity, ribbon-style oreshoots, shallow but laterally extensive, are found; but at deeper levels there is a tendency for the vertical dimension to exceed the horizontal.

Area 15 may be regarded as forming the northern margin of the Craven Basin, which since 1973 has been investigated in connexion with the BGS Mineral Reconnaissance Programme on behalf of the Department of Industry, using geochemical and geophysical methods. The results are available in an open-file report by Wadge, Bateson and Evans (1984) and where these cover parts of the present area they are noted in the text.

There is little doubt that galena was exposed at surface in some of the veins in the limestones before mining commenced, and Raistrick's suggestion (1973, p. 16) that there may have been some working in Iron Age or even Bronze Age times is a reasonable one. That the Romans were in the area is very probable since they left behind the pigs of lead found at Heyshaw Bank and (possibly) Nussey Knot. The first documentary evidence comes from the 12th century (Raistrick, 1927, 1973; Raistrick and Jennings, 1965) but though some activity has continued through the past 800 years, the industry was at its height in the 18th and 19th centuries, almost dying out in the depression of the 1880s. That documentation of the workings is better here than in some other parts of the Yorkshire Pennines is in no small measure due to M. Newbould, agent to the Sunside Mining Company in the mid-19th century, whose plans remain the prime source of information on the underground workings. The driving of an aqueduct tunnel for Bradford Corporation to communicate with its Scar House reservoir in the closing years of the 19th century, passing beneath one of the most highly mineralised parts of Greenhow Hill, revealed that ore-bearing veins continued below the deepest workings of the time, and though production of lead ore during the present century has been small, the district has remained of some interest. In 1941 the possibility that it might provide short-term needs of lead ore and fluorspar led to a revision of the unsatisfactory primary survey, mapping at the 1:5000 scale being undertaken. The principal geological results of this work were published (Dunham and Stubblefield, 1945) but not the geological details as applied to individual workings; the present chapter is designed to supply this information. Meanwhile Jennings (1967) and Dickinson (1964a,

1964b, 1967, 1969, summed up in his 1970 memoir) have made valuable contributions by bringing together the known facts and local tradition on mining at and around Greenhow. The important collection of plans made by Varvill, which formed the basis for his researches (1920, 1937), were made available to the Geological Survey in 1941, and are now in the safe keeping of the Department of Mining at the University of Leeds, which also maintains the Gillfield Adit at Greenhow for teaching purposes. In recent years some drilling has been done both at Ashfold Side Beck and Greenhow Hill by Bewerley Mines Ltd, the one concern left of the 47 listed by Dickinson (*op. cit.* Appendix A) as having been active at Greenhow since 1585. Some of the boreholes provide new stratigraphical and structural data (Wilson and Dunham, in preparation).

Description in this chapter proceeds southwards from Blayshaw Gill and Lolly to Ashfold Side Beck and the east end of Greenhow Hill; then westward along the Greenhow and Skyreholme anticlines to the Craven knoll-reefs, and concludes with Malham and Settle.

Limley Veins Lead ore

SE 17 NW; Yorkshire 100 SW
Four veins, three E – W, one N80° W, the outer pair downthrowing S (Figure 15).

Towards the head of Nidderdale, limestones including the Middle, Five Yard and Three Yard form a triangular outcrop fault-bounded to the north, and with Limley at the southern point. Considerable sinks, known as the Manchester and Goydin Pot-holes conduct the River Nidd underground in this area, into the Middle Limestone. In a second inlier south of Limley, four small mineral veins crossing the Nidd [1015 7556, 1016 7542, 1016 7537 and 1014 7519] were recognised by the primary surveyors. The workings are now largely obscured, but the third of these is a N80°W fault downthrowing S not more than about 10 ft (3 m). The Three Yard Limestone is in the river, overlain by Grassington Grit. Two short opencuts were worked on the northernmost fracture in the grit; a level was driven on the east bank to test the second; several pits into grit tested the third, while a level was driven on the west bank into the fourth. This may be the drift referred to in N. Yorkshire Record Office plan ZLB 41/185 marked 'Limley', showing 1150 ft (351 m) of drivage to the west, with small stopes in the western part 30 to 50 ft (9–15 m) above the drift. These workings, if they are correctly identified, would be in Grassington Grit and probably yielded the 316 tons of ore produced between 1894 and 1900. In Geological Survey records there is also a plan (6606) of a small trial level at Middlesmoor, surveyed by Littlewood and Crossland of Leeds in June 1916, the probable, though not certain, location of which was on the south bank of the Nidd [0928 7337] near How Stean cottages. The level ran 175 ft (53 m) SSW, and explored several fractures in a drive of about 250 ft (75 m) to the E.

Blayshaw Gill Veins Lead ore

SE 07 SE; Yorkshire 117 NW
Four veins coursing N85°E, downthrowing S (Figure 15).

The deposits seem to have been discovered by means of pits into the Grassington Grit in Stean Pasture, south of Hard Gap Lane; workings appear to have found four parallel veins in a width of about 275 ft (84 m). The two southern veins, that might usefully be named the Main and Sun veins respectively continue to the east, with a total prospected length exceeding 4000 ft (1219 m) and the Main Vein can be followed farther east as a fault to beyond the Nidd. On the north side, from Blayshaw Gill eastward, the surface formations include the Three Yard, Five Yard and Middle limestones; the Grassington Grit appears on the hangingwall of the Sun Vein. An extensive exploration of the complex was undertaken in 1901–06 by means of Blayshaw Gill Level [0975 7280] at 640 ft (195 m) OD. The plan (ZLB 41/186, by A. Rodwell dated 1881) shows that this followed Sun Vein for 625 ft (191 m) then turned NW to reach the Main Vein which was followed for another 1400 ft (427 m) with a 30 ft (9 m) shift to the N at 425 ft (130 m) from the crosscut, suggesting that the fracture is probably a compound one. The 1881 section shows a sublevel 50 ft (15 m) above the adit for part of the course. The following notes regarding development after 1901 are taken from a record by T. Eastwood of data supplied by E. Cherry of Fremington: near the site of Water Shaft, approximately 1000 ft (305 m) W of the shift mentioned above, the adit was turned WNW for 550 ft (152 m) cutting through the two small veins north of Main Vein. A crosscut was then driven 800 ft (244 m) S, mainly in shale; a vein was probably cut at 450 ft (137 m) from the origin of the crosscut, for a drift was carried W for 1170 ft (357 m) and rises were put up from it, no doubt into the Grassington Grit. Eastwood remarks that if this vein was the Sun Vein, it had suffered a shift to the S of at least 360 ft (110 m). The Sun Vein which throws 15 ft (4.6 m) in the eastern part of the adit has not been detected by surface mapping in this western area. Unfortunately the whole exploration was a failure, yielding in the period 1876 to 1910 when it was abandoned, only 83 tons of lead ore. A little baryte was present, but no workable amount is in evidence. The original bell-pit workings may well have yielded much more lead ore, but at some remote time for which no statistics are available.

Silver Hill Vein

SE 07 SE, 17 SW; Yorkshire 117 NW
Direction N85°E

A shaft 850 ft (259 m) SW of the portal of Blayshaw Gill Level on the east bank of the gill is said to have cut a vein in Grassington Grit. Silver Hill Level [1023 7272] was driven 280 ft (85 m) SSW in the grit; a fracture found at 180 ft (55 m) from the mouth was tested by means of a drift 240 ft (72 m) long to the WNW, with a rise of 24 ft (7 m) and a sump of 6 ft (1.8 m), but no driving was done on Silver Hill Vein, if it was reached. However a shaft on the line 725 ft (221 m) farther E may have tested the fracture, but only sandstone occurs on the dump. The level was driven in 1905–6 but it must be assumed that no productive ground was found.

Lolly Scar Vein Lead ore Witherite

SE 07 SE, 17 SW; Yorkshire 117 NW
Direction N83°E, possibly downthrowing N

Lolly South Vein
Direction near E – W

These were the veins worked at *Lolly Scar Mine*, approximately 1 mile (1.6 km) NNW of Ramsgill, on the west bank of the Nidd. Near Lolly Scar the main vein cuts through the Grassington Grit, and probably the discovery was made here, for to the west it passes beneath Nidderdale Shales. A level was started, according to Raistrick (1973, p. 59) in 1866. The main period of work however was between 1894 and 1910, under J. Cradock of Stockton on Tees, who also worked the Limley and Blayshaw Gill prospects. Low Level, [1073 7251] driven S at 525 ft (160 m) OD cut the main vein in Grassington Grit at 280 ft (85 m), and subsidiary veins or strings at 350 and 400 ft (107 and 122 m) from the portal. The plan in the N. Yorkshire Records Office (ZLB 41/184) shows Low Level driven 1205 ft (367 m) W from the adit crosscut but information obtained by T. Eastwood from E. Cherry suggests that it was continued to 1330 ft (405 m) where it was linked by rises with sublevels ('durk drifts') above. A second adit, Top or High Level, reached the vein 800 ft (244 m) W of the first, at 655 ft (200 m) OD, having been driven above the Grassington Grit close to the horizon of the Cockhill Marine Band. Harker's sump connected the two levels, and to the west a sublevel was driven 70 ft (21 m) above Low Level, whereas east of the sump, sublevels ranged from 30 to 45 ft (9–14 m) above. These workings almost certainly reflect the fact that there is, for this district, a rather steep westward rise in the beds, as illustrated by the fact that the third adit, Bents Level [0945 7246], in Blayshaw Gill, is at 800 ft (244 m) OD but at the same stratigraphical horizon as Top Level. It is probable, though not certain, that the sublevel 70 ft (21 m) above Low Level continued to within 300 ft (91 m) of the crosscut adit of Bents Level, but unfortunately no stope section appears to have been preserved, and it can be no more than a conjecture that the orebody worked here was a patchy ribbon, in Grassington Grit dipping east at about 4½°. Raistrick (*op. cit.*) records that the vein was 3 to 4 ft (0.9 – 1.2 m) wide, of baryte with some yellow fluorite. A little witherite was present, and some is said to have been shipped to Germany.

From Low Level a test was made into the beds below the unconformity at the east end of the mine, near Thompson Shaft [1084 7240], where 12 ft (3.6 m) of chert, presumably in the Five Yard Limestone, were encountered, but no workable ground on the vein. At this end the vein probably fingers out. North of the vein the outcrop of the Three Yard Limestone may be terminated as a result of its presence, but this is the only evidence for assigning a N downthrow to it. Also from Low Level, a crosscut was driven S starting 1175 ft (358 m) W of the level head. This found a second vein at 495 ft (151 m) S of the Main Vein, and it was evidently considered sufficiently promising to drive a second crosscut from the higher sublevel; about 900 ft (274 m) of ground was tried laterally on this vein, with what result is not known. The output of Lolly Mine, 1894 to 1910 when it closed, was 5747 tons of concentrates yielding 3846 tons of lead worth £50 400.

Kidd's House Vein

SE 16 NW; Yorkshire 117 SE
Direction N85°W, throw about 50 ft (15 m) N

On either side of the tributary now running into Gouthwaite Reservoir near the former site of the Hall, there are remains of substantial shafts [1307 6835 and 1319 6826] which may perhaps have been sunk to test a fracture running through them in a WNW direction. The second is known as Leyfields Shaft, and was collared at 520 ft (158 m) OD; an adit from the gill drained it to about 20 ft (6 m) deep. Before 1840 a drive at the adit level, known as Burn Gill Low Level, had been carried S17°W to test a fault that had been found passing through springs to the south of Eanings, and a shaft (Kidd's House) [1310 6798][1] sunk. In 1846 Kidd's Shaft was connected to Low Level 216 ft (66 m) below surface, close to 500 ft (152 m) OD. From near Kidd's House Shaft a drift was carried eastward in the vein 12 to 18 ft (4–5 m) above Low Level for between 70 and 80 fathoms (thus about 135 m); Newbould remarks: 'Sometimes we have had a fine strong-looking vein with some pieces of ore in it, but in the last 10 to 15 fathoms it was filled with black rubbish or plate and no ore'. Subsequently a trial was made west of the shaft driving at least 150 ft (46 m), but though some ore was found particularly in rise workings about 12 ft (4 m) above Low Level, much of the ground was gouge-filled and barren. After more than ten years work was abandoned, even though at times the vein looked as though it ought to produce a mine. Upstream from the adit portal, bullions from the Cockhill Marine Band [1279 6818] (p. 23) indicate that the top of the Grassington Grit Formation is close by. The exploratory drives would thus be near the top of the grit on the hangingwall of the vein, and well into it on the footwall. Nothing is known of the fracture—if one exists— between the original pair of shafts but both were sunk into the grit and originally had an objective different from the Kidd's House Vein.

[1]The work after 1840 is documented in reports by M. Newbould to Sir John Yorke preserved in papers on the case Yorke *v.* Bradford Corporation in the High Court, anno 1906.

Stony Grooves Vein — Lead ore (barytes)

SE 06 NE, 16 NW; Yorkshire 135 NW
Direction N68°W, throw about 250 ft (76 m)

Goodhamsyke Vein — Lead ore

SE 16 NW; Yorkshire 135 NW
Direction N58°W, throw to NE perhaps 200 ft (60 m) at Goodhamsyke Level.

Merryfield Hole Veins — Lead ore

SE 16 NW; Yorkshire 135 NW
Direction N78°E, aggregate throw approximately 350 ft (107 m) SSE.

Providence Veins — Lead ore

SE 16 NW, SW; Yorkshire 135 NW, NE
Direction N65–35°W, throw 295 to 375 ft (90–114 m) SW.

The supposed continuation of the Bycliffe vein-system from Hebden Moor enters the Nidd drainage at High Stony Groove Mine, in the headwater of Ashfold Side Beck (Figure 36). Although having regard to the persistent strong SW downthrow it is not unreasonable to suppose, as most previous writers have done, that Stony Grooves Vein is directly linked to Bycliffe; the connexion between Speculation Level, (p.200)where Bycliffe Vein was last tested, and Stony Grooves remains unproved and the extrapolation of the direction of the latter vein would not take it into Speculation Level, but considerably to the north. Thus it is feasible that Bycliffe Vein dies out SE of Speculation Level while Stony Grooves takes over *en échelon* to the north, and there was nothing in the peat-covered ground between to enable a choice between these alternatives to be made, nor is there any sign of the swing to an E – W direction which would be needed to join the exposures. Such a swing is, of course, known elsewhere on Bycliffe Vein. Geophysical measurements carried out in connexion with the BGS Mineral Reconnaisance Programme, reported by Wadge, Bateson and Evans (1984, pp. 30–31) throw considerable light on this problem, Airborne EM showed a conductor continuing westward on the line of Stony Grooves Vein, while ground follow-up using VLF showed a double belt of anomalies, swinging to E–W and suggesting continuity to and beyond Speculation Level, and suggesting that both *en échelon* and looping elements are present in the Bycliffe–Providence vein-zone. These anomalies have not yet (1984) been investigated by drilling.

At Stony Grooves the Red Scar Grit and overlying beds are at the surface on the footwall of the vein, and the Upper Follifoot Grit is the hangingwall formation, indicating a SW downthrow of the order of 250 ft (76 m). Shafts mark the course of the vein for 3300 ft (1.006 km) WNW of Low Stony Grooves Shaft [1001 6655] and tailings containing abundant baryte (analysis, Table 17, p. 100) remain from the hand dressing of lead ore. Sinkings of over 500 ft (152 m) would be required here to command the Grassington Grit even on the footwall, and it seems certain from the shaft heaps that this depth was not attained; in short the evidence points to oreshoots at the unusually high stratigraphical position of the Red Scar and Follifoot grits. The workings appear to have been fairly continuous. At Low Stony Grooves Shaft, a drive 75 ft (23 m) below the collar, at about 1000 ft (305 m) OD was carried 1250 ft (381 m) to the

WNW, but the plan shows that this diverged steadily north of the surface position of the vein and the numerous irregularities suggest that a compound system of fractures was being explored. With beds rising in the direction of the level at 10°, this drive must have passed into the footwall position of the Nidderdale Shales and it is assumed that it was below the worked oreshoots, and probably followed strings diverging into the footwall.

Low Stony Grooves Shaft was sunk to connect with Storey's Level (described below) which here formed the 42 Fathom Level, 252 ft (77 m) below the collar. This was still probably above the footwall position of the top of the Grassington Grit, but no geological section of the shaft is extant. Stony Grooves Vein was apparently not developed at this horizon. Near the shaft the main throw of the fault proved to be transferred to two subparallel veins coursing a little N of E, which crossed the stream from the Bewerley Royalty into the Yorke Royalty to the NE, where they were worked by Merryfield Hole Mine. On the south-west side of the stream, the direct line of Stony Grooves Vein is continued by Goodhamsyke Vein, throwing NE. The 42-fathom level was driven 1340 ft (406 m) SE in this vein, but there is no information as to what was found. Goodhamsyke Level [1093 6603] reached the vein from surface beneath the hangingwall position of the Cayton Gill Beds, which dip NE from the vein at 20°, but appears to have found little; a short level from Cross Gill Dyke also seems to have found nothing. Towards the SE, the throw of Goodhamsyke Vein dies away and is nil at Brandstone Beck.

The Merryfield Hole veins were at first worked opencast (in the Hole), an operation that could well date back to the time before Sir John Yorke purchased Heathfield Moor at the dissolution of Bylands Abbey, whose property it had been since the 12th century (see Raistrick, 1973, p. 18). The section exposed on the footwall of North Vein in the Hole is as follows:

		ft	m
Sandstone, brown, micaceous, flaggy with shale partings		15	4.6
Shale, black, micaceous;	seen	20	6.1
Colsterdale Marine Beds			
Shale, fossiliferous		5	1.5
Limestone, tesselated; goniatites		1	0.3
Shale, fossiliferous		6	1.8
Red Scar Grit			
Grit, coarse, pebbly;	seen	4	1.2

and on the hangingwall side, probably of the South Vein, typical green cherty shale with brachiopods, and softer bluish shale of the Cayton Gill Beds are seen overlying sandstone referred to as the Upper Follifoot Grit. Mineralisation was still reaching an usually high stratigraphical position. Five shafts, in a line on the north-east side of Ashfold Side Beck, were subsequently sunk to work the veins in depth; from west to east: New 150 ft (46 m), communicating with levels on North Lode at 72 ft (22 m) and at the bottom, and on South Lode at 72 ft(22 M); Engine [1013 6657] sunk to the 42 Fathom or Storey Level, and continued to 52 Fathom, 312 ft (95 m); Horner's [1021 6660] (150 ft, 46 m); an unnamed shallow shaft; and Borehole Shaft [1033 6665] (240 ft, 73 m). On Bradford Corporation Waterworks plan Z.1545 stoping is indicated down to the 52-Fathom Level; even so, the Grassington Grit on the footwall (up) side of the lodes can hardly have been reached. The 'hade' (presumably the dip) of North Lode is shown as varying between 60 and 87°, and of South Lode, 85°; the two veins are 50 to 70 ft (15–21 m) apart.

At Borehole Shaft another dramatic change in structure occurs. The Providence Vein, running SE, now takes over and the Hole veins, as indicated by a trial shaft east of the junction, effectively cease to exist.

From this point Providence Vein pursues a gently sinuous SE course for 1½ miles (2.4 km) until it begins to finger out as it approaches the North Craven Fault. The beds continue to dip SE at 10°, so that when the vein crosses Ashfold Side Beck from the Yorke into the Bewerley royalty, the footwall formation is still the Red Scar Grit; however, along much of this first stretch, the Libishaw Sandstone is present on the hangingwall and this is still the position at Prosperous – Providence Mine. The extensive workings on Providence Vein are shown in section on Figure 36, and the geology of the footwall, necessarily somewhat generalised, is indicated. *Merryfield Mine* (not the same as Merryfield Hole Mine) worked the vein in the Yorke Royalty, *Prosperous – Providence Mine* in Bewerley ground.

Throughout the worked portion of Providence Vein there is evidence that it is a broad zone of fracturing, up to 125 ft (38 m) wide. The footwall portion, dipping at about 85° was known as Black Vein, perhaps a reference to shale gouge on this side. Some ore may have been got from this at New West Gin [1070 6654], in the stretch 500 ft (152 m) ESE of this shaft, but otherwise the evidence suggests that it was unproductive at Merryfield. The principal orebodies were on the near-vertical hangingwall region of the vein. There is no certain evidence as to where the main displacement occurred, but it is most likely to have been associated with the footwall. Merryfield Mine was extensively developed during the 18th century though the opencut [around 1100 6635] may be much older. By 1757 the working of the vein from surface shafts was becoming impeded by drainage difficulties and according to Raistrick (1973) Yorke Level [1162 6613], (Figure 36, loc. 1), starting a short distance below the point where the vein crosses Ashfold Side Beck, was begun. This is at 740 ft (226 m) OD, with Red Scar Grit on the footwall, and Cayton Gill Beds on the hangingwall at the start. About 30 years later, Storey's or Jossey (?Joicey) Level [1140 6610] (Figure 36, loc. 2) was driven as a crosscut adit to reach the vein in its middle stretch beneath the Libishaw Sandstone, 60 ft (18 m) above Yorke Level. This became the principal working level of the mine, and was, in the 19th century, driven up to and beyond Low Stony Grooves Shaft, after serving as the 42-Fathom Level at Merryfield Hole. The courses of Yorke and Storey's levels, as shown by Raistrick (1973, diagram 3) intertwine, again indicating that different parts of a wide channel were being tried and worked. In 1800, according to Dickinson (1970) another crosscut adit, College Level (Figure 36, loc. 3) was driven to the Old West Gin [1084 6647] (also known as Varty's or College Shaft). This adit reaches the vein 62 ft (19 m) above Storey's Level; it is a curious fact that Merryfield Mine seems to have been developed in the reverse order of levels to usual practice. Old West Gin was now sunk to Chalder's Level at 650 ft (198 m) OD, which was driven 700 ft (213 m) to connect with East Gin. These workings were now over 100 ft (30 m) below the Yorke Level, and were drained by an underground water wheel. By 1814, when Dyson's plan (Bradford Corporation Z.1551) was made, the substantial area 1440 x 380 ft (439 x 116 m) shown on his section had been worked out. It is not unfortunately certain that this was continuously stoped, but it is clear that an oreshoot of unusual size for the district was extracted here. It also appears from the extant sections that little if any ore was mined from the section of the vein between the western end of this oreshoot and the Borehole Shaft of Merryfield Hole Mine. Further preparations were made for working in depth when, by agreement between John Horner, the operator of Merryfield Mine, and John Wood of Prosperous – Providence, Wood's Shaft [1166 6615] was sunk on the north side of Ashfold Side Beck close to the royalty boundary to 305 ft (93 m) depth, reaching approximately 500 ft (153 m) OD to serve as a joint pumping shaft. On the Merryfield side the main development was the Low Level, close to 590 ft (190 m) OD, which gave access to the bottom of the oreshoot already described. Below this, Bell's Sump, 925 ft (287 m) WNW of Wood's Shaft was sunk to approximately 400 ft (122 m) above OD, with a shallower sump nearby, but no development was

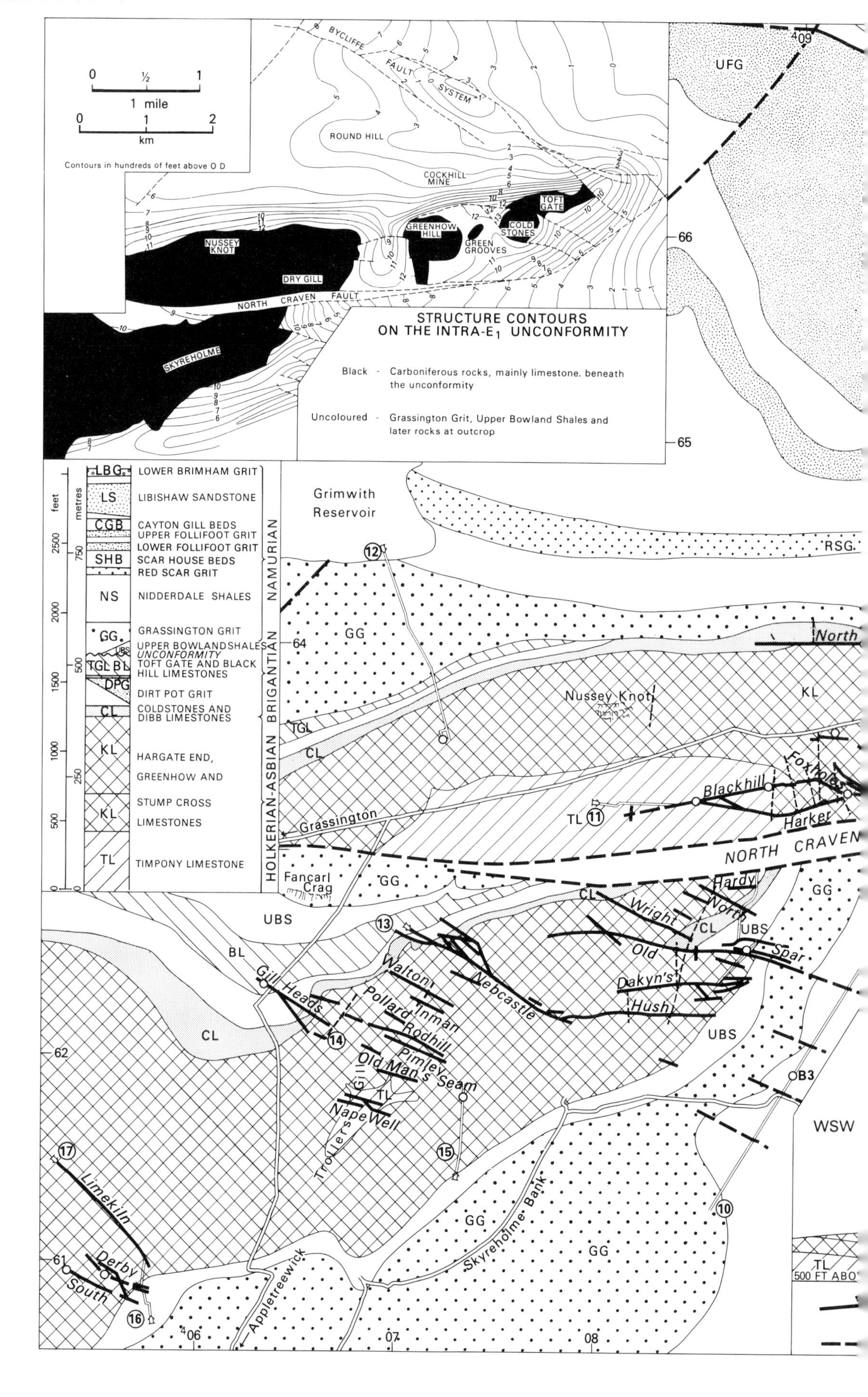

Figure 36
Map to show the veins in the Greenhow and Skyreholme anticlines.

Key to numbered levels and mines **1** Yorke Level, Merryfield–Providence Mine; **2** Jossey or Storey's Level, Merryfield; **3** College Level, Merryfield; **4** Wonderful Level, Merryfield–Providence; **5** Perseverance Level; **6** Bale Bank Level; **7** Eagle Level; **8** Gillfield Adit, Sunside Mines; **9** Cockhill Adit, West Waterhole Mine; **10** Bradford Aqueduct; **11** Blackhill Adit; **12** Kelshaw or California Level; **13** Hopewell Level, Nebcastle Rake; **14** Gill Heads Level; **15** Glory Level; **16** Wellington Adit, Appletreewick Mine; **17** Limekiln Vein Level

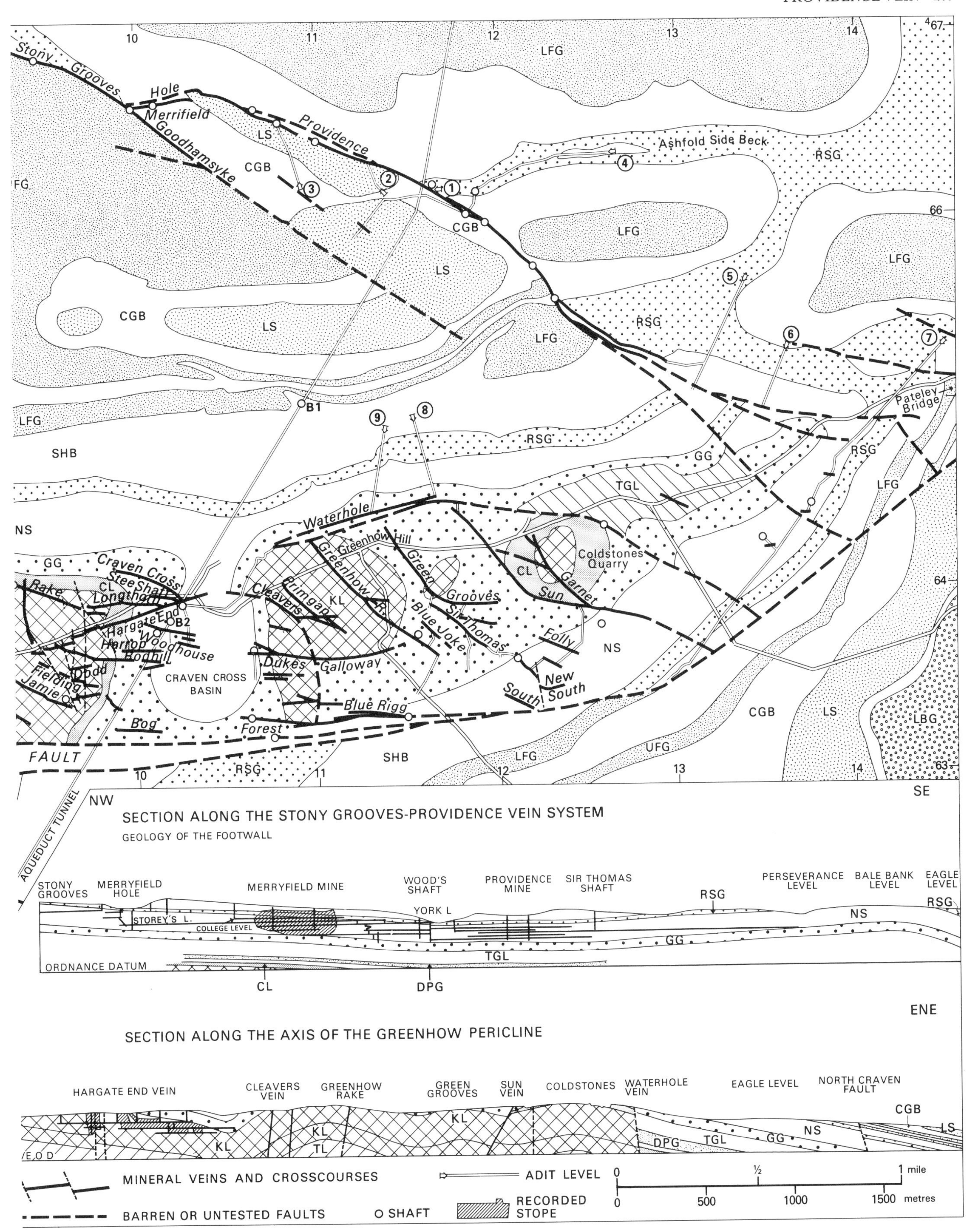
Stony Grooves
Hole
Merrifield
Goodhamsyke
Providence
Ashfold Side Beck
LFG
RSG
CGB
LS
Coldstones Quarry
Waterhole
Greenhow Hill
Greenhow R.
Green Grooves
Sun
Garnet
Blue Joke
Sir Thomas
Folly
New South
South
Blue Rigg
Forest
Bog
Dukes
Galloway
Cleavers
Primgap
Craven Cross
Stee Shaft
Longthorn
Hargate End
Woodhouse
Hartop
Rodhill
Dodd
Fielding
Jamie
Rake
CRAVEN CROSS BASIN
Pateley Bridge
FAULT
SHB
NS
GG
CL
KL
TGL
UFG
LBG
B1
B2
AQUEDUCT TUNNEL
NW
SE
SECTION ALONG THE STONY GROOVES-PROVIDENCE VEIN SYSTEM
GEOLOGY OF THE FOOTWALL
STONY GROOVES
MERRYFIELD HOLE
MERRYFIELD MINE
WOOD'S SHAFT
PROVIDENCE MINE
SIR THOMAS SHAFT
PERSEVERANCE LEVEL
BALE BANK LEVEL
EAGLE LEVEL
STOREY'S L.
COLLEGE LEVEL
YORK L
ORDNANCE DATUM
DPG
ENE
SECTION ALONG THE AXIS OF THE GREENHOW PERICLINE
HARGATE END VEIN
CLEAVERS VEIN
GREENHOW RAKE
GREEN GROOVES
SUN VEIN
COLDSTONES
WATERHOLE VEIN
EAGLE LEVEL
NORTH CRAVEN FAULT
TL
MINERAL VEINS AND CROSSCOURSES
BARREN OR UNTESTED FAULTS
ADIT LEVEL
SHAFT
RECORDED STOPE
0 ½ 1 mile
0 500 1000 1500 metres

done—whether because of lack of ore or inability to deal with the water is not known. The reason why the two lower levels of Wood's Shaft were not extended into Merryfield is also unknown; initially the shaft had a water wheel to power the pumps, but a steam engine was installed in 1802; how adequately this was able to keep the water down is in doubt.

The stratigraphical position along the Providence Vein has been illuminated, in a way that was impossible from surface mapping alone, by the drilling of a series of seven inclined cored boreholes by Bewerley Mines Ltd, the present holders of the royalty, and one vertical hole by Claro Water Board, of Harrogate.[1] An account of these is being prepared by Wilson, Dunham and Pattison and graphic logs are included in Figures 5, 7, 8 and 14, but it may assist the present account of the workings if a statement of the sequence is included here:

	ft	m
Libishaw Sandstone; surface formation against the hangingwall of Providence Vein for much of its length	Not drilled	
Libishaw Shale	do	
Cayton Gill beds; green, hard, fossiliferous shales	78	24
Upper Follifoot Grit; sandstone with seatearths	100	30
Shale	32	10
Lower Follifoot Grit	30	9
Scar House Beds; shale, sandy shale, minor sandstone	160–180	48–58
Colsterdale Marine Beds; fossiliferous shale, thin limestone	42	13
Red Scar Grit; pebbly sandstone	40	12
Nidderdale Shales; shale, sandy shale, thin sandstones	275	84
Cockhill Marine Band; limestone and fossiliferous shale	10	3
Grassington Grit Formation; pebbly sandstone with shale beds and thin coals	180	55
Intra-Pendleian Unconformity		
Three Yard Limestone (Merryfield only)	21	5.4
Shale	3	0.9
Five Yard Limestone	17–24	5–9
Shale	15	4.5
Middle Limestone } Toft Gate Limestone	68	21
Thin limestones and shales } Toft Gate Limestone	41	12
Simonstone Limestone } Toft Gate Limestone	40	12
Dirt Pot Grit; medium-grained sandstone	58–68	18–21
Shale	65	20
Coldstones Limestone, equivalent to Hardrow Scar, Gayle and Hawes limestones of Wensleydale; Girvanella Band near base	68–80	21–24
Hargate End Limestone	130+	40+
Greenhow Limestone	Not drilled	

Applying this information to the Merryfields oreshoot, it emerges that on the footwall side of the structure, the beds present ranged from the lowest part of the Scar House Beds down to the Three Yard or Five Yard Limestone, but since no mention of the limestones appears in the records, it may be taken as fairly certain that the footwall was not in fact penetrated here. If, as suggested, the main movement was on the footwall fracture (the Black Vein), the beds on either side of the oreshoot itself ranged from Libishaw Sandstone down to a little below the Red Scar Grit, and the Grassington Grit and underlying limestones remain far below, untested by mining.

Crossing Ashfold Side Beck to Prosperous – Providence Mine, virtually no record remains of workings above 650 ft (198 m) OD, the horizon of Wonderful Level, but the mine was certainly operating long before this level was driven, and it must be assumed that extensive stopes probably existed in the upper 200 ft (61 m) of the ground, related to Prosperous [1188 6602] and Providence [1203 6589] shafts. Wonderful Level [1265 6630] (Figure 36, loc. 4) was started by Sir Thomas White at the opening of the 19th century. It runs from a point on the south bank of Ashfold Side Beck, 3375 ft (1.03 km) E of Wood's Shaft, but it was driven to Smelt Mill Shaft [1192 6613] where water was pumped up to it from the 15-Fathom Level at 555 ft (169 m) OD; subsequently a branch connected it to Prosperous Shaft, and this was later continued NW along the vein to near the river, then due W for 1000 ft (305 m) and thereafter 1000 ft (305 m) SW, to search for parallel veins on the hangingwall side of Providence Vein. Several strings were cut; one was tried with a short drive, but nothing productive was found; it is uncertain whether Goodhamsyke Vein was reached. As, however, the whole of this drive must have been in Nidderdale Shales, the level having been started under the Red Scar Grit, it was perhaps not decisive. The closely spaced levels down to 38 fathoms (70 m) below Wonderful Level shown on Figure 36 were all driven from Providence and Prosperous shafts during the years 1803–37. No stope section unfortunately remains, but it is difficult to believe that Raistrick's figure of 900 ft (274 m) for the length of the oreshoot is correct; an excessive amount of dead work would be implied. It is more reasonable to allow twice this figure, and to recognise a suggestion of a plunge towards the SE. By 1837 Providence Shaft had been deepened to the 38-Fathom Level, 456 ft (139 m) below surface and here at 430 ft (131 m) OD. A level was being driven when the mine was flooded out, though not before limestone had been noticed on the footwall of the vein. It has been assumed by all writers on the mine that this was of Lower Carboniferous (Dinantian) age, but in fact the recent boring programme has proved this to be the Cockhill Limestone of Dunham and Stubblefield (1945), M. Newbould's 'Top Limestone'; the implication is that the whole of the Grassington Grit on both walls of the vein is intact below the 38-Fathom Level. A width of 42 ft (13 m) for the vein channel, quoted in reports of the time, is quite credible having regard to the distance between the footwall and hangingwall fractures proved at Prosperous Shaft and at Merryfield.

Information on production from both mines is scanty for the years when they were doing well. Production from Prosperous – Providence formed a substantial part of the 11 268 tons of lead metal produced on Bewerley royalties between 1806 and 1817 (p. 221); 1825 was a year of great prosperity according to Raistrick (1973); the only other figures remaining refer to the period 1837–42, during which the operator, Watson, became bankrupt and the Yorkshire District Banking Co. took over; 1314 tons of lead were made, equivalent at 67 per cent recovery to about 2000 tons of concentrates. Though this was from clean-up operations, it may give some indication of smelting capacity which would ultimately control rate of production. If this rate is applied to the whole period of working below Wonderful Level, 1803–42, an output of the order of 13 000 tons would be suggested. If the stoped area was 1400 × 225 ft (427 × 69 m) this would represent 1.28 tons per fathom^2 of stope, a figure which can be used to provide a check by comparison with contemporary lead mines of the London Lead Co. (Dunham, 1944, p. 201). In these the minimum yield for the top class stope areas was 0.8 tons, corresponding to a grade of 6.6 per cent Pb for a vein width of 4 ft (1.2 m). The figure of 13 000 tons, if high, is not unreasonably so, but a report by M. Colling[2] in which

[1]These boreholes were logged for the Survey by Dr Wilson and we are very much indebted to the owners for permission to quote the stratigraphical results.

[2]Kindly made available by the late Mr D. Gill, with other data on Greenhow area.

he states 'In sinking 8 fathoms below the 30-Fathom Level at Providence Shaft no ore yet discovered which will pay for working; a level is there driving by the side of the vein and crosscuts occasionally through it'; this was written only a year before the 38 Fathom Level was drowned out and suggests that 175 ft (53 m) ought to be taken rather than 225 as the vertical component of the oreshoot. This would lead to rescaling of the output to the order of 10 000 tons. It must however be recalled that Colling also remarks that 'probably no piece of ground in Yorkshire has been more productive' than Merryfield – Providence. A figure of 25 000 tons concentrates to include extraction above and below Wonderful Level at Prosperous – Providence, the Merryfield oreshoot, Merryfield Hole and Stony Grooves is probably conservative.

With the loss of the deep levels at Prosperous – Providence, the mine was flooded at least to the 15-Fathom Level. Sir Thomas Shaft 1250 ft (381 m) SE of Providence Shaft, sunk after the lessors had seized the property, was down 90 ft (27 m) and a borehole was put down from the bottom in 1840; however a prospectus issued by the Nidderdale Mining Co. (1859) Ltd. in 1863 shows the shaft down to the 15-Fathom Level. Ore worth £850 (about 85 tons of concentrates) was obtained above Wonderful Level, and inevitably there were reports of rich ore left below.

Still farther south-east, Perseverance Level [1337 6560] (Figure 36, loc. 5) had been begun in 1825 as a SSW crosscut adit, cutting the north fracture of Providence Vein 2610 ft (796 m) from the entrance, 55 ft (17 m) lower than the outfall of Wonderful Level. This level starts in Red Scar Grit and enters lower beds as the rise towards the Greenhow Anticline affects the structure. The result did not encourage further work at the time, but the Nidderdale Company was formed in 1859 to resume this drive and to carry it beneath Wonderful Level. A weak branch fracture that had been cut 1570 ft (479 m) from the entrance was followed W and WNW until the main footwall fracture was encountered and on the west side of Round Hill an old shaft at Hole Bottom [1263 6528] was reopened and deepened for ventilation, but hereabouts extensive old workings were encountered. A S crosscut was put out to the Sun or hangingwall fracture, on which the orebodies at Prosperous – Providence had been worked, but this was the only test south-east of Ebenezer Level, a short adit from Brandstone Beck at 800 ft (244 m) OD. The evidence of the underground workings coupled with surface mapping shows that in this eastern stretch, the Sun Vein fracture diverges rapidly from the footwall fracture, so that it was at least 400 ft (122 m) ahead of the end of the direct drive of Perseverance Level, and thereafter it has remained untested until Eagle Level, discussed below, is reached. The Providence Vein had thus split up into at least three divergent fractures at this end. The North Branch from Perseverance Level was driven to connect with Sir Thomas Shaft, but the whole of these extensive operations from Perseverance Level yielded only 336 tons of concentrates between 1861 and 1873. At the end there were still ambitious plans to sink a new shaft to 240 ft (73 m) below Perseverance Level, but this would only have been a short distance below the old 38-fathom Level of Providence Shaft, and it proved impossible to raise the necessary capital. Meanwhile a small resurgence of interest at Merryfield had led to the partial reopening of Storey's Level, and the production of 273 tons of concentrates in 1861–67. Some ore from Prosperous was returned amongst an output of 803 tons from Stony Grooves between 1861 and 1867; Stony Grooves alone produced 299 tons between 1878 and 1887. Including a final 41 tons from Providence, the later 19th century workings along Ashfold Side Beck added only 1752 tons to the total.

The eastern fingering-out of Providence Vein was explored by Bale Bank Level [1362 6526] (Figure 36, loc. 6), 1350 ft (411) m SE of Perseverance Level, at 800 ft (244 m) OD, which cut and found unproductive yet another N branch of the system. The final test, however, was made by the remarkable Eagle Level [1451 6530] (Figure 36, loc. 7), at 620 ft (189 m) OD started from the grounds of Eagle Hall in the prosperous year 1825, and driven at intervals during the next 25 years in a south-westward direction to reach a total length of 8000 ft (2.43 km). Some details of the several veins and faults cut by the level are available from reports by M. Colling for the period up to 1843, and Geological Survey records indicate that the level entered limestone in the core of the plunging nose of the Greenhow Anticline near Toft Riggs, about 3000 ft (0.91 km) from the portal. With the aid of surface mapping it is possible to reconstruct the geology in more detail and this is summarised in the following table:

From Portal ft	m	
50	15	Top of Red Scar Grit, dip 20°N
150	45	Base of Red Scar Grit
810	245	Fault, throw about 30 ft (9 m) S; in Nidderdale Shales; barren
1880	573	Fault, throw perhaps 50 ft (15 m) N; branch of Providence Vein; in Nidderdale Shales; barren
2150	655	Top of Grassington Grit, dip 10°N, direction near E – W
2380	725	Fault, small N throw; branch of Providence Vein; in grit; barren
3300	1005	Enters limestone, gentle N dip
3830	1167	Shaft 102 ft (31 m) from surface and borehole for ventilation; foul gas released from grit
3955	1205	Fracture, in limestone; barren
4130	1259	Fault, in limestone, throw N; branch of Providence Vein, probably Sun Vein; barren
4730	1442	Vein crossing WNW in limestone, 'water burst out overnight showing lead ore among great quantity of sludge' (Colling). Butts Shaft sunk off tunnel line from surface, no mine found
5050	1539	Much water from joint in limestone
5244	1598	'In shale with limestone above and below' (Colling)
5700	1737	Several fractures; Grassington Grit base about here, dipping S
5900	1798	Fault, probable continuation of Waterhole Vein; barren.
7350	2240	Sun Vein, followed by drift 160 ft (49 m) to NW
8000	2438	Tunnel ends at or near North Craven Fault

As far south-west as Waterhole 'Vein' there is adequate written information and, taken in conjunction with surface exposures, a consistent picture can be built up with some confidence. For the last 2550 ft (777 m) the only documentary evidence is M. Newbould's plan of 1848, and beyond Waterhole Vein the ground is entirely drift and peat covered. There are two possible interpretations of this part of the level: (1) Waterhole Vein throws NE as at Gillfield Adit (p. 212), in which case the drive probably continued to the end in limestone, or (2) the throw of Waterhole Vein reverses, in which case the drive would be in Grassington Grit. The geometry of the two alternatives, as it affects the workings on Waterhole Vein, is shown on Figure 17, pp. 78–79; no critical evidence at present exists which would enable a choice to be made. The level has long since ceased to be accessible owing to bad ventilation resulting from the filling in of the air shafts; it is now used as a source of water for public supply. As a mining venture, it was a failure; it tested Providence Vein in the act of fingering out and turning barren; and the two major Greenhow veins were equally disappointing at the horizons cut.

The minerals in the Providence and Merryfield Hole veins include those typical of the field: baryte predominates, galena is the sole primary lead ore but some cerussite may be present in shallow workings. A little sphalerite is generally present. Fluorite, colourless or pale yellow is present throughout the oreshoots, but its abundance relative to baryte varies in an interesting way (analyses, Table 17). Expressed as $BaSO_4/CaF_2$ ratios, Prosperous – Providence tailings (about 30 000 tons) 2.40; Merryfield, tailings around West and East Gin shafts, 3.01; Merryfield Hole 18.2; and Stony Grooves, 23.6. Baryte becomes relatively more abundant proceeding north-west from Prosperous – Providence. Some attempt was made to recover barytes from Stony Grooves late last century, but the remote situation has discouraged further efforts.

East Waterhole Vein Lead ore Fluorspar

SE 16 SW; Yorkshire 135 NW, NE
Direction N82°W, continuing as a fault running N50°W, with maximum throw about 135 ft (41 m) N

Garnet Vein Lead ore

SE 16 SW; Yorkshire 135 NW, NE
Direction N45°W, branching N to N70°W and N8°W

Sun Vein = Coldstones Sun Vein Lead ore Fluorspar

SE 16 SW; Yorkshire 135 NW, SE
Direction N35°W, turning S to N82°W, throw perhaps 25 ft (8 m) SW

Toft Gate Veins and Strings

SE 16 SW; Yorkshire 135 NE

Eagle Level, described in the previous section, cuts through the gently-plunging nose of the Greenhow Anticline, the mineralisation of which will be discussed in the sections which now follow. The most recent description of the geology is that of Dunham and Stubblefield (1945) supplemented to take account of more recent data from boreholes by Wilson, Dunham and Pattison (in preparation). A general résumé is included here since it aids description of the veins. The presence of the structure first becomes apparent about ¼-mile (0.4 km) E of Eagle Hall. From here the main axis may be considered as pursuing a direction slightly S of W for a distance of 6 miles (9.7 km) until it noses out on the south-east side of Grimwith Reservoir, beyond the area where California Level (p. 175) is driven through the north flank. It is periclinal in the sense that a series of axial culminations and depressions occur along its length; these are, from east to west, the Coldstones dome, the gentle Green Grooves basin, the Greenhow Hill dome, the Craven Cross basin and the generally monoclinal Craven Cross – Blackhill – Grimwith region, where the North Craven Fault, downthrowing S, has taken the place of the southern limb of the structure. West of Coldstones, the crestal region is cut by numerous veins throughout, a number of them showing curviform trends, swinging from NNW almost to E – W as they cross the anticline.

As first demonstrated by Chubb and Hudson (1925) and worked out in detail by the mapping of Dunham and Stubblefield (*op. cit.*), the intra-Pendleian unconformity, upon which rests the Grassington Grit Formation, has its maximum effect within the area of the anticline. At the eastern end, the grit rests on high beds of the Toft Gate Limestone, believed to be equivalent to the top beds of the Middle Limestone farther north and west. Towards the west, the unconformity successively cuts out the whole of the Toft Gate Limestone and the underlying Dirt Pot Grit so that, at Coldstones Quarry and around the Coldstones dome, the Coldstones Limestone, its upper part the equivalent of the Hardrow Scar Limestone, is the formation beneath the erosion surface. In the Green Grooves basin, the grit transgresses on to Hargate End Limestone, and on the north side of Greenhow Hill dome, near the church, it must rest on Greenhow Limestone, equivalent to the middle part of the Great Scar Limestone (Asbian). The Hargate End Limestone reappears on the east side of the Craven Cross basin, while on the west side, Coldstones Limestone is again present, to be succeeded by Dirt Pot Grit and Toft Gate Limestone before California Level is reached. If the Grassington Grit was deposited on a more or less flat surface, as seems possible from the consistent thickness of the formation in this area, the summit of Greenhow Hill must have coincided with a pre-Namurian elevation from which about 750 ft (225 m) of beds were eroded prior to the arrival in the area of the Grassington Grit delta.

The following general section of the full sequence in the Coldstones dome is based upon measurements in Gillfield Adit made in 1941, and on Bewerley Mines Ltd inclined boreholes Nos. 3 to 5 (Coldstones) drilled in 1971 (for full details see Wilson, Dunham and Pattison, in preparation):

	ft	m
Colsterdale Marine Beds; dark fossiliferous shale, limestone		
Red Scar Grit; pebbly sandstone	32	10
Nidderdale Shales; shale with siderite layers, sandy shale, thin sandstones	351	107
Cockhill or Top Limestone	2	0.6
Grassington Grit Formation; pebbly sandstone with seatearths, thin coals and shale beds	176	54
Intra Pendleian Unconformity		
Toft Gate Limestone; pale crinoidal limestone with chert;		
Upper Section = Middle Limestone	176	54
Shale, shaly limestone	16	5
Lower Section = Simonstone Limestone	101	31
Dirt Pot Grit; sandstone	62	19
sandy shale and sandstone	58	18
mainly sandstone	128	39
Coldstones Limestone; Girvanella Band near base	40	12
Hargate End Limestone	160 +	49 +
Greenhow Limestone; pale, well bedded, some shale partings	490	128
Stump Cross Limestone; pale, structureless	270	82
Timpony Limestone; dark grey, bedded	55 +	17 +
(Base not reached)		

Unlike the veins along Ashfold Side Beck, those of the Greenhow district in general carry oreshoots in the limestones only, and almost without exception are barren above the unconformity.

The veins listed at the head of the present paragraphs are those of the Coldstones dome and its extension towards the nose of the anticline. The most important of them were worked through Gillfield Adit [1154 6487] (Figure 36, loc. 8) starting at 960 ft (293 m) OD, the principal entrance of the *Sunside Mine* (Leeds University Department of Mining). The level was commenced shortly after 1782, no doubt at a time when the veins to which it gives access had been worked as far down as possible from surface shafts. It runs SSE through the steep north limb of the anticline, cutting through beds from the Red Scar Grit downwards, and passes through the unconformity 1216 ft (371 m) from the portal, to reach West Waterhole North Vein at 1380 ft (421 m) in Toft Gate Limestone. Nathan Newbould's sections of this level and of the adjacent Cockhill Level are classics of Yorkshire geology, and have been quoted by several authors from Phillips (1836) on. Figure 17, pp. 78 – 79 is however based on new detailed measurements in 1941. West Waterhole veins are discussed in connexion with Cockhill Mine; Gillfield North Vein, which contained only small amounts of galena, has been stoped 200 ft (61 m) W and 130 ft (40 m) E of the level head for fluorspar by

Caldbeck Mining Co., by 1941. The W drive was carried about 400 ft (122 m) from the level head, entering a gulf which separated it from the Cockhill workings. Crosscuts S give access, in Greenhow Limestone, to East Waterholes Vein, making an angle of about 20° with the strike of the West veins, and to West Waterholes South Vein. A rise of 60 ft (18 m) gave access to stopes on the East Vein 4 to 12 ft (1.2–3.6 m) wide and 20 to 40 ft (6–12 m) high on East Waterholes vein, worked for fluorspar following previous slitting for lead ore. The late Mr D. Gill (personal communication) was of the opinion that the oreshoot here was determined by the intersection of two fractures with the vein. It is also controlled by overlying shale, where the hade of the vein increases from 10–20° to about 45°. West Waterholes South Vein does not appear promising at Gillfield Level. Gillfield Level runs 260 ft (79 m) to a crosscut leading S to Sun Vein, and was continued along East Waterhole Vein for another 500 ft (152 m). Close to the Sun Vein crosscut a working, suggesting a flat beneath a shale parting, could be seen and nearby a sump was sunk to 72 ft (22 m), but there is no evidence of much working from it. The East Waterhole Vein can be followed as a fault by surface mapping as far as the intersection in Eagle Level (p. 211) but there is little evidence of prospecting along it and present evidence suggests that it turned barren after the Sun Vein intersection, which possibly also terminated West Waterhole North Vein.

Sun Vein is reached by Gillfield Level 260 ft (79 m) along the crosscut from East Waterhole Vein, and from this point the next 870 ft (265 m) is open, to a fall which blocks the level; of this, 640 ft (135 m) had been stoped at widths of about 3 ft (1 m). Dickinson (1970) however records that stoping here was carried 80 ft (24 m) above the level and in places reached 10 ft (3 m) wide. Judging from the extensive heaps which continue for 2600 ft (792 m) along the curviform course of the vein beyond the blockage, the most productive part of the vein still lay ahead. As far east as Sun Forefield Shaft [1263 6380] it appears that all the working lay above Gillfield Adit, but no record of the shape of the stopes has survived. A comparison of the surface line of the vein with that of the adit indicates that the hade of the vein increases from about 10° in the N part to nearly 45° at the SE, also that at least one subparallel branch is present in depth. Beyond Forefield Shaft, ore was followed in rich ground beneath the adit (Figure 17, pp. 78–79) to about 800 ft (244 m) OD. The underground shaft shown on Figure 17, 185 ft (56 m) deep was equipped with a steam pump driven from underground boilers installed in 1839 (Dickinson, *op. cit.*) but after Eagle Level intersected the vein the water drained away, even though no direct connexion was made to this level. M. Colling's report of 1844 remarks: 'It was also observed in working the vein eastward and underlevel that its productiveness was very much influenced by the junctions of strings or thin veins from the south; whenever they were numerous, the vein was rich'. He also adds: 'The Dun Vein and Garnet Vein form a junction near Forefield Engine; and this is considered to be the cause of the great productiveness of the Sun Vein in that part'. The plan shows a fracture corresponding with Garnet Vein cutting through 220 ft (67 m) W of Forefield Shaft, and the underlevel oreshoot extended about 600 ft (183 m) E of this. The minerals in Sun Vein, in addition to galena were, in order of abundance, fluorite, calcite and baryte. It is possible, though not certain, that resources of fluorspar remain *in situ* or as fill in the workings now inaccessible.

Garnet Vein forms a loop with Sun Vein on the north-east side of the latter. It cuts through the crest of the Coldstones dome and was mainly worked from a large opencut and shafts in the Hargate End Limestone, pinching along strike to the SE in the Coldstones Limestone. The width varied up to 6 ft (1.8 m) in places. The WNW branch must diverge from Sun Vein north of the main road, near the cemetery but it is not recorded on the plan; the junction near Forefield Shaft 2150 ft (655 m) to the SE has already been mentioned. Some 1150 ft (350 m) SE of the main road, a flat known as Pendleton Pipe was worked in Hargate End Limestone on both sides of Garnet Vein (Figure 36). Under a dolomitised roof, the flat contained sandy banded fluorite with calcite, baryte and a little galena. It appears to be related to small E – W strings crossing the vein. It was being worked for fluorspar in 1942 by Mackwell and Storey, the total accessible length of ground at that time being 380 ft (116 m), the maximum width 40 ft (12 m) and the height 15 ft (4.6 m); the end of the deposit appeared to have been reached to the east, but to the west it may persist, from the evidence at surface, for 300 ft (91 m) further.

East of Coldstones, around Toft Gate, a few minor veins and strings have been found carrying calcite, baryte and traces of galena, but none appears to be of any significance. One of these, tried from two shafts on either side of the main road, lines up approximately with the 4730 ft (1442 m) vein in Eagle Level.

A north crosscut from the elaborate system of levels related to Cockhill Adit (described below) eventually reached Sun Vein from Folly Vein, 345 ft (105 m) W of Forefield Shaft and was continued to test Garnet Vein to the north. Folly Vein, about 550 ft (168 m) S of Sun Vein was exploited, with the upper levels at Sun Vein, from a shaft at [1236 6375] as well as from Cockhill Level. Fluorite and baryte are present in the tailings near the shaft; no doubt both veins contributed to these.

West Waterhole North Vein — Lead ore Fluorspar

SE 16 SW; Yorkshire 135 NW, SW
Direction N75°E, considerable N downthrow, possibly pre-Namurian

West Waterhole South Vein = Lum Vein
Direction N75°E, small N downthrow

Green Grooves Vein
Direction N48°W, turning S to near E – W

Moor Vein,? = Folly Vein
Direction N75°W, N68°W

Sir Thomas Vein
General direction, N50°W

Haiding Vein ? = Blue Joke Vein = New South Vein
General Direction NW to WNW

Cleavers ? = Hezle Vein
Direction N75°W, N60°W

Great Cross Vein
Direction N10°W

Banks Vein
Direction N40°W

Greenhow Rake = Moss Vein — Lead ore Fluorspar
Direction N40–55°W, probably no throw

Primgap Vein
Direction N45°W

The veins listed above are the principal veins of *Cockhill Mine*, situated, with the remains of its smelt mill (Clough, 1980, pl. 3) ½-mile (0.8 km) N of Greenhow village. All but the last and first two veins were worked beneath the cover of Pendleian strata in the Green Grooves basin, and it may be doubted whether any of these were discovered at surface. Greenhow Rake and Primgap veins cut across the crest of the Greenhow Hill dome and the superficial workings on these must be ancient; the Waterholes veins were likewise discovered at an early stage, and the two very old levels,

Sam Oon and Jackass (Raistrick, 1973) were driven in from the N-facing slope below the village in search of these veins, which had probably already been opencut between Hargate End Limestone and Grassington Grit near their western extremity. The development of the mine really began with the establishment of the partnership of Cleaver, Hutchinson and Hutchinson in 1782 (Raistrick, *op. cit.*, p. 49). The first step was the driving of Cockhill Level (Figure 36, loc. 9) from Brandstone Beck at 978 ft (298 m) OD, a crosscut adit reaching West Waterhole North Vein 1480 ft (451 m) S8°W of the entrance. Nathan Newbould made a section of this level in 1805 and new measurements bed by bed through the steeply-dipping strata made in 1942 confirmed his observations, led to the first record of *Cravenoceras cowlingense* (E_2) in the area and brought out the highly angular nature of the intra-Pendleian unconformity (Dunham and Stubblefield, 1945, fig. 2). Cockhill Level starts only 575 ft (175 m) from Gillfield Adit, which was apparently being driven at the same time, but Cockhill was used to exploit the southern and western part of the White royalty, while Gillfield gave access to Coldstones. At the level head, the North Vein brings Toft Gate Limestone on the hangingwall into contact with Greenhow Limestone, implying a substantial N throw which, since it is not reflected in the grit beds above the unconformity, suggests that North Vein predates the Grassington Grit as a fault; there is no evidence that it had already been mineralised at this early stage. The W branch of the adit follows North Vein for 1070 ft (326 m). East Waterhole Vein possibly united with North Vein at 176 ft (54 m) W of the level head, but the first 400 ft (122 m) of the drift show a poor vein with only 1 to 2 ft (0.3–0.6 m) of calcite with a little fluorite, hading at 20–25°. The vein then widens out and here a stope, in places 10 ft (3 m) wide had been carried up to 40 ft (12 m) above the level by G. Boddy by 1942, producing good-quality fluorspar, the lens appearing to pitch E. The vein continued strong in fluorite for 280 ft (85 m) W of the rise to this stope, beyond which extensive old stoping could be seen above the level for 390 ft (119 m) farther. The level then turns into the footwall and continues along South Vein, which is poor, to give access to Greenhow Rake. Surface shaft heaps which are numerous along the worked stretch on North Vein show considerably more baryte than can be seen underground. South Vein is marked by old shafts only at the east end, where some fluorite is present.

Greenhow Rake, one of the major veins of the district, crosses Greenhow Hill north-east of the summit; its total length including SE branches exceeds 3000 ft (914 m). It may be followed on surface from the large opencut north of the main road, 900 ft (274 m) W of Greenhow church through a continuous line of surface cuts and shafts to Greenhow Quarry where it splits into three branches. Of these the middle branch, which has been worked intermittently for fluorspar, appears to be the strongest; it probably continues under the Green Grooves basin to join up with Blue Joke Vein to the SE. The southern branch, which is only feeble as exposed in the quarry, continues to the SE as Moss Vein. In 1939 J. Busfield sank or reopened an old shaft from surface on the Rake, 420 ft (128 m) S of the main road, and developed the vein in whole ground 4 to 5 ft (1.2–1.6 m) wide of fluorite at 70 ft (21 m) depth, but by 1942 the shaft had collapsed and spar was being worked from the quarry. Underground, the Cockhill Level gave access to Greenhow Rake both from the west end of West Waterhole Vein, and from the south by crosscuts joining Blue Joke and Moss veins; the workings were not, however, joined through and it appears that some barren ground at this horizon occurred beneath the summit of the hill. However, the north part of the Rake was productive down to and considerably below Cockhill Level, which is here at about 990 ft (302 m) OD. No details of the stopes above the level remain, but this part of the mine was accessible in 1942 and the impression was gained that streaks of lead ore had been slit out, leaving pillars of fluoritic veinstuff; an assay of a sample across one of these is given in Table 17. An Engine Shaft was sunk to a depth of 180 ft (55 m) below Cockhill Level from a point 180 ft (55 m) S of South Vein with levels at 9-Fathoms (934 ft (285 m) OD) and 20-Fathoms (870 ft (265 m) OD). An old section in the Varvill Collection (C50) now at Leeds University shows workings below the 20-Fathom Level at 280, 320 and 405 ft (85, 98 and 123 m) SE of Engine Shaft, the first being sunk to 816 ft (249 m) OD. The surface being at 1250 to 1350 ft (381–412 m) OD, the vertical range of workings on Greenhow Rake exceeds 500 ft (152 m) extending substantially below the horizon of the Bradford Aqueduct Tunnel, to which reference is made below. Stratigraphically they extended through the Greenhow Limestone and may have entered Stump Cross Limestone. The old section shows 'Bearing Beds', dipping through the workings beneath Cockhill Level; these may have been divided by the marly partings characteristic of the Greenhow Limestone, and may have influenced the distribution of ore. The Engine Shaft, originally sunk in 1859–62 was served by steam pumps and winding gear operated from five underground boilers (Dickinson, 1970). When it was reopened by Bewerley Mines Ltd in 1926–27, it was found to have been sunk to the 30-Fathom level, and that stoping had been carried down to the bottom. Unfortunately before any sampling or trial of the 30-fathom sole had been made, it proved impossible to control the water, and the attempt was abandoned; a heap of good ore left from the earlier workings is said to have been noticed.

Beyond Greenhow Rake, the main Cockhill Level continued 730 ft (223 m) farther W, partly in South Vein to the point where North and South veins unite before dying out. Thence the drive was continued southward along the east side of the deep Craven Cross basin and eventually joined up with the Craven Cross mines, referred to later. A second southward crosscut turned off 450 ft (137 m) W of Greenhow Rake to give access to Primgap Vein, and beyond that to Low Row and the Galloway veins. North of High View a stretch of 500 to 600 ft (152–183 m) of Primgap Vein seems to have been highly mineralised and to have contained much fluorite, which has been recovered from surface dumps. The workings extended down to Cockhill Level but not below. To the SE it converges on Cleaver's Vein and passes into minor veinlets with baryte and traces only of fluorite. Cleaver's Vein runs through the large shaft [1074 6394] opposite Low Row, and here surface mapping indicates that it has a considerable N downthrow but as it continues and turns towards the SE, the throw dies away perhaps by virgation with the cross vein of Galloway gulf. Underground the vein was worked for a length of nearly 500 ft (152 m) at Cockhill Level horizon, SE of its intersection with Primgap Vein, fluorite, baryte and dolomitised limestone being present on the shaft heaps. On the south-west slopes of Greenhow Hill the vein deteriorates into feeble strings with baryte and fluorite. In the Green Grooves basin, what might be the continuation of the same fracture causes a sudden change in direction in the workings of Sir Thomas Vein.

In the ground extending about 500 ft (152 m) N of Low Row cottages, a great number of small veins are present, trending approximately E–W, south of Lum Vein, the title given here to West Waterhole South Vein. According to a plan dated 1804 (Mines Department R. 206D) the following have been worked: North and South Maiden veins, Lord Howe Vein, Old Man's Mistress Vein, Sir Jon Jervis Vein, Lord Nelson Vein, Captain Bralap Vein. None of these extends more than 400 ft (122 m) from the unconformity at the eastern margin of the Craven Cross basin, and only the Jervis Vein can be identified at surface, a shaft on this having been descended by members of the British Speleological Association. Since so little is known about this group, they have not been included in the main list; local tradition has it that this ground was very 'gulfy'. Karst action is to be expected with so many close-spaced fractures available for attack.

The workings beneath Green Grooves basin were reached by a drive which followed a barren fracture ('Sunside Vein' of Dickinson, 1970) from the Cockhill level head SE for a distance of 1550 ft

(472 m) to [1162 6400] where the productive Green Grooves Vein, curving away to the E was encountered. This was worked for about 1500 ft (457 m) to its junction with Moor Vein, the probable NW continuation of Folly Vein west of the Great Cross Vein, near Moor Vein Engine Shaft. Green Grooves Vein was highly mineralised with fluorite, which is abundant on the spoil heaps. There is no record of the stoping but the horizon of working was the Hargate End Limestone, which forms a small inlier, perhaps due to a shallow flexure, where the vein was first found. In this area a belt of small shafts crossing the NW line of the vein in an E – W direction suggests that a flat comparable with the Pendleton Pipe may have been found. From here the Cockhill Level gives access to an elaborate system of drives and vein workings, illustrated in outline on Figure 36, where the vein workings at Cockhill horizon are shown in solid black and the ramifying continuations of the various fractures are interpreted in accordance with the prevailing opinion in the early 19th century, when the Newboulds made their plans. The only sections which have survived are those on Sir Thomas and Blue Joke veins; true scale sections based on these and incorporating the approximate wallrock geology have been published by Dunham and Stubblefield (1945, fig. 6) while the original 1818 sections by Newbould have been reproduced by Dickinson (1970). The oreshoots are excellent examples of elongate shallow ribbon orebodies, related to the unconformity and following Hargate End Limestone wallrocks; they also illustrate the effect of karst features on the orebodies. Many joints were cut and recorded in the crosscuts, which must be regarded as having explored the basin with considerable thoroughness. Almost all the ore lay above the Cockhill Level, as shown on the Sir Thomas and Blue Joke sections, save on Haiding Vein (representing the combined continuation of Sir Thomas and Blue Joke veins) where Colling's report mentions in 1844 'moderate underlevel workings' west of the Great Cross Vein. With regard to the cross vein, he remarks 'when in driving the Haiding, the Hezle or the Folly Vein the Great Cross Vein is come at, the vein is lost and the drift has to be turned about 10 fathoms to the right before the vein can be found on the other side of the cross vein'. Haiding Vein seems to have been named from its considerable N hade, 42° according to a cross section by Nathan Newbould (Bradford Corporation Z.1571). On Figure 36 the principal shafts (though by no means all) have been shown; many of them probably communicated directly or indirectly with the Cockhill system of drives. Fluorite is common in the shaft heaps and the possibility that resources of this mineral exist in slit ground or in fill in the stopes cannot be dismissed. At the same time it must be added that all experience of Greenhow Veins shows them to be very subject to swells and pinches.

The farthest extension of the Cockhill workings reached the South and New South veins, considered to be near or perhaps part of the North Craven Fault. On the latter the workings again encountered the base of the Grassington Grit on the south side.

West Galloway = Duke's Vein — Lead ore Fluorspar

SE 16 SW; Yorkshire 135 SW
Direction near E – W, turning to N75°E at E end; S downthrow, probably small

Noway's Vein
Direction N75°W, branching from Duke's

Old Galloway Vein — Lead ore
Direction E – W

Trott Vein and Strings
Direction E – W

North Forest Vein = Blue Rigg Veins — Lead ore
Direction N80°W – N85°E

South Forest Vein
Direction N85°E, throw S

South Vein
Direction N60°E

This group of veins lies south of Greenhow Hill on the flank of the dome, extending towards the southern side of the Green Grooves basin; the general area is known as Galloway Pasture and Galloway Rigg. South Vein was the southernmost mineralised fracture cut by the Sunside crosscut at Cockhill horizon, which reached it by way of Green Grooves, Sir Thomas and Great Cross veins. The Newbould plan shows two branches uniting eastward; both were tested but do not appear to have been developed. They may be interpreted as flyers from the North Craven Fault (Dunham and Stubblefield, 1945, pl.XII). A heap of tailings almost above them, near Cadger Beck, contains fluorite, calcite and baryte from hand dressing of lead ore, but since there is no shaft on South Vein, these must have been brought here from elsewhere, possibly from the Blue Rigg veins. The main Blue Rigg Vein is marked by a line of nine shafts running W from a shaft [1152 6326] which marked the end of the crosscut from Cockhill Level via Blue Joke and Moss veins. Two subparallel branches up to 50 ft (15 m) apart were tested by drives extending 1330 ft (403 m) W of this shaft, at about 1000 ft (305 m) OD, approximately 300 ft (90 m) below surface. The depths of the other shafts to the Blue Rigg veins are indicated on Newbould's plan; they vary between 120 and 160 ft (37 – 49 m), with one sump reaching 198 ft (60 m) below surface. The worked ground was thus substantially above the Cockhill horizon. All the shafts were sunk through sandstone and shale into crinoidal limestone. These workings continue westward into the Duchy of Lancaster royalty as North Forest Vein, and here at least one shaft shows evidence of having encountered Coldstones Limestone, which has been present above the pale Hargate End Limestone, coming on westward beneath the unconformity. Several shafts mark the line of South Forest Vein, including one with a large spoil heap containing fluorite, calcite, limestone and sandstone. A second small array of shafts suggests another vein or branch about 150 ft (45 m) to the south. The ground here is heavily covered with peat, but the slope rising southward towards Simonseat exposes Red Scar Grit, indicating that the line of the North Craven Fault has been crossed. The alignment of South Forest Vein with the footwall fracture of this fault as cut in the Bradford Aqueduct tunnel leads us to consider that this vein is part of the fault zone, and that most or all of the more or less E – W veins listed above are sympathetic fractures related to it. The net southward displacement by the Craven Fault may be no more than about 250 ft (76 m) here but in the aqueduct tunnel there is some evidence of a trough-like structure which might persist as far east as Galloway Pasture. All that can be said with certainty here is that a number of small orebodies were present on veins in the footwall and on the footwall side, in limestone beneath the unconformity; the south side of the fault has not yet been tested at this stratigraphical level.

Galloway Vein, and its continuation in the Duchy of Lancaster royalty known as Duke's Vein, has been worked for a total length exceeding 3000 ft (914 m), probably up to its junction with Moss Vein. Duke Shaft [1071 6362], sunk between this vein and the branch known as Noway's Vein, is reputed to be about 350 ft (107 m) deep, connecting with the end of the third drive into this area from Cockhill Adit which reached it from Primgap Vein. A descent was made in this shaft during the resurvey in 1942, but it was found to be blocked at a depth of 124 ft (38 m), exposing only part of the Grassington Grit Formation which dips W into the Craven Cross basin at 24°. From Duke's Shaft crosscuts not only at Cockhill horizon, about 1000 ft (305 m) OD but also at some higher level, perhaps only a short distance below the unconformity, ran to the Duke's and Noway's veins; at this higher level, a drift was carried 1370 ft (418 m) SSE beneath the western edge of Galloway Pasture, cutting through Old Galloway Vein, a weak double belt of fractures probably with flats in the Hargate End Limestone on the south side and reaching Trott Vein, another weak structure north of the North Forest Vein. The Trott Vein and strings continue west in the exposed Hargate End Limestone of Galloway Pasture. All these veins contain fluorite, calcite and baryte (in that relative order of abundance), in addition to galena, but only Galloway Vein has yielded much fluorspar production. Here, 1100 ft (335 m) E of Duke Shaft is the centre of a large gulf, 80 ft (24 m) in diameter which has developed at the junction of the vein with a N – S cross fault downthrowing E. The swallow hole has been filled as it formed with residual clay containing large patches of excellent quality fluorspar, lumps of galena with cerussite coatings, and pieces of water-worn limestone. In 1942 this deposit was being worked successfully for fluorspar by G. Mackwell. East of the cross vein, the surface formation is Grassington Grit which is faulted in along Galloway Vein; west of the cross vein a salient of Hargate End Limestone continues at surface as far south as North Forest Vein. There are many mineralised strings other than those already mentioned in this limestone, tested by shallow cuts and pits; and a crosscut probably at the 1000 ft (305 m) OD level which followed the edge of the Craven Cross basin south to beyond Old Galloway Vein found seven such strings, but nothing worth working. Dickinson (1970) and Raistrick (1973) both give historical information about the partnerships working the Galloway, Forest and Black Rigg veins from 1757, when the Duke of Devonshire acquired rights in the area and following promising discoveries at Black Rigg erected a small smelt-mill at Hoodstorth in the Upper Washburn. From 1764 to 1829 when this mill was closed, Raistrick's figures, not quite complete, indicate an output of 1555 tons lead equivalent at the recovery normal in the district at that time to 2321 tons of concentrates. These veins related to the Craven Fault are not attractive as lead producers; some fluorite no doubt remains, but in patches not easy to reach. It is not known whether they contributed significantly to output from Cockhill Mine after they were reached, between 1830 and 1844, by the crosscuts from that mine, but had much ore been left after 1829 it would surely have been more economical to raise it to surface at Galloway Rigg than to tram it through a substantial part of the 4.197 miles (6.7 km) length that according to Colling, the Cockhill adit and related crosscuts had reached by 1844. Colling's reports for 1840 give a glimpse of what was found: 'The vein, supposed to be one of the Blue Riggs veins, continues unpromising though sometimes giving encouraging samples of ore' and later 'South or south-west level going towards Blue Riggs or Galloways, cut a vein some time since, in which a considerable trial made. At times promising with fine ore, but strata and vein much confused and vein lost at present. If level were continued south it would cut Blue Riggs veins, in which it is supposed good ore was left under water.' The fact that a drainage level started from the Washburn [1070 6232] was commenced and driven N by Asquith and Co. in 1784 (Dickinson, *op. cit.*) suggests that by that time water was already a problem. There are shafts on what appears to be the line of the drift at 400 and 800 ft (122 and 244 m) N of the portal, and it may have been driven beyond the northern one before being abandoned, but the dump is not large enough to indicate that it reached a third shaft on the line, 2500 ft (762 m) from the portal, and 525 ft (160 m) S of the centre of South Forest Vein, which the level would have drained to over 200 ft (61 m) below surface.

A test of Galloway Vein in depth was made in 1971 by Bewerley Mines Ltd by means of their No. 5 Borehole. Drilled from [1128 6347] at an inclination of 47°N, this passed through Grassington Grit Formation to enter Hargate End Limestone at 331 ft (100.9 m) and Greenhow Limestone at 400 ft (121.9 m); stratigraphical results of this and other boreholes mentioned below are to be given by Wilson, Dunham and Pattison (in preparation). Narrow mineralised zones were intersected in No. 5 Borehole at 635 to 637 ft (193.5–194 m), 672 to 674 ft (204.8–205.4 m) and near 835 ft (254.5 m), perhaps indicating that the veins split. Ankerite at the dolomite end of the series is present in each case, with silica in the first case. Zinc values respectively of 11 per cent, 2 per cent and trace were found, the zinc mineral present being smithsonite; traces of zinc, none of them significant, were also found by the M.3182 Mineral Analyser at twelve other depths in the Greenhow Limestone[1]. The zinc carbonate is almost certainly secondary in origin, indicating circulating oxygenated waters to considerable depths in the vicinity of the Craven Fault zone. Farther north-east, No. 4 Borehole [1178 6401], inclined 47½° SSW into the Green Grooves basin, found hemimorphite and smithsonite in a 4-inch intersection at 311 ft (95 m) rod length, and a 1-inch vein of hemimorphite at 1061 ft (323 m) the zinc values being 4.7 and 3.6 per cent respectively. Bewerley No. 3 Borehole [1161 6460], drilled into the Waterhole veins from the north, found narrow veinstuff with fluorite, calcite, baryte and sphalerite at 894 ft (272 m). The evidence of zinc mineralisation here is unimpressive compared with that adjacent to the western part of the Craven fault system (p. 228).

Craven Cross Vein Lead ore

SE 06 SE, 16 SW; Yorkshire 135 NW
Direction N75–63°W

Stee Shaft Vein

SE 16 SW; Yorkshire 135 NW
Direction sinuous about E – W

Longthorn Vein

SE 16 SW; Yorkshire 135 NW
Direction N50°W

Hargate End Vein = Willie Waters Vein

SE 06 SE, 16 SW; Yorkshire 135 SW, NW
Direction N75°E, throw 20 ft (6 m) N

Woodhouse Vein

SE 16 SW; Yorkshire 135 SW
Direction N78°W

Harrop Vein

SE 16 SW; Yorkshire 135 SW
Direction E – W

Rodhill Vein

SE 16 SW; Yorkshire 135 SW
Direction near E – W, displacement 40 ft (12 m) N

Of the veins impinging on the west side of the Craven Cross basin, Hargate End is the most important. It may be regarded as the central element in an *en échelon* series including also Blackhill Vein to the west, and the Waterhole veins to the east, exhibiting an E side N shift as one element dies out and another takes up. Running along the south side of the Hebden to Pateley Bridge main road, the worked length of the vein amounts to 4640 ft (1.41 km) if the western or Willie Waters part of the zone is included. Surface mapping of the old opencast workings in the Greenhow and Hargate End limestones show that it is shifted by at least nine of the cross frac-

[1]The examination of the cores and the follow-up laboratory work were carried out as part of a Department of Industry mineral reconnaissance project in 1973, by I. R. Wilson, Aynsley M. Shilston and D. J. Bland.

tures on Craven Moor, shifts of 10 to 20 ft (3–6 m) being characteristic; these may be either E side N or E side S. All the workings in Greenhow Limestone appear to have been shallow and not very productive, but at Donk Shaft [0975 6373] the vein enters Hargate End Limestone dipping east, and deeper workings ensue. Here an oreshoot with maximum strike length 200 ft (61 m) was followed down to 900 ft (279 m) OD, 400 ft (122 m) below surface. The opencut, with walls of solid Greenhow Limestone, is up to 10 ft (3 m) wide west of the shaft. The oreshoot was clearly formed where a cross fault, bringing in Hargate End Limestone at surface to the east, passed through the vein. East of it there is a barren patch 250 ft (76 m) long, succeeded by the main oreshoot, extending down from the surface to the 70-Fathom Level, following the dip of the measures. The maximum vertical dimension is about 200 ft (61 m) below the Coldstones Limestone which, generally was unproductive. The longitudinal section (Dunham and Stubblefield, 1945, pl. XII) shows that the oreshoot was patchy, the most productive portions being related to minor cross veins. On that part which we were able to examine on the 70-Fathom Level in 1942, these shift the vein a few inches to a few feet E side S. The old main or Derby Shaft, 1345 ft (410 m) ENE of Donk Shaft, collared at 1245 ft (379 m) OD gave access to levels at 30-Fathoms (about 1180 ft 360 m OD) and 42-Fathoms (1100 ft 335 m OD), the lower level being carried back to link with Donk Shaft. Below this, an extension of Cockhill Level from the end of the Lum workings on Waterhole Vein had been driven across the basin through Grassington Grit (with, according to local tradition, a coal seam) to become the 56-Fathom Level on Hargate End Vein. This 'Joint Level' was driven by agreement between the Cockhill Company (Cleaver), Wood at Craven Cross and John Yorke as owner of the Craven Moor royalty, the last-mentioned contributing substantially to the cost; it passed through the royalty boundary in 1801 and Bell's Gin (mentioned below) was sunk to it in 1803 (Dickinson, 1970; Raistrick, 1973). Hargate End Vein and its branches were thus drained to near 1000 ft (305 m) at the beginning of the 19th century and mined out in the succeeding years, but work did not proceed below this level and probably nothing further would have been done but for the driving of the Bradford Corporation Waterworks aqueduct tunnel in the closing years of the century. This passed directly beneath the Hargate End workings at 911 ft (277.7 m) OD and the Corporation's No. 2 Shaft was sunk nearby [1021 6382]. The tunnel revealed Hargate End Vein as well mineralised with, according to Varvill (1920) galena 1 ft (0.3 m) wide which yielded 11 tons of ore during the driving. In 1927 the Greenhaugh Mining Company acquired Bradford No. 2 Shaft, and developed the 70-Fathom levels on Hargate End and Woodhouse veins.

The records of Bradford Corporation Waterworks[1] include a log of the strata and faults encountered during the construction of the tunnel (Figure 36, loc. 10) which, while it may have been adequate for the engineering works, was unfortunately not examined by the Geological Survey while the work was in progress, nor as far as can be ascertained, was any of the material seen by a palaeontologist. In 1942 the data was used to construct a section (Dunham and Stubblefield, 1945, fig. 5) which, in conjunction with the surface mapping, produces a consistent picture of the formations and structures encountered but which would have been more satisfactory had it been possible to identify the positions of key horizons such as the Cockhill Limestone and the Coldstones Limestone with certainty. It would nevertheless be difficult to produce a picture very different from the one published, which brings out the relationship of the Greenhow and Skyreholme anticlines; it requires a thickening of the Grassington Grit Formation south of the North Craven Fault, but this is consistent with the general trend in Namurian strata to the south of the present area. It would have been particularly interesting in connexion with the establishment of continuity of mineralisation in depth (even though it is recognised that orebodies cannot be worked if they endanger the tunnel) if a single detailed account of the various veins seen in the tunnel had been kept. In fact, three mutually inconsistent versions are in Geological Survey records and form the basis of the following table:

[1]Generously made available to Mr L. H. Tonks of the Geological Survey in 1939–40.

Lead veins and faults cut in aqueduct tunnel, Greenhow Anticline
(Measurement in feet N or S of Bradford No. 2 Shaft)

Bradford Waterworks Tunnel Section	*J. Pounder, 1899, per Walker Hardwick*	*D. Gill, 1934, from Barton and Stancliff*
1232N Enters limestone		
497N Watercourse		
402N Lead vein	440N Stee Shaft Vein	430N Junction, Stee Shaft Vein and cross vein
	400N Hargate End Vein	
372N Lead vein, 3 in		
342N Watercourse		
172N Lead vein, 12 in in roof, 17 to 20 in in floor		100N Hargate End Vein
0 BRADFORD SHAFT No. 2		
	32S String dips S	27S Small vein
180S Lead vein, steep S dip	178S Woodhouse Vein	171S Woodhouse Vein, ore-rib 3–3½ in, dip 75°S
	290–300S Strings	300S Strings
	480S NW vein, dip S	520S String, mineralised
	605S NW vein, dip S	604S String, little mineral
	665S NW vein, dip S	694S String
	1350–1380S Broken ground	1393S Cavern, sludge
	1540–1570 concreted	1561½–1611½ cavern
	1680S String, NE, 12 in	1708S Vein 12 in, dip 64°S
	2300S Calcite vein 14 in	2366S Calcite vein, dip 55°S
2438S Lead vein	2390S Blackhill Vein 66 in	2444S Blackhill Vein, dip 56°N
		2569S Water
2808S Fault, dip 50°S		2797S Fault 66 in wide, dip 62°S
2843S Fault	3100S Faults dipping N and S	2800S Fault, dip 71°S

In addition to these versions, there are other and more optimistic accounts of the veins. Varvill (1920) mentions fourteen veins, of which six were said to be payable where cut. In private reports, notes said to have been taken from engineer's reports are quoted: Blackhill and Harker veins are given as 12 in wide each, Woodhouse 6 in, Hargate End '3 ft solid'; Fielding, Jamie and Stee are said to have been cut in a workable condition, the last containing fluorite; a cross vein with baryte was seen, and four other veins are mentioned. It will be obvious that no really satisfactory account of what was seen in the tunnel exists. Hargate End Vein, as proved in the 72-Fathom Level, is actually 390 ft (119 m) N of the centre of No. 2 Shaft, and Woodhouse was 158 ft (48 m) S; neither figure agrees with any previous record. In the absence of an accurate account by an experienced observer, it does not appear that more is warranted (in addition to what is known from the 70-Fathom Level) than the following: (i) At least one mineralised vein exists north of Hargate End Vein, possibly the Craven Cross Vein, described below; (ii) Harrop Vein was probably represented by strings with a little mineralisation (iii) Lodge Vein (p. 219) may persist to tunnel level; (iv) The so-called Blackhill Vein may be the continuation of Foxholes Vein (pp. 219–220), or it may be a vein unknown at surface; it is wide but not necessarily well mineralised.

From Bradford No. 2 Shaft, the Greenhaugh Mining Company drove crosscuts N to Hargate End and S to Woodhouse Vein. These veins were then stoped, the yield being 1300 tons of concentrates up to the time of the liquidation of the company in 1929-30. Subsequently the mine was acquired by Caldbeck Mining Co., and by 1942 four sumps and an inclined drift had been sunk below the 70-Fathom Level, the deepest being 68 to 875 ft (21–267 m) OD. These proved the continuation of values below the level, and sampling of the level sole in 1937 showed, west of the line of the aqueduct, 350 ft (107 m) long with average vein-width 13.5 in (0.35 m) with 15.6 per cent Pb, 55.7 per cent CaF_2 while the corresponding figures for 400 ft (122 m) E of the line, were 23.2 in (0.59 m), 10.3 per cent Pb, 60.7 per cent CaF_2. The vein is subject to pinches and swells, and the values are better in the wider parts. At 470 ft (143 m) W of the tunnel line, the vein becomes completely barren and continues so to the forehead at 640 ft (195 m). The mineralisation dies out exactly 200 ft (66 m) below the top of the Hargate End Limestone.

Craven Cross Vein branches off Hargate End Vein near Bell's Gin Shaft [1025 6392] and has been worked for 1250 ft (381 m) to the NW. A section by N. Newbould dated 1824 (Z1555) indicates that it was wholly worked under grit cover, the other shafts being Gulf Shaft [1011 6402], 550 ft (168 m) N of Derby Shaft, Steam Engine Shaft and West Gin, respectively 250 and 450 ft (76 and 137 m) W of Gulf Shaft. In these three shafts the base of the Grassington Grit was proved at 125, 115 and 112 ft (38, 35 and 34 m) respectively below surface. The deepest working level lay nearly 300 ft (91 m) below surface at Gulf Shaft, at 1024 ft (312 m) OD; others were at 1062, 1084 and 1125 ft (324, 330 and 342 m) OD; the oreshoot thus extended almost 200 ft (67 m) below the unconformity. There was apparently no inducement to drive up the 56-Fathom Joint Level on this vein. Fluorite and baryte are common on the shaft heaps.

Stee Shaft Vein has been worked from the shaft of this name, 240 ft (73 m) N of Derby Shaft for 400 ft (122 m) to Bell's Gin. There were also workings from a 40-Fathom Level at Bell's Gin, but no details are available. Tested at the 56-Fathom Level, nothing appears to have been found. Longthorn Vein, a NW branch from Hargate End Vein near Derby Shaft was worked mainly about the 30-Fathom Level (Z1560, dated 1858) and a stretch 250 ft (76 m) long was extracted to 45 ft (14 m) below the level.

Woodhouse Vein branches ESE from the south side of Hargate End Vein about 400 ft (122 m) W of Derby Shaft; Woodhouse Shaft, sunk to the 42-Fathom Level was still open and in good condition in 1942. East of this shaft the vein splits into three branches, worked on the 56-Fathom Level from Bell's Gin and what is probably the northern branch was worked from the 70-Fathom Level of Bradford No. 2 Shaft. As seen in 1942, an oreshoot approximately 200 ft (67 m) long had been stoped above this level west of the shaft crosscut. Sampling of the sole here in 1937 gave the following results: Average vein width, 6.3 in (0.16 m), 8.8 per cent Pb, 47.5 per cent CaF_2. A sump near the eastern end 48 ft deep, to 893 ft (272 m) OD showed vein up to 10 in (0.3 m) wide with up to 33 per cent Pb, turning barren at the bottom. The height of the worked oreshoot is not exactly known, but it coincides in position with workings shown on a section dated 1858 (Z.1562) below the 42-Fathom Level, so it probably exceeded 150 ft (45 m). This section shows the vein continuously stoped up to the unconformity. As already noted, the vein was still ore-bearing in the aqueduct tunnel at 911 ft (278 m) OD.

From Hargate End Vein 30-Fathom Level (1180 ft, 354 m) 120 ft (37 m) W of Derby Shaft, a crosscut runs SSW for 2000 ft (610 m), its course marked by a line of old shafts with sandstone, shale and Coldstones Limestone on their tips; the drive must have been in the limestone. Harrop Vein was cut 640 ft (195 m) from Hargate End Vein and was worked for a length of about 160 ft (49 m) between this drive and a nearly parallel crosscut at the 42-Fathom Level from Woodhouse Shaft. The southward extension of the 56-Fathom Level which found the three branches of Woodhouse Vein failed to find Harrop Vein; perhaps it passed through a barren patch, for it seems to have been present in the aqueduct tunnel, and it is the only vein on the west side of the Craven Cross basin that lines up with one (in this case Duke's Vein) on the east side. Rodhill Vein, 200 ft (67 m) farther south and parallel to Harrop appears not to have been productive where cut by the 30-Fathom crosscut (section Z.1554, dated 1858) but carried an oreshoot farther east, worked from the 42-Fathom Level of Moss Shaft [1013 6357], where flats were associated with a southern branch of the vein. The gangue was fluorite. West of Rodhill Shaft to the 30-Fathom crosscut, the Rodhill Vein turns NW and may be followed through small opencasts in white limestone back to Hargate End Vein, but it was evidently too small to develop in depth.

North Rake Lead ore (fluorspar)

SE 06 SE, 16 SW; Yorkshire 135 NW
Direction N72°W turning W to E – W

Pity Me Vein
Direction N10–35°W

North Rake, the northernmost vein of the exposed limestone area of Craven Moor can be traced from the shaft [0986 6390], for almost a mile (1.6 km) to the WNW and W, following the outcrop of the Hargate End Limestone north of the Hebden – Pateley Bridge road. At the east end it is probably split into several weak strings, though the heaps at the shaft mentioned indicate fairly strong mineralisation with fluorite. After encountering the Pity Me cross vein, 470 ft (143 m) W of the shaft, the line appears to be shifted W side S, but thereafter the fracture appears stronger and more collected. According to a section by D. Williams for the East Grassington Mining Co. dated 1891, stoping began 240 ft (73 m) beyond Pity Me Vein and extended 1760 ft (536 m) to the W. Harris Shaft [0948 6397], with levels at 27-Fathoms (1105 ft, 337 m OD), 37-Fathoms (1045 ft, 319 m OD) and 47-Fathoms (994 ft, 303 m OD), had been deepened to the 60-Fathom at the time the section was made and consideration was being given to a further deepening to 70-Fathoms. The ground above the 27-Fathom is shown as stoped out, with about two-thirds of the ground extracted down to the 47-Fathom Level. Dickinson (1970, p. 22) reproduces a later section, showing Harris Shaft down to the 70-Fathom and an oreshoot 220 ft (67 m) long continuing into the sole; he also mentions having seen one plan showing this shaft down to the

82-Fathom Level. Two other shafts were working during the second half of the 19th century, Hammond, 500 ft (152 m) WNW of Harris, and West Boundary, 700 ft (213 m) W of Hammond; the former went down to a 50-Fathom Level, and there were sump workings on an oreshoot about 250 ft (76 m) long down to the 60-Fathom; while West Boundary Shaft according to Williams' section went to the 27-Fathom, but the later section shows it down to the 47-Fathom. The impression gained from these sections is of a substantial orebody with two discrete roots below the 47 or 50-Fathom Level, but it is not possible to be certain of the real continuity of the workings between surface and 27-Fathom Level, or of the validity of the later section in view of the lack of returns of production in Hunt's statistics. The only return by the East Grassington Mining Company was of 88 tons of concentrates in 1886, and the company went out of existence in the early 1890s. The output of the Craven Moor operations as a whole from 1861 to 1887 amounted to only 3620 tons of concentrates, and this includes contributions from veins already described, and some, like Blackhill Vein, yet to be considered. A possible explanation would be that the development of Harris and Hammond shafts down to the 47-Fathom, and the former to 60-Fathom predated the official records, and that the deeper extensions were anticipated but not in practice realised. The dumps remaining in the central part of North Rake are consistent with fairly substantial workings; more baryte is present than farther east, but some fluorite was removed by Greenhaugh Mining Co. West of the Boundary Shaft, a N – S row of shafts indicates that a cross vein has been tested, and still farther west the North Rake continues into a cavernised area said to join up with Stump Cross caverns. On the heaps from shallow shafts, *Girvanella* algal limestone, typical of the Coldstones Beds is noticeable, but only a little baryte and calcite indicate the presence of weak mineralisation.

Before leaving this area, the contact between the linear traces of the veins and cross-fractures and the ramifying outline of the caverns at Stump Cross may be noted. The veins and that part of the caverns explored up to 1942 are shown in Dunham and Stubblefield (1945, pl. XII).

Dodd Vein

SE 06 SE; Yorkshire 135 SW
Direction N82°E

Fielding Rake? = Lodge Vein Lead ore
Direction N70°W, throw possibly N

Jamie Vein (Lonsdale Vein, Lupton String) Lead ore
Direction N75°W to N85°E; throw possibly N

Bathole Veins Lead ore
Direction N62°W

The plateau of Greenhow Limestone lying between North Rake and Blackhill Vein is traversed by a reticulated pattern of weak mineralised veins running WNW (of which those listed above are the chief) and N – S to NNW cross fractures, many of which have been tested but none has been a producer. No attempt has been made to name the latter, but the pattern is displayed in Figure 36. Fielding Vein may be seen in opencuts 6 to 8 ft (1.8–2.4 m) wide on either side of the track leading from the main road to Jamie Shaft. It was strongest on the west side of the cross fracture which passes through this shaft and can be followed for 1000 ft (304 m) on surface to its junction with the Lonsdale branch of Jamie Vein. Beyond that it may be represented by Lodge Vein, the last vein cut by the 30-Fathom S crosscut from Hargate End Vein. No record has been found of the vein and the nearby shaft from surface appears only to have found sandstone, shale and Coldstones Beds. Jamie Vein can also be followed through shallow surface cuts, the length of which is at least 1300 ft (336 m). Near Jamie Shaft [0964 6366] a southern flyer, Lupton String is given off, and the shaft was sunk upon this. Collared at 1234 ft (376 m), it is 8 ft (2.4 m) in diameter, with a working level at 1080 ft (329 m) OD though the shaft is supposed to continue to 1048 ft (319 m) OD, with a 38 ft (11.60 m) winze below this. According to a report to the Greenhaugh Mining Co. by H. Louis in 1917 the level runs 120 ft (37 m) SE in Lupton String until this is cut off by a cross fault; this was then followed N for 100 ft (30 m) until the main vein was found, and continued to pick up Lonsdale Vein, here regarded as a north branch of Jamie Vein. In the main vein sumps 28 ft (8.5 m) deep are stated to have showed good ore. Jamie and Lonsdale veins are recorded from the 1917 reopening as 1 to 1½ ft (0.3–0.45 m) wide carrying calcite, baryte, a little fluorite and galena valued at 20 to 25 cwts per square fathom (approximately 8 to 10 per cent Pb). The mine is said to be in very broken ground, liable to flooding; according to a report by R. Salveson in 1916, the water stood originally at 87 ft (27 m) from surface in the shaft, but the driving of the aqueduct tunnel lowered it by 50 ft (15 m). Lonsdale Vein is said to have been left in ore, going east, presumably when cessation of pumping led to the closure of the mine. The apparent shift of these veins by the cross fracture is 100 ft E side S; east of this fracture, Lonsdale Vein continues through surface workings for 700 ft (213 m) to its junction with Fielding Rake. The cross fracture at Jamie Shaft is one of the most persistent in the area; it can be followed through Fielding Rake, where it imparts a similar shift, through the confused western reaches of Hargate End Vein and thence by shallow opencuts showing a little fluorite and baryte to North Rake. The dumps at Jamie Shaft were worked over several times for fluorspar before being removed for ballast and the shaft, open in 1942, has now collapsed. Other minerals present in this complex of veins are a coarse 'columnar' variety of calcite, discoloured baryte; and in galena from Jamie Shaft occurred the only record of gold from the Northern Pennines. This has not been confirmed.

Jamie Vein runs NW towards Willie Waters Vein, a branch from which suggests that it may join up near Bell Shaft [0924 6355] (not to be confused with Bell's Gin, mentioned earlier). According to a plan by Salveson (1917), Waters Vein was worked to a depth of 360 ft (109 m) but S. Miall's report to Greenhaugh Mining Co. of the same date gives the depth as 48 to 60 ft (15–18 m), which from the surface appearance seems more probable. A continuation of Waters Vein to the west may be the vein cut in the Gill flats, as suggested by Varvill (1937). According to him, northward dipping flats extended up to 200 ft (61 m). These were worked for lead ore from two inclines (now collapsed) and from Gill Shaft [0922 6343] which also gave access to a vertical pipe deposit figured by Varvill (*op. cit.*, p. 32), communicating with the flat. The stratigraphical horizon here is the Stump Cross Limestone, and the Gill flat has possibly developed at the same horizon as the Middle Flat of Nussey Knot, described below. The pale limestone has been dolomitised and the minerals present are baryte, calcite and subordinate fluorite in addition to galena. The vertical pipe had a maximum diameter near the bottom of 60 ft (18 m), was tapered upwards, and was worked from surface to a depth of 132 ft (40 m). Mapping suggests that it was controlled by the crossing of a N – S fracture through the combined continuation towards the NW of the two Bathole veins, two feeble veins worked over a length of 830 ft (253 m) SE from the outcrop of the base of the Greenhow Limestone, in the lower beds of that formation. Baryte is the principal mineral, with only traces of fluorite.

Blackhill Vein Lead ore

SE 06 SE; Yorkshire 135 SW
Direction N75°W turning abruptly to N75°E

Harker Vein
Direction N80°E turning to N82°W

Foxholes Vein
Direction N70°W, throw SW

Nussey Knot Cross Fractures
Direction N – S

Blackhill Mine, the westernmost mine of the Greenhow Area, consists of a main level [0802 6322] (Figure 36, loc. 11) driven eastward at 980 ft (299 m) OD for 3755 ft (1.14 km) to the point where Foxholes Vein is given off to the SE, and thence a further 460 ft (140 m) along this vein. There were three important shafts, Engine [0862 6323], Ashworth, 1115 ft (340 m) ENE, and New Blackhill 475 ft (148 m) beyond this; the first two communicated with the level, but New Blackhill was carried on to 90 ft (27 m) below the level, with a drift of 42-Fathoms (about 915 ft (279 m) OD). As far east as New Blackhill Shaft, the workings are entirely in the Timpony Limestone of Holkerian age, a grey to dark grey series of well bedded limestone with carbonaceous shaly films and partings. No information is available as to the depth to Lower Palaeozoic rocks beneath this; a report that slate had been found in Blackhill Level could not be confirmed. The first 1210 ft (369 m) of the level was driven off vein; the best of the old plans shows the vein changing direction from NE to ENE at the point where the level cuts it, but the fact that a crosscut was made S from the adit suggests that the direct fracture may also have continued. North of this first section of the level, the lowest of three horizons of flatting associated with Nussey Knot (Nursery Knot of modern six-inch maps) was worked from numerous shallow shafts in Drygill, covering an area perhaps 200 × 400 ft (61 × 122 m), and these workings extended into the level. The beds dip north here at 25° and the northern limit of workings in the Low Flat was probably where they were drowned out by the sub-surface drainage beneath Drygill, which often has no surface water in it. The Blackhill Vein carried three oreshoots according to D. Williams' section of 1891 (Dickinson, 1971), separated by barren ground: (i) with the old Engine Shaft as centre, 700 ft long × 125 ft (213 × 38 m) vertically, tapering upwards; (ii) at Ashworth Shaft, 300 × 150 ft (91 × 46 m); and (iii) at New Blackhill, 450 × 240 ft (137 × 73 m) at its maximum dimensions. A considerable subsidence marks the site of the first of these. There is evidence that ore occurs in the vicinity of intersecting cross fractures or oblique flyers; for example, near Engine Shaft two SE-trending flyers, regarded as part of the Harker vein-system, can be followed for 1300 ft (396 m) to a point where they become united. Blackhill Vein is shifted E side S by a N – S cross fracture near Ashworth Shaft, and several more cut through it near New Blackhill. Fluorite was plentiful in the heaps around Engine Shaft, where it was being extracted; since then the dump at Ashworth has been worked over by Clay Cross Co. for this mineral. Baryte is widespread and was abundant at Ashworth Shaft.

After the abrupt change of direction to SE, the vein is called Foxholes Vein; this portion was worked through West Foxholes Shaft, 340 ft (104 m) beyond the turn, steeply inclined west, and East Foxholes Shaft [0930 6330]. The working level was the 20-Fathom (37 m) and the two oreshoots, respectively 200 and 300 ft (61 and 92 m) long were separated by 75 ft (23 m) of barren ground. Both continued below the level and were worked from sumps 25 ft (8 m) deep; the section suggests that ore was left in the soles, but none is shown in the foremost workings of Blackhill Level which reach under the western but not the eastern oreshoot at Foxholes. A crosscut from Blackhill Level, the exact course of which is not known, runs to the Gill Shaft pipe. At least four cross fractures cut through Foxholes Vein and on one of these a drive went south from Foxholes Shaft to link with Harker Shaft [0930 6324]. The Harker Vein forms, in effect a large but weak loop between Blackhill Vein and the North Craven Fault; the workings are generally shallow, and little development was done even at the 20-Fathom Level. The dumps at Foxholes contained abundant colourless fluorite, yellowish baryte, calcite and much dolomitised limestone. A shaft 520 ft (158 m) E of East Foxholes shaft on the line of the vein failed to find it, in spite of crosscutting for it, and no justification is forthcoming from surface mapping for the view that the fracture cut in the aqueduct tunnel at 2800 ft S (p. 217) was Blackhill Vein; alternative possibilities are discussed later.

Returning to the prominent crag of Greenhow Limestone at Nussey Knot, two additional flat deposits were worked beneath the crag, one within Stump Cross Limestone and one at the top of this limestone, under the strongly-bedded Greenhow Limestone. Varvill (1937) has illustrated how these were related to a series of cross fractures trending N – S; the workings are scattered along strike for about 1000 ft (305 m) but they are not continuous. The mineralisation is rather pink baryte with a little galena and some cubic pseudomorphs indicating the former presence of either pyrite or fluorite. The deposits have been worked down dip by inclines following the pipes formed adjacent to cross fractures for upwards of 100 ft (30 m); the most recent operation, by Greenhaugh Mining Co., was an incline directly north of Drygill House, which enters the deposit through the roof at an angle greater than the angle of dip. A shaft on the north side of The Knot may be on the feeder to this flat. Barytes was the object of these workings.

The Nussey Knot belt of flats continues to, or is resumed at, Croft Gill, ½-mile (0.8 km) farther west, where there are trials on the top flat. From a point [0689 6435] south-west of Grimwith Reservoir, Kelshaw or California Level (Figure 36, loc. 12) was driven 2850 ft (869 m) in an average direction S21°E to intercept the belt down dip on the steep north flank of the monocline. A section of this level by J. R. Dakyns made during the primary geological survey in the 1870s when the level was still fully accessible shows that it starts in Nidderdale Shales and passes through the Grassington Grit Formation, dipping 20°N, reaching the unconformity at the base at approximately 850 ft (259 m) from the portal. The section in this formation resembles that at the Mire Shaft, Grassington (p. 198) in containing a thick grit in the upper part, separated by shale from a thinner sandstone in the lower part. Three coal seams were cut, one near the top of the formation, the second in the shale and the third on top of the lower sandstone; these were worked by the Aket Coal Pits which extend as far south as the outcrop of the lowest of the three seams. Kelshaw Level may have been driven in the first case to drain these. The first shaft shown on the section as connecting with the level is Lambert's, some 940 ft (288 m) from the portal; here the level is believed to be in beds equivalent to the Dirt Pot Grit. The final shaft, Wood's, is 2675 ft (815 m) SSE of the portal, starting near the top of the Greenhow Limestone. Beyond this the level entered the top series of flats at a depth of about 260 ft (79 m) below surface. Dickinson (1970, p. 16) records that the Yorkshire Lead Mining Company were driving the level in 1863 when a 'pipe' of baryte with two ribs of galena was cut. Before the expiry of their lease in 1874 the level had been abandoned due, it is said, to the hardness of the ground. In 1942 the level was blocked 545 ft (166 m) from the portal. At 916 ft (279 m) OD, it lies below the new level of Grimwith Reservoir and will never open again, having been sealed off.

The only separate production figures for the Blackhill mines (listed as Craven Moor East) are for 1875–84 and 1894, when a total of 1074 tons of concentrates was produced; the Craven Moor West area yielded 803 tons in 1875–84. The figures however give no satisfactory picture of the output of these mines and in this respect they are typical of the whole field. Data prior to the early 19th century are scanty and no continuous record is available before the commencement of official statistics in 1845, by which time production was past its zenith. Excluding the Ashfold Side Beck mines, the available information is as follows:

Years		*Long Tons* *Lead*	*Lead ore*	*per year*
1706–1732	Cockhill, Coldstones, Galloway, Lum[a]	2503	(4170)[f]	160
1764–1829	Forest Moor[b]	1555	(2321)	36
1806–1817	Cockhill, Sunside, N. Coldstones, Black Rigg[c]	5634	(8409)	701
1839–1844	S. Coldstones[d]	1528	(2281)	380
1845–1854	Pateley Bridge mines[e]	6015	8940	894
1855	Craven Moor	229	342	342
1855	Sunside and Cockhill	184	274	274
1856–1860	Pateley Bridge mines	2518	4168	834
1861–1887	Sunside and Cockhill	3406	4687	293
1861–1866	Forest Moor	66	118	20
1861–1873, 1875–1884, 1894	Craven Cross, Craven Moor	2603	3627	165
1915–1936	Craven Cross, Craven Moor, Cockhill	1400	2009*	

Sources: [a]Jennings, 1967, p. 159 and Raistrick, 1973, p. 31 give a total of 3178 fothers including Stony Grooves; the fother is calculated at 21 cwt and Stony Grooves is assumed to have contributed one-quarter; [b]Raistrick, 1973, p. 58; [c]Mr D. Gill's figure, communicated in 1939, gave 11 268 tons as the metal production from the Bewerley Royalty for this period; half is assumed to have come from the Providence–Prosperous mines. If the figures used were the same as the Eagle Hall duty lead figures mentioned by Jennings *op. cit.*, p. 285, he expresses the view that an exact output cannot be derived from them, but interprets them as indicating 1000 tons per year average lead production, consistent with Gill's figure; [d]Bewerley Royalty records per Mr Gill; [e]Official statistics provide the data for this and subsequent entries. [f]Concentrates are calculated from metal production on the basis of 60 per cent recovered prior to 1800, 67 per cent thereafter.

*This figure includes a few tons from Lunehead Mine, near Brough.

Though these data, leading to a recorded total of 41 346 long tons of concentrates, give some idea of the scale of operations at Greenhow, they are incomplete in several important respects; for example, there is no record for the years 1806–44 for Craven Cross when extraction was proceeding down to the 56-Fathom Level, or for the years between 1818 and 1844 when exploitation of the Green Grooves Basin was at its most extensive. Nor are the 500 or more years prior to 1705–6 represented, though it may be inferred that lead ore was already being mined at the rate of at least 500 tons per year by that year. An attempt—admittedly unsatisfactory—may be made to derive a total picture from the geological and mining evidence, bearing in mind the probable smelting capacity, and using the few historical clues available. During the period prior to 1600, smelting was by bole hill and demand small. Although about 13 tons were sent to roof Windsor Castle, and larger quantities were used on abbeys and churches, it is unlikely that concentrates output greatly exceeded 2000 tons. At the beginning of the 17th century, practice began to improve, partly as a result of the spread of technology brought to Cumberland by the German miners of Elizabeth I. Sir Stephen Proctor erected a bellows-blow smelt mill, probably at Cockhill, and soon afterwards Sir John Yorke's Heathfield Mill came into existence. Considering what is known about contemporary smelting practice in Wharfedale, it is safe to conclude that at least 100 tons per year capacity was now available to smelt Greenhow ores, and allowing for poor years, a total of 7500 tons ore may have been smelted from the beginning of operations up to Sir Thomas White's probable reconstruction of Cockhill Mill prior to or during his vigorous period beginning in 1705. Capacity for making over 400 tons metal now existed, and though White's mines went through a poor period from 1718–32, it is difficult to believe that production at Greenhow as a whole fell below 200 tons of ore per year, implying 10 000 tons in 1733–82. For 1783–1805, prosperous conditions had led to the driving of Cockhill and Gillfield adits and to the sinking of shafts at Craven Cross, where production rose to 300 tons lead per year (Jennings, *op. cit.*,). A figure, probably conservative, of 500 tons ore per year is adopted for this period, implying 17 500 tons output. Now followed the peak of lead production at Greenhow. A committee set up to promote a railway from Knaresborough to Pateley Bridge in 1818 reported, according to Jennings (pp. 285–286), that lead output was 3000 tons per year, but he considers it unlikely that the Yorke royalties produced more than Bewerley, and postulates 2000 tons as a more likely figure, including the Ashfold Side Beck mines. The fall in the price of lead after 1829 had a discouraging effect, and Jennings cites a correspondent in the Mining Journal for 1837 who stated that the output from the whole field had been 1200 tons lead 'for some years', and by the end of the 1830s Jennings estimates that output had fallen to 700 tons/year. The appropriate average figure for Greenhow, excluding Ashfold Side for 1818–44 appears to be 500 tons ore per year, adding 13 500; 2000 tons more may be added to cover the missing Yorke royalty figures for 1806–17. In conjunction with the recorded figures given in the table above this produces a highly tentative estimate of about 95 000 tons lead concentrates for the total yield of Greenhow Mining District, excluding Prosperous – Providence, Merryfield, Merryfield Hole and Stony Grooves, for which an estimated total of 25 000 tons should be added (p. 229) bringing the total to an order of magnitude similar to that of Grassington.

North Craven Fault = Foxholes South Vein

SE 06 SE, SW, SD 96 SE; Yorkshire 135 SW, 134 SE, SW
Direction N82°E (Pateley Bridge to River Dibb); N77°W (to Austwick); N50°W (to junction with Dent Fault). Variable but substantial S displacement.

Middle Craven Fault
Direction approximately E – W from near Threshfield to Settle, where it joins the South Craven Fault

Separating the Greenhow Mining District from that of Skyreholme runs the North Craven Fault, a broad zone of disturbed strata. An account of its structural significance has already been given, noting that east of Grassington it lies close to the belt of intra-Pendleian uplift marking the boundary of the Askrigg Block to the north and the Craven Basin to the south, though it does not coincide in detail with the change of facies in the Carboniferous. Reference has already been made to the Green Grooves South Vein as a possible branch of the fault (pp. 213–215) and to the Blue or Black Rigg and Forest Moor veins as sympathetic fractures if not actually parts of the fault zone (p. 215). Bog Vein, an E – W fracture found in a single trial shaft in Green Bog, 2250 ft (686 m) SSW of Bradford No. 2 Shaft, is probably in the same category, but little is known about it. In the Aqueduct Tunnel, the fault cut at 2800 ft (953 m) S of Bradford No. 2 Shaft (p. 217) and recorded in some descriptions as Blackhill Vein we regard as the footwall of the North Craven Fault; one record places it at 2808 ft (956 m), dipping 50°S, another at 2797 ft (952 m), dipping 62°S, but there is little doubt that both refer to the same fault. On the north side it terminates the limestone a short distance beneath the unconformity; to the south the tunnel passed into thick shale dipping north, interpreted by Dunham and Stubblefield (1945, fig. 5) as Nidderdale Shales, the displacement being of the order of 800 ft (244 m) S, accentuated beyond the stratigraphical separation because the beds dip steeply towards one another, north and south of the fault. Farther south other, probably smaller faults were cut, dipping both N and S and eventually the Grassington Grit is believed to have reappeared in the tunnel forming a broad anticline with its axis at about 2 miles

(3.2 km) S of Bradford No. 2 Shaft. The 2800 ft (953 m) S fault we consider to be continuous with Foxholes South Vein, worked from a S crosscut 1020 ft (311 m) long at the 20-Fathom Level from Foxholes East Shaft (p. 220); this would be consistent with a 50°S dip on the fault. The vein was followed for 550 ft (168 m) E from the crosscut and exploited from seven shallow shafts, the tips of which display grit, limestone, baryte and a little galena, the limestone being typical of the Coldstones Beds. No other penetration of the North Craven Fault has been made from the Greenhow side.

As demonstrated by F. W. Anderson (1928) the presence of the fault is nevertheless evident from surface mapping between Blackhill, where the surface formation is the lower division of the Carboniferous Limestone (the Timpony Limestone), and the lower slopes of Rear Clouts, where the northernmost of the Skyreholme group of veins were worked in strata equivalent to the Greenhow and Coldstones limestones, beneath Grassington Grit. Here a crosscut N from Hardy Vein (p. 223) is believed to have reached the hangingwall fracture of the North Craven Fault at shallow depth; a winze 48 ft (15 m) deep was sunk here [0868 6289] but no development was done. It should be emphasised that the ground mapped as limestone for nearly ½-mile (0.8 km) W of this position along the south side of the fault does not coincide with the axis of the Skyreholme anticline, but lies on the NW flank of this structure, with the base of the Coldstones Beds dropping more than 200 ft (61 m) to the NW across it. The axis in fact turns more or less parallel to and ¼-mile (0.4 km) S of the fault, to coincide approximately with the position of Wright Vein (p. 223) as it cuts the Grassington Grit, continuing eastward to the position established by the Aqueduct Tunnel. From the Hardy Crosscut Winze, the fault zone passes westward north of the outlier of Grassington Grit exposed in Fancarl Crag [065 625] where existing geological maps show only a single fault; however, apart from the crag itself, the outlier is very poorly exposed, and a fault zone as much as 400 ft (182 m) wide could well pass through here. A shaft at [0668 6289] is considered to have been sunk to the footwall fracture since highly dolomitised limestone is present on the dump. Beneath the unconformity under the outlier, the Dibb and Black Hill limestones of Black and Bond (1952), equivalent respectively to the Simonstone and part of the Middle limestones have come in, and Upper Bowland Shales probably occur between the Black Hill Limestone and the base of the Grassington Grit. The position of the North Craven Fault can next be fixed in the River Dibb, where at 1570 ft (479 m) SW of Dibbles Bridge the outcrop of the Greenhow Limestone is terminated, strongly dolomitised, against the footwall fracture. A faulted-in wedge of grit is exposed immediately to the south; the ground is then concealed for 450 ft (137 m) to the point where Dirt Pot Grit is exposed in the stream, overlying Coldstones Limestone, indicating that the total displacement by the fault zone is at least 550 ft (168 m) S; if the grit against the footwall fracture is Grassington Grit, the throw at that point could be substantially greater. Trial shafts in the fields ½-mile (0.8 km) E of the Dibb may have entered the zone near the horizon of the Simonstone Limestone; the debris shows the limestone veined with limonite and calcite.

During the IGS mineral reconnaissance survey, begun in 1973, geophysical measurements were made over the North Graven Fault between Drygill House and Skyreholme. A VLF anomaly was found to coincide with the hangingwall side of the fault-zone, in the vicinity of the Hardy sump, but to diverge westwards from the course of the fault as mapped from surface, eventually crossing diagonally the line of Wright Vein. Reporting this work, Wadge, Bateson and Evans (1984, pp. 20–22) note 'the possibility that disseminated sulphides were trapped against tight Lower Palaeozoic basement on the downthrow side of the . . . fault'. Unfortunately the geophysical data yields no indication of the possible depth, but comparison with the succession at Malham suggests that 250 to 350 m of beds might be present beneath the Skyreholme Anticline. The prospect, though speculative, is not without interest.

Between the River Dibb and Hebden Beck we follow the view of Black (1950) that the belt of disturbed ground widens out so that the distance between the northern fracture, which crosses the beck below Town Hill, and the southern fracture at Hebden Mill has increased to 3000 ft (0.91 km). The effect of the footwall fracture has meanwhile lessened, for the Grassington Grit outlier of Ratlock Hill, south of this fracture, shows the base of the grit only 150 ft (46 m) lower than in the main outcrop on Knot Hill. It is suggested that a link crossing the zone from the grit wedge in the Dibb to West Gate Lathe to join the southern element of the zone transfers part of the throw, for around Bent's Lathe, south of Ratlock Hill, Grassington Grit is faulted against Great Scar Limestone making up the Burnsall area. The throw on the southern fracture could thus exceed 500 ft (150 m). Hebden Beck exposes Upper Bowland Shales, including the Scale Haw Limestone of Black (*op. cit.*), the base of the limestone stepping down from 700 ft (213 m) OD north of the footwall fracture to below 550 ft (168 m) OD before reaching the southern fractures. At Hebden the Trial Level [0284 6298] at 595 ft (181 m) OD (see also p. 201) is considered to have started in Upper Bowland Shales, but these beds are not well seen owing to the arching and timbering of the level to 318 ft (100 m) from the portal. From here to 422 ft (129 m) the walls are clear and the south-dipping footwall fractures, with anticlinally-folded limestone, sandstone and shale between them, were measured in detail in 1942. North of the fault, a limestone rolls over and dips steeply north, overlain by a sandstone, in part shaly, at least 15 ft (4.6 m) thick. The sandstone can only be the Dirt Pot Grit, the underlying limestone being the top part of the Coldstones (= Gayle) Limestone, and the limestone above the sandstone just seen before the arching recommenced to the north is probably the Simonstone. Unfortunately the unconformity beneath the Bowland Shales was not exposed. No mineralisation was found in any of these fractures, the filling being shale gouge.

West of Hebden, there is a middle fracture within the zone, as Black (1950) has demonstrated, throwing N, and bringing Upper Bowland Shales, identified from the presence of *Eumorphoceras pseudobilingue* in Isingdale Beck, against Great Scar Limestone to the south, so that here the North Craven zone is a trough with a net north downthrow. At the Sewage Works at Linton, a segment of Grassington Grit is faulted in, and at least four fractures must cross the Wharfe, including the footwall fracture at Linton Tin Bridge, giving the disturbed zone a width of 1200 ft (365 m). The fractures are very poorly exposed, if at all, but had any float ore been found, they would have been opened up; on existing evidence they must be regarded, as at Hebden Trial Level, as unmineralised. The North Craven zone narrows to the west and at Skythorns Great Scar Limestone to the north is faulted by it against Grassington Grit to the south.

A subparallel fracture, the Middle Craven Fault, appears westward of Threshfield and forms the southern boundary of the impressive Great Scar Limestone area of Malham and Gordale Scar. The two Craven faults pass into the Settle (60) Sheet, now (1981) re-surveyed, and except for a reference later in this chapter to mineralisation associated with the Middle Craven Fault at Stockdale, east of Settle, no further description will be attempted here. Detailed information, such as is available, has however been given for the North Craven Fault in its passage through the Pateley Bridge (61) Sheet because of its possible interest as a prospect for mineral discovery following the discoveries from 1959–72 of major lead-zinc, copper and barytes deposits in the Carboniferous Limestone of the Irish Republic, adjacent to the downthrow sides of faults of similar direction and magnitude.

In this connexion, the possibility that the North Craven Fault continues as a pre-Namurian structure east-north-east of Pateley Bridge must also be considered. Petroleum exploration boreholes at Aldfield [2406 6812] and Sawley [2449 6502] (Falcon and Kent,

1960, p. 32 and pl. II), about 6 miles (10 km) beyond the last known position of the fault at Pateley Bridge, both drilled on small folds, gave remarkably different stratigraphical results. At Aldfield, the lower part of the Arnsbergian and the Pendleian succession, from the Lower Follifoot Grit down to the base of the Grassington Grit, is missing, so that Upper Follifoot Grit[1] rests directly on a limestone with 'a normal D_2' (that is, Brigantian) 'fauna'. It might be suggested that the intra-Pendleian movements left high ground here at the beginning of the Namurian. Only 3 miles (4.8 km) away, the Sawley Borehole entered Colsterdale Marine Beds at 425 ft (129.50 m), Grassington Grit at 937 ft (326 m) and 'Great Scar Limestone' at 1268 ft (441 m). The existence of a penecontemporaneous fault between these boreholes, which could be of interest as an ore prospect, is implied.

Hardy Vein Lead ore

SE 06 SE; Yorkshire 134 SE, 135 SW
Direction N75°E

Burhill North Vein and String
Direction N60°W, N70°W

Wright Vein
Direction N65–80°W, with short E – W section, throws N

Spar Vein Lead ore Fluorspar
Direction N70°W

Burhill Old Vein Lead ore
Direction variable between E – W and N70°W, throws N, perhaps 40 ft (12 m)

Dakyn's String
Direction E – W

Nebcastle Rake = Hopewell Vein Lead ore Fluorspar
Direction N65°W, breaking up to NW into five divergent veins

Hush or Rush Vein Lead ore
Direction N80°E to E – W

Walton Vein Lead ore Fluorspar
Direction N60°W

Inman Vein
Direction N62°W

This group of veins was worked at the *Burhill Mines*, active during the middle years of last century for lead ore, and during recent years for fluorspar (Figure 36). They constitute the northern group of deposits in the limestone core of the Skyreholme Anticline, the axis of which runs NE through Troller's Gill and then turns eastward through the eastern workings of Wright and Old Veins to reach the Aqueduct Tunnel as already described (p. 217). On the south flank of the plunging anticline, the Fancarl Series of Anderson (1928), equivalent to the Coldstones Limestone of Greenhow, are present beneath the intra-Pendleian unconformity as far south as Old Vein, but south of this the Grassington Grit and underlying shale rest directly on Great Scar Limestone, probably on beds equivalent to the Hargate End Limestone. All three divisions of the limestone have yielded some production, but a case can be made for the view that the upper part of the Great Scar Limestone was most favourable to mineralisation. Some of the veins cross the north-west flank of the structure, these running NW heading towards the North Craven Fault, but in no case has a vein been followed to its intersection with the fault, all having apparently become unproductive in that direction. The axial plunge of the anticline can be judged from the fact that the base of the Grassington Grit on the footwall of Old Vein is at 1340 ft (908 m) OD, while the corresponding position above the Aqueduct Tunnel, almost 1200 ft (366 m) to the east, is a little below 1200 ft (366 m) OD.

Most of the veins come to outcrop in the bare or thinly-covered limestones and must therefore have been worked from very early times. Their courses are marked, as elsewhere in Wharfedale, by numerous bell-pits and shallow shafts left by small partnerships, a situation that at Greenhow only applies to Craven Moor. Wright Vein was worked opencast from [0802 6268] for 1550 ft (472 m) to the SE in early times, the positions of associated shafts indicating that the vein dips NE. These workings were in Hargate End and Greenhow limestones, but Virgin Shaft, 250 ft (76 m) NW of the end of the opencast was sunk, apparently unsuccessfully, in calcite mudstone of the Fancarl (= Coldstones) Series. The vein was not worth following for the remaining few hundreds of feet to its junction with the North Craven Fault. The opencast, 6 to 10 ft (2–3 m) wide, reveals dolomitised limestone on the walls, with 'Blue John' type dark purple fluorite, unusual for the present region, but only in small quantities. Dickinson (1970) considers that the shafts must have reached depths over 100 ft (30 m); Rodhill Shaft, midway along the opencast seems to have been the most important, but the general impression given by this part of the vein was that any workings were shallow. A cross fracture trending NNE apparently terminates this part of the vein, shifting it 20 ft (6 m) E side S. At 300 ft (91 m) E of this a second cross fracture came through, apparently with beneficial effects upon the vein, for there are now a number of deeper shafts. Gulf Shaft [0868 6247] was 210 ft (64 m) deep according to Newbould's plan of 1860, but later records indicate that the vein was worked to 300 ft (91 m) deep from a level at 270 ft (82 m) and an incline dipping SE. It is uncertain whether the 270 ft level was connected with Gulf Shaft, or Whim Shaft close to Spar Vein, 380 ft (116 m) SE of Gulf Shaft. At this end of the mine the oreshoots were plunging SE with the dip of the shales overlying the limestone, Wright Vein being abandoned after it had been followed 800 ft (244 m) from Gulf Shaft. Gulf Shaft was collared at about 1230 ft (395 m) OD, so that the deepest workings here reached 930 ft (283 m) OD, only a short distance above the horizon of the Bradford Aqueduct Tunnel, which lies 1120 ft (341 m) ESE of the forehead, and is here driven, according to Dunham and Stubblefield's (1945) interpretation, in the lower part of the Grassington Grit. Measured from Bradford No. 3 Shaft [0899 6188], fractures at 2090 (ft) N and 1600 N might correspond with Wright and Old veins respectively, assuming that these have a strong hade towards the N, as is likely. No mineralisation was recorded nor is rumoured from the driving of the tunnel, but none was found in any of the Burhill veins at surface in this formation. The last working of Wright Vein was in 1964–66, when according to Dickinson (*op. cit.*) the Clay Cross Company got good fluorspar from opencast working, and were sinking an incline in 1967 when they ceased operations. Underground examination to 50 ft (15 m) depth at Dross Shaft, 325 ft (99 m) W of Gulf Shaft had shown that flats carrying fluorite, baryte and some galena had been worked, according to dates carved on boulders, between 1833 and 1857.

From Gulf Shaft an exploratory level, driven before 1860, ran N to Burhill North Vein and Hardy Vein, terminating, beyond Shack or Shake Shaft, at the winze already mentioned as probably being in the hangingwall fracture of the North Craven Fault (p. 222). The crosscut, altogether 2075 ft (632 m) long from Gulf Shaft was at only about 50 ft (15 m) below surface if it connected with shafts on North and Hardy veins shown by Newbould as only 8-Fathoms deep. North Vein was worked for a length of at least 750 ft (229 m) from six shafts and there was a small parallel vein 120 ft (37 m) to the SW. Fluorite and baryte were present, but these veins have not

[1]Reinterpreted by I. C. Burgess as the Red Scar Grit.

recently been worked; probably they are narrow. Similar considerations apply to Hardy Vein.

Burhill Old Vein pursues a sinuous course, running from the north-west flank of the anticline into the axial region, the worked length being 4300 ft (1.3 km). Altogether the remains of 22 shafts can be identified along this vein and the strings which parallel it. Matthew Newbould's plan identifies Whitehead, Harker, Pimley and Lonsdale shafts in the middle stretch of the vein, the last-mentioned [0831 6242] being sunk to 240 ft (73 m) according to a plan dated 1858 cited by Dickinson (1967, 1970) Farther east, the two Mecca Shafts were sunk either side of the NNE cross fracture mentioned above as shifting Wright Vein; the shift on Old Vein is in the opposite sense, about 40 ft (12 m) E side N. The deeper eastern workings, following oreshoots down the plunge of the structure, were probably from Whim Shaft on the parallel vein north-east of Old Vein called Spar Vein. As already noted, the deepest level here was the 45-Fathom (82 m) but an incline on Old Vein was carried down 54 ft (16 m) below this and a series of sumps were sunk on both veins from the 45-Fathom Level. The ESE forehead on Old Vein stands 790 ft (240 m) beyond the crosscut from Whim Shaft, no doubt at the effective limit of working; but whether ore was continuing down the plunge into the sole is not known. On the downthrow (north) side of the vein, Fancarl Series debris can be identified in all the shaft heaps, but not to the south, where the upper beds of the Great Scar Limestone seem to underlie the intra-E_1 unconformity. Thus the Old Vein fault may have been active prior to the deposition of the Upper Bowland Shales and Grassington Grit, though in the inaccessible state of the mines, this cannot be regarded as certain. In the western two-thirds of the Old Vein workings, the gangue mineral is calcite, but fluorite and baryte are both present in the eastern workings. South-west of Old Vein, the crosscut from Whim shaft continues on a winding course to link up with the weak but persistent vein known here as Dakyn's String, a branch from which, near Green Groves Well, links with Old Vein. The considerable heap of tailings near the eastern end of these workings, containing calcite, limestone, fluorite and a little baryte probably came mainly from Old Vein; it was being worked over for fluorspar in 1942.

During the BGS Mineral Reconnaissance Programme, the discovery of an EM-anomaly near Stump Cross, at the western end of the Greenhow field, and north of Skyreholme led to detailed IP and VLF work on the ground (Wadge, Bateson and Evans, 1984, pp. 20–27). In spite of the use of a large array, the IP results were not altogether satisfactory in this complex ground, but two anomalies were identified by VLF methods. The larger of these, about 3500 ft (1.07 km) long corresponds with Burnhill Old Vein in its eastern course (that is, where the vein was most productive) but turns away to the west. An ENE branch is also suggested, corresponding with no known feature. There is thus clear evidence of a conductor here, but the anomaly has not yet (1984) been drilled. Farther north, a second, weaker VLF anomaly corresponds in part with the course of the North Craven Fault; this is discussed below.

Nebcastle Rake has been the most extensively worked of the Burhill veins in recent years and Dickinson (1970, pp. 10–12 and pl. 1) has given a valuable account of operations by the Clay Cross Company between 1965 and 1967. The vein had already been worked for lead ore from the upper part of Trollers Gill for 2600 ft (793 m) to its junction with Hush Vein, some of the shallow shafts probably being very old. Hopewell Level [0712 6240] (Figure 36, loc. 13) starts from the east side of the Gill close to 875 ft (267 m) OD, cuts the vein obliquely from the N at 90 ft (27 m) and was driven in vein to 790 ft (241 m), where the forehead is 130 ft (40 m) below surface. The level, according to Dickinson was driven in the 18th century, and the block of ground above it was stoped out to surface for lead ore. A winze 200 ft (61 m) from the portal, 54 ft (16 m) deep enabled an oreshoot plunging WNW, parallel to the bedding of the Hargate End Limestone on the flank of the anticline, to be extracted over a horizontal length of about 150 ft (46 m). The vein width in the Hopewell stopes varied between 1½ and 3 ft (0.45 and 0.9 m), calcite occurring near the walls and fluorite in the centre. A sublevel from White's Shaft, 90 ft (27 m) deep near the east end of the Hopewell stopes enabled working to be extended 180 ft (55 m) further. At 1220 ft (372 m) ESE of Hopewell adit portal the vein gives off a NW branch, Nebcastle North Vein, from the north side, with several strings between the vein and branch. The intersection had a beneficial effect upon the vein, and a substantial oreshoot composed mainly of fluorite, 200 ft (61 m) long had already been worked opencast to 60 ft (18 m) depth when the Clay Cross Company sank an incline to 130 ft (40 m) vertical depth and proved that the shoot plunges W at 40°, but narrows downward. It appears to be following the plunge of the intersection between main and North veins rather than the bedding here. At the bottom or 22-Fathom Level, the vein, consisting of soft and friable fluorite in a matrix of dry clay with strings of calcite and of galena with cerussite (Dickinson, *op. cit.*), was workable only over a length of 70 ft (27 m), the maximum width being 10 ft (3 m). At the east end of the 22-Fathom Level pure white calcite filled the vein, and was extracted from an inclined raise. Exploration of the ground between main and North veins west of the intersection showed it to be highly mineralised with bands of calcite and fluorite, but not at the time workable for the latter mineral. Several old shafts to the ESE were investigated for fluorspar and Skip Shaft, 965 ft (294 m) from the incline was equipped to 40 ft (12 m) depth; the vein which had been 12 ft (4 m) wide to the west, nipped down to 9 in (0.22 m). A gulf following a NE-trending cross fracture occurs near 1100 ft (335 m) SE of the incline, and beyond this workable fluorspar was again found; and at 1180 ft (360 m) from the first incline, a second inclined drift exploited a deposit 250 ft (75 m) long × 4 ft (1.3 m) wide which cut out at 90 ft (27 m) depth. A drive from the incline foot at approximately 1000 ft (305 m) OD found the vein nipped to the west, and mainly calcite with 'odd strings of galena and fluorite' (Dickinson, *op. cit.*) to the east. Throughout the ground SE of the first incline, the vein tended to be split into several strings. All the workings here were in the Great Scar Limestone below the Davidsonina septosa Band, thus in the equivalent of the Greenhow Limestone, crossing the axis of the anticline in the vicinity of the Blackhill Road. Trials on the northern part of Nebcastle North Vein and associated strings made by Clay Cross Company are described by Dickinson (*op. cit.*, p. 12).

Hush or Rush Vein can be followed through 16 old shafts to the contact of the Upper Bowland Shales to the east; the old plan mentions the names of Newbould, Marshall, Walton, Green, Broadley, Press and Hannam in connexion with them. Three shafts near the west end worked an oreshoot up to 12 ft (4 m) wide, 150 ft (46 m) long but otherwise, where it has been investigated recently, widths of 2 ft (0.6 m) are general, notwithstanding which extensive stoping was done by the lead miners. There is no evidence that the vein was followed beneath the unconformity. Walton Vein, the next vein south of Nebcastle Rake, is crossed by a conspicuous NE fracture [0711 6216], south-east of which an opencast yielding good fluorspar to F. C. Walker was developed by means of a 50 ft (15 m) shaft by Clay Cross Company; the deposit was 8 ft (2.4 m) wide, but it terminated against the cross fault. Inman Vein also contained good grade fluorspar, but under barren limestone 10 to 15 ft (3–4.5 m) thick. Walker's workings here are said to have bottomed the deposit at 40 ft (12 m) below surface, winzes beneath this being barren. In the past Inman Vein yielded lead ore from Morlands Shaft.

Returns from the Burhill Mining Company, formed in 1857, appear in official statistics from 1861–81 (no output 1874–76). The total for the period is 1432 tons of concentrates yielding 1006 tons of metal, an average recovery of 70 per cent. The principal operations were probably the 45-Fathom levels on Old and adjacent veins from Whim Shaft, though many other places were probably tried. No data remain for the earlier workings, which certainly went back to 1675, the date on a penny found underground at Nebcastle (Dickinson, *op. cit.*). It is not certain whether the system of crosscuts from Gulf and Whim shafts belongs to this period. Dickinson mentions a Green Grooves Mining Co. as involved in the eastern development to depth, but this does not figure in official records, though a Burhill No. 2 company involving people different from those in the Lead Company, does appear, though with no production. Consideration of all the evidence suggests that a reasonable global figure for lead output from Burhill might be 6000 tons of concentrates. Fluorspar production, up to the present, can be stated in a round figure as 20 000 tons and some hundreds of tons of white calcite have been sold. There can be little doubt that fluorspar ore remains, but the experience of the Clay Cross Company showed that the ground is very pockety and to get a steady output of feed would demand considerable development in advance, with a substantial part of it through barren ground. Changing economics of fluorspar mining might someday bring about a reconsideration of this area.

Pollard Vein

SE 06 SE; Yorkshire 134 SE
Direction N60°W

Rodwell or Rodhill Vein Lead ore
Pimley Vein
Two veins forming an elongate loop with average trend N75°W

Gill Heads Veins Fluorspar (lead ore)
Principal vein trending N63°W,
branch called Joyce Vein, N33°W
Main vein throws 8 ft (2.4 m) NE

Old Man's Scar Vein ? = Stubbs Vein
Direction N77°W

Nape Well Strings

The spectacular Trollers Gill part of the Skyreholme Anticline exposes a series of veins among which Gill Heads Vein has been a significant producer of fluorspar. The working level here [0663 6203] (Figure 36, loc. 14) is close to 840 ft (256 m) OD, and was the Middle Level of the old lead mine. Above it the vein was exposed in the steep western cliff of the gill, and has been opencast for 375 ft (114 m) up to the point where the oreshoot is cut off against overlying shale of the Fancarl Series. Beneath the shale, the opencut exposes 25 ft (7.6 m) of grey calcite mudstone and crinoidal limestone, with a shale parting and with the Girvanella Band in the lower part; these beds also belong to the Fancarl Series of Anderson (1928), and rest upon the pale grey to white Great Scar Limestone equivalent to the Hargate End Limestone of Greenhow. The throw of the vein here is 8 ft (2.4 m) NE. A section published by Dickinson (1970, pl. 2) shows Middle Level driven 1310 ft (399 m) WNW to the shale contact, indicating an average dip of the strata of 10° in this direction; this is however less by 5° than the dip observed at surface in the bearing limestone. The point is of some importance because, near the mouth of Middle Level, a low dipping plane, inclined at 20°W cut off the bottom of the orebody. W. W. Varvill, who was managing the mine when this was discovered, wrote as though he regarded it as a bedding plane (1937, p. 10); later opinion has tended to regard it as a fault, since the vein has not been found beneath it. Dickinson takes this view and gives the strike as 'S45' (presumably S45°W), which would correspond with the strike of the bedding. It is possible that the dip gradually increases eastward towards the axis of the anticline, but it remains uncertain whether the fault is truly discordant, or is a bedding plane with some lateral movement on it. From Middle Level, at 50 ft (16 m) inbye, a 20° incline was sunk by Greenhaugh Mining Company to a depth of 580 ft (147 m) on the dip, and from this, fluorspar was stoped 6 to 10 ft (1.8–3 m) wide, the oreshoot being 430 ft (131 m) long at Middle Level, and the maximum depth below surface 330 ft (101 m). This is the largest single fluorite body yet discovered in the area covered by the present volume; using the minimum width, the tonnage at 14 cu.ft/ton would be 12 000, so that the round figure assigned by Dickinson, 15 000 tons, is a reasonable one. Assays of recent crude are given in Table 17. In sinking the incline, an ancient level was discovered below Middle Level at 800 ft (244 m) OD which had been driven 90 ft (27 m) below the low-angle fault before cutting the vein; it is said to continue through ground stoped for lead ore to the contact with the Fancarl Shale; a third level had also been driven 42 ft (13 m) above Middle Level. The branch known as Joyce Vein was discovered by Greenhaugh Mining Company in a SW crosscut 450 ft (137 m) from the entrance to Middle Level; it was proved according to Dickinson (*op. cit.*) over a length of 1000 ft (305 m) and carried 1 to 6 ft (0.3–1.8 m) of white fluorite with clay selvedges. It had been located in depth from the incline when the mine was closed in 1924. Recent investigations by F. C. Walker and the present (1979) operator, C. R. Schwartz have established not only that the oreshoot is continuing down dip, but that a vein coming in from the SW side carries good quality fluorspar. The problem of the whereabouts of Gill Heads Vein under the bottom fault, if it exists, remains. One possibility is that it is represented on the east side of the Gill by the combined Rodwell and Pimley veins; if this is so, it would imply a NE shift on the footwall side of the fault of 100 to 120 ft (30–36 m). It has, however, not yet been proved that the fault has moved since the mineralisation was introduced, indeed Varvill's description would imply a pre-mineralisation age for it; the continuation of the strong fluorspar oreshoot may not, therefore, exist.

The existing workings on Gill Heads Vein may extend 1680 ft (512 m) NW of the adit. A shaft 230 ft (70 m) and others 1400 to 1950 ft (427–594 m) farther ahead (the latter group known as the Hen and Chicken workings) were sunk into Brigantian strata on the line of the vein, but with little success.

The remaining veins mentioned above are exposed in the Great Scar Limestone outcrop in Trollers Gill; all are small, carrying at most a few inches of fluorite, with calcite.

None of the veins of the Gill Heads group has been worked as far east as the shale on the eastern flank of the anticline. However, a crosscut adit known as Glory Level [0731 6131] (Figure 36, loc. 15) east of Percival Hall was begun, it is said during times of depression in the 19th century, near the unconformity on the south-east flank at 780 ft (238 m) OD and driven northwards into the limestone. The course was NNE for 225 ft (69 m) to an air shaft, then 860 ft (262 m) N to a second shaft. The exact length of the drive is not known, but there is no evidence that it reached Rodwell – Pimley vein. It would go into progressively lower beds in the succession, and the dump shows that Fancarl Series and Great Scar Limestone were cut. Dickinson records (1970, p. 9) that an E – W vein was cut 700 ft (213 m) inbye, carrying calcite, vertical on the W side of the level, but hading S on the E side. The position could correpond with Nape Well Vein. Some dolomite is present on the dump, but no galena was found. Glory Level, though it starts very little below Low Level at Gill Heads Mine, would have given extra working depth on Rodwell, Inman, Walton and Nebcastle veins had it been driven through. On the last-mentioned vein it could have provided 80 ft (24 m) extra depth below Hopewell Adit, but the oreshoot had already been extracted to 54 ft (16 m) below this adit (p. 224).

Limekiln Vein Lead ore

SE 06 SE; Yorkshire 134 SE
Direction N55°W

Derby Veins
Direction N55–75°W

Appletreewick South Vein Calcite
Direction N60°W

The veins of *Appletreewick Mine* (Figure 36) cross the exposed Great Scar Limestone about ¼-mile (0.4 km) N of the village, on the SE side of the Skyreholme Anticline, and are the last group to be described within this structure. Raistrick (1973, p. 74) identifies the first lease of this part of the Yorke estate as 1709 and no doubt the shallow workings belong to the 18th century. Wellington Adit [0578 6054] (Figure 36, loc. 16) at 625 ft (191 m) OD, its name indicating early 19th century work, follows a NNW course, connecting with South Vein at 300 ft (91 m), a flyer called Napoleon Vein at 430 ft (131 m) and Derby Vein at 20 ft (6 m) beyond this; the crosscut was afterwards driven N to reach Limekiln Vein 500 ft (152 m) farther N. The principal producer was Derby Vein, which split into two branches, both workable, 340 ft (104 m) W of Wellington Level; Engine and Cabin shafts were sunk near this point after the formation of the Appletreewick Lead Mining Company in 1857, with levels at 4-, 15-, 20- and 25-Fathoms (7, 27, 37 and 46 m), a level at 33-Fathoms (60 m) being added after the reconstruction of the company in 1875. The predominant mineral in all the veins here is white calcite, but Derby Vein carried enough galena to make it worth working, and output 1861–72 was 4627 tons concentrates, 1875–83, 1568 tons. Limekiln Vein proved disappointing where cut by Wellington Level, though Raistrick (*op. cit.*) reports fluorite in this area. South Vein was mainly calcite, and when Greenhaugh Mining Company in 1923 sank the Varvill Pit [0538 6075] to 246 ft (75 m) depth on it, little galena was found, but calcite was produced in quantity for a few years. A test of the Derby Vein, presumably the south branch, failed to reveal workable ground. Towards the end of their period, the second Appletreewick Lead Company had begun a level running NE from near the entrance to Wellington Level, with the intention of testing Derby Vein against the unconformity, but this project was never completed. Work was also done at the north-west end of Limekiln Vein by the Appletreewick Gill Mining Company, who drove a level [0518 6130] (Figure 36, loc. 17) at about 700 ft (273 m) OD into the outcrop of the vein, but only a few tons of ore were mined. Including early work, an estimate of 7500 tons of lead concentrates for Appletreewick is probably conservative. Some contribution to the total was made by working a flat discovered in the Village Level [0525 6021] at 550 ft (168 m) OD in the course of a drive through Great Scar Limestone at least 625 ft (191 m) to the NE.

Thorpe Kail Veins Lead ore

SE 06 SW, SD 96 SE; Yorkshire 134 SW
Six short veins, trending between N80°W and E – W

Elbolton Veins
At least 14 veins and strings, trending mainly between N60° and 80°W

Stebden Veins
Three short veins trending E – W

Carden Vein
Direction E – W; branch may link N80°E with Stebden

Langerton Hill Veins
Three short veins, N60°W; one N80°E

Swinden Vein
Direction N50°W

West of Appletreewick, the broad outcrop of the Great Scar Limestone through Burnsall, though in part drift covered, is not known to carry any mineral deposits. At Thorpe Kail, however, in the main belt of limestone reef-knolls often referred to as the Cracoe Reefs (Bond, 1950), mineralisation reappears (Figure 6). The largest mineralised Knoll, Elbolton, is nearly 1 mile (0.8 km) in diameter and rises over 300 ft (91 m) above the surrounding limestone plateau. The reefs, consisting of limestone and shell debris, form roughly conical mounds that have been buried by Bowland Shales and re-excavated, though not completely, in the present cycle of erosion. Structurally and texturally, it might have been expected that they would provide excellent environments for ore deposition. Of the ten knolls, four are mineralised, but only with weak veins; no large replacements such as characterise some of the Irish lead-zinc deposits, or those of Pine Point, N. W. T., Canada, have been found. The stages in the formation of the knolls have been very well illustrated by Ramsbottom (1974, p. 63) following the ideas of Black (1958) and others. Bedding is generally poorly developed or absent in the knolls, but where it can be seen, as at Stebden and Butter Haw, the dip is outward from the centre of the hill and the surface of the knoll is almost certainly an original surface of deposition.

The small veins on Thorpe Kail [016 614] lie roughly at right angles to the long axis of the knoll and can be followed by lines of shallow pits and surface cuts, from which the only debris is, in many cases, reef limestone, though calcite was also present in places. The impression is given of narrow stringers like those forming the scrin belts in Swaledale and Wensleydale, but here the spacing is 100 to 400 ft (30–120 m) between stringers. The second vein south of the summit has been followed almost across the knoll, and has been tried by three shafts through shale to the east, near Micklefoots Lathe, but no mineral can be found on the dumps from these. On Elbolton the orientation has changed to predominantly WNW and the veins are both stronger and more persistent. *Elbolton Mine* consisted of a level [0097 6147] and a shaft on the next vein, 175 ft (53 m) to the south. This shaft passed through shale, as did several bell-pits to the south, and a little galena is present on the heaps. Although the bulk of the debris from workings at Elbolton is limestone showing little or no alteration, fluorite and calcite can be found here, notably on the tips from shafts on the west and north-west sides of the hill. The official statistics list Cracoe and Elbolton (with one reference to Thorpe in 1862) from 1860 to 1898, the total production for this period being 244 tons of concentrates, a clear indication of the very small scale of the operations. On Stebden [002 608], limonite, not seen at the other localities, is present probably representing former pyrite. Two shafts on the south-west side of the hill show massive purple fluorite with galena and calcite, and the vein here apparently extends westward under the Bowland Shales to Carden Knoll, at least one shaft having found similar material between the knolls on the line of the vein. Dark purple fluorite also occurs with calcite and galena in a vein worked north-west of the summit of Carden Hill. At the *Langerton Hill Mine* [994 612] the working of a small vein has yielded galena with calcite and dolomite; while at *Swinden Lime Works Quarry* [983 616] a NW-trending zone of tiny fractures containing purple fluorite with calcite and bitumen cut through the centre of the main face as exposed in 1942. Since then the quarrying operations have been greatly expanded. Dickinson (1972) has placed on record the remains of former lead prospects that have been or will be obliterated as quarrying proceeds. He has identified three veins running with the dominant joint-direction (NW) and five running WNW, of which the northernmost is the zone seen in 1942. Apart from this one, the veins carry little other than calcite, and residual clay from solution by groundwater. Traces of galena could be found in debris from one of the NW veins, and the fluorite-bearing vein, now quarried away, also contained galena, with malachite and aragonite. That there were prospect pits here at all shows how persistent was the search for lead ore in former times; these veins must have proved entirely unrewarding.

The thickest vein at Swinden [9776 6135] trends north-west and corresponds to Dickinson's Vein 4 which crosses the hill from side to side via a shallow col, currently being quarried away. This vein, tested by two oblique boreholes sunk for Tilcon Ltd takes the form of a 100 ft (30 m) wide fracture belt with the main veins totalling 10 to 15 ft (3–4.50 m) in width, concentrated towards the edges. The central zone contains numerous clay-filled fissures. Analyses of the cores showed the veins to consist of almost pure calcite with only trace amounts of fluorine (up to 210 ppm) and barium (less than 300 ppm). Rare malachite and pyrite were noted in hand specimens. Recent quarry exposures (1980) show high slickensided faces on the margins of veins with the striae running nearly horizontally.

Dickinson's Vein 6, trending WNW was well exposed in the quarry in 1980 [9783 6153]. This 2 ft (0.60 m) wide vein shows a distinctive layering of pale pink calcite with pale grey calcite, but only the latter contains disseminated chalcopyrite. KCD

Middle Craven Fault

SD 86 SE, SW; Yorkshire 132 NE, 133 NW, SW
General direction E – W, varying between N70°W and N65°E; downthrow S

Pikedaw – Grizedales Veins Copper ore Lead ore

SD 86 SE; Yorkshire 132 NE
Direction N35–45°W, some with small W downthrow

Malham Calamine Deposit Zinc ore

SD 86 SE; Yorkshire 132 NE

Great Scar Veins

SD 86 SE; Yorkshire 132 NE
Small veins trending N – S, N30°W, N45°E

Settle Scar – Back Scar Veins Lead ore Copper ore

SD 86 SW, SE; Yorkshire 132 NE
Direction N50–68°E, one with W downthrow

Benscar Veins Lead ore
SD 86 SW, Yorkshire 132 NE
Direction N30°W and E – W

Between Threshfield and Malham, the ground between the North and Middle Craven faults (see also p. 73) is crossed by a number of faults running NNW and NW, some of which link the two major faults. These affect beds ranging from the Gordale Limestone up to the Hardrow Scar Limestone and the unconformably overlying grits and shales of the Grassington Grit Formation, but no mineralisation is known in them. The displacements of most of them are greater than those of the mineral veins in Areas 14 and 15, a factor possibly unfavourable to the passage of mineralising fluids; also, the ground is not well exposed. At Pikedaw, west of Malham Cove occurs a belt of NNW and NNE mineral veins starting on the north side of the Middle Craven Fault, and extending as much as 0.8 mile (1.3 km) towards the North Craven Fault though never reaching it. This area of mineralisation continues for 3 miles (4.8 km) towards Settle (Figure 6).

The only previous general descriptions of this mining area, detached by barren ground from the nearest veins at Cracoe and High Mark, are those of Raistrick (1938 a, b); there are historical references by T. Hurtley (1786) and accounts of his 'calamine' (smithsonite) operations by Lord Ribblesdale (1807 a, b). Raistrick has also written on the history of the area (1953, 1973) and described the calamine mine (1954). Many of the small veins outcrop on the limestone scars, but the ores are pyritic and therefore gossan-forming, and not much lead may have been visible in the backs. It is nevertheless difficult to believe that the veins had not been seen by the miners at Arncliffe, no great distance away, who left behind them there a coin of the period 1272–1327. Copper smelting was already going on at the end of the 17th century, the smelters, according to Hurtley, living in caves. By 1786 the remains of the copper mill could be seen, but evidently it was derelict. However, about this time a search of the old copper workings was being made at the instance of Birmingham brass-founders, who wished to break the monopoly of the copper industry established by 1785 by John Williams as a result of his opencast operations at Parys Mountain, Anglesey (Hornshaw, 1975, p. 116). About 1788, in the course of this work, the unique Malham cave-deposit of smithsonite was discovered. Mining activity for this and sporadically for lead continued until about 1830, the lead ores being smelted at a mill the ruins of which can still be seen 1½ miles (2.4 km) to the north of Pikedaw, but the Birmingham men do not appear to have been able to revive the mining of copper. From 1966 to 1970 there was interest in a part of the area north of Stockdale Farm but after a series of preliminary borings, the Canadian company concerned, at a public enquiry was refused permission to develop on the grounds that this is an area of natural beauty.

The question as to whether zinc orebodies of 'Irish Style' occur nevertheless remains of some interest, especially since, were such discoveries to be made, exploitation might well be possible without jeopardising the natural beauty of the area. Reference should be made to Wadge, Bateson and Evans' (1984) account of the Settle – Malham area (pp. 19–20), who conclude 'The possibility of large disseminations of sulphides along the Mid-Craven Fault, trapped in the basinal carbonates against the Lower Palaeozoic basement of the Askrigg Block, remains as yet untested'.

The mineral suite here differs from that of the deposits in Area 14 and in the previously-described parts of Area 15 in some important respects. Chalcopyrite and its oxidation products malachite and azurite were present in sufficient amount to make some copper mining possible, though it may be suspected that the ores were lean. Concentrated smithsonite was successfully worked here for shipment as calamine. The gangue of most veins includes quartz (conspicuous by its absence in the Askrigg orefield save at Mallerstang) and there is extensive silicification of limestone wallrocks. Dolomite, in both discordant and concordant bodies, has extensively replaced limestone to a greater extent than in Wharfedale or Nidderdale. The remaining suite is normal, including galena and its oxidation products anglesite and cerussite; sphalerite at one locality at least, baryte and calcite. We have no record of fluorite.

The Pikedaw – Grizedales group of veins (Figure 6, loc. 1) includes at least fourteen NNW fractures in a belt up to 3000 ft (0.9 km) wide. Some of these carry only siliceous dolomite, and stand up as well marked ridges. The mining developments include bell-pits, shafts, and an adit [8800 6376]. The veins cut through Gordale Limestone and, to the north and west Hawes Limestone with the Girvanella Band. No underground plans are available, but it is doubtful whether much stoping has been done. Baryte and malachite can be seen at [8839 6455] and the same minerals, with galena, from shafts [8830 6466]. The area of intersection of some of the Grizedales fractures with the Middle Craven Fault may have been investigated by an adit driven westward beneath Namurian [8821 6363] but little spoil remains. East of this adit a shaft has investigated two E – W veins that may be regarded as flyers from the Middle Fault; baryte is present here.

The smithsonite deposit (Figure 6, loc. 2) was discovered from an old copper mine shaft [approximately 8757 6401] at a depth below surface of about 48 ft (14.6 m). From the shaft a short eastward working broke into a pothole, blind upwards, and from this access was gained to three caverns, named by Lord Ribblesdale from their lengths in yards the 104, running more or less W – E, the 44 (SSW) and the 84 (SSW). The smithsonite was mostly in the form of a white powder filling the bottoms of these caverns. The original entrance proved too difficult for extraction of the ore, and after accurate survey a new shaft [8757 6400] was sunk in 1806 into the junction of the 104 and 44 caverns. After the cessation of mining, though a local tradition remained, the site of the operation was lost for over a century, until the historical researches of Raistrick led to its rediscovery and entry in 1953. His description (1954) includes a plan and sections of the caverns. Some of the white powder or silt still remained in small pockets associated with the caverns in 1979, and an XRD analysis by A. C. Dunham showed it to be smithsonite, not hemimorphite. A chemical analysis is given in Table 20. Deposition is probably not still in progress since Raistrick recorded calcite stalagmite overlying it. The origin of this smithsonite silt is an interesting question. The fact that zinc moves freely in solution in carbonated phreatic waters has been well known since Loughlin's work at Tintic, Utah (see p. 107), and it is normal for it to be exported some distance from its primary sphalerite source during the oxidation stage. However, it normally accumulates as incrustations of 'dry bone' ore on the walls of caverns or fissures. Two possibilities in the present case are (a) that the silt results from subsequent underground-stream erosion of a massive smithsonite deposit; or (b) that it represents a direct chemical precipitate from carbonated water carrying zinc, perhaps as a result of a change in oxygen fugacity as in the case of calcite. It is difficult to understand why a solid crystalline deposit was not, however, produced in the latter case. An estimate of the quantity in the three caverns may be obtained by assuming that the average thickness of the deposit was 3 ft (0.91 m), and that 18 cu.ft is required for 1 ton: roughly, this yields a figure of 15 000 tons and this may be regarded as a maximum. The only guide to actual production is that in 1800, 144 tons and in 1811, 268 tons were obtained. It may be suggested that the figure of 5000 tons adopted here for the 30 years production is reasonably consistent with the facts, and it has Dr Raistrick's approval. The 1806 shaft was sunk through Lower Hawes Limestone and the cavern system is in Gordale Limestone, evidently near the top of the section; spoil includes baryte, malachite, azurite and quartz. One of Raistrick's sections (1954, fig. 3) shows the cavern cutting through a lead vein.

The smithsonite deposit suggests that a fairly substantial sphalerite deposit is, or was, available in the vicinity. Workings [8688 6379] on the Middle Craven Fault where several NNW fractures impinge on it show sphalerite, baryte and malachite in the dumps. Around [862 637] the Middle Fault is again mineralised, here with quartz, baryte and malachite, with accompanying silicification of wallrocks, while a short distance to the north a N – S vein of the Great Scar group (Figure 6, loc. 3) carries abundant baryte, with galena and malachite. Other small veins of this group to the north and west also carry baryte and galena. Massive 'dry bone' smithsonite can also be picked up at surface in this general area between Stockdale and Back Scar, and it was here that, as already noted, Canadian interests began a drilling programme in the hope of finding a large, 'Irish-style' sulphide body related to the Middle Craven Fault. The programme was not continued to a conclusion and the question remains open. The stratigraphical data obtained from the boreholes will be included in the forthcoming memoir on Settle (Sheet 60).

The belt of fracturing here called the Settle Scar – Back Scar veins (Figure 6, loc. 4) includes at least four subparallel NE veins, cutting through strata from the Cove Limestone (Holkerian) to the Hardrow Scar Limestone of the Brigantian. Baryte, malachite and galena are present in the spoil from trials, but in no considerable amount. Farther west, the Benscar veins (Figure 6, loc. 5) worked from shafts around [841 645] behind Attermire Scar contained baryte and calcite gangue with a little smithsonite in silicified limestone. Quartz also occurs as a vein mineral and the amount of work done suggests that galena was obtained. This is the westernmost occurrence in the Malham – Settle belt.

Mineralisation south of the Middle Craven Fault

SD 86 SW; Yorkshire 132 SE

On High Hills and Scaleber, overlooking Settle, there are considerable areas of discordant dolomitisation, and some concordant replacement at the junction of the Gordale and Lower Hawes limestones. Reef limestone at [830 635] is partly dolomitised and cut by veins with calcite, vuggy quartz and traces of chalcopyrite, azurite and malachite, the only evidence of mineralisation (other than dolomitisation) found in the reefs of Settle and Maham, save on Scaleber [841 630], where quartz again occurs in cavities in dolomite. Also on Scaleber, [8414 6270] sphalerite is disseminated in the Scaleber Quarry Limestone near its junction with the Boulder Bed.

Mineralisation north and west of Settle, between the South and North Craven Faults.

SD 86 NW, 76 SE; Yorkshire 114 SW, 132 NW

Nothing comparable with the Malham – Settle belt is known, but concordant and discordant dolomitisation is seen at various places and there are a few minor trials of veins. A calcite vein [8257 6644] about 3 ft (1.0 m) wide trending NW has been tested, while a short distance to the west concordant dolomite is found at the junction of the Kilnsey and Cove limestones of Holkerian age. The Cove Limestone is dolomitised [8085 6478] in contact with the South Craven Fault (here trending NW) and cut by calcite veins. At [7925 6620] WNW-trending veins have been tried in the Upper Hawes and Gayle limestones near the same fault, but these and a number of shallow pits to the NE and SE appear to have found nothing workable. A shaft [7990 6764] has tested a 3 ft 3 in (1.0 m) wide E – W calcite vein carrying a little baryte. To the north, extensive concordant dolomitisation of the Gordale Limestone lies between this trial and the position of the North Craven Fault.

DJCM, KCD

OUTPUT AND RESOURCES

Although mining in this area has a very long history from which a few clues as to output can be gained the earliest specific record begins in 1706, and a complete account dates only from the commencement of official statistics in 1845.

Very tentative estimates covering gaps in the record above, and the long period prior to the record add 52 750 tons for the Greenhow Mining Area (p. 221), some 13 475 for the mines along Ashfold Side Beck, 6000 for Appletreewick and Burhill, and perhaps 3000 tons for the reef-knolls and the Malham area, making an estimated total production of lead concentrates of the order of 145 000 long tons.

Production of fluorspar from the area began in 1920 and up to 1938 it is probable that nearly the whole of the Yorkshire production of 40 188 tons came from Greenhow and Skyreholme, partly from underground workings such as Gill Heads and Cockhill, partly from opencast operations, for example at Galloway Pasture, and also from washing of dumps. Detailed official records are no longer available but general figures suggest that Yorkshire produced about 88 500 tons from 1939 to 1968, after which even general figures are not separated from those for Durham. The contributors to the Yorkshire total since World War II have included the Old Gang – Arkengarthdale area, Wet Grooves in Wensleydale, and the Dales Chemicals operations in Grassington Moor; allowing for 22 591 tons of outside (presumably Greenhow) production milled at Grassington[1], Area 15 may have contributed five-sixths of the total. Thus allowing for a small continuing output since 1968, the total figure for this area may be of the order of 120 000 long tons.

There is believed to have been a small production of barytes, principally from the Nussey Knot flats, but no output figures are available. Probably the output was no more than a few thousand tons, for in spite of the interest of the occurrence as described by Varvill (1937) it has never become a continuing producer. An estimate of a reserve of 70 000 tons by an owner in the 1930s has been quoted (Dickinson, 1970, p. 15) but it is unlikely that sufficient work to justify such a figure has been done.

A total of 720 tons of calcite were produced at Appletreewick during 1925–27.

Table 25 Recorded production, lead metal and concentrates, Area 15

Mine or Group	Years	Lead metal long tons	Concentrates long tons	Lead %
Greenhow Hill and Stony Grooves	1706–1732	3 339†	5 562	60‖
Forest Moor	1764–1829	1 555†	2 321	67‖
Ashfold Side and Greenhow Hill	1806–1817	11 268*	16 818	67‖
Providence–Prosperous	1837–1842	1 314†	2 000	67‖
South Coldstones	1839–1844	1 528‡	2 281	67‖
Greenhow Hill, Craven Moor	1845–1936	16 813§	24 165	67
Blayshaw Gill, Limley, Lolly Scar	1876–1910	4 159§	6 149	68.6
Stony Grooves, Merryfield, Providence	1861–1887, 1855	856§	1 236	69.2
Appletreewick, Burhill, Gill Heads	1861–1888	5 252§	7 715	68.1
Grimwith Kelshaw	1861–1881, 1894	859§	1 248	68.8
Cracoe, Elbolton	1860–1876	170§	244	68.3
		47 113	69 739	67.5¶

Sources
* Stony Grooves, Providence, Prosperous, Cockhill, Sunside, Black Rigg, North Coldstones, Thornhill Meres: estate records communicated by Mr D. Gill.
† Raistrick (1973, pp.58, 31, 39 and 40); Jennings (1967, p.285).
‡ Estate records per Mr Gill.
§ Official statistics, *Mem. Geol. Surv.*, 1849–1881; Home Office 1882–1919; Mines Department, 1920–1936.
‖ Assumed as basis for calculation of concentrates production.
¶ Average value.

[1]Information supplied by the Agent to the Trustees of the Chatsworth Settlement

A small reserve of lead ore with fluorspar matrix is in place at Craven Cross Mine, but the two narrow veins in which it occurs would not be workable under present conditions. The northern flank of the Greenhow Anticline offers a prospect for fluorspar in depth, for example on Waterhole North Vein, while the North Craven Fault-system along the southern flank deserves serious attention. There are resources of fluorspar, perhaps with some recoverable barytes, both at Greenhow and at Skyreholme, but the patchy and short-lived nature of the oreshoots has discouraged previous operators. The quality of the raw spar is unusually good in that the silica content is negligible. Since the Grassington Grit and underlying limestones have nowhere been reached by mining in the Ashfold Side Beck mines, oreshoots could exist in depth, but water might be a serious problem. Geophysical measurements (Wadge, Bateson and Evans, 1984) clearly indicate the structural continuity of the Bycliffe vein-system from Grassington to Ashfold Side Beck; the existence of conductors along this virgin stretch may indicate that resources of lead ore, fluorspar and barytes comparable with the orebodies already worked to the west and east exist in depth, but drilling is required. There could also be concealed orebodies at Burhill Old Vein and perhaps related to the North Craven Fault nearby.

Production of zinc ore in the form of 'calamine' (smithsonite) is estimated in round figures as 5000 tons (p. 228); undiscovered resources may remain. The 18th century output of copper ore from Malham is represented by a nominal figure of 500 tons.

At some future time the Craven Fault-system may be considered worthy of systematic prospecting for concealed massive sulphide deposits comparable with those in Eire. The fact that movements contemporaneous with Carboniferous sedimentation occurred within the zone, particularly on the Middle Fault (Dixon and Hudson, 1931) would be regarded by some as a favourable factor, as would the widespread presence of oxidised zinc minerals. KCD

REFERENCES

ANDERSON, F. W. 1928. The Lower Carboniferous of the Skyreholme Anticline, Yorkshire. *Geol. Mag.*, Vol. 65, pp. 518–527.

BLACK, W. W. 1950. The Carboniferous geology of the Grassington area, Yorkshire. *Proc. Yorkshire Geol. Soc.*, Vol. 28, pp. 29–42.

— 1958. The structure of the Burnsall – Cracoe district, and its bearing on the origin of the Cracoe knoll-reefs. *Proc. Yorkshire Geol. Soc.*, Vol. 31, pp. 391–414.

— and BOND, G. 1952. The Yoredales succession on the northern flank of the Skyreholme Anticline, Yorkshire. *Proc. Yorkshire Geol. Soc.*, Vol. 28, pp. 180–187.

BOND, G. 1950. The Lower Carboniferous reef limestones of Cracoe, Yorkshire. *Q.J. Geol. Soc. London*, Vol. 105, pp. 157–188.

CHUBB, L. J. and HUDSON, R. G. S. 1925. The nature of the junction between the Lower Carboniferous and the Millstone Grit of North-West Yorkshire. *Proc. Yorkshire Geol. Soc.*, Vol. 20, pp. 257–291.

CLOUGH, R. T. 1967. Catalogue of a collection of Yorkshire mine plans relative to the Greenhow area. *Mem. North. Cavern Mine Res. Soc.*, pp. 11–15.

— 1980. The lead smelting mills of the Yorkshire Dales and Northern Pennines. 2nd Edition. (Keighley: Clough.) 332 pp.

DICKINSON, J. M. 1964a. Some notes on the lead mines of Greenhow Hill. *Trans. North. Cavern Mine Res. Soc.*, Vol. 1, pp. 34–47.

— 1964b. The Appletreewick Lead Mining Company 1870–72. *Mem. North. Cavern Mine Res. Soc.*

— 1967. The Burhill mines. *Mem. North. Cavern Mine Res. Soc.*, pp. 1–10.

— 1969. The Greenhow lead mining field. *Mem. North. Cavern Mine Res. Soc.*, pp. 35–41.

— 1970. The Greenhow lead mining field (A historical survey). *North. Cavern Mine Res. Soc.*, Individ. Surv. Ser. No. 4, 50 pp.

— 1971. Dry Gill Mill. *Trans. North. Cavern Mine Res. Soc.*, Vol. 2, pp. 31–36.

— 1972. The mineral veins of Swinden Knoll. *Trans. North. Cavern Mine Res. Soc..*, Vol. 2, pp. 98–99.

DIXON, E. E. L. and HUDSON, R. G. S. 1931. A mid-Carboniferous boulder bed near Settle. *Geol. Mag.*, Vol. 68, pp.81–92.

DUNHAM, K. C. 1944. The production of galena and associated minerals in the Northern Pennines, with comparative statistics for Great Britain. *Trans. Inst. Min. Metall.*, Vol. 53, pp. 181–252.

— and STUBBLEFIELD, C. J. 1945. The stratigraphy, structure and mineralisation of the Greenhow mining area, Yorkshire. *Q. J. Geol. Soc. London*, Vol. 100, pp. 209–268.

FALCON, N. L. and KENT, P. E. 1960. Geological results of petroleum exploration in Britain 1945–1957. *Mem. Geol. Soc. London*, No. 2.

HORNSHAW, T. R. 1975. Copper mining in Middleton Tyas. *N. Yorkshire County Record Office*, Publ. No. 6.

HURTLEY, T. 1786. *A concise account of some natural curiosities in the environs of Malham, in Craven, Yorkshire.* (London.) 267 pp.

JENNINGS, B. (Editor). 1967. *A history of Nidderdale.* (Huddersfield: Advertiser Press.) 504 pp. 2nd. Ed., 1983.

PHILLIPS, J. 1836. *Illustrations of the geology of Yorkshire*, Part II. The Mountain Limestone District. (London: John Murray.) 253 pp.

RAISTRICK, A. 1927. Lead mining and smelting in West Yorkshire. *Trans. Newcomen Soc.*, Vol. 7, pp. 81–97.

— 1938a. Mineral deposits in the Settle – Malham district, Yorkshire. *Naturalist*, pp. 119–125.

— 1938b. The mineral deposits. Pp. 343–349 in *Geology of the country around Harrogate.* HUDSON, R. G. S. (Editor). *Proc. Geol. Assoc.*, Vol. 26.

— 1953. The Malham Moor Mines, Yorkshire, 1790–1830. *Trans. Newcomen Soc.*, Vol. 26, pp. 69–73.

— 1954. The calamine mines, Malham, Yorks. *Proc. Univ. Durham Philos. Soc.*, Vol. 11, pp. 125–130.

— 1973. *Lead mining in the Mid-Pennines.* (Truro: Bradford Barton.) 172 pp.

— and JENNINGS, B. 1965. *A history of lead mining in the Pennines.* (London: Longmans.) 347 pp.

RAMSBOTTOM, W. H. C. 1974. Dinantian. Pp. 47–73 in *The geology and mineral resources of Yorkshire.* RAYNER, D. H. and HEMINGWAY, J. E. (Editors). (Leeds: Yorkshire Geological Society.) 405 pp.

RIBBLESDALE, LORD. 1807a. Three letters giving an account of a mine of zinc ore found in caverns at Malham Moors, and its application as a paint. *Trans. Soc. Arts*, Vol. 35, pp. 35–38.

— 1807b. Account of a mine of zinc ore, and its application as a paint. *Nat. Philos. Chem.*, Ser. 2, Vol. 21, pp. 12–14.

VARVILL, W. W. 1920. Greenhow Hill lead mines. *Mining Mag.*, Vol. 22, pp. 275–282.

— 1937. A study of the shapes and distribution of the lead deposits in the Pennine limestones in relation to economic mining. *Trans. Inst. Min. Metall.*, Vol. 46, pp. 463–559.

WADGE, A. J., BATESON, J. H. and EVANS, A. D. 1984. Mineral reconnaissance surveys in the Craven Basin. *Mineral Reconnaissance Programme Rep. Inst. Geol. Sci.*, No. 66.

WILSON, A. A., DUNHAM, K. C. and PATTISON, J. (in preparation). The stratigraphy of recent boreholes in the Greenhow and Grimwith areas, Yorkshire.

CHAPTER 14

Future prospects

The southern half of the Northern Pennine metalliferous orefield now lies in part within the boundaries of the Yorkshire Dales National Park (Figure 1). Part of the eastern fringe lies within military ranges, but those areas in any case appear to offer no prospect for further subsurface development. It is hoped that this volume will provide, for those visitors to the National Park and for residents who are interested in the remains of the ancient mining industry which form a characteristic part of the Dales landscape, accurate information about the significance of the remains. Beyond this, however, it is pertinent to ask whether a district which has yielded metals and associated minerals worth at 1980 prices at least £200 million still contains subsurface resources of interest to the nation. It may be noted that some planning permissions for mining operations continue to exist within the Park; that some interesting parts of the orefield such as the Greenhow – Skyreholme district are not within the Park, and that, in the West Craven hills near Settle, permission for mining development was refused on the grounds that this is an area of outstanding natural beauty.

LEAD AND ZINC ORES

Output of lead ore from the North Swaledale Belt (Chapter 9) was already in marked decline, presumably due to exhaustion of reserves, before the sharp drop in the price of lead in the 1880s which effectively ended profitable operations in this formerly highly productive zone. During the previous century, a series of trial levels designed to test the continuation of payable ground below the main bearing beds (Crow Beds down to base of Main or base of Underset Limestone) had met with no success except beneath Gunnerside Gill, where ore was left in the sole at the horizon of the Middle Limestone. This root, which we regard as one of the feeders of the Belt, was tested by one borehole (Gunnerside G1) in 1966, but this was not sufficient to give a decisive result. Nor can the amount of driving done from the deep Mill Level (p. 134) be regarded as sufficient to establish that no extension occurs below the very highly mineralised ground between Turf Moor and Dam Rigg. Neverthless, the general point made by Carruthers and Strahan (1923) that in (North) Swaledale the valleys have cut through the mineralised productive zone into unfavourable beds is, in the main, justified. The last underground working, near Hurst (pp. 147 – 149) shortly before World War II, found some ore, and it is likely that on the down-dip side of the Hurst monocline some ground remains intact.

The Swale – Ure watershed, though extensively mined in the past, appears to have little to offer except that a geophysical anomaly near Oxnop, lying in one of the principal mineralised trends, remains to be explained. In Wensleydale a number of trials and small mines have been worked in beds almost down to the Great Scar Limestone, but with generally poor results. Only the old Keld Heads Mine near Wensley offers much interest.

West of Upper Wharfedale lies the great area of Geological Sheet 50 (Hawes) occupying the plateau region of the Askrigg Block, underlain by the Wensleydale Granite and yet almost completely devoid of known mineral deposits. This area provides the greatest contrast between this block and the Alston Block, which carries mineral veins throughout the half-dome. Were the oreshoots present in the now largely eroded Main Limestone and associated bearing beds over this area? As Raistrick (1938) has suggested, the numerous but feeble High Mark veins may represent the feeble extensions in depth of such oreshoots. This question is not, of course, answerable but it is interesting that the exhaustive chemical survey of groundwaters in the area by Ineson and Al-Badri (1980) revealed metal anomalies in the Litton and Dent area that are difficult to explain on the basis of known mineralisation. Not far away, the Beckermonds Scar Borehole (Wilson and Cornwell, 1982) encountered subeconomic fluorite flats with low copper values where no deposit was expected. At the same time, the West Upper Wharfedale – Dent region could not be regarded as highly promising.

In connexion with the BGS mineral reconnaissance survey a thorough geochemical drainage basin survey of the marginal area from Dent to the Ingleton Coalfield (Bateson and Johnson, 1983) including stream sediment, panned concentrate and water sampling work has failed to indicate any major mineral deposit, though a few new instances of minor mineralisation have been found. Similar considerations apply farther north (Bateson and Evans, 1982).

Since the publication of Dunham (1948), the remarkable discoveries of large zinc-lead deposits in the Carboniferous of the Irish Republic have taken place. This raises the question whether, in the English Carboniferous, the mineralising fluids available have been dispersed into a very large number of relatively small oreshoots without forming orebodies comparable with those in Ireland, or whether such deposits still remain to be found, as they did in Eire until 1959. Nearly all the Irish-style deposits are in Tournaisian host-rocks, and the two largest, Silvermines (Taylor and Andrew, 1978) and Navan (Byrne, Downing and Romer, 1971) have features which most observers interpret as indicating that mineralisation was in part contemporary with sedimentation, particularly of the 'Waulsortian reef' calcilutite facies. Pre-Viséan fault movements are implied at both places and also at the third major deposit, Tynagh. In the present area, Tournaisian sediments are present only doubtfully in the Stainmore Trough; they are not known to extend to the mineralised areas and only an arenaceous facies is known. There is no evidence of pre-Viséan, late Tournaisian faulting in the Pennines. Thus a strict parallel cannot be established. Nevertheless, the concept of contemporaneous faults releasing mineralising fluids into euxinic pools in the sea might be applicable along the Craven Fault System, where the Middle Craven Fault was certainly moving before the deposition of the Pendleian Upper Bowland Shales (Hudson, 1944), and where the black shales suggest near-euxinic conditions. Calcilutite reef-mounds are also present nearby. It is this combination of facts that lent interest to the explorations around Stockdale Farm, near Settle (p. 227), but it should be added that the North Craven Fault at Greenhow, and perhaps beyond Pateley Bridge if it continues as a pre-late Pendleian structure, may offer comparable conditions (pp. 221–223).

West of Pateley Bridge, the Perseverance – Providence – Prosperous – Merryfield vein system diverges west-north-west from the North Craven Fault; it has now been demonstrated geophysically that this continues to Grassington High Moor as the Bycliffe Vein (Wadge, Bateson and Evans, 1984). A substantial gap, untested at the critical Grassington Grit horizon, exists between the two sets of workings. The eastern end, from Providence to Stony Grooves in Ashfold Side Beck, carried substantial lead oreshoots against wallrocks ranging from Kinderscoutian down to Arnsbergian, the highest stratigraphical level reached by mineralisation anywhere in the Askrigg Orefield. It was formerly believed that the top of the Lower Carboniferous limestone had been touched in the deepest workings, but borings (Wilson, Dunham and Pattison, in preparation) have now proved decisively that the limestone encountered was the small Cockhill bed. Thus the bearing bed of the Grassington field was not reached by the workings, and remains, with the underlying Brigantian and Asbian limestones, an interesting target for exploration in depth. Greenhow Hill itself is probably of more interest for its fluorspar potential, and is mentioned below.

Study of the positions of the numerous oreshoots worked in the Grassington district (pp. 192–197) indicates a clear stratigraphical control by particular divisions of the Grassington Grit, but to the east and south-east this formation was about to dip beneath the workings drained by Duke's Level at 864 ft (263 m) OD when mining came to an end. The fact that the eastern and south-eastern foreheads had passed out of ore by no means proves that new oreshoots do not lie ahead in these directions. With the driving of the Hebden Trial Level at 595 ft (181 m) OD the next logical step had been taken, for a structural analysis shows that this could have controlled the base of the Grassington Grit over most of Hebden Moor and Grimwith. It was, however, driven when the price of lead had reached its lowest point in the 19th century, and it is perhaps surprising that it was pursued to almost 2½ km before abandonment. The next 2½ km would have proved or disproved the down-dip recurrence of the Grassington vein-complex (Dunham, 1959, p. 144). Now, however, the area of interest partly underlies the catchment of the enlarged Grimwith Reservoir and this would have to be taken carefully into account.

Concealed oreshoots almost certainly exist between upper Wharfedale and Nidderdale, as witness the reappearance of mineralisation where the Grassington Grit re-emerges from below the cover of later Namurian strata around Lolly in upper Nidderdale. These, however, would be difficult and expensive to locate by any method at present available, and the production from the oreshoots north of the Grassington field is insufficient to justify high hopes for this remote ground. When, in valleys farther north such as Coverdale and Walden, the bearing beds, now including the Main Limestone, again make their appearance east of the main orefield, only a few small deposits have been found (pp. 181–182) though there is reasonably good exposure.

It remains to mention the western boundary structure of the orefield, the Dent Fault and monocline. In spite of the intense disturbance of the strata, or perhaps because of this, the southern stretch from the Ingleton Coalfield up to Clouds, west of Wild Boar Fell, is devoid of known mineralisation other than the ubiquitous dolomitisation. Geochemical work, already mentioned, may suggest some potential around Dent village, but nothing has previously been found here and more recent BGS geochemical work failed to find anything significant (Bateson and Johnson, 1983). From Clouds northward to Hartley Birkett there are some small deposits which have yielded small tonnages of lead and (probably) copper ores. This area was selected, with the south-west marginal area of the Alston Block, for investigation by applied geophysical methods in connexion with the Mineral Reconnaissance Programme of the IGS (Cornwell, Patrick and Hudson, 1978), the methods used being airborne electro-magnetic, radiometric and magnetic surveys over selected parts of the areas, followed by ground follow-up of the most promising anomalies using VLF, Turam and Slingram methods. Even in remote country such as this, interference affected the work; some anomalies were traced to shales in the stratigraphical sequence, and only in one area, Kitchen Gill [around 790 037], east of Birkett Common (see Great Bell, p. 116) were promising anomalies found, and these were not considerd to indicate strong mineralisation. The results of this work, on open file at BGS, are an interesting illustration of the difficulties of this work in the Pennine setting. From surface evidence, Hartley Birkett (p. 115) appears the most mineralised part of the Dent zone and perhaps at some time it might be worth drilling.

It should be stressed that the BGS Mineral Reconnaissance Programme did not include a comprehensive geophysical or geochemical survey of the area covered by the present memoir. The intention was to test the applicability of various indirect methods of investigation to the Pennine situation, it being recognised from the outset that the cyclothemic sedimentation (including some notably pyritic shales) are not easy subjects for geophysical discovery of new orebodies, and that contamination at surface by lead is widespread as a result of the numerous smelt mills formerly operated, making Pb and probably Zn trace element detection of little value. Evans, Patrick, Wadge and Hudson (1983) probably give voice to a general conclusion when they say that the airborne EM method employed is not an effective tool in the northern Pennines. Nevertheless, two out of the eleven anomalies followed up on the ground proved to show strong VLF-EM indications and these, respectively at Oxnop and Brownfield, may yet repay further investigation. In Areas 13 and 14, the ground anomalies indicating the structural continuity of the Bycliffe and Merryfield veins (p. 97) and those on and near the North Craven Fault at Skyreholme (pp. 221–223) deserve testing by drilling, particularly since should sulphide deposits be found to be the conductors, VLF-EM could be regarded as a valuable tool for the discovery of new deposits in the orefield.

Two attempts have been made to forecast the potential of the southern half of the Northern Pennine Orefield by statistical methods, but neither appears to have taken the realities of the geological situation sufficiently into account. Bozdar and Kitchenham (1972) assume that the Askrigg Block is in effect a mirror image of the Alston Block, and that the distribution of mines (negative binomial) and of values (lognormal) in the latter can be applied to determine the potential of the former. On this basis they conclude that one new mine should be found in the present region for every 20 km^2 of area. A comparison of Volumes I and II of the present memoir will show that this assumption is by no means a valid one, the most obvious difference being the large blank central area of the Askrigg Block, already referred to above compared with the fairly even distribution of veins through the Alston Block. There are other, more subtle differences. The study of the base-metal potential of northwest England by Cruzat and Meyer (1974) used eighteen stratigraphical and four structural parameters derived from the Geological Survey one-inch maps in conjunction with the results of the Wolfson stream-sediment survey carried out by the Applied Geochemistry Research Group at Imperial College, forecasting being based upon multiple regression of the data against production indexes. Altogether thirty-four new lead mines and three new zinc mines are forecast in the present area, but not in places consistent with the geological evidence as it appears to us. It may further be questioned whether the use of stream sediment geochemistry in regions like the present, well known for the extensive contamination from ancient smelting and widespread mine waste, can possibly be admitted in any forecasting process.

The view adopted here is that the Craven fault-belt holds the highest potential for new lead-zinc discoveries, with Ashfold Side to Grassington next, followed by Gunnerside Gill – Surrender, and perhaps Keldheads.

FLUORSPAR

The main fluorspar producing area has been Greenhow – Skyreholme, and this occupies a poor third place in this country to the much more productive fields of North Derbyshire and West Durham. Small tonnages only have so far come from Grassington, Wensleydale and Arkengarthdale. Only in Area 15 do fluorite-dominated veins comparable with Hucklow Edge and High Rakes in Derbyshire or the quarter-point veins of Durham occur, and so far there are only two good examples: Waterhole Vein and the Gill Heads deposit. Other veins at Skyreholme opened up during the Clay Cross Company's operations between 1963 and 1967 proved to contain shoots of fluorite with calcite up to 12 ft (3.65 m) wide, but their lengths are so short that mining of them became uneconomic and the operations were short-lived. At Greenhow at least one 'gulf' (on Galloway Vein) contained excellent crude fluorspar and contributed substantially to the district's output. Karst action here had probably concentrated veinstuff from high levels, now eroded, into the sinkhole. Other Greenhow 'gulfs' might be worth investigation, for example those on Sir Thomas and Blue Joke veins, illustrated in Dunham and Stubblefield (1945, fig. 6, and Dickinson, 1970, pls. 2 and 4). Opportunities for further prospecting also remain on the Waterhole veins, and Gill Heads is being actively mined on a small scale at present. Most of the dump material readily capable of being treated at Greenhow – Skyreholme was cleaned up by the Greenhaugh and Clay Cross companies. At Grassington, the clean fluorite-bearing tailings sampled during the mineral survey (Dunham, 1952, pp. 71, 72) formed the basis for flotation milling by Dales Chemicals Ltd, but the yield, only a little over 10 000 tonnes was disappointing, and later production was based on material brought mainly from Greenhow. At Grassington the spoil heaps, scattered over a wide area, contain fluorite and baryte, often mixed with coarse dead rock; a substantial amount of fluorite and baryte must remain when the lead production of the district is recalled and a stage may some day be reached when a thorough clean-up of this area might be worth while. No reserves of fluorite-dominated, or indeed any other ore are known to exist underground and the veins were generally narrow, perhaps averaging little more than 3 to 4 ft (say 1 m). Though many of the Upper Wharfedale mines contained fluorite in their mineral assemblage, it nowhere appears to have been sufficiently abundant to offer the possibility of production. This applies also to such mines in Wensleydale and the Ure – Swale watershed as carry the mineral; save that at Seata (pp. 180 – 181), Raygill House (p. 166) and Worton Hall (p. 167) the presence of strongly fluoritised limestone on the dumps suggests the possibility that fluorite-bearing flats might be found at low limestone horizons in the Brigantian; the evidence is, however, tenuous. The tailings at Wet Grooves have mostly been removed and the mine offers little attraction for reopening. In the North Swaledale Belt only one tailings heap, Dam Rigg West, is said to have proved worth processing for fluorspar. The mineral is widely present between Gunnerside Gill and Langthwaite but forms only a few per cent of the extensive tailings. If these were systematically treated, as suggested below, it should be possible to take off a fluorspar product.

BARIUM MINERALS

The results of sampling the very extensive tailings heaps of Old Gang, Friarfold and Surrender, first published by Dunham and Dines (1945, table 16) showed 44 to 72 per cent $BaSO_4$ with 1 to 4 per cent $BaCO_3$. These have been augmented by recent sampling of heaps in Arkengarthdale North Side, Sun Side and East Side (Table 15). Exact estimation of tonnage would require boring or trenching of many scattered heaps and this has been undertaken for Arkengarthdale and Hurst by private interests. There may be as much as ½-million tons of barium mineral scattered along the belt as a whole. At present the small plant in Hard Level Gill is stated (Mitchell, 1977) to be producing 50 tons per week of barium minerals for drilling mud. The resource of tailings in the Old Gang area, if combined with Surrender and Arkengarthdale, could provide the basis for a larger operation in which fluorspar and lead (a surprising amount of which remains, as Table 15 shows) would also be recovered. The countryside would benefit from the cleaning-up operation. The possibility that unmined areas with substantial resources of barium minerals remain here cannot be ruled out, but it is understood that an examination of those headings accessible through Hard Level in the early 1950's failed to reveal such places. No serious reopening was, however, undertaken at this time. Witherite has previously been widespread in the North Swaledale Belt, but much of it has been converted into secondary baryte in the oxidation zone. While it is possible that virgin witherite exists, the only places known are the deepest workings below Sir Francis Level, beneath Gunnerside Gill (p. 126) and the small flots at the very inaccessible Virgin Moss Mine (p. 172).

No vein or tailings area likely to be of interest for baryte alone is known in Wensleydale, and here, witherite begins to die out southward. It is absent in Wharfedale save as a mineralogical rarity, and the only tips interesting for barytes recovery are those of Cam Pasture (p. 189) and Grassington Moor, particularly Yarnbury, where intermittent working is taking place. Substantial tonnages remain, but without deposits underground. Finally the Nussey Knot flats at Greenhow have formerly yielded a small production of barytes and may contain some resources; but the free drainage of the lower run of these by way of Kelshaw Level is no longer possible because the mouth of this adit is now submerged under Grimwith Reservoir.

COPPER ORE

The outer-zone copper deposits of this part of the orefield have yielded small tonnages only, but one area—Middleton Tyas—was for about 30 years extremely profitable because of the very high grade of the ore obtainable at shallow depth. While it is well realised that in modern times, copper comes from large, very low-grade deposits, there may still be room for small operations if sufficiently high-grade. This was the justification for the inclusion of the Richmond area in the IGS Mineral Reconnaissance Programme (Chapter 10). No workable extension of the high-grade deposits has been found, and the small prospects in Areas 10 (Mallerstang), 11 (Great Sleddale) and 15 (Malham – Settle) cannot be regarded as showing any promise.

KCD

REFERENCES

Bateson, J. H. and Evans, A. D. 1982. Mineral exploration in the Ravenstonedale area. *Mineral Reconnaissance Programme Rep. Inst. Geol. Sci.*, No. 57.

— and Johnson, C. C. 1983. A mineral reconnaissance of the Dent–Ingleton area of the Askrigg Block, northern England. *Mineral Reconnaissance Programme Rep. Inst. Geol. Sci.*, No. 64.

Bozdar, L. B. and Kitchenham, B. A. 1972. Statistical appraisal of the occurrences of lead mines in the northern Pennines. *Trans. Inst. Min. Metall.*, Vol. 81, pp. B183–188.

Byrne, B., Downing, D. and Romer, D. 1971. Some aspects of the genesis of the zinc-lead orebody at Navan. *Proc. Galway Symp. Ir. Geol. Assoc.*

Carruthers, R. G. and Strahan, A. 1923. Lead and zinc ores of Durham, Yorkshire and Derbyshire, with notes on the Isle of Man. *Spec. Rep. Miner. Resour. Mem. Geol. Surv. G. B.*, Vol. 26.

Cornwell, J. D., Patrick, D. J. and Hudson, J. M. 1978. Geophysical investigations along parts of the Dent and Augill faults. *Mineral Reconnaissance Programme Rep. Inst. Geol. Sci.*, No. 24.

Cruzat, A. C. E. and Meyer, W. T. 1974. Predicted base-metal resources of north-west England. *Trans. Inst. Min. Metall.*, Vol. 83, pp. B131–134.

Dickinson, J. M. 1970. The Greenhow lead mining field (A historical survey). *North Cavern Mine Res. Soc.*, Individ. Surv. Ser. No. 4.

Dunham, K. C. 1948. Geology of the Northern Pennine Orefield: Vol. 1, Tyne to Stainmore. *Mem. Geol. Surv. G. B.*, 357 pp.

— 1952. Fluorspar. *Spec. Rep. Miner. Resour. Mem. Geol. Surv. G. B.*, 4th edition, Vol. 4.

— 1959. Non-ferrous mining potentialities of the northern Pennines. Proc. Symposium on Future of non-ferrous mining in Great Britain and Ireland. *Inst. Min. Metall., London*, pp. 115–147.

— and Dines, H. G. 1945. Barium minerals in England and Wales. *Wartime Pamphlet Geol. Surv. G. B.*, No. 46.

— and Stubblefield, C. J. 1945. The stratigraphy, structure and mineralisation of the Greenhow mining area, Yorkshire. *Q. J. Geol. Soc. London*, Vol. 100, pp. 209–264.

Evans, A. D., Patrick, D. J., Wadge, A. J. and Hudson, J. M. 1983. Geophysical investigations in Swaledale. *Mineral Reconnaissance Programme Rep. Inst. Geol. Sci.*, No. 65.

Hudson, R. G. S. 1944. A pre-Namurian fault-scarp at Malham. *Proc. Leeds Philos. Lit. Soc.* (Science Ser.), Vol. 4, pp. 226–232.

Ineson, P. R. and Al-Badri, A. S. 1980. Hydrochemically anomalous areas in the Askrigg Block of the northern Pennine Orefield. *Proc. Yorkshire Geol. Soc.*, Vol. 43, pp. 17–29.

Mitchell, W. R. 1977. Old Gang and beyond. *Dalesman*, Vol. 38, pp. 800–803.

Raistrick, A. 1938. The mineral deposits. Pp. 343–349 in *Geology of the country around Harrogate*. Hudson, R. G. S. (Editor). *Proc. Geol. Assoc.*, Vol. 49.

Taylor, S. and Andrew, C. J. 1978. Silvermines orebodies, County Tipperary, Ireland. *Trans. Inst. Min. Metall.*, Vol. 87, pp. B111–124.

Wadge, A. J., Bateson, J. H., and Evans, A. D. 1984. Mineral reconnaissance surveys in the Craven Basin. *Mineral Reconnaissance Programme Rep. Inst. Geol. Sci.*, No. 66.

Wilson, A. A. and Cornwell, J. D. 1982, The IGS borehole at Beckermonds Scar, North Yorkshire. *Proc. Yorkshire Geol. Soc.*, Vol. 44, pp. 59–88.

— Dunham, K. C. and Pattison, J. In preparation. The stratigraphy of recent boreholes in the Greenhow and Grimwith areas, Yorkshire.

INDEX

HER MAJESTY'S STATIONERY OFFICE

Government Bookshops
49 High Holborn, London WC1V 6HB
13a Castle Street, Edinburgh EH2 3AR
Brazennose Street, Manchester M60 8AS
Southey House, Wine Street, Bristol BS1 2BQ
258 Broad Street, Birmingham B1 2HE
IDB House, Chichester Street, Belfast BT1 4JY

Government publications are also available through booksellers

BRITISH GEOLOGICAL SURVEY

Keyworth, Nottingham NG12 5GG

Murchison House, West Mains Road,
Edinburgh EH9 3LA

Exhibition Road, London SW7 2DE

The full range of Survey publications is displayed and sold in London (until 31 March 1985 at the Bookshop at the Geological Museum, Exhibition Road, London SW7 2DE; thereafter at the BGS Information Point adjacent to the Geology Library), and through the Sales Desks at Keyworth and Murchison House. All the books are listed in HMSO's Sectional List 45. Maps are listed in the BGS Map Catalogue and Ordnance Survey's Trade Catalogue. They can be bought from Ordnance Survey Agents as well as from BGS.

On 1 January 1984 the Institute of Geological Sciences was renamed the British Geological Survey. It continues to carry out the geological survey of Great Britain and Northern Ireland (the latter as an agency service for the government of Northern Ireland), and of the surrounding continental shelf, as well as its basic research projects. It also undertakes programmes of British technical aid in geology in developing countries as arranged by the Overseas Development Administration.

The British Geological Survey is a component body of the Natural Environment Research Council.